NASA STI Program ... in Profile

Since its founding, NASA has been dedicated to the advancement of aeronautics and space science. The NASA scientific and technical information (STI) program plays a key part in helping NASA maintain this important role.

The NASA STI program operates under the auspices of the Agency Chief Information Officer. It collects, organizes, provides for archiving, and disseminates NASA's STI. The NASA STI program provides access to the NTRS® and its public interface, the NASA Technical Reports Server, thus providing one of the largest collections of aeronautical and space science STI in the world. Results are published in both non-NASA channels and by NASA in the NASA STI Report Series, which includes the following report types:

- TECHNICAL PUBLICATION. Reports of completed research or a major significant phase of research that present the results of NASA Programs and include extensive data or theoretical analysis. Includes compilations of significant scientific and technical data and information deemed to be of continuing reference value. NASA counter-part of peer-reviewed formal professional papers but has less stringent limitations on manuscript length and extent of graphic presentations.

- TECHNICAL MEMORANDUM. Scientific and technical findings that are preliminary or of specialized interest, e.g., quick release reports, working papers, and bibliographies that contain minimal annotation. Does not contain extensive analysis.

- CONTRACTOR REPORT. Scientific and technical findings by NASA-sponsored contractors and grantees.

- CONFERENCE PUBLICATION. Collected papers from scientific and technical conferences, symposia, seminars, or other meetings sponsored or co-sponsored by NASA.

- SPECIAL PUBLICATION. Scientific, technical, or historical information from NASA programs, projects, and missions, often concerned with subjects having substantial public interest.

- TECHNICAL TRANSLATION. English-language translations of foreign scientific and technical material pertinent to NASA's mission.

Specialized services also include organizing and publishing research results, distributing specialized research announcements and feeds, providing information desk and personal search support, and enabling data exchange services.

For more information about the NASA STI program, see the following:

- Access the NASA STI program home page at http://www.sti.nasa.gov

- E-mail your question to help@sti.nasa.gov

- Phone the NASA STI Information Desk at 757-864-9658

- Write to:

NASA STI Information Desk
Mail Stop 148
NASA Langley Research Center
Hampton, VA 23681-2199

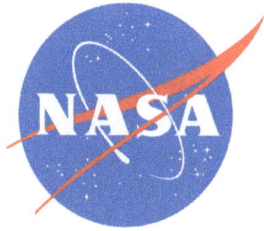

NASA/SP-2014-3705

NASA Space Flight Program and Project Management Handbook

National Aeronautics and Space Administration Headquarters
Office of the Chief Engineer
Washington, D.C. 20546

September 2014

Given ships or sails adapted to the breezes of heaven, there will be those who will not shrink from even that vast expanse.

– Johannes Kepler
Letter to Galileo, 1610

To request copies or provide comments,
contact the Office of the Chief Engineer at NASA Headquarters

Electronic copies are available at

http://ntrs.nasa.gov/

Table of Contents

4   Project Life Cycle, Oversight, and Activities by Phase 109

Boxes

Figures

Tables

Preface

This handbook is the companion document to NPR 7120.5E and represents the accumulation of knowledge NASA gleaned on managing program and projects coming out of NASA's human, robotic, and scientific missions of the last decade. At the end of the historic Shuttle program, the United States entered a new era that includes commercial missions to low-earth orbit as well as new multi-national exploration missions deeper into space. This handbook is a codification of the "corporate knowledge" for existing and future NASA space flight programs and projects. These practices have evolved as a function of NASA's core values on safety, integrity, team work, and excellence, and may also prove a resource for other agencies, the private sector, and academia. The knowledge gained from the victories and defeats of that era, including the checks and balances and initiatives to better control cost and risk, provides a foundation to launch us into an exciting and healthy space program of the future.

This handbook provides implementation guidance for *NPR 7120.5E, NASA Space Flight Program and Project Management Requirements*, which changed and streamlined key procedural requirements across the Agency for space flight program and project management. The goal of the NPR requirements is to ensure programs and projects are developed and successfully executed in the most cost-effective and efficient manner possible. This handbook provides context, rationale, and explanation to facilitate the application of requirements so that they make sense and to pass on some of the hard-won best practices and lessons learned.

While thoughtful planning and execution is important in all phases of a program or project, the Agency is placing particular emphasis on activities during the Formulation Phase of the life cycle. This focus is needed to:

- Accurately characterize the complexity and scope of the program or project;
- Increase understanding of programmatic requirements;

- Better identify and mitigate high safety, technical, acquisition, cost, and schedule risks; and

- Improve the fidelity and realism of cost and schedule commitments made when the program or project is approved to transition from Formulation to Implementation.

NPR 7120.5E changes also reflect a strengthening of the elements of Governance, which include the independent role of the Technical Authority and an expanded role for Center Directors as full partners. To support these policy changes, NPR 7120.5E defined a new structure for formulating, baselining, and rebaselining (if necessary) the agreements that set the parameters within which programs and projects work. These concepts are explained in more detail in this handbook.

The handbook also provides additional details on the cost and schedule estimates, probabilistic assessments, and earned value management as it applies at NASA. As applicable to some programs and projects, it also provides information and discussions on developing confidence levels leading to the formal joint cost and schedule confidence level established at the transition from Formulation to Implementation.

The life cycles for programs and projects have been refined and are explained in more detail in Chapter 3 (programs) and Chapter 4 (projects) including the timing of the life cycle transitions, clarifying the reviews leading to launch, and describing the added decommissioning review. Additionally, this handbook describes the expected maturity of products at reviews and the elements of program and project plans.

Finally, there is also increased emphasis on the ability and need to properly tailor these requirements to fit the size, complexity, cost and risk of the program/project. Tailoring can be expeditiously captured for approval in the Compliance Matrix that accompanies the Program or Project Plan.

The information, techniques, methodologies, and practices described in this handbook are the compilation of best practices and lessons learned from some of the best program and project managers, systems engineers, technical teams, procurement specialists, scientists, financial managers, and leadership within the Agency, academia, commercial organizations and other government agencies. Thank you for your dedication and insight.

Mike Ryschkewitsch
NASA Chief Engineer

Acknowledgments

Envisioned by NASA Chief Engineer Mike Ryschkewitsch in conjunction with streamlining NPR 7120.5E, the handbook effort was led by Michael Blythe for Sandra Smalley and Ellen Stigberg of the Office of the Chief Engineer. The core team members on the project were Linda Bromley, Robert Moreland, David Pye, Mark Saunders, Kathy Symons, and Linda Voss. Additional editorial team members included Brenda Bailey, Steve Caporaletti, Nita Congress, Diedtra Henderson, Keith Maynard, Brenna McErlean, Rob Traister, Steven Waxman, and Grace Wiedeman.

The team would like to acknowledge in appreciation the time, effort, and expertise contributed by many to draft, develop, improve, and complete the content of this handbook. Many discussions sharpened the precision of the language. This body of knowledge is made possible through the efforts of those who live it and work to shape it for the better. It is not possible to list all the formal and informal contributions. Apologies and thanks to those who contributed but are not listed below.

Subject matter expertise was contributed by:
Sue Aleman, HQ
Omar Baez, KSC
Greg Blaney, IV & V
Roger Galpin, JSC
Johanna Gunderson, HQ
Charles Hunt, HQ
Tupper Hyde, HQ
Jim Lawrence, HQ
Ken Ledbetter, HQ
Jeff Leising, JPL
David Liskowsky, HQ
Cynthia Lodge, HQ
John Lyver, JPL
Bryan O'Connor, HQ
James Ortiz, JSC
Eric Plumer, HQ
Anne Sweet, HQ
Anna Tavornmina, JPL
Richard Williams, HQ
Robert Woods, HQ
Mary Beth Zimmerman, HQ

Inputs were also contributed by:
Tahani Amer, LARC
Ronald Baker, LARC
Simon Chung, LARC
Alfredo Colon, HQ
John Decker, GSFC
Homayoon Dezfuli, HQ
James Kyle Johnson, HQ

John Kelly, HQ
Mary Kerwin, HQ
Arthur Maples, HQ
Trisha Pengra, HQ
Harry Ryan, SSC
Charles Smith, ARC
Kathleen Teale, HQ
Marc Timm, HQ
J.K. Watson, HQ
Terrence Wilcutt, HQ
Donna Wilson, HQ

Review comments were coordinated by:
Dan Blackwood, GSFC
Kris Brown, HQ
Mina Cappuccio, ARC
Diane Clayton, HQ
Dan Dittman, ARC
Lonnie Dutriex, SSC
John Gagosian, HQ
Paul Gilbert, MSFC
Eve Lyon, HQ
Mike McNeill, HQ
Constance Milton, KSC
Ray Morris, JPL
Lara Petze, HQ
Nang Pham, GRC
Irene Piatek, JSC
Kevin Power, SSC
Jennifer Rochlis, JSC
Jan Rogers, MSFC
Joseph Smith, HQ
Kevin Weinert, DFRC

Other reviewers included:
George Albright, HQ
David Anderson, GRC
Rob Anderson, HQ
Dan Andrews, ARC
Melissa Ashe, LARC
David Beals, LARC
Hal Bell, HQ
Christine Bonniksen, HQ
Jack Bullman, HQ

Madeline Butler, HQ
Edgar Castro, JSC
Carolyn Dent, GSFC
Eric Eberly, MSFC
Michele Gates, HQ
Helen Grant, HQ
David Hamilton, JSC
Robert Hammond, SSC
Bob Hodson, LARC
Steve Kapurch, HQ
Jerald Kerby, MSFC
William Knopf, HQ
Trudy Kortes, GRC
Jeri Law, MSFC
Alan Little, LARC
Kelly Looney, MSFC
Bill Luck, LARC
Bill Marinelli, HQ
Paul McConnaughey, MSFC
Steven McDaniel, MSFC
Dave Mobley, MSFC
Luat Nguyen, LARC
Louis Ostrach, HQ
Todd Peterson, GRC
Charles Polen, LARC
James Price, LARC
Eric Rissling, LARC
Kevin Rivers, LARC
Steven Robbins, MSFC
Kathy Roeske, MSFC
Monserrate Roman, MSFC
Robert Ross, SSC
Jim Schier, HQ
Marshall Smith, HQ
Carie Sorrels, HQ
Van Strickland, MSFC
Barmac Taleghani, LARC
Sandeep Wilkhu, KSC

Scientific and Technical Information review and publication was supported by:
Darline Brown, HQ
Aaron Goad, HQ

1 Introduction

1.1 Purpose

This handbook is a companion to *NPR 7120.5E, NASA Space Flight Program and Project Management Requirements* and supports the implementation of the requirements by which NASA formulates and implements space flight programs and projects. Its focus is on what the program or project manager needs to know to accomplish the mission, but it also contains guidance that enhances the understanding of the high-level procedural requirements. (See Appendix C for NPR 7120.5E requirements with rationale.) As such, it starts with the same basic concepts but provides context, rationale, guidance, and a greater depth of detail for the fundamental principles of program and project management. This handbook also explores some of the nuances and implications of applying the procedural requirements, for example, how the Agency Baseline Commitment agreement evolves over time as a program or project moves through its life cycle.

1.2 Document Structure

Guidance begins in Chapter 2 with a high-level overview of NASA's space flight program and project management structure and references to specific topics elsewhere in the document that provide greater levels of detail. The overview also includes NASA's Governance structure and a description of the program and project life cycles and management decision points.

Details of the activities in the phases of the life cycle begin in Chapter 3 with programs and continue in Chapter 4 with projects. These chapters capture the flow of program and project activities and give a perspective on what needs to be accomplished while progressing through the phases of the program and project life cycles. Chapter 3 describes the four different program types, their common activities, and how they differ. Chapter 4 covers activities for all categories of projects, with a greater focus on Category 1 projects. All the activities to meet the requirements of a Category 1 project are detailed, including activities that may not be applicable to Category 2 or 3 projects. To help Category 2 and 3 projects, a section in Chapter 4 describes how projects can and typically do tailor the NPR 7120.5E requirements to meet their specific needs. A specific example of tailoring for small projects is provided in Table 4-2. (See Section 4.1.5.)

Chapter 5 delves into special topics to explain important concepts from NPR 7120.5E in more detail. It explains the nuances and implications of Governance, Technical Authority, tailoring principles, and the Dissenting Opinion process and how they are implemented in specific situations such as a project being developed in a multi-Center environment. Key program and project documentation is explored in more detail in the section on maturing, approving, and maintaining baselines that include the Agency Baseline Commitment and the Management Agreement. Other special topics include:

- Earned value management;
- Analyses/work supporting decisions, including joint cost and schedule confidence level analysis;
- The Federal budgeting process;
- The independent Standing Review Boards and life-cycle reviews;
- Other reviews such as the Termination Review;
- Requirements for external reporting;
- Program or project management selection and certification;
- Leading indicator guidance; and
- The work breakdown structure and its relationship to Agency financial processes.

1.3 How to Use This Handbook

This handbook was structured as a reference document to make it useful from the perspective of the practitioner. The focus is on the activities a program or project manager needs to perform with context and explanation for the requirements. Rather than reading the handbook as a chronological narrative, the program or project manager can go to a specific section to learn about a particular area of interest, i.e., Section 5.3 on Dissenting Opinion. Chapter 3 on programs and Chapter 4 on projects stand on their own, so a project manager can go to Chapter 4 and determine what is required in one place. That means that some of the material that is common between chapters and phases is duplicated to be complete. When a particular topic such as the work breakdown structure is introduced, it is defined in margin text. If the topic is discussed in greater depth in this handbook, the reader is referred to that location. On occasion, the reader will be referred to another handbook or a community of practice for more in-depth knowledge.

Additional margin text contains content about key concepts, including points of elucidation or emphasis on best practices as well as rationales or principles behind some of the requirements. In addition, required products are bolded in the text, so content about them can be more easily located.

Though the content of this handbook was intended to stand the test of time, the electronic version of the handbook is subject to revision as NPR 7120.5 evolves. However, dynamic content was reserved for online forums. For example, information supplemental to policy documents can be found in the Office of the Chief Engineer listing under the "Other Policy Documents" tab in the NASA Online Directives Information System (NODIS) library and the NASA Engineering Network (NEN) Program and Project Management Community of Practice (PM CoP). While the handbook presents core information, it also references extended content with pointers to various NASA communities of practice which contain additional guidance, best practices, and templates that are updated to be current with latest practice. Also, additional information in other handbooks, websites, and policy documents is liberally referenced rather than duplicated.

2 High-Level Overview of Program and Project Management

Space flight programs and projects are often the most visible and complex of NASA's strategic investments. These programs and projects flow from the implementation of national priorities, defined in the Agency's Strategic Plan, through the Agency's Mission Directorates as part of the Agency's programmatic organizational hierarchy shown in Figure 2-1.

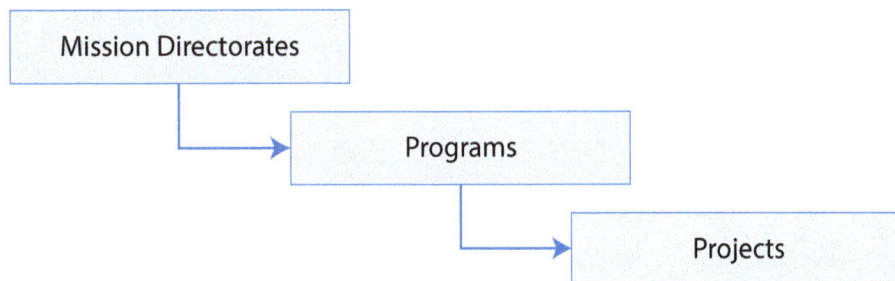

```
┌─────────────────────────┐
│   Mission Directorates   │
└─────────────────────────┘
          │
          └──────────▶ ┌─────────────────────────┐
                       │        Programs          │
                       └─────────────────────────┘
                                  │
                                  └──────────▶ ┌─────────────────────────┐
                                               │        Projects          │
                                               └─────────────────────────┘
```

Figure 2-1 Programmatic Authority Organizational Hierarchy

This hierarchical relationship of programs to projects shows that programs and projects are different and their management involves different activities and focus. The following definitions are used to distinguish the two:

- **Program**—a strategic investment by a Mission Directorate or Mission Support Office that has a defined architecture, and/or technical approach, requirements, funding level, and a management structure that initiates and directs one or more projects. A program implements a strategic direction that the Agency has identified as needed to accomplish Agency goals and objectives.

Architecture is the structure of components, their relationships, and the principles and guidelines governing their design and evolution over time.

- **Project**—space flight projects are a specific investment identified in a Program Plan having defined requirements, a life-cycle cost,[1] a beginning, and an end. A project also has a management structure and may have interfaces to other projects, agencies, and international partners. A project yields new or revised products that directly address NASA's strategic goals.

All NASA space flight programs and projects are subject to NPR 7120.5 requirements. NPR 7120.5 requirements apply to contractors, grant recipients, or parties to agreements only to the extent specified or referenced in the appropriate contracts, grants, or agreements.

2.1 Overview of Program and Project Life Cycles

NASA manages programs and projects to life cycles that include the systems engineering processes described in *NPR 7123.1, NASA Systems Engineering Processes and Requirements*. These life cycles are divided into defined phases that correspond to specific activities and increasing levels of expected maturity of information and products. A program or project moves through the life-cycle phases as it progresses from concept to operations, and ultimately to decommissioning. Programs and projects are periodically evaluated at specific points to gain formal approval to progress through their life cycle.

At the top level, program and project life cycles are divided into two phases, Formulation and Implementation. (See Section 2.6 and Figure 2-4 for a description of the activities of these phases.) The activities and work to be accomplished in these phases are as follows:

- **Formulation**—identifying how the program or project supports the Agency's strategic goals; developing and allocating program requirements to initial projects; deriving a technical approach from an analysis of alternatives; assessing risk and possible risk mitigations; conducting engineering and technology risk reduction activities; developing organizational structures and building teams; developing operations concepts and acquisition strategies; developing preliminary cost and schedule

---

[1] The life-cycle cost (LCC) is the total cost of the program or project over its planned life cycle from Formulation (excluding Pre–Phase A) through Implementation (excluding extended operations). For long-duration (decades) programs such as human space flight programs, it is difficult to establish the duration of the life cycle for the purposes of determining the LCC. Under these circumstances, programs define their life-cycle scope in the Formulation Authorization Document (FAD) or Program Commitment Agreement (PCA). Projects that are part of these programs document their LCC in accordance with the life-cycle scope defined in their program's Program Plan, PCA or FAD, or the project's FAD. The life-cycle cost is discussed in more detail in Section 5.5.

estimates; establishing high-level requirements, requirements flow down, and success criteria; developing preliminary designs; assessing the relevant industrial base/supply; preparing plans essential to the success of a program or project; and establishing control systems to ensure accurate execution of those plans.

- **Implementation**—executing approved plans for the development and operation of the program and/or project; using control systems to ensure performance to approved plans and requirements and continued alignment with the Agency's strategic goals; performing acquisition, detailed design, manufacturing, integration, and test; conducting operations; implementing sustainment during which programs' constituent projects are initiated and their formulation, approval, implementation, integration, operation, and ultimate decommissioning are constantly monitored; and adjusting the program and/or project as resources and requirements change.

There are three different life cycles for four different types of programs (see Chapter 3) and one life cycle for three categories of projects (see Chapter 4). The life cycles are divided into phases. Transition from one phase to another requires management approval at Key Decision Points (KDPs). (See Section 2.2.3.) The phases in program and project life cycles include one or more life-cycle reviews, which are considered major milestone events. A life-cycle review is designed to provide the program or project with an opportunity to ensure that it has completed the work of that phase and an independent assessment of a program's or project's technical and programmatic status and health. The final life-cycle review in a given life-cycle phase provides essential information for the KDP that marks the end of that life-cycle phase and transition to the next phase if successfully passed. As such, KDPs serve as gates through which programs and projects must pass to continue.

KDPs for projects are designated with capital letters, e.g., KDP A. The letter corresponds to the project phase that will be entered after successfully passing through the gate. Program KDPs and life-cycle reviews are analogous to project KDPs and life-cycle reviews. KDPs for single-project programs are designated with letters as are projects, i.e., KDP A, KDP B, etc. KDPs associated with other types of programs (i.e., uncoupled, loosely coupled, and tightly coupled) are designated with Roman numerals and zero. The first KDP is KDP 0, the second is KDP I, etc.

Life-cycle reviews are essential elements of conducting, managing, evaluating, and approving space flight programs and projects and are an important part of NASA's system of checks and balances. Life-cycle reviews are conducted by the program and project and often an independent Standing Review Board (SRB). (SRBs are defined and discussed further in

NASA defines acquisition as the process for obtaining the systems, research, services, construction, and supplies that the Agency needs to fulfill its mission. Acquisition—which may include procurement (contracting for products and services)—begins with an idea or proposal that aligns with the NASA Strategic Plan and fulfills an identified need and ends with the completion of the program or project or the final disposition of the product or service. (The definition of acquisition in accordance with *NPD 1000.5, Policy for NASA Acquisition* is used in a broader context than the FAR definition to encompass strategic acquisition planning and the full spectrum of various NASA acquisition authorities and approaches to achieve the Agency's mission and activities.)

Section 3.1.1, Section 4.1.1, and Section 5.10.) NASA accords special importance to maintaining the integrity of its independent review process. Life-cycle reviews provide the program or project and NASA senior management with a credible, objective assessment of how the program or project is progressing. The independent review also provides vital assurance to external stakeholders that NASA's basis for proceeding is sound.

The KDP decision to authorize a program or project's transition to the next life-cycle phase is made by the program or project's Decision Authority. (See Section 2.2.1.) This decision is based on a number of factors, including technical maturity; continued relevance to Agency strategic goals; adequacy of cost and schedule estimates; associated probabilities of meeting those estimates (confidence levels); continued affordability with respect to the Agency's resources; maturity and the readiness to proceed to the next phase; and remaining program or project risk (safety, cost, schedule, technical, management, and programmatic). At the KDP, the key program or project cost, schedule, and content parameters that govern the remaining life-cycle activities are established.

Figure 2-2 shows a simplified, high-level version of the NASA project life cycle to illustrate the relationship between the phases, gates, and major events, including KDPs and major life-cycle reviews. Note that the program life cycles (discussed in Chapter 3) vary from this simplified life cycle depending on the program type.

2.2 Oversight and Approval

NASA has established a program and project management oversight process to ensure that the experience, diverse perspectives, and thoughtful programmatic and technical judgment at all levels are accessible, available, and applied to program and project activities. The Agency employs management councils and independent review boards, including the SRB, to provide the Decision Authority and upper management with insight on the status and progress of programs and projects and their alignment with Agency goals. This process enables a disciplined approach for developing the Agency's assessment, which informs the Decision Authority's KDP determination of program or project readiness to proceed to the next life-cycle phase.

This section describes NASA's oversight approach and the process by which programs and projects are approved to move forward through their life cycle. It defines and describes NASA's Decision Authority, management councils, and KDPs. (See Sections 3.2 and 4.2 for more detailed information on these topics.)

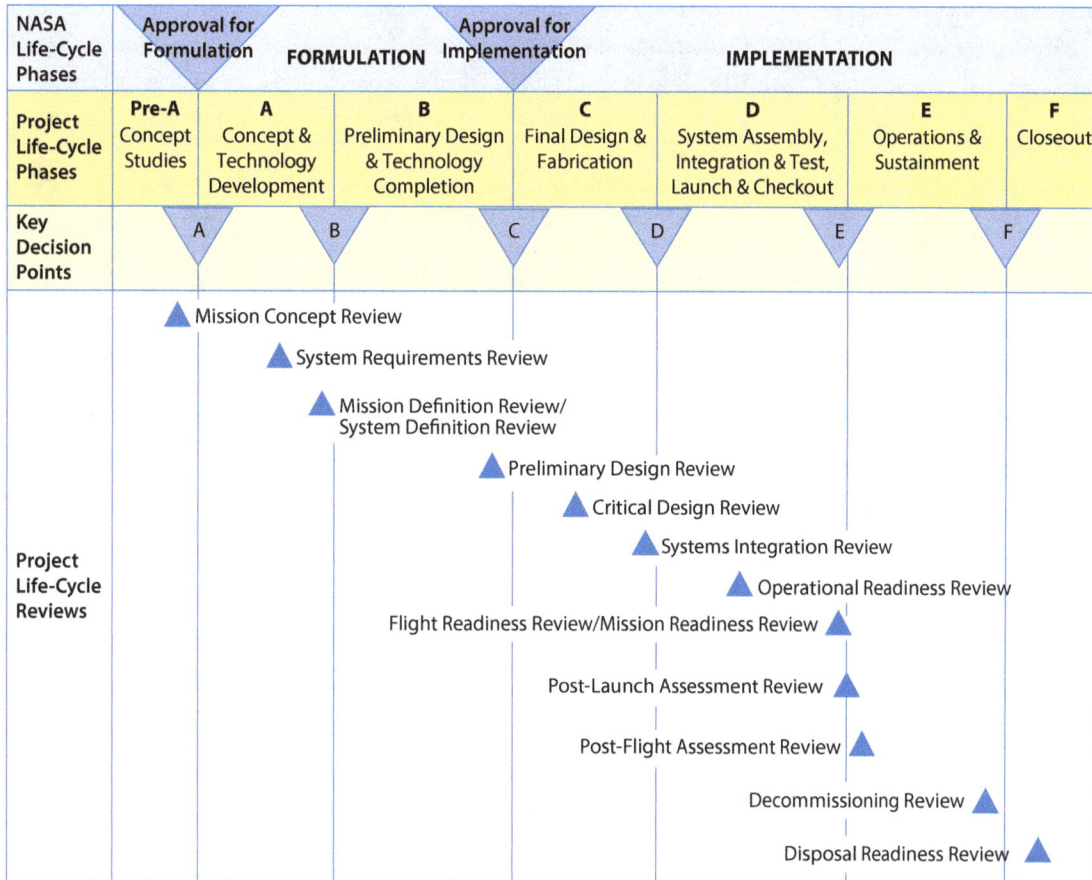

Figure 2-2 Simplified Project Life Cycle

2.2.1 Decision Authority

The Decision Authority is the Agency individual who is responsible for making the KDP determination on whether and how a program or project proceeds through the life cycle and for authorizing the key program cost, schedule, and content parameters that govern the remaining life-cycle activities.

For programs and Category 1 projects, the Decision Authority is the NASA Associate Administrator (AA). The NASA AA may delegate this authority to the Mission Directorate Associate Administrator (MDAA) for Category 1 projects. For Category 2 and 3 projects, the Decision Authority is the MDAA. (See Sections 2.4 and 2.5 for more information on program and project categories.)

The MDAA's limitation on delegation is necessary to preserve the separation of the roles of the Programmatic and Institutional Authorities as required by NASA Governance. (See Section 2.3.)

The PCA (see *NPR 7120.5*, Appendix D) is an agreement between the MDAA and the NASA AA (the Decision Authority) that authorizes program transition from Formulation to Implementation. The PCA is prepared by the Mission Directorate and documents Agency and Mission Directorate requirements that flow down to the program; program objectives, management, and technical approach and associated architecture; program technical performance, schedule, time-phased cost plans, safety and risk factors; internal and external agreements; life-cycle reviews; and all attendant top-level program requirements.

The MDAA may delegate to a Center Director the Decision Authority to determine whether Category 2 and 3 projects may proceed through KDPs into the next phase of the life cycle. However, the MDAA will retain authority for all program-level requirements, funding limits, launch dates, and any external commitments.

All delegations are documented and approved in either the Program Commitment Agreement (PCA) or the Program Plan (see NPR 7120.5E, Appendix G) depending on which Decision Authority is delegating.

The Decision Authority's role during the life cycle of a program and project is covered in more detail in NPR 7120.5E, Section 2.3 Program and Project Oversight and Approval, and in Chapters 3 and 4 of this handbook.

2.2.2 Management Councils

At the Agency level, NASA Headquarters has two levels of program management councils (PMCs): the Agency PMC (APMC) and the Mission Directorate PMC (DPMC). The PMCs evaluate the safety, technical, and programmatic performance and content of a program or project under their purview for the entire life cycle. These evaluations focus on whether the program or project is meeting its commitments to the Agency and on ensuring successful achievement of NASA strategic goals. For all programs and Category 1 projects, the governing PMC is the APMC. The NASA AA chairs the APMC. For all Category 2 and 3 projects, the governing PMC is the DPMC. The MDAA chairs the DPMC.

The governing PMC conducts reviews to evaluate programs and projects in support of KDPs; makes a recommendation to the Decision Authority on a program or project's readiness to progress in its life cycle; and provides an assessment of the program or project's proposed cost, schedule, and content parameters. A KDP normally occurs at the governing PMC review. Prior to the governing PMC review, the program or project is reviewed by the responsible Center Director and/or Center Management Council (CMC), which provides its findings and recommendations to the MDAA/DPMC. In cases where the governing PMC is the APMC, the responsible MDAA and/ or DPMC also conduct an in-depth assessment of the program or project. The Center Director/CMC and MDAA/DPMC provide their findings and recommendations to the APMC.

2.2.3 Key Decision Points

At KDPs, the Decision Authority reviews all the materials and briefings at hand to make a decision about the program or project's maturity and readiness to progress through the life cycle and authorizes the content, cost, and schedule parameters for the ensuing phase(s). The materials and briefings include findings and recommendations from the program manager, the project manager (if applicable), the SRB, the CMC, the DPMC, the MDAA (if applicable), and the governing PMC. KDPs conclude the life-cycle review at the end of a life-cycle phase. A KDP is a mandatory gate through which a program or project must pass to proceed to the next life-cycle phase.

The potential outcomes at a KDP include approval or disapproval to enter the next program or project phase, with or without actions for follow-up activities.

The KDP decision is summarized and recorded in the Decision Memorandum. The Decision Authority completes the KDP process by signing the Decision Memorandum. The expectation is to have the Decision Memorandum signed by concurring members as well as the Decision Authority at the conclusion of the governing PMC KDP meeting. (See more information on the Decision Memorandum, including signatories and their respective responsibilities, in Section 5.5.6, Decision Memorandum.)

A life-cycle review is complete when the governing PMC and Decision Authority complete their assessment and sign the Decision Memorandum.

2.3 Governance

To successfully implement space flight programs and projects, NASA's management focuses on mission success across a challenging portfolio of high-risk, complex endeavors, many of which are executed over long periods of time. *NPD 1000.0, NASA Governance and Strategic Management Handbook* sets forth the Governance framework through which the Agency manages its missions and executes its responsibilities. The Governance model provides for mission success by balancing different perspectives from different elements of the organization and is also fundamental to NASA's system of checks and balances.

The cornerstone of this organizational structure is the separation of the Programmatic and Institutional Authorities. The separation of authorities is illustrated in Figure 2-3.

Programmatic Authority resides within the Mission Directorates and their respective programs and projects. (Appendix D provides a summary of the roles and responsibilities for key program and project management officials.)

Institutional Authority encompasses all organizations and authorities not in Programmatic Authority. This includes the Mission Support Directorate and Mission Support Offices at Headquarters and associated organizations at the Centers; other mission support organizations; Center Directors; and

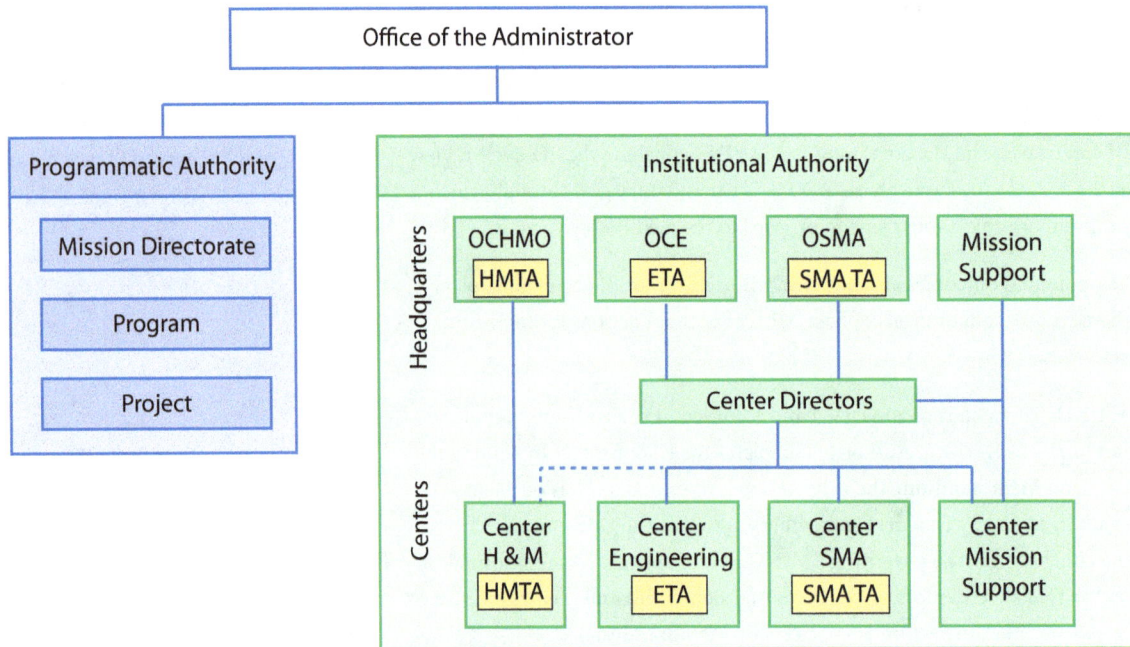

Legend: ---- indicates that not all Centers have HMTA. Sometimes that function is served by Engineering and SMA TAs.

Acronyms: OCE = Office of the Chief Engineer; OCHMO = Office of the Chief Health and Medical Officer; OSMA = Office of Safety and Mission Assurance; TA = Technical Authority.

Figure 2-3 Separation of Programmatic and Institutional Authorities

the Technical Authorities, who are individuals with specifically delegated authority in Engineering, Safety and Mission Assurance, and Health and Medical.

The Engineering, Safety and Mission Assurance, and Health and Medical organizations are a unique segment of the Institutional Authority. They support programs and projects in two ways:

- They provide technical personnel and support and oversee the technical work of personnel who provide the technical expertise to accomplish the program or project mission.

- They provide Technical Authorities, who independently oversee programs and projects. These individuals have a formally delegated Technical Authority role traceable to the Administrator and are funded independent of programs and projects.

(See Section 5.2 for more detail on the Technical Authorities.)

Each of these authorities plays a unique role in the execution of programs and projects. For example, with respect to requirements:

- Programmatic Authorities are responsible for "programmatic requirements" and focus on the products to be developed and delivered that specifically relate to the goals and objectives of a particular NASA program or project. These programmatic requirements flow down from the Agency's strategic planning process.

- Institutional Authorities are responsible for "institutional requirements." They focus on how NASA does business and are independent of any particular program or project. These requirements are issued by NASA Headquarters (including the Office of the Administrator and Mission Support Offices) and by Center organizations. Institutional requirements may respond to Federal statute, regulation, treaty, or Executive order.

For more information on the roles and responsibilities of these authorities see Appendix D.

The "Types of Requirements" box provides definitions for some basic types of requirements. See Appendix A for definitions of these and other types of requirements.

The Programmatic and Institutional Authorities are further explained in Section 5.1, NASA Governance.

2.4 NASA Programs

As a strategic management structure, the program construct is extremely important within NASA. Programs provide the critically important linkage between the Agency's strategic goals and the projects that are the specific means for achieving them.

NASA space flight programs are initiated and implemented to accomplish scientific or exploration goals that generally require a collection of mutually supporting projects. Programs integrate and manage these projects over time and provide ongoing enabling systems, activities, methods, technology developments, and feedback to projects and stakeholders. Programs are generally created by a Mission Directorate with a long-term time horizon in mind. Programs are generally executed at NASA Centers under the direction of the Mission Directorate and are assigned to Centers based on decisions made by Agency senior management consistent with the results of the Agency's strategic acquisition process.

The strategic acquisition process is the Agency process for ensuring that NASA's strategic vision, programs, projects, and resources are properly developed and aligned throughout the mission and life cycle. (See *NPD 1000.0, NASA Governance and Strategic Management Handbook*, and *NPD 1000.5, Policy for NASA Acquisition*, for additional information on the strategic acquisition process.)

Types of Requirements

Programmatic Requirements—focus on space flight products to be developed and delivered that specifically relate to the goals and objectives of a particular program or project. They are the responsibility of the Programmatic Authority.

Institutional Requirements—focus on how NASA does business independent of the particular program or project. They are the responsibility of the applicable Institutional Authority.

Allocated Requirements—established by dividing or otherwise allocating a high-level requirement into lower level requirements.

Derived Requirements—arise from:

• Constraints or consideration of issues implied but not explicitly stated in the higher level direction originating in Headquarters and Center institutional requirements or

• Factors introduced by the architecture and/or the design.

These requirements are finalized through requirements analysis as part of the overall systems engineering process and become part of the program/project requirements baseline.

Technical Authority Requirements—a subset of institutional requirements invoked by Office of the Chief Engineer, Office of Safety and Mission Assurance, and Office of the Chief Health and Medical Officer documents (e.g., NASA Procedural Requirements (NPRs) or technical standards cited as program or project requirements or contained in Center documents). These requirements are the responsibility of the office or organization that established the requirement unless delegated elsewhere.

Additional types of requirements are defined in Appendix A.

For additional information on the strategic acquisition process, refer to Sections 3.3.1 and 4.3.1.1. (See also *NPD 1000.5, Policy for NASA Acquisition.*)

Because the scientific and exploration goals of programs vary significantly in scope, complexity, cost, and criticality, different program management strategies are required ranging from simple to complex. As a result, the Agency has developed three different life cycles for four different program types: uncoupled, loosely coupled, tightly coupled, and single-project programs. These life cycles are illustrated in figures in Chapter 3 and show the program life-cycle phases; program life-cycle gates and major events, including KDPs; major program life-cycle reviews; and the process of recycling through Formulation when program changes warrant such action.

NASA programs have a common life-cycle management process, regardless of the type of program:

- **Program Formulation** is designed to establish a cost-effective program that is demonstrably capable of meeting Agency and Mission Directorate goals and objectives. During Formulation, the program team derives a technical approach, develops and allocates program requirements to initial projects and initiates project pre-Formulation activities, develops preliminary designs (when applicable), develops organizational structures and management systems, defines the program acquisition strategies, establishes required annual funding levels, and develops preliminary cost and schedule estimates.

- **Program Implementation** begins when the program receives approval to proceed to Implementation with the successful completion of KDP I (KDP C for single-project programs). Implementation encompasses program acquisition, operations, and sustainment, during which constituent projects are initiated. Constituent projects' formulation, approval, implementation, integration, operation, and ultimate decommissioning are constantly monitored. The program is adjusted to respond as needs, risks, opportunities, constraints, resources, and requirements change, managing technical and programmatic margins and resources to ensure successful completion of Implementation.

Independent evaluation activities occur throughout all phases.

2.5 NASA Projects

As with programs, projects vary in scope and complexity and, thus, require varying levels of management requirements and Agency attention and oversight. Consequently, project categorization defines Agency expectations of project managers by determining both the oversight council and the specific approval requirements of each category. Projects are Category 1, 2, or 3 and are assigned to a category based initially on: (1) the project life-cycle cost estimate, the inclusion of significant radioactive material,[2] and whether the system being developed is for human space flight; and (2) the priority level, which is related to the importance of the activity to NASA, the extent of international participation (or joint effort with other government agencies), the degree of uncertainty surrounding the application of new or untested technologies, and spacecraft/payload development risk classification. (See

[2] Nuclear safety launch approval is required by the Administrator or Executive Office of the President when significant radioactive materials are included onboard the spacecraft and/or launch vehicle. (Levels are defined in *NPR 8715.3, NASA General Safety Program Requirements*. See also Section 4.4.3.3 in this handbook.)

The life-cycle cost of the project includes all costs, including all unallocated future expenses and funded schedule margins for formulation and development through prime mission operations (the mission operations as defined to accomplish the prime mission objectives) to disposal, excluding extended operations.

Tightly coupled and single-project programs also have life-cycle costs. These programs document their life-cycle cost estimate in accordance with the life-cycle scope defined in the Formulation Authorization Document (FAD) or Program Commitment Agreement (PCA). Projects that are part of these programs document their life-cycle cost estimate in accordance with the life-cycle scope defined in their program's Program Plan, FAD or PCA, or the project's FAD.

NPR 7120.5E, NASA Space Flight Program and Project Management Requirements, Section 2.1 and Table 2-1, and Table 4-1 in this Handbook for a table of project categorization guidelines and *NPR 8705.4, Risk Classification for NASA Payloads* for payload risk classification guidelines.)

NASA projects have a common life cycle regardless of the category of the project: The project life cycle is explained in detail in Chapter 4:

- Although not part of the project life cycle, a Mission Directorate, typically supported by a program office, provides resources for concept studies (i.e., Pre–Phase A Concept Studies) prior to initiating a new project. These pre-Formulation activities involve Design Reference Mission analysis, feasibility studies, technology needs analyses, engineering systems assessments, and analyses of alternatives that typically are performed before a specific project concept emerges. Pre-Formulation activities include identifying risks that are likely to drive the project's cost and schedule and developing mitigation plans for those risks. Note that pre-Formulation costs are not included in life-cycle cost estimates.

- Project Formulation consists of two sequential phases, denoted as Phase A (Concept and Technology Development) and Phase B (Preliminary Design and Technology Completion). NASA places significant emphasis on project pre-Formulation and Formulation to ensure adequate preparation of project concepts and plans and mitigation of high-risk aspects of the project essential to position the project for the highest probability of mission success. During Formulation, the project explores the full range of implementation options, defines an affordable project concept to meet requirements, and develops needed technologies. The activities in these phases include developing the system architecture; completing mission and preliminary system designs; acquisition planning; conducting safety, technical, cost, and schedule risk trades; developing time-phased cost and schedule estimates and documenting the basis of these estimates; and preparing the Project Plan for Implementation. For projects with a life-cycle cost greater than $250 million, these activities allow the Agency to present to external stakeholders time-phased cost plans and schedule range estimates at KDP B and high-confidence cost and schedule commitments at KDP C.

- At KDP C, Project Approval for Implementation, the Decision Authority approves or disapproves the transition to Implementation and the technical scope, cost estimate, and schedule estimate.

- Project Implementation consists of Phases C, D, E, and F. During Phase C (Final Design and Fabrication) and Phase D (System Assembly, Integration and Test, Launch and Checkout), the primary activities are developmental in nature, including acquisition contract execution. Phase C includes completion of final system design and the fabrication, assembly, and test of components, assemblies, and subsystems. Phase D includes system assembly, integration, and test; prelaunch activities; launch; and on-orbit checkout (robotic projects) or initial operations (human space flight projects). All activities are executed according to the Project Plan developed during Formulation. KDP E marks approval to launch. After successful on-orbit checkout or initial operations, the project transitions to Phase E. The start of Phase E (Operations and Sustainment) marks the transition from system development and acquisition activities to primarily systems operations and sustainment activities. In Phase F (Closeout), project space flight and associated ground systems are taken out of service and safely disposed of, although scientific and other analyses might continue under project funding. Independent evaluation activities occur throughout all phases.

2.6 Interrelationships Between NASA Programs and Projects

Figure 2-4 is a summary of the NASA life cycles for space flight programs and projects and provides an overview of their interrelated life-cycle management processes.

PROGRAM LIFE CYCLE

| PROGRAM PRE- FORMULATION | Program Initiation | PROGRAM FORMULATION | A (Imple |

A (includes Decision Authority & AA) establishes strategic goals and outcomes for new work based on external direction or guidance.
A revises Strategic Plan or issues guidance document.

MD(s) issue program FAD(s) to initiate program development based on results of strategic planning meetings.
AA approves ACD, if appropriate.

MD establishes program office.
Program develops preliminary technical and management approach including work assignments, acquisition strategies, budget and schedules, requirements, interfaces, etc.

Program conducts SRR).
Program conducts ASM at appropriate time.
Program conducts SDR.

A and affected MDAAs and other senior mgmt conduct strategic planning to assess approaches, organizational structures, initial work assignments, acquisition strategies, etc.

Program prepares budget and submits for approval as part of annual budget process.
Program updates annually as required, ensuring program content and budget match.

If desired by AA, AA conducts KDP 0 and provides guidance and/or approval.

Program resolves any issues

Program decomposes program requirements to specific project requirements, including project-to-project interfaces
Program may begin initiation of project pre-formulation and may authorize project formulation to coincide with program formulation, particularly for tightly coupled programs.

Tightly Coupled Program conducts PDR.
AA conducts KDP I and approves PCA.
MDAA approves Program Plan.

Key

A	Administrator
AA	Associate Administrator
ACD	Architectural Control Document
ASM	Acquisition Strategy Meeting
DRR	Disposal Readiness Review
FAD	Formulation Authorization Document
KDP	Key Decision Point
MCR	Mission Concept Review
MD	Mission Directorate
MDAA	Mission Directorate Associate Administrator
MDR	Mission Definition Review
PCA	Program Commitment Agreement
PDR	Preliminary Design Review
SDR	System Definition Review
SRR	System Requirements Review

PROJECT PRE-FORMULATI

Pre-Phase A

Program establishes team to conduct broad range of concept studies that meet Agency goals and program requirements; defines management and technical approaches, and selects acceptable alternatives.
Pre-Project Team conducts MCRs.
AA/MDAA conducts project ASMs, as required.
MD/Program develops FAD.
Project develops Formulation Agreement.

Decision Authority conducts KDP A.
MDAA approves FAD and Formulation Agreement.
Decision Authority approves entry to Phase A.

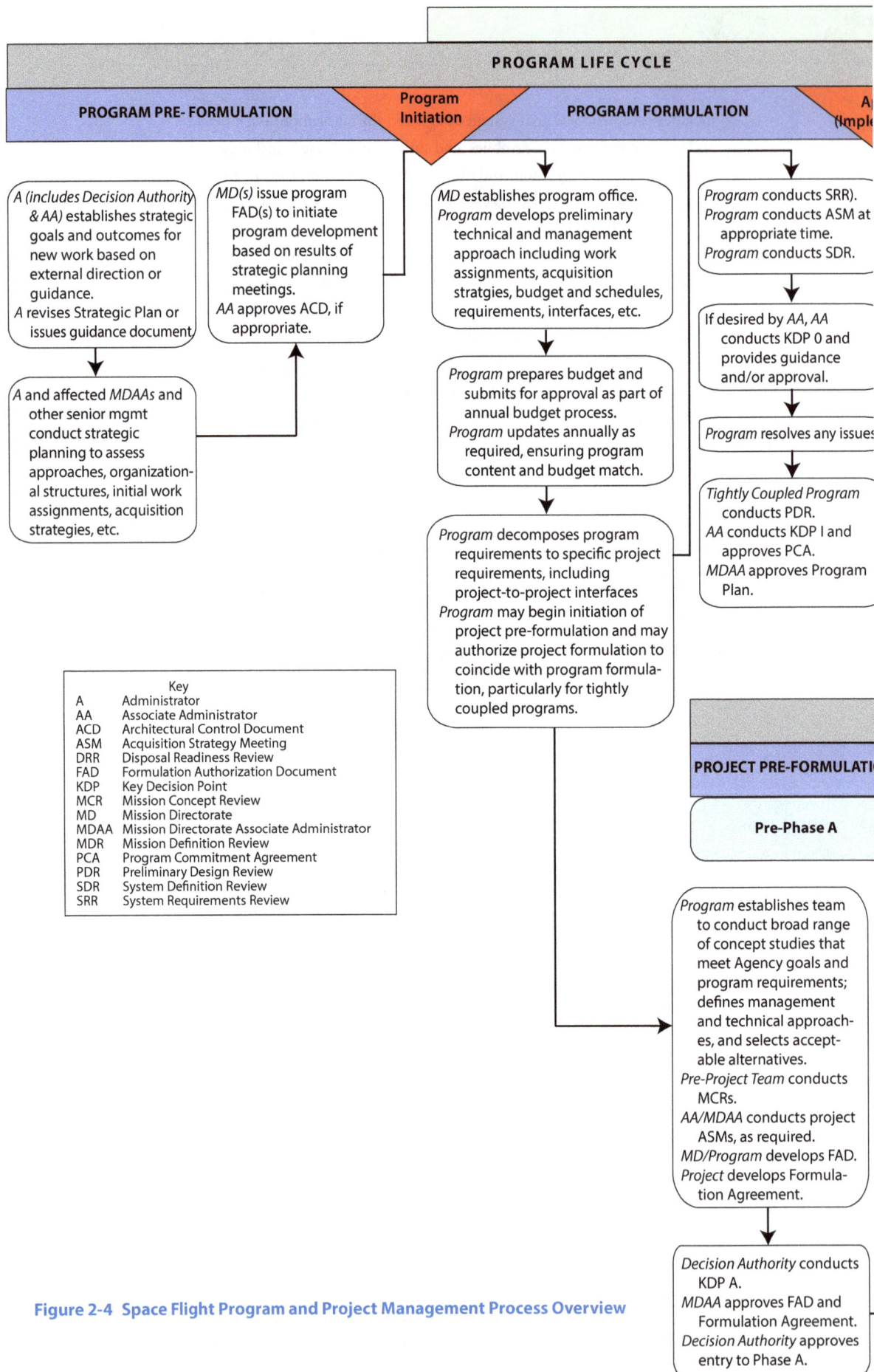

Figure 2-4 Space Flight Program and Project Management Process Overview

pproval
ementation)

PROGRAM IMPLEMENTATION

Program implements in accordance with Program Plan and program life cycle:
• Updates program approach, PCA, Program Plan, and budget when major budget or strategic issues require such changes.
• Conducts program reviews and KDPs as shown in Figures 3-1, 3-2, and 3-3.

Program integrates projects as necessary

Program initiates new projects, as required, and *MD/Program* approve project FADs.
Program oversees project formulation, approval, implementation and closeout.

PROJECT LIFE CYCLE

ON	PROJECT FORMULATION	Approval (Implementation)	PROJECT IMPLEMENTATION			
	Phase A	Phase B	Phase C	Phase D	Phase E	Phase F

Project Evaluation

MD/Program establish project office and conduct ASM.
Project develops concept, management and technical approaches, requirements, etc.; conducts SRR and refines technical approach.
Project conducts SDR or MDR, and develops preliminary Project Plan.

Project develops baseline design to meet requirements with acceptable risk within cost and schedule constraints, completes technology development, conducts PDR, and completes baseline Project Plan.

Project implements in accordance with Project Plan and project life cycle:
• Updates project approach, PCA, Program Plan, Project Plan and budget when major budget or content issues require such changes.
• Conducts project reviews.
• Supports special reviews and KDPs as required.

Decision Authority conducts KDP B and approves entry to Phase B.

Decision Authority conducts KDP C and approves Agency Baseline Commitment and entry to Phase C.
MDAA approves Project Plan.

Project conducts Decommissioning Review.
Decision Authority conducts KDP F and approves decommissioning.

Project conducts DRR.

Project executes Decommissioning/Disposal Plan.

3 Program Life Cycles, Oversight, and Activities by Phase

3.1 NASA Programs

A program implements a strategic direction that the Agency has identified as needed to accomplish Agency goals and objectives. Because the scientific and exploration goals of programs vary significantly, in scope, complexity, cost, and criticality, different program management strategies are required ranging from simple to complex. To accommodate these differences, NASA identifies four basic types of programs that may be employed:

- Single-project programs (e.g., James Webb Space Telescope (JWST)) tend to have long development and operational lifetimes and represent a large investment of Agency resources. Multiple organizations or agencies contribute to them. Generally, single-project programs have one project, but the program and project may be combined into a program structure.

- Uncoupled programs (e.g., Discovery Program) are implemented under a broad theme (like planetary science) and/or a common program implementation mechanism, such as providing flight opportunities for formally competed cost-capped projects or Principal Investigator (PI)-led missions and investigations. Each project in an uncoupled program is independent of the other projects within the program.

- Loosely coupled programs (e.g., Mars Exploration Program) address specific objectives through multiple space flight projects of varied scope. While each project has an independent set of mission objectives, the projects as a whole have architectural and technological synergies and strategies that benefit the program. For instance, Mars orbiters designed for more than one Mars year in orbit are required to carry a communication system to support present and future landers.

- Tightly coupled programs have multiple projects that execute portions of a mission or missions. No single project is capable of implementing a complete mission. Typically, multiple NASA Centers contribute to the program. Individual projects may be managed at different Centers. The program may also include other agency or international partner contributions.

3.1.1 Program Life Cycles

Programs follow a life cycle that matches their program type. The different life cycles formalize the program management process. The life cycles for uncoupled and loosely coupled programs, tightly coupled programs, and single-project programs are shown in Figures 3-1, 3-2, and 3-3, respectively. These life-cycle figures illustrate the different life-cycle phases, gates, and major events, including Key Decision Points (KDPs); major life-cycle reviews; and principal documents that govern the conduct of each phase. They also show how programs recycle through Formulation when program changes warrant such action.

Each program life-cycle phase includes one or more life-cycle reviews. A life-cycle review is designed to provide a periodic assessment of a program's technical and programmatic status and health at a key point in the life cycle. Life-cycle reviews are essential elements of conducting, managing, evaluating, and approving space flight programs and are an important part of NASA's system of checks and balances. Most life-cycle reviews are conducted by the program and an independent Standing Review Board (SRB). NASA accords special importance to maintaining the integrity of its independent review process to gain the value of an independent technical and programmatic perspective.

Life-cycle reviews provide the program and NASA senior management with a credible, objective assessment of how the program is doing. The final life-cycle review in a given program life-cycle phase provides essential information for the KDP, which marks the end of that life-cycle phase. A KDP is the point at which a Decision Authority determines whether and how a program proceeds through the life cycle, and authorizes key program cost, schedule, and content parameters that govern the remaining life-cycle activities. For programs, the Decision Authority is the NASA Associate Administrator. A KDP serves as a mandatory gate through which a program must pass to proceed to the next life-cycle phase. During the period between the life-cycle review and the KDP, the program continues its planned activities unless otherwise directed by the Decision Authority.

KDPs associated with uncoupled, loosely coupled, and tightly coupled programs are designated with Roman numerals except for the potential first KDP, which is KDP 0. Because of the close correlation of steps between a single-project program and project life cycles, KDPs for single-project programs are designated by letters (KDP A, etc.).

The Standing Review Board (SRB) is a group of independent experts who assess and evaluate program and project activities, advise programs and convening authorities, and report their evaluations to the responsible organizations, as identified in Figure 3-6. They are responsible for conducting independent reviews (life cycle and special) of a program and providing objective, expert judgments to the convening authorities. The reviews are conducted in accordance with approved Terms of Reference and life-cycle requirements per *NPR 7120.5* and *NPR 7123.1, NASA Systems Engineering Processes and Requirements.* For more detail, see Section 5.10 of this handbook and the *NASA Standing Review Board Handbook.*

For uncoupled and loosely coupled programs, the Formulation Phase is completed at KDP I after the program System Definition Review (SDR). Program approval for Implementation occurs at KDP I. After that, as depicted in Figure 3-1, Program Implementation Reviews (PIRs) are conducted during the Implementation Phase. (See Section 5.11.3 in this handbook for guidance on PIRs.) The Decision Authority determines on an annual basis the need for PIRs to assess the program's performance, evaluate its continuing relevance to the Agency's Strategic Plan, and authorize its continuation.

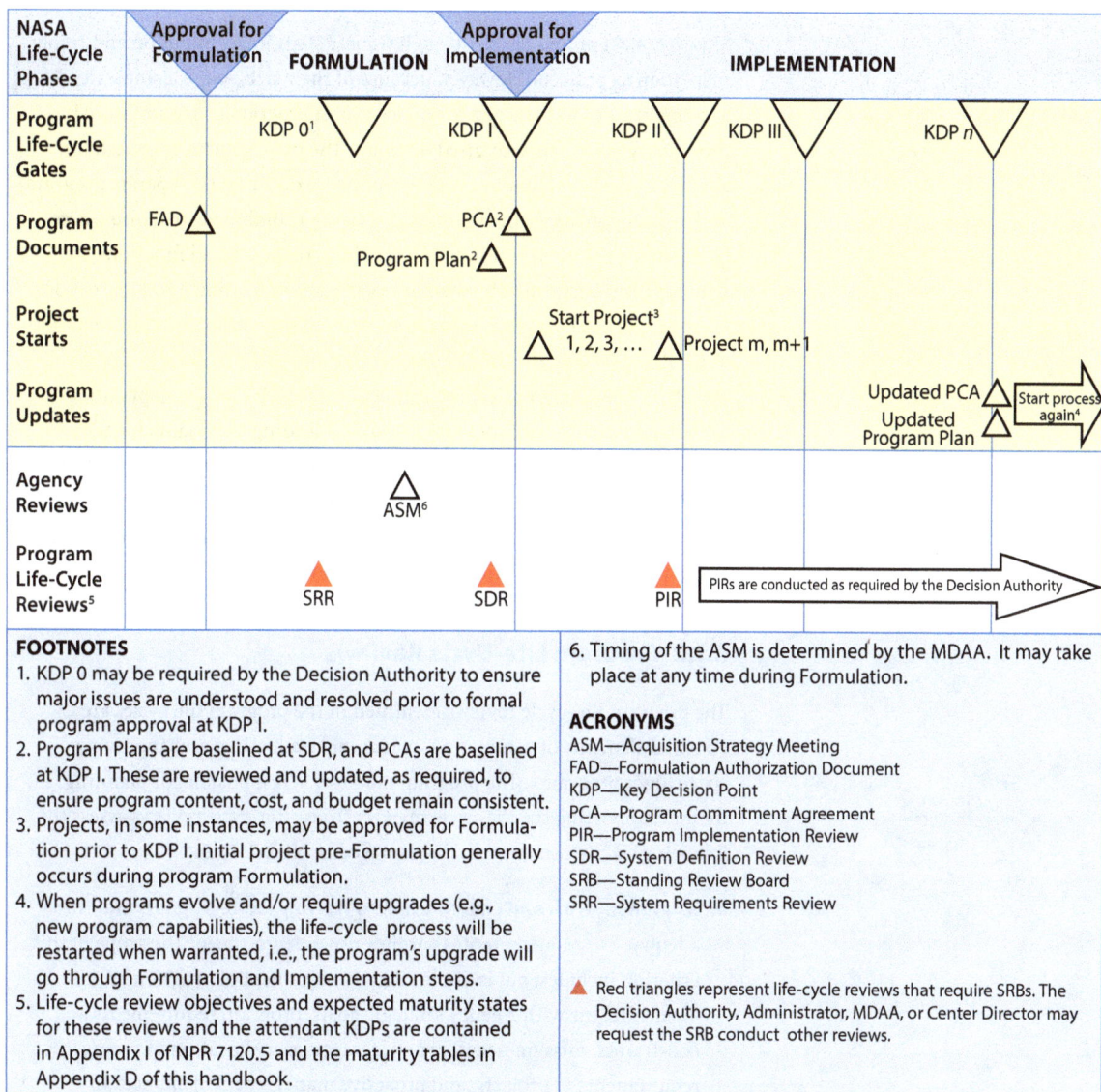

FOOTNOTES

1. KDP 0 may be required by the Decision Authority to ensure major issues are understood and resolved prior to formal program approval at KDP I.
2. Program Plans are baselined at SDR, and PCAs are baselined at KDP I. These are reviewed and updated, as required, to ensure program content, cost, and budget remain consistent.
3. Projects, in some instances, may be approved for Formulation prior to KDP I. Initial project pre-Formulation generally occurs during program Formulation.
4. When programs evolve and/or require upgrades (e.g., new program capabilities), the life-cycle process will be restarted when warranted, i.e., the program's upgrade will go through Formulation and Implementation steps.
5. Life-cycle review objectives and expected maturity states for these reviews and the attendant KDPs are contained in Appendix I of NPR 7120.5 and the maturity tables in Appendix D of this handbook.

6. Timing of the ASM is determined by the MDAA. It may take place at any time during Formulation.

ACRONYMS

ASM—Acquisition Strategy Meeting
FAD—Formulation Authorization Document
KDP—Key Decision Point
PCA—Program Commitment Agreement
PIR—Program Implementation Review
SDR—System Definition Review
SRB—Standing Review Board
SRR—System Requirements Review

▲ Red triangles represent life-cycle reviews that require SRBs. The Decision Authority, Administrator, MDAA, or Center Director may request the SRB conduct other reviews.

Figure 3-1 NASA Uncoupled and Loosely Coupled Program Life Cycle

Tightly coupled programs are more complex as shown in Figure 3-2. Since the program is intimately tied to its projects, the Formulation Phase mirrors the single-project program life cycle shown in Figure 3-3, and program approval for Implementation occurs at KDP I after the program-level Preliminary Design Review (PDR). In the Implementation Phase, program life-cycle reviews continue to be tied to the project life-cycle reviews to ensure the proper integration of projects into the larger system. Once a tightly coupled program is in operations, the Decision Authority determines on an annual basis the need for PIRs to assess the program's performance, evaluate its continuing relevance to the Agency's Strategic Plan, and authorize its continuation.

Single-project programs go through similar steps in Formulation and Implementation as projects. However, because of their size, scope, complexity, and importance to the Agency, single-project programs have additional program requirements imposed on them. The management approach for single-project programs can take one of two structures: (1) separate program and project management organizations or (2) a combined structure where both program and project functions are integrated, and all functions are managed and performed by the one organization. As shown in Figure 3-3, the single-project program transitions from Formulation to Implementation at KDP C following the single-project program's PDR. Following approval at KDP C, the single-project program continues with design, fabrication/manufacturing, system integration, and test leading up to launch and checkout following KDP E. Once a single-project program is in operations, the Decision Authority determines on an annual basis the need for PIRs to assess the program's performance, evaluate its continuing relevance to the Agency's Strategic Plan, and authorize its continuation.

3.1.2 Program Life-Cycle Reviews

The program life-cycle reviews identified in the program life cycles are essential elements of conducting, managing, evaluating, and approving space flight programs. The program manager is responsible for planning for and supporting the life-cycle reviews. These life-cycle reviews assess the following six assessment criteria identified in NPR 7120.5:

- **Alignment with and contribution to Agency strategic goals and the adequacy of requirements that flow down from those.** The scope of this criterion includes, but is not limited to, alignment of program requirements/designs with Agency strategic goals, program requirements and constraints, mission needs and success criteria; allocation of program requirements to projects; and proactive management of changes in program scope and shortfalls.

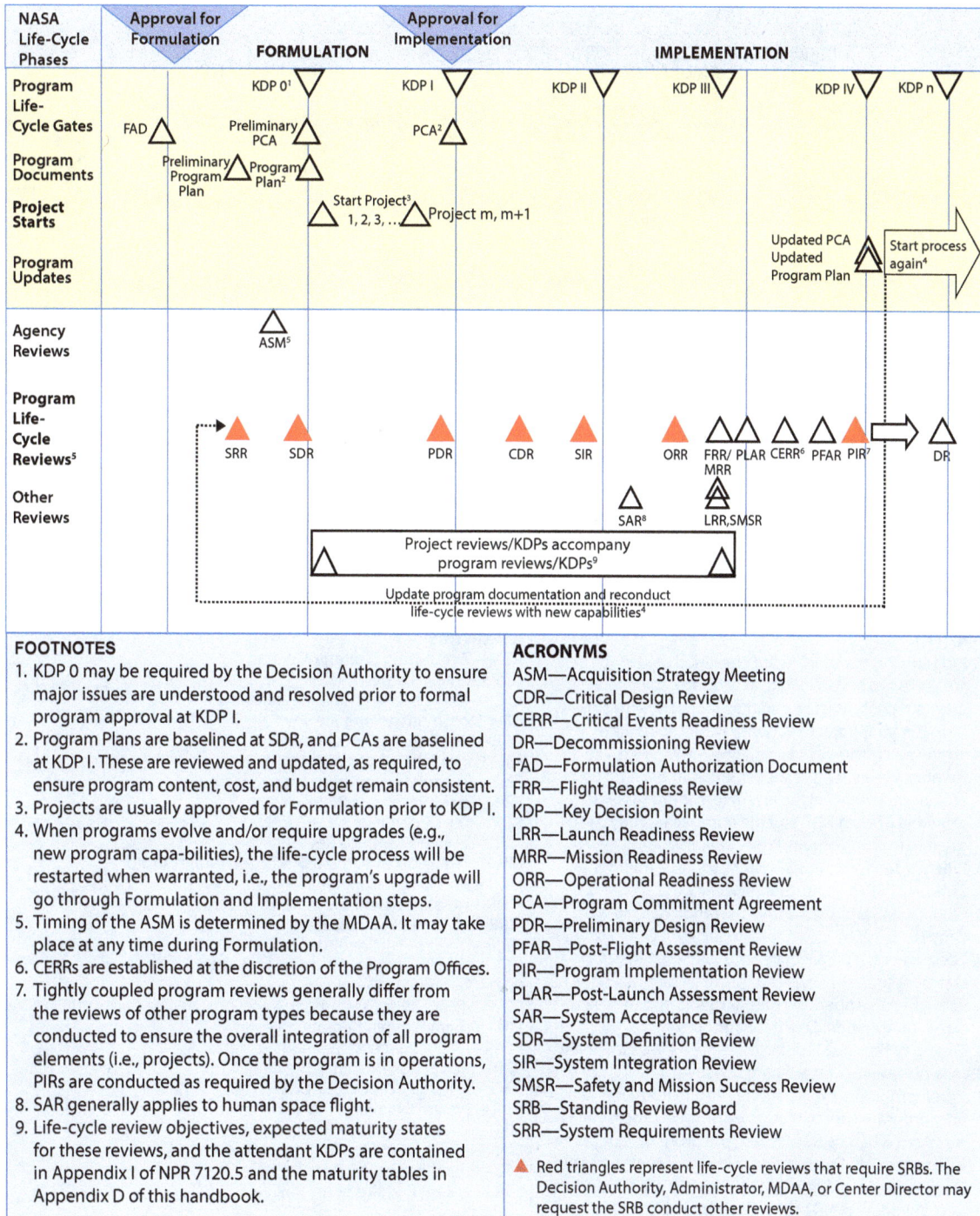

Figure 3-2 NASA Tightly Coupled Program Life Cycle

FOOTNOTES

1. KDP 0 may be required by the Decision Authority to ensure major issues are understood and resolved prior to formal program approval at KDP I.
2. Program Plans are baselined at SDR, and PCAs are baselined at KDP I. These are reviewed and updated, as required, to ensure program content, cost, and budget remain consistent.
3. Projects are usually approved for Formulation prior to KDP I.
4. When programs evolve and/or require upgrades (e.g., new program capa-bilities), the life-cycle process will be restarted when warranted, i.e., the program's upgrade will go through Formulation and Implementation steps.
5. Timing of the ASM is determined by the MDAA. It may take place at any time during Formulation.
6. CERRs are established at the discretion of the Program Offices.
7. Tightly coupled program reviews generally differ from the reviews of other program types because they are conducted to ensure the overall integration of all program elements (i.e., projects). Once the program is in operations, PIRs are conducted as required by the Decision Authority.
8. SAR generally applies to human space flight.
9. Life-cycle review objectives, expected maturity states for these reviews, and the attendant KDPs are contained in Appendix I of NPR 7120.5 and the maturity tables in Appendix D of this handbook.

ACRONYMS

ASM—Acquisition Strategy Meeting
CDR—Critical Design Review
CERR—Critical Events Readiness Review
DR—Decommissioning Review
FAD—Formulation Authorization Document
FRR—Flight Readiness Review
KDP—Key Decision Point
LRR—Launch Readiness Review
MRR—Mission Readiness Review
ORR—Operational Readiness Review
PCA—Program Commitment Agreement
PDR—Preliminary Design Review
PFAR—Post-Flight Assessment Review
PIR—Program Implementation Review
PLAR—Post-Launch Assessment Review
SAR—System Acceptance Review
SDR—System Definition Review
SIR—System Integration Review
SMSR—Safety and Mission Success Review
SRB—Standing Review Board
SRR—System Requirements Review

▲ Red triangles represent life-cycle reviews that require SRBs. The Decision Authority, Administrator, MDAA, or Center Director may request the SRB conduct other reviews.

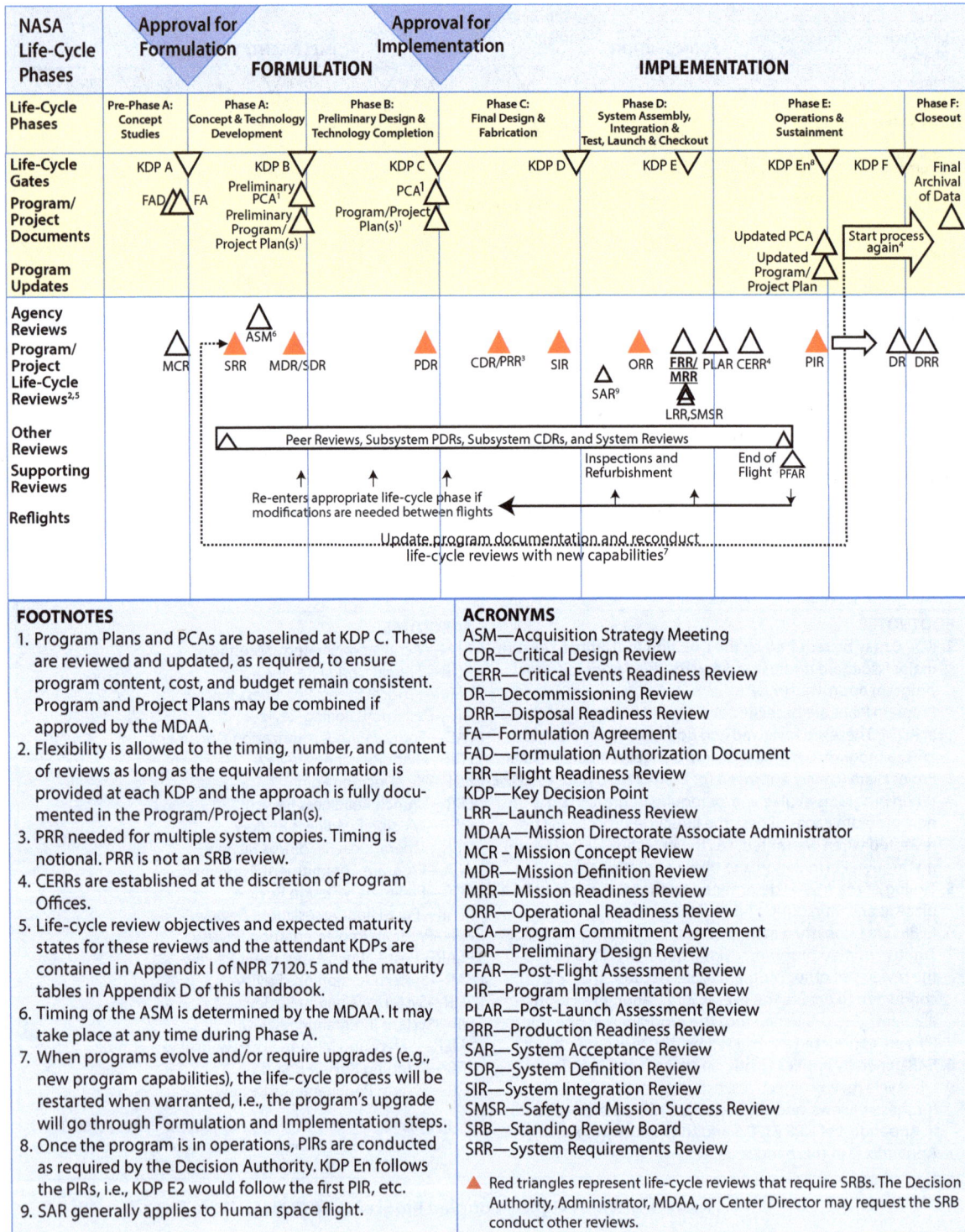

Figure 3-3 NASA Single-Project Program Life Cycle

FOOTNOTES

1. Program Plans and PCAs are baselined at KDP C. These are reviewed and updated, as required, to ensure program content, cost, and budget remain consistent. Program and Project Plans may be combined if approved by the MDAA.
2. Flexibility is allowed to the timing, number, and content of reviews as long as the equivalent information is provided at each KDP and the approach is fully documented in the Program/Project Plan(s).
3. PRR needed for multiple system copies. Timing is notional. PRR is not an SRB review.
4. CERRs are established at the discretion of Program Offices.
5. Life-cycle review objectives and expected maturity states for these reviews and the attendant KDPs are contained in Appendix I of NPR 7120.5 and the maturity tables in Appendix D of this handbook.
6. Timing of the ASM is determined by the MDAA. It may take place at any time during Phase A.
7. When programs evolve and/or require upgrades (e.g., new program capabilities), the life-cycle process will be restarted when warranted, i.e., the program's upgrade will go through Formulation and Implementation steps.
8. Once the program is in operations, PIRs are conducted as required by the Decision Authority. KDP En follows the PIRs, i.e., KDP E2 would follow the first PIR, etc.
9. SAR generally applies to human space flight.

ACRONYMS

ASM—Acquisition Strategy Meeting
CDR—Critical Design Review
CERR—Critical Events Readiness Review
DR—Decommissioning Review
DRR—Disposal Readiness Review
FA—Formulation Agreement
FAD—Formulation Authorization Document
FRR—Flight Readiness Review
KDP—Key Decision Point
LRR—Launch Readiness Review
MDAA—Mission Directorate Associate Administrator
MCR – Mission Concept Review
MDR—Mission Definition Review
MRR—Mission Readiness Review
ORR—Operational Readiness Review
PCA—Program Commitment Agreement
PDR—Preliminary Design Review
PFAR—Post-Flight Assessment Review
PIR—Program Implementation Review
PLAR—Post-Launch Assessment Review
PRR—Production Readiness Review
SAR—System Acceptance Review
SDR—System Definition Review
SIR—System Integration Review
SMSR—Safety and Mission Success Review
SRB—Standing Review Board
SRR—System Requirements Review

▲ Red triangles represent life-cycle reviews that require SRBs. The Decision Authority, Administrator, MDAA, or Center Director may request the SRB conduct other reviews.

- **Adequacy of management approach.** The scope of this criterion includes, but is not limited to, program authorization, management framework and plans, acquisition strategies, and internal and external agreements.

- **Adequacy of technical approach**, as defined by *NPR 7123.1, NASA Systems Engineering Processes and Requirements* entrance and success criteria. The scope of this criterion includes, but is not limited to, flow down of project requirements to systems/subsystems, architecture and design, and operations concepts that respond to and satisfy requirements and mission needs.

- **Adequacy of the integrated cost and schedule estimate and funding strategy** in accordance with *NPD 1000.5, Policy for NASA Acquisition*. The scope of this criterion includes, but is not limited to, cost and schedule control plans; cost and schedule estimates (prior to KDP I (KDP C for single-project programs)) and baselines (at KDP I (KDP C for single-project programs)) that are consistent with the program requirements, assumptions, risks, and margins; Basis of Estimate (BoE); Joint Cost and Schedule Confidence Level (JCL) (when required); and alignment with planned budgets.

- **Adequacy and availability of resources other than budget.** The scope of this criterion includes, but is not limited to, planning, availability, competency and stability of staffing, infrastructure, and the industrial base/supplier chain requirements.

- **Adequacy of the risk management approach and risk identification and mitigation** in accordance with *NPR 8000.4, Agency Risk Management Procedural Requirements* and *NASA/SP-2011-3422, NASA Risk Management Handbook*. The scope of this criterion includes, but is not limited to, risk-management plans, processes (e.g., Risk-Informed Decision Making (RIDM) and Continuous Risk Management (CRM)), open and accepted risks, risk assessments, risk mitigation plans, and resources for managing/mitigating risks.

Life-cycle reviews are designed to provide the program an opportunity to ensure that it has completed the work of that phase and an independent assessment of the program's technical and programmatic status and health. Life-cycle reviews are conducted under documented Agency and Center review processes. (See Section 5.10 and the *NASA Standing Review Board Handbook*.)

The life-cycle review process provides:

- The program with a credible, objective independent assessment of how it is doing.

The joint cost and schedule confidence level is the product of a probabilistic analysis of the coupled cost and schedule to measure the likelihood of completing all remaining work at or below the budgeted levels and on or before the planned completion of the development phase. A JCL is required for all tightly coupled and single-project programs, and for all projects with an LCC greater than $250 million. The JCL calculation includes consideration of the risk associated with all elements, regardless of whether or not they are funded from appropriations or managed outside of the program or project. JCL calculations include the period from approval for Implementation (KDP I for tightly coupled programs, KDP C for projects and single-project programs) through the handover to operations. Per NPR 7120.5, Mission Directorates plan and budget tightly coupled and single-project programs (regardless of life-cycle cost) and projects with an estimated life-cycle cost greater than $250 million based on a 70 percent JCL or as approved by the Decision Authority. Mission Directorates ensure funding for these projects is consistent with the Management Agreement and in no case less than the equivalent of a 50 percent JCL.

A life-cycle review is complete when the governing PMC and Decision Authority complete their assessment and sign the Decision Memorandum.

- NASA senior management with an understanding of whether:
  - The program is on track to meet objectives,
  - The program is performing according to plan, and
  - Impediments to program success are addressed.
- For a life-cycle review that immediately precedes a KDP, a credible basis for the Decision Authority to approve or disapprove the transition of the program at the KDP to the next life-cycle phase.

The independent review also provides vital assurance to external stakeholders that NASA's basis for proceeding is sound.

The program finalizes its work for the current phase during the life-cycle review. In some cases, the program uses the life-cycle review meeting(s) to make formal programmatic and technical decisions necessary to complete its work. In all cases, the program utilizes the results of the independent assessment and the resulting management decisions to finalize its work. In addition, the independent assessment serves as a basis for the program and management to determine if the program's work has been satisfactorily completed, and if the plans for the following life-cycle phases are acceptable. If the program's work has not been satisfactorily completed, or its plans are not acceptable, the program addresses the issues identified during the life-cycle review, or puts in place the action plans necessary to resolve the issues.

Prior to life-cycle reviews, programs conduct internal reviews in accordance with NPR 7123.1, Center practices, and NPR 7120.5. These internal reviews are key components of the process used by programs to solidify their plans, technical approaches, and programmatic commitments and are part of the normal systems engineering work processes defined in NPR 7123.1. Internal reviews assess major technical and programmatic requirements along with the system design and other implementation plans. Major technical and programmatic performance metrics are reported and assessed against predictions.

The program manager has the authority to determine whether to hold a one-step review or a two-step review. This determination usually depends on the state of the program's cost and schedule maturity as described below. Any life-cycle review can be either a one-step review or a two-step review. The program manager documents the program's review approach in the Program Review Plan.

Descriptions of the one-step and two-step life-cycle review processes are provided in Figures 3-4 and 3-5. These descriptions are written from the perspective of life-cycle reviews conducted by a program and an SRB. For life-cycle reviews that do not require an Agency-led SRB, the program manager will work with the Center Director or designee to prepare for and conduct the life-cycle review in accordance with Center practices and a Center-assigned independent review team. When the life-cycle review is conducted by the program and a Center independent review team rather than an Agency-led SRB, the remaining references to SRB are replaced with Center independent review team:

- In a one-step review, the program's technical maturity and programmatic posture are assessed together against the six assessment criteria. In this case, the program has typically completed all of its required technical work as defined in NPR 7123.1 life-cycle review entrance criteria and has aligned the scope of this work with its cost estimate, schedule, and risk posture before the life-cycle review. The life-cycle review is then focused on presenting this work to the SRB. Except in special cases, a one-step review is chaired by the SRB. The SRB assesses the work against the six assessment criteria and then provides an independent assessment of whether or not the program has met these criteria. Figure 3-4 illustrates the one-step life-cycle review process. (A one-step review for a program is analogous to a one-step review for a project.)

- In a two-step review, the program typically has not fully integrated the program's cost and schedule with the technical work. In this case, the first step of the life-cycle review is focused on finalizing and assessing the technical work described in NPR 7123.1. However as noted in Figure 3-5, which illustrates the two-step life-cycle review process, the first step does consider the preliminary cost, schedule, and risk as known at the time of the review. This first step is only one half of the life-cycle review. At the end of the first step, the SRB will have fully assessed the technical approach criteria but will only be able to determine preliminary findings on the remaining criteria since the program has not yet finalized its work. Thus, the second step is conducted after the program has taken the results of the first step and fully integrated the technical scope with the cost, schedule, and risk, and has resolved any issues that may have arisen as a result of this integration. The period between steps may take up to six months depending on the complexity of the program. In the second step, which may be referred to as the Independent Integrated Life-Cycle Review Assessment, the program typically presents the integrated technical, cost, schedule, and risk, just as is done for a one-step review, but the technical presentations may simply update information provided during the first step. The SRB then completes its assessment of whether or not the program has met the six assessment criteria. In a two-step life-cycle review, both steps are necessary to fulfill the life-cycle review requirements. Except in special cases, the SRB chairs both steps of the life-cycle review. (A two-step review for a program is analogous to a two-step review for a project.)

There are special cases, particularly for human space flight programs, where the program uses the life-cycle review to make formal decisions to complete the program's technical work and align it with the cost and schedule. In these cases, the program manager may co-chair the life-cycle review since the program manager is using this forum to make program decisions, and the SRB will conduct the independent assessment concurrently. The program manager will need to work with the SRB chair to develop the life-cycle review agenda and agree on how the life-cycle review will be conducted to ensure that it enables the SRB to fully accomplish the independent assessment. The program manager and the SRB chair work together to ensure that the life-cycle review Terms of Reference (ToR) reflect their agreement and the convening authorities approve the approach.

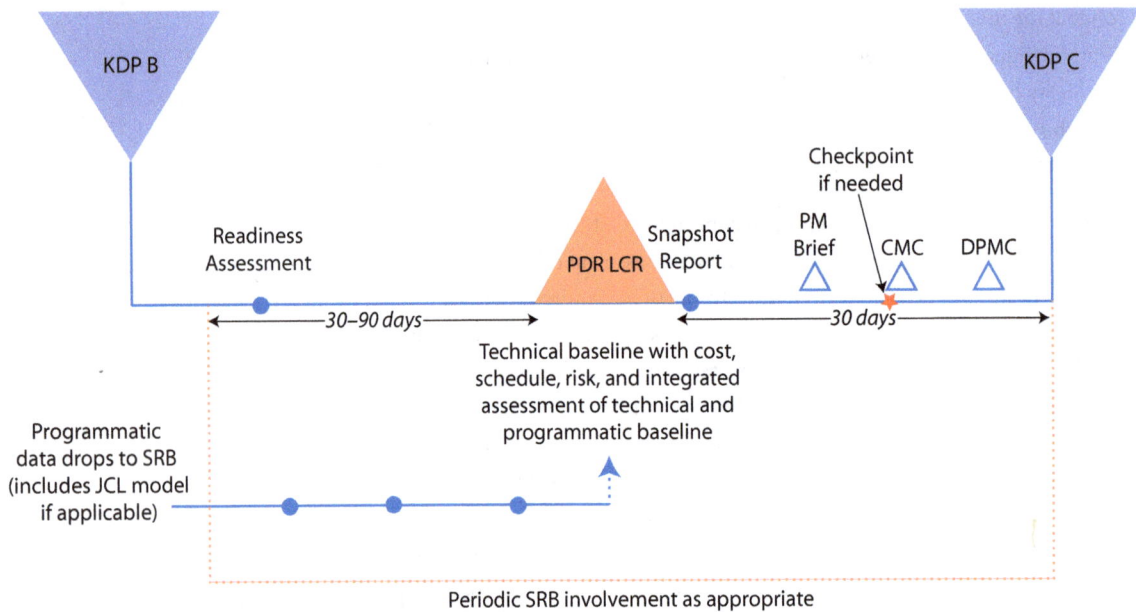

Notes: A one-or two-step review may be used for any life-cycle review. Section 5.10 and the *NASA Standing Review Board Handbook* provide information on the readiness assessment, snapshot reports, and checkpoints associated with life-cycle reviews. Time is not to scale.

Figure 3-4 One-Step PDR Life-Cycle Review Overview

Details on program review activities by life-cycle phase are provided in the sections below. The *NASA Standing Review Board Handbook* and Section 5.10 in this handbook also contain more detailed information on conducting life-cycle reviews. NPR 7123.1 provides life-cycle review entrance and success criteria, and Appendix I in NPR 7120.5E and Appendix E in this handbook provide specifics for addressing the six assessment criteria required to demonstrate that the program has met the expected maturity state for the KDP.

3.1.3 Other Reviews and Resources

Special reviews may be convened by the Office of the Administrator, the Mission Directorate Associate Administrator (MDAA), Center Director, the Technical Authorities (TAs),[1] or other convening authority. Special reviews may be warranted for programs not meeting expectations for achieving safety, technical, cost, or schedule requirements; not being able to develop

---

[1] That is, individuals with specifically delegated authority in Engineering (ETA), Safety and Mission Assurance (SMA TA), and Health and Medical (HMTA). See Section 5.2 for more information on Technical Authorities.

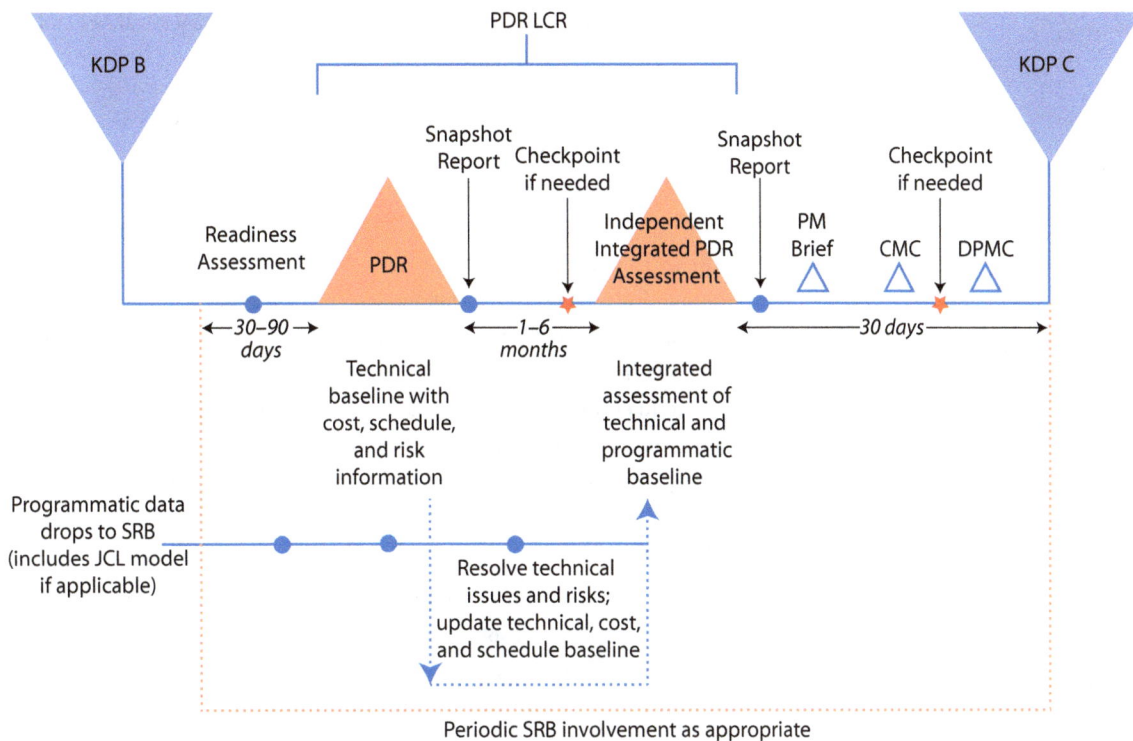

Figure 3-5 Two-Step PDR Life-Cycle Review Overview

Notes: A one-or two-step review may be used for any life-cycle review. Section 5.10 and the *NASA Standing Review Board Handbook* provide information on the readiness assessment, snapshot reports, and checkpoints associated with life-cycle reviews. Time is not to scale.

an enabling technology; or experiencing some unanticipated change to the program baseline. Special reviews include a Rebaseline Review and Termination Review. Rebaseline reviews are conducted when the Decision Authority determines the Agency Baseline Commitment (ABC) needs to be changed. (For more detail on Rebaseline Reviews, see Section 5.5.4.1. For more detail on the ABC, see Sections 3.2.4 and 5.5.1.) A Termination Review may be recommended by a Decision Authority, MDAA, or program executive if he or she believes it may not be in the Government's best interest to continue funding a program.

Other reviews, such as Safety and Mission Assurance (SMA) reviews, are part of the regular management process. For example, SMA Compliance/Verification reviews are spot reviews that occur on a regular basis to ensure programs are complying with NASA safety principles and requirements. For more detail on Termination Reviews and SMA reviews, see Section 5.11.

Other resources are also available to help a program manager evaluate and improve program performance. These resources include:

- The NASA Engineering and Safety Center (NESC), an independently funded organization with a dedicated team of technical experts, provides objective engineering and safety assessments of critical, high-risk programs. NESC is a resource to benefit programs and organizations within the Agency, the Centers, and the people who work there by promoting safety through engineering excellence that is unaffected and unbiased by the programs it is evaluating. The NESC mission is to proactively perform value-added independent testing, analysis, and assessments to ensure safety and mission success and help NASA avoid future problems. Programs seeking an independent assessment or expert advice on a particular technical problem can contact the NESC at http://www.nasa.gov/offices/nesc/contacts/index.html or the NESC chief engineer at their Center.

- The NASA Independent Verification and Validation (IV&V) Facility strives to improve the software safety, reliability, and quality of NASA programs and missions through effective applications of systems and software IV&V methods, practices, and techniques. The NASA IV&V Facility applies software engineering best practices to evaluate the correctness and quality of critical and complex software systems. When applying systems and software IV&V, the NASA IV&V Facility seeks to ensure that the software exhibits behaviors exactly as intended, does not exhibit behaviors that were not intended, and exhibits expected behaviors under adverse conditions. Software IV&V has been demonstrated to be an effective technique on large, complex software systems to increase the probability that software is delivered within cost and schedule, and that software meets requirements and is safe. When performed in parallel with systems development, software IV&V provides for the early detection and identification of risk elements, enabling early mitigation of the risk elements. For projects that are required, or desire to, do software IV&V, go to the "Contact Us" link at http://www.nasa.gov/centers/ivv/home/index.html. (All Category 1 projects; all Category 2 projects that have Class A or Class B payload risk classification per *NPR 8705.4, Risk Classification for NASA Payloads*; and projects specifically selected by the NASA Chief, Safety and Mission Assurance are required to do software IV&V. See NPR 7120.5E and Section 4.1 in this handbook for project categorization guidelines.)

3.1.4 Program Evolution and Recycling

A program may evolve over time in ways that require it to go back and restart parts of its life cycle. A program may evolve as a result of a planned series of upgrades, with the addition of new projects, when the need for new capabilities is identified, or when a new mission is assigned to the program. This can happen in a number of ways.

For tightly coupled and single-project programs, when the requirements imposed on a program significantly change, the program typically evaluates whether the changes impact the program's current approved approach and/or system design and performance. In these cases, the Decision Authority may ask the program to go back through the necessary life-cycle phases and reviews, and to update program documentation, to ensure that the changes have been properly considered in light of the overall program/system performance. Each case is likely to be different and thus may not require completely restarting the process at the beginning. The decision on when and where to recycle through the life-cycle reviews will be based on a discussion between the program, the Mission Directorate, and the Decision Authority. This case is depicted as the "Start process again" arrows on Figures 3-2 and 3-3. As an example, after the Hubble Space Telescope (HST) was deployed in April 1990 and was in operations, a component for the HST started back through the life cycle. The Corrective Optics Space Telescope Axial Replacement (COSTAR) program for correcting the optics of the HST was required to repeat a concept definition phase after approval in January 1991 and start back through the life cycle at the PDR.

There are also cases of evolution for a single-project program where operational reusable systems are refurbished after each flight or modifications are required between flights. A program going back through a part of its life cycle is depicted on the single-project program life-cycle figure on the "Reflight" line (Figure 3-3).

For uncoupled and loosely coupled programs, program evolution is also possible. An example of a simple change to an uncoupled program that might warrant performing another SDR and subsequent program reapproval might be the addition of a new science discipline to the program that requires a totally different implementation approach. In this case, the Decision Authority may wish to have the program evaluated to ensure the program's approach is satisfactory.

3.2 Program Oversight and Approval

NASA has established a program management oversight process to ensure that the experience, diverse perspectives, and thoughtful programmatic and technical judgment at all levels is available and applied to program activities. The Agency employs management councils and management forums, such as the Baseline Performance Review (BPR) (see Section 3.2.5), to provide insight to upper management on the status and progress of programs and their alignment with Agency goals. This section describes NASA's oversight approach and the process by which programs are approved to move forward through their life cycles. It defines and describes NASA's Decision Authority, KDPs, management councils, and the BPR. (See Section 4.2.2 for more information about management councils for projects.)

The general flows of the program oversight and approval process for life-cycle reviews that require SRBs and of the periodic reporting activity for programs are shown in Figure 3-6. Prior to the life-cycle review, the program conducts its internal reviews. Then the program and the SRB conduct the life-cycle review. Finally, the results are reported to senior management via the management councils.

Additional insight is provided by the independent perspective of SRBs at life-cycle reviews identified in Figures 3-1, 3-2, and 3-3. Following each life-cycle review, the independent SRB chair and the program manager brief the applicable management councils on the results of the life-cycle review to support the councils' assessments. These briefings are completed within 30 days of the life-cycle review. The 30 days ensures that the Decision Authority is informed in a timely manner as the program moves forward to preclude the program from taking action that the Decision Authority does not approve. These briefings cover the objectives of the review; the maturity expected at that point in the life cycle; findings and recommendations to rectify issues or improve mission success; the program's response to these findings; and the program's proposed cost, schedule, safety, and technical plans for the follow-on life-cycle phases. This process enables a disciplined approach for developing the Agency's assessment, which informs the Decision Authority's KDP determination of program readiness to proceed to the next life-cycle phase. Life-cycle reviews are conducted under documented Agency and Center review processes.

3.2.1 Decision Authority

The Decision Authority is the Agency individual who is responsible for making the KDP determination on whether and how the program proceeds

> The Decision Authority is the individual authorized by the Agency to make important decisions on programs and projects under their purview. The Decision Authority makes the KDP decision by considering a number of factors, including technical maturity; continued relevance to Agency strategic goals; adequacy of cost and schedule estimates; associated probabilities of meeting those estimates (confidence levels); continued affordability with respect to the Agency's resources; maturity and the readiness to proceed to the next phase; and remaining program risk (safety, cost, schedule, technical, management, and programmatic). The NASA AA signs the Decision Memorandum as the Decision Authority for programs at the KDP. This signature signifies that the Decision Authority, as the approving official, has been made aware of the technical and programmatic issues within the program, approves the mitigation strategies as presented or with noted changes requested, and accepts technical and programmatic risk on behalf of the Agency.

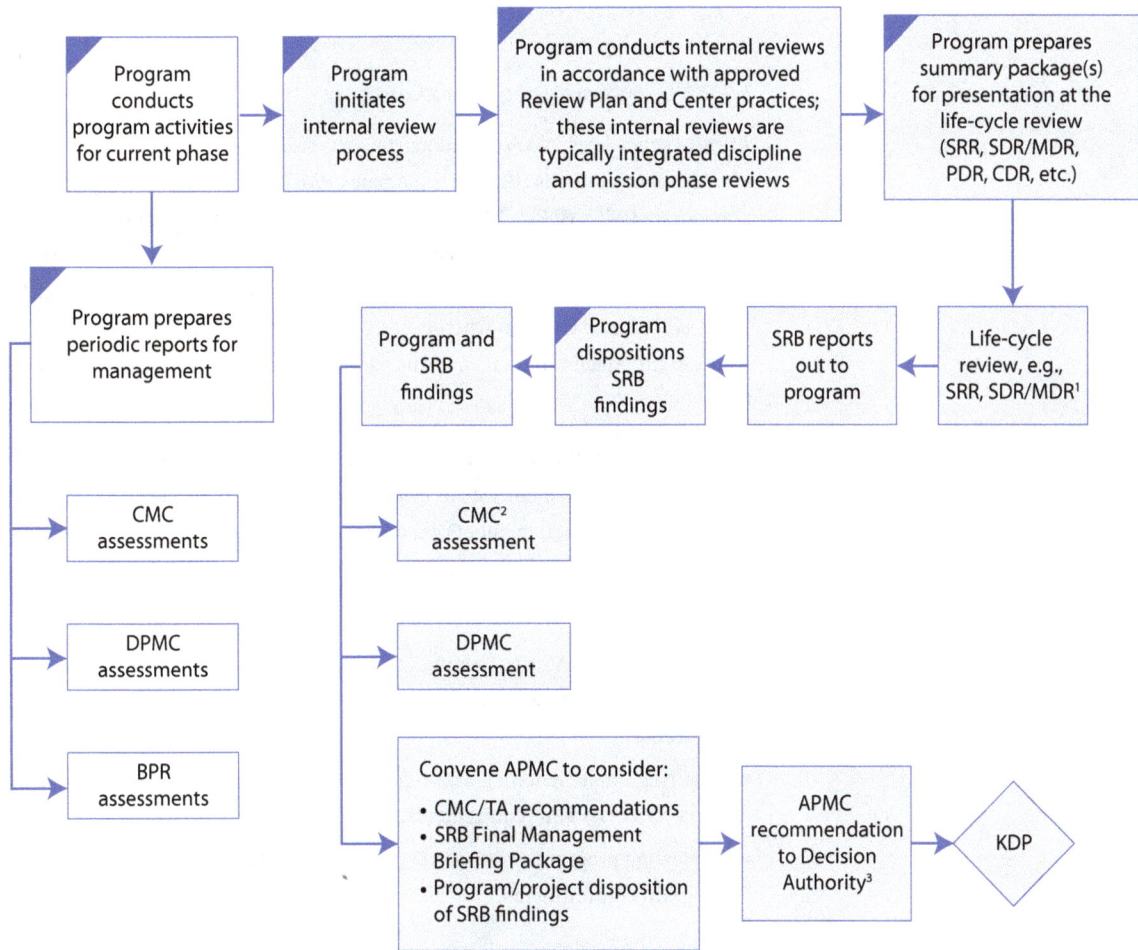

Legend: ▼ Program activity ☐ Periodic reporting activity ☐ Life-cycle review activity

[1] See Section 5.10 and the *NASA Standing Review Board Handbook* for details.

[2] May be an Integrated Center Management Council when multiple Centers are involved.

[3] Life-cycle review is complete when the governing PMC and Decision Authority complete their assessment.

Figure 3-6 Program Life-Cycle Review Process and Periodic Reporting Activity

through the life cycle and for authorizing the key program cost, schedule, and content parameters that govern the remaining life-cycle activities. The NASA AA is the Decision Authority for all programs.

3.2.2 Management Councils

3.2.2.1 Program Management Councils

At the Agency level, NASA Headquarters has two levels of Program Management Councils (PMCs): the Agency PMC (APMC) and the Mission Directorate PMCs (DPMCs). The PMCs evaluate the safety, technical, and programmatic performance (including cost, schedule, risk, and risk mitigation) and content of a program under their purview for the entire life cycle. These evaluations focus on whether the program is meeting its commitments to the Agency and on ensuring successful achievement of NASA strategic goals.

For all programs, the governing PMC is the APMC. The APMC is chaired by the NASA AA, and consists of Headquarters senior managers and Center Directors. The council members are advisors to the AA in the capacity as the PMC Chair and Decision Authority. The APMC is responsible for the following:

- Ensuring that NASA is meeting the commitments specified in the relevant management documents for program performance and mission assurance,
- Ensuring implementation and compliance with NASA program management processes and requirements,
- Reviewing programs routinely, including institutional ability to support program commitments,
- Approving PCAs,
- Reviewing special and out-of-cycle assessments, and
- Approving the Mission Directorate strategic portfolio and its associated risk.

As the governing PMC for programs, the APMC evaluates programs in support of KDPs. A KDP normally occurs at the APMC review as depicted in Figure 3-6. The APMC makes a recommendation to the NASA AA on a program's readiness to progress in its life cycle and provides an assessment of the program's proposed cost, schedule, and content parameters. The NASA AA, as the Decision Authority for programs, makes the KDP determination on whether and how the program progresses in its life cycle and authorizes the key program cost, schedule, and content parameters that govern the remaining life-cycle activities. Decisions are documented in a formal Decision Memorandum, and actions are tracked in a Headquarters tracking system (e.g., the Headquarters Action Tracking System (HATS)). See Section 3.2.4 and Section 5.5.6 for a description of the Decision Memorandum.

A Directorate PMC (DPMC) provides oversight for the MDAA and evaluates all programs executed within that Mission Directorate. The DPMC is usually chaired by the MDAA and is composed of senior Headquarters executives from that Mission Directorate. The MDAA may delegate the chairmanship to one of the senior executives. The activities of the DPMC are directed toward periodically (usually monthly) assessing programs' performance and conducting in-depth assessments of programs at critical milestones. The DPMC makes recommendations regarding the following:

- Initiation of new programs based on the results from advanced studies,
- Transition of ongoing programs from one phase of the program life cycle to the next, and
- Action on the results of periodic or special reviews, including rebaselining or terminating programs.

The results of the DPMC are documented and include decisions made and actions to be addressed. The MDAA may determine that a program is not ready to proceed to the APMC and may direct corrective action. If the program is ready to proceed, the MDAA carries forward the DPMC findings and recommendations to the APMC.

3.2.2.2 Center Management Council

In addition to the APMCs, Centers have a Center Management Council (CMC) that provides oversight and insight for the Center Director (or designee) for all program work executed at that Center. The CMC evaluation focuses on whether Center engineering, SMA, health and medical, and management best practices (e.g., program management, resource management, procurement, institutional) are being followed by the program under review; whether Center resources support program requirements; and whether the program is meeting its approved plans successfully. As chair of the CMC, the Center Director or designated chair may provide direction to the program manager to correct program deficiencies with respect to these areas. However, with respect to programmatic requirements, budgets, and schedules, the Center Director does not provide direction, but only recommendations to the program manager, Mission Directorate, or Agency leadership. The CMC also assesses program risk and evaluates the status and progress of activities to identify and report trends and provide guidance to the Agency and affected programs. For example, the CMC may note a trend of increasing risk that potentially indicates a bow wave of accumulating work or may communicate industrial base issues to other programs that might be affected. Prior to KDPs, the Center Director/CMC chair provides the Center's findings and recommendations to program managers and to the DPMC and APMC regarding the performance, technical, and management

In accordance with NPR 7120.5: "Center Directors are responsible and accountable for all activities assigned to their Center. They are responsible for the institutional activities and for ensuring the proper planning for and assuring the proper execution of programs and projects assigned to the Center." This means that the Center Director is responsible for ensuring that programs develop plans that are executable within the guidelines from the Mission Directorate and for assuring that these programs are executed within the approved plans. In cases where the Center Director believes a program cannot be executed within approved guidelines and plans, the Center Director works with the program and Mission Directorate to resolve the problem. (See Section 5.1.2 for additional information on Center Directors' responsibilities.)

viability of the program. This includes making recommendations to the Decision Authority at KDPs regarding the ability of the program to execute successfully. (Figure 3-6 shows this process.) These recommendations consider all aspects, including safety, technical, programmatic, and major risks and strategy for their mitigation and are supported by independent analyses, when appropriate.

The relationship of the various management councils to each other is shown in Figure 3-7.

3.2.2.3 Integrated Center Management Councils

An Integrated Center Management Council (ICMC) is generally used for any program conducted by multiple Centers. This is particularly true for tightly coupled programs. The ICMC performs the same functions as the CMC but includes the Center Director (or representative) from each Center responsible for management of a project within the program and each Center with a substantial program development role. The ICMC is chaired by the Center Director (or representative) of the Center responsible for program management.

When an ICMC is used to oversee the program, the participating Centers work together to define how the ICMC will operate, when it will meet, who

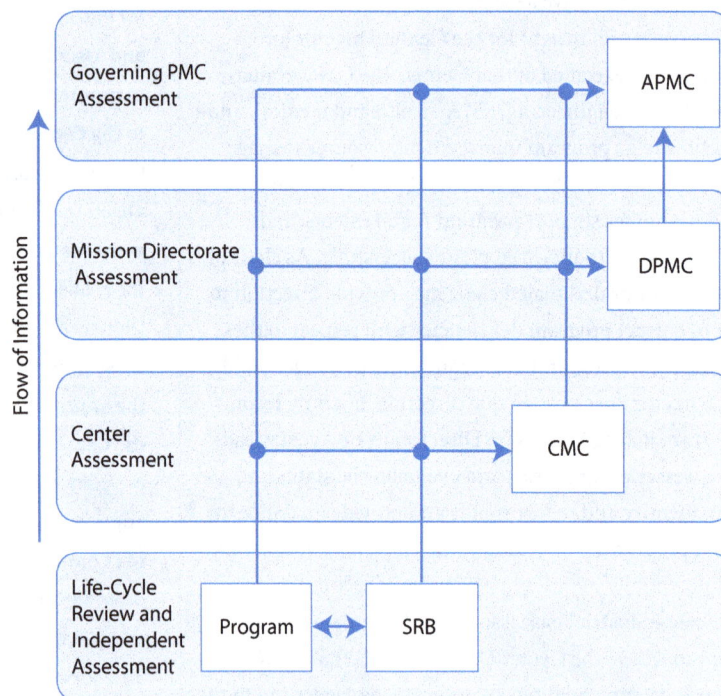

Figure 3-7 Management Council Reviews in Support of KDPs

will participate, how decisions will be made, and how Dissenting Opinions will be resolved. (See Section 5.3 on Dissenting Opinion.) In general, final decisions are made by the chair of the ICMC. When a participating Center Director disagrees with a decision made at the ICMC, the standard Dissenting Opinion process is used. As an example, this would generally require that the NASA Chief Engineer resolve disagreements for engineering or program management policy issues.

3.2.3 Key Decision Points

At Key Decision Points (KDPs), the Decision Authority reviews all the materials and briefings at hand to make a decision about the program's maturity and readiness to progress through the life cycle and authorizes the content, cost, and schedule parameters for the ensuing phase(s). KDPs conclude the life-cycle review at the end of a life-cycle phase. A KDP is a mandatory gate through which a program must pass to proceed to the next life-cycle phase.

The potential outcomes at a KDP include the following:

- Approval to enter the next program phase, with or without actions.
- Approval to enter the next phase, pending resolution of actions.
- Disapproval for continuation to the next phase. In such cases, follow-up actions may include:
  - A request for more information and/or a follow-up review that addresses significant deficiencies identified as part of the life-cycle review preceding the KDP;
  - A request for a Termination Review;
  - Direction to continue in the current phase; or
  - Redirection of the program.

The KDP decision process is supported by submitting the appropriate KDP readiness products to the Decision Authority and APMC members. This material includes the following:

- The program's proposed cost, schedule, safety, and technical plans for their follow-on phases. This includes the proposed preliminary and final baselines.
- Summary of accepted risks and waivers.
- Program documents or updates signed or ready for signature (e.g., the program Formulation Authorization Document (FAD), Program Plan, Program Commitment Agreement (PCA), Formulation Agreement (single-project programs), Memoranda of Understanding (MOUs), and Memoranda of Agreement (MOAs)).

> A life-cycle review is complete when the governing PMC and Decision Authority complete their assessment and sign the Decision Memorandum.

- Summary status of action items from the previous KDP (with the exception of KDP 0/A).
- Draft Decision Memorandum and supporting data. (See Section 3.2.4.)
- The program manager recommendation.
- The SRB Final Management Briefing Package.
- The CMC/ICMC recommendation.
- The MDAA recommendation.
- The governing PMC review recommendation.

After reviewing the supporting material and completing discussions with all parties, the Decision Authority determines whether and how the program proceeds and approves any additional actions. These decisions are summarized and recorded in the Decision Memorandum. The Decision Authority completes the KDP process by signing the Decision Memorandum. The expectation is to have the Decision Memorandum signed by concurring members as well as the Decision Authority at the conclusion of the governing PMC KDP meeting. (See more information on the Decision Memorandum, including signatories and their respective responsibilities in Section 5.5.6.

The Decision Authority archives the KDP documents with the Agency Chief Financial Officer and the program manager attaches the approved Decision Memorandum to the Program Plan. Any appeals of the Decision Authority's decisions go to the next higher Decision Authority, who (for programs) is the NASA Administrator.

3.2.4 Decision Memorandum, Management Agreement, and Agency Baseline Commitment

The Decision Memorandum is a summary of key decisions made by the Decision Authority at a KDP, or, as necessary, in between KDPs. Its purpose is to ensure that major program decisions and their basis are clearly documented and become part of the retrievable records. The Decision Memorandum supports the clearly defined roles and responsibilities and a clear line of decision making and reporting documented in the official program documentation.

When the Decision Authority approves the program's entry into the next phase of the life cycle at a KDP, the Decision Memorandum describes this approval, and the key program cost, schedule, and content parameters authorized by the Decision Authority that govern the remaining life-cycle activities. The Decision Memorandum also describes the constraints and parameters within which the Agency and the program manager will operate, i.e., the Management Agreement, the extent to which changes in plans may be made without additional approval, and any additional actions from the KDP.

The Management Agreement contained within the Decision Memorandum defines the parameters and authorities over which the program manager has management control. A program manager has the authority to manage within the Management Agreement and is accountable for compliance with the terms of the agreement. The Management Agreement, which is documented at every KDP, may be changed between KDPs as the program matures with approval from the Decision Authority. The Management Agreement typically is viewed as a contract between the Agency and the program manager and requires renegotiation and acceptance if it changes.

During Formulation, the Decision Memorandum documents the key parameters related to work to be accomplished during each phase of Formulation. It also documents a target Life-Cycle Cost (LCC) range (and schedule range, if applicable) that the Decision Authority determines is reasonable to accomplish the program. (For uncoupled and loosely coupled programs, the LCC range may be represented merely as a single annual funding limit consistent with the budget.) Given the program's lack of maturity during Formulation, the LCC range reflects the broad uncertainties regarding the program's scope, technical approach, safety objectives, acquisition strategy, implementation schedule, and associated costs. When applicable, the range is also the basis for coordination with the Agency's stakeholders, including the White House and Congress. Tightly coupled and single-project programs document their Life-Cycle Cost Estimate (LCCE) in accordance with the life-cycle scope defined in their FAD or PCA. (Projects that are part of these programs document their LCCE in accordance with the life-cycle scope defined in their program's Program Plan, PCA or FAD, or the project's FAD.)

During Implementation, the Decision Memorandum documents the parameters for the entire life cycle of the program. At this point, the approved LCCE of the program is no longer documented as a range but instead as a single number. The LCCE includes all costs, including all Unallocated Future Expenses (UFE) and funded schedule margins, for development through prime mission operation to disposal, excluding extended operations. (See "Extended Operations" box.)

Unallocated Future Expenses (UFE) are the portion of estimated cost required to meet the specified confidence level that cannot yet be allocated to the specific Work Breakdown Structure (WBS) sub-elements because the estimate includes probabilistic risks and specific needs that are not known until these risks are realized. (For programs and projects that are not required to perform probabilistic analysis, the UFE should be informed by the program or projects unique risk posture in accordance with Mission Directorate and Center guidance and requirements. The rationale for the UFE, if not conducted via a probabilistic analysis, should be appropriately documented and be traceable, repeatable, and defendable.) UFE may be held at the program level and the Mission Directorate level.

Extended Operations

Extended operations are operations conducted after the planned prime mission operations are complete. (The planned prime mission operations period is defined in a program's FAD or PCA and in a project's FAD.) Extended operations may be anticipated when the PCA or FAD is approved, but the complexity and duration of the extended operations cannot be characterized. Examples of this case include long-duration programs, such as the space shuttle and space station programs. Alternatively, the need for extended operations may be identified later, as the program or project is nearing the completion of its planned prime mission operations period. Examples include cases when extended operations contribute to the best interests of the Nation and NASA. For example, a mission may become vital to the success of programs run by another Federal agency, such as the need for mission data for terrestrial or space weather predictions by the National Oceanic and Atmospheric Administration. NASA's best interest may include continuing value to compelling science investigations that contribute to NASA's strategic goals. All extended operations periods need to be approved. The approval process is determined by the program or project's Mission Directorate and may require Agency-level approval. Program or project documentation, such as the Program or Project Plan, needs to be revised to continue the mission into extended operations.

The Agency Baseline Commitment (ABC) is an integrated set of program requirements, cost, schedule, technical content, and JCL. The ABC cost is equal to the program LCC approved by the Agency at approval for Implementation. The ABC is the baseline against which the Agency's performance is measured during the Implementation Phase of a program. Only one official baseline exists for a program, and it is the ABC. The ABC for tightly coupled and single-project programs forms the basis for the Agency's external commitment to the Office of Management and Budget (OMB) and Congress, and serves as the basis by which external stakeholders measure NASA's performance for these programs. Changes to the ABC are controlled through a formal approval process. An ABC is not required for loosely coupled and uncoupled programs.

Tightly coupled programs and single-project programs establish a program baseline, called the Agency Baseline Commitment (ABC), at approval for Implementation (KDP I for tightly coupled programs and KDP C for single-project programs). The ABC and other key parameters are documented in the Decision Memorandum.

See Section 5.5 for a detailed description of maturing, approving, and maintaining program plans, LCCs, baselines, and commitments and for additional information on the Decision Memorandum and Management Agreement.

3.2.5 Management Forum—Baseline Performance Review

NASA's Baseline Performance Review (BPR) serves as NASA's monthly, internal senior performance management review, integrating Agency-wide communication of performance metrics, analysis, and independent assessment for both mission and mission support programs, projects and activities. While not a council, the Baseline Performance Review (BPR) is closely linked with the councils and integral to council operations. As an integrated review of institutional, program, and project activities, the BPR highlights interrelated issues that impact performance and program and project risk enabling senior management to quickly address issues, including referral to the governing councils for decision, if needed. The BPR forum fosters communication across organizational boundaries to identify systemic issues and address mutual concerns and risks. The BPR is the culmination of all of the Agency's regular business rhythm performance monitoring activities, providing ongoing performance assessment between KDPs. The BPR is also used to meet requirements for quarterly progress reviews contained in the Government Performance Reporting and Accountability Modernization Act of 2010 (GPRAMA) and OMB Circular A-11 Section 6.[2]

The NASA Associate Administrator and Associate Deputy Administrator cochair the BPR. Membership includes Agency senior management and Center Directors. The Office of the Chief Engineer (OCE) leads the program and project performance assessment process conducted by a team of independent assessors drawn from OCE, the Office of the Chief Financial Officer (OCFO), and the Office of Safety and Mission Assurance (OSMA).

A typical BPR agenda includes an assessment of each Mission Directorate's program and project performance, including performance against

---

[2] Additional information on GPRAMA can be found at http://www.gpo.gov/fdsys/pkg/PLAW-111publ352/pdf/PLAW-111publ352.pdf. Additional information on A-11 Section 6 can be found at http://www.whitehouse.gov/sites/default/files/omb/assets/a11_current_year/s200.pdf.

Management Agreements and ABCs (if applicable), with rotating in-depth reviews of specific mission areas. The schedule ensures that each mission area is reviewed on a quarterly basis. Mission support functions are included in the BPR. Assessors use existing materials when possible. Table 3-1 shows typical information sources that may be used by the BPR assessors. Different emphasis may be placed on different sources depending on which mission is being assessed.

Table 3-1 Typical Information Sources Used for BPR Assessment

Program/Project Documents	FAD, Formulation Agreement, PCAs, and Program and Project Plans
Reviews	Life-cycle reviews
	Monthly, quarterly, midyear, and end-of-year Mission Directorate reviews
	Other special reviews (see Section 3.1.3)
	Monthly Center status reviews
Meetings	APMC (presentations and decision memorandums)
	DPMC (presentations and decision memorandums)
	Recurring staff/status meetings including project monthly status
	Program Control Board (meetings and weekly status reports)
	Biweekly tag-ups with the SMA TAs supporting and overseeing the program
Reports	Annual Performance Goals (for programs)
	Reports from Agency assessment studies (CAD, IPAO, etc.)
	PPBE presentations
	Quarterly cost and schedule reports on major programs/projects delivered to OCFO
	Center summaries presentations at BPR
	Weekly Mission Directorate report
	Weekly project reports
	Weekly reports from the NESC
	Monthly EVM data
	Project anomaly reports
	Center SMA reports
	Technical Authority reports
Databases	N2 budget database
	SAP and Business Warehouse financial databases
	OMB/Congressional cost/schedule data

3.3 Program Formulation

3.3.1 Program Activities Leading to the Start of Formulation

The process for initiating programs begins at the senior NASA management level with strategic acquisition planning. When a need for a program is first identified, the Agency examines and considers acquisition alternatives from several perspectives. This process enables NASA management to consider the full spectrum of acquisition approaches for its programs—from commercial off-the-shelf buys to in-house design and build efforts. For a "make or buy" decision, the Agency considers whether to acquire the capability in-house, where NASA has a unique capability and capacity or the need to maintain or develop such capability and capacity; to acquire it from outside the Agency; or to acquire it through some combination of the two. Other than preservation of core competencies and unique facilities, considerations include maturity of technologies affecting the technical approach, priorities from the White House and Congress, and commercialization goals. Strategic acquisition at the Agency level promotes best-value approaches by taking into account the Agency as a whole.

Many processes support acquisition, including the program management system, the budget process, and the procurement system. The NASA Planning, Programming, Budgeting, and Execution (PPBE) process supports allocating the resources of programs through the Agency's annual budgeting process. (See Section 5.8, Federal Budgeting Process; *NPR 9420.1, Budget Formulation*; and *NPR 9470.1, Budget Execution*.) The NASA procurement system supports the acquisition of assets and services from external sources. See NPD 1000.5, the Federal Acquisition Regulation (FAR), and the NASA FAR Supplement (NFS) for NASA's specific implementation of the FAR.

3.3.2 Program Formulation Activities

Programs provide the critically important linkage between the Agency's strategic goals and the projects that are the specific means for achieving them. The purpose of program Formulation activities is to establish a cost-effective program that is demonstrably capable of meeting Agency and Mission Directorate goals and objectives. During Formulation, the program team:

- Derives a technical approach from an analysis of alternatives;
- Develops and allocates program requirements to initial projects;

NASA defines acquisition as the process for obtaining the systems, research, services, construction, and supplies that the Agency needs to fulfill its mission. Acquisition—which may include procurement (contracting for products and services)—begins with an idea or proposal that aligns with the NASA Strategic Plan and fulfills an identified need and ends with the completion of the program or project or the final disposition of the product or service. (The definition of acquisition in accordance with *NPD 1000.5* is used in a broader context than the FAR definition to encompass strategic acquisition planning and the full spectrum of various NASA acquisition authorities and approaches to achieve the Agency's mission and activities.)

- Initiates project pre-Formulation activities;

- Develops organizational structures and initiates work assignments;

- Defines and gains approval for program acquisition strategies;

- Develops interfaces to other programs;

- Establishes required annual funding levels and develops preliminary cost and schedule estimates;

- Develops products required during Formulation in accordance with the Program Product Maturity tables at the end of this chapter;

- Designs a plan for Implementation;

- Puts in place management systems; and

- Obtains approval of formal program documentation, all consistent with the NASA Strategic Plan and other higher level requirements.

Official program Formulation begins with a FAD that authorizes a program manager to initiate the planning of a new program and to perform the analyses of alternatives required to formulate a sound Program Plan. However, in many cases, Mission Directorates engage in pre-Formulation activities prior to the development of a FAD to develop the basic program concept and have it approved by NASA's senior management.

One of the first activities is to select the management team. The program managers are recommended by the Center Director with approval for appointment by the MDAA.

3.3.2.1 Program Formulation Activities Across Program Types

The following paragraphs describe the activities all program types must accomplish to develop a sound Program Plan. However, programs vary significantly in scope, complexity, cost, and criticality, and the activities vary as a result. The differences in activities are described by program type in Section 3.3.2.2.

Program Formulation is initiated at approval for Formulation and completes when the Decision Authority approves the program's transition from Formulation to Implementation at KDP I (KDP C for single-project programs). Authorization of program transition from Formulation to Implementation is documented in the Program Commitment Agreement (PCA) and other retrievable program records. The program assists the Mission Directorate in preparing this agreement, as requested. A draft PCA is prepared by KDP 0 and baselined by KDP I. (Single-project programs are the exception: they follow a life cycle similar to projects, so they are approved at KDP C. However, single-project programs are also required to

The FAD is issued by the MDAA to authorize the formulation of a program whose goals will fulfill part of the Agency's Strategic Plan and Mission Directorate strategies. The FAD describes the purpose of the program, including a clear traceability from the goals and objectives in the Mission Directorate strategies. It describes the level or scope of work, and the goals and objectives to be accomplished in the Formulation Phase. It also describes the NASA organizational structure for managing the formulation process from the Mission Directorate Associate Administrator (MDAA) to the NASA Center program or project managers, as applicable, and includes lines of authority, coordination, and reporting. It identifies Mission Directorates, Mission Support Offices, and Centers to be involved in the activity, their scope of work, and any known constraints related to their efforts (e.g., the program is cofunded by a different Mission Directorate). It identifies any known participation by other organizations external to NASA that are to be involved in the activity, their scope of work, and any known constraints related to their efforts (e.g., the program or project must be cofunded by the external participant). It identifies the funding that will be committed to the program during each year of Formulation. Finally, it specifies the program life-cycle reviews planned during the Formulation Phase.

The PCA (see *NPR 7120.5*, Appendix D) is an agreement between the MDAA and the NASA AA (the Decision Authority) that authorizes program transition from Formulation to Implementation. The PCA is prepared by the Mission Directorate and documents Agency and Mission Directorate requirements that flow down to the program; program objectives, management and technical approach and associated architecture; program technical performance, schedule, time-phased cost plans, safety and risk factors; internal and external agreements; life-cycle reviews; and all attendant top-level program requirements.

Major Acquisitions are directed at and critical to fulfilling the Agency's mission, entail the allocation of relatively large resources, or warrant special management attention.

develop a Program Plan and have a PCA, unless the Mission Directorate approves otherwise.)

Major program and life-cycle reviews leading to approval at KDP I (KDP C for single-project programs) are the Acquisition Strategy Meeting (ASM); the System Requirements Review (SRR); the System Definition Review (SDR)/Mission Definition Review (MDR);[3] the governing PMC review; and for single-project programs and tightly coupled programs, the PDR.

Acquisition Strategy. As early as possible in Formulation, all program types begin to define the program's acquisition strategy. The Acquisition Strategy is the plan or approach for using NASA's acquisition authorities to achieve the program's mission. The strategy includes recommendations from make/buy analyses, the recommendations from competed/directed analyses, proposed partnerships and contributions, proposed infrastructure use and needs, budget, and any other applicable considerations. This strategy addresses the program's initial plans for obtaining the systems, research, services, construction, and supplies that it needs to fulfill its mission, including any known procurement(s); the availability of the industrial base capability and supply chain needed to design, develop, produce, and support the program and its planned projects; identifying risks associated with single source or critical suppliers; and attendant mitigation plans.

The program develops their preliminary strategy, which is informed by the Agency's strategic planning process, prior to the SRR. The MDAA and AA determine whether an ASM is required. If an ASM is required, the team plans, prepares for, and supports the ASM as part of the formulation of its acquisition strategy. The ASM is typically held early in Formulation and precedes making partnership commitments, but the timing is determined by the Mission Directorate. The results of this meeting are used to finalize the Acquisition Plan. (See Section 3.3.3.5.)

The purpose of the ASM is for senior Agency management to review and agree upon the acquisition strategy before authorizing resource expenditures for major acquisitions. This includes implementation of the decisions and guidance that flowed out of the previous Agency Strategic Implementation Planning (SIP) process and consideration of resource availability, impact on the Agency workforce, maintaining core capabilities, make-or-buy planning, supporting Center assignments, and the potential for partnerships. (See Section 5.8.3.1 for information on the SIP process.) The development of an acquisition strategy also includes an analysis of the industrial base capability to design, develop, produce, support, and even possibly

---

[3] The SDR and the MDR are the same review. Robotic programs tend to use the terminology MDR and human space flight programs tend to use SDR.

restart an acquisition program or project. The plan also includes the mechanisms used to identify, monitor, and mitigate industrial base and supply chain risks. The ASM review is based on information provided by the associated Mission Directorate or Mission Support Office, and results in the approval of plans for Formulation and Implementation. Decisions are documented in the ASM meeting minutes. The results of the ASM are used to finalize the Acquisition Plan. (See Section 3.3.2.)

System Requirements Review. The purpose of the SRR, regardless of program type, is to evaluate whether the program functional and performance requirements are properly formulated and correlated with the Agency and Mission Directorate strategic objectives and to assess the credibility of the program's estimated budget and schedule. For uncoupled and loosely coupled programs a KDP 0 may be required, at the discretion of the Decision Authority, to ensure that major issues are understood and resolved prior to proceeding to SDR and KDP I. At a KDP 0, the program shows how it meets critical NASA needs and proves it has a good chance of succeeding as conceived.

System Definition Review/Mission Definition Review. The purpose of the SDR/MDR for uncoupled and loosely coupled programs is to evaluate the proposed program requirements/architecture and allocation of requirements to initial projects, to assess the adequacy of project pre-Formulation efforts, and to determine whether the maturity of the program's definition and associated plans is sufficient to begin Implementation. After a successful SDR/MDR, the program proceeds to KDP I. The program is expected to demonstrate that it (1) is in place and stable, (2) addresses critical NASA needs, (3) has adequately completed Formulation activities, (4) has an acceptable plan for Implementation that leads to mission success, (5) has proposed projects that are feasible within available resources, and (6) has a level of risk that is commensurate with the Agency's risk tolerance.

The purpose of the SDR/MDR for tightly coupled and single-project programs is to evaluate the credibility and responsiveness of the proposed program requirements/architecture to the Mission Directorate requirements and constraints, including available resources and allocation of requirements to projects. The SDR/MDR also determines whether the maturity of the program's mission/system definition and associated plans is sufficient to begin preliminary design. For tightly coupled programs a KDP 0 may be required, at the discretion of the Decision Authority, to ensure that major issues are understood and resolved prior to proceeding to PDR and KDP I. If the KDP 0 is held, the program will be expected to demonstrate how it meets critical NASA needs and that projects are feasible within available resources. For single-project programs, the program proceeds to KDP B, where the program is expected to demonstrate that (1) the proposed mission/system architecture is credible and responsive to program requirements and constraints, including resources; (2) the maturity of the mission/system definition and associated plans is sufficient to begin Phase B; and (3) the mission can probably be achieved within available resources with acceptable risk.

Preliminary Design Review. The purpose of the PDR for tightly coupled and single-project programs is to evaluate the completeness/consistency of the program's preliminary design, including its projects, in meeting all requirements with appropriate margins, acceptable risk, and within cost and schedule constraints, and to determine the program's readiness to proceed with the detailed design phase of the program.[4] After the PDR, the program proceeds to KDP I (KDP C for single-project programs). The program is expected to demonstrate that (1) it is in place and stable, (2) it addresses critical NASA needs, (3) it has adequately completed Formulation activities, (4) it has an acceptable plan for Implementation that leads to mission success, (5) the proposed projects are feasible within available resources, and (6) the program's level of risk is commensurate with the Agency's risk tolerance. The decisions made at KDP I (KDP C for single-project programs) establish the ABC for the program. (See Section 5.5.1.)

The general flow of activities for the various program types in Formulation is shown in Figures 3-8, 3-9, and 3-10.

While not part of Formulation, some Implementation activities such as initiating project Pre–Phase A may occur during Formulation.

Program Formulation is a recursive and iterative process that requires concurrent development of the program organization, structure, management approach, management processes and the technical and management products required for program implementation. The level of maturity of each of these items continues to evolve and each item becomes more mature as the program goes through the formulation process. Each of the life-cycle milestones and associated KDPs provides an opportunity for the program and its management to review and assess the program's progress.

3.3.2.2 Program Activities in Formulation by Program Type

The different program types require different levels of management and planning in Formulation.

Uncoupled and Loosely Coupled Program Formulation. As a result of the loose affiliation between the projects in these programs, the program does

---

[4] Uncoupled and loosely coupled programs do not have a PDR.

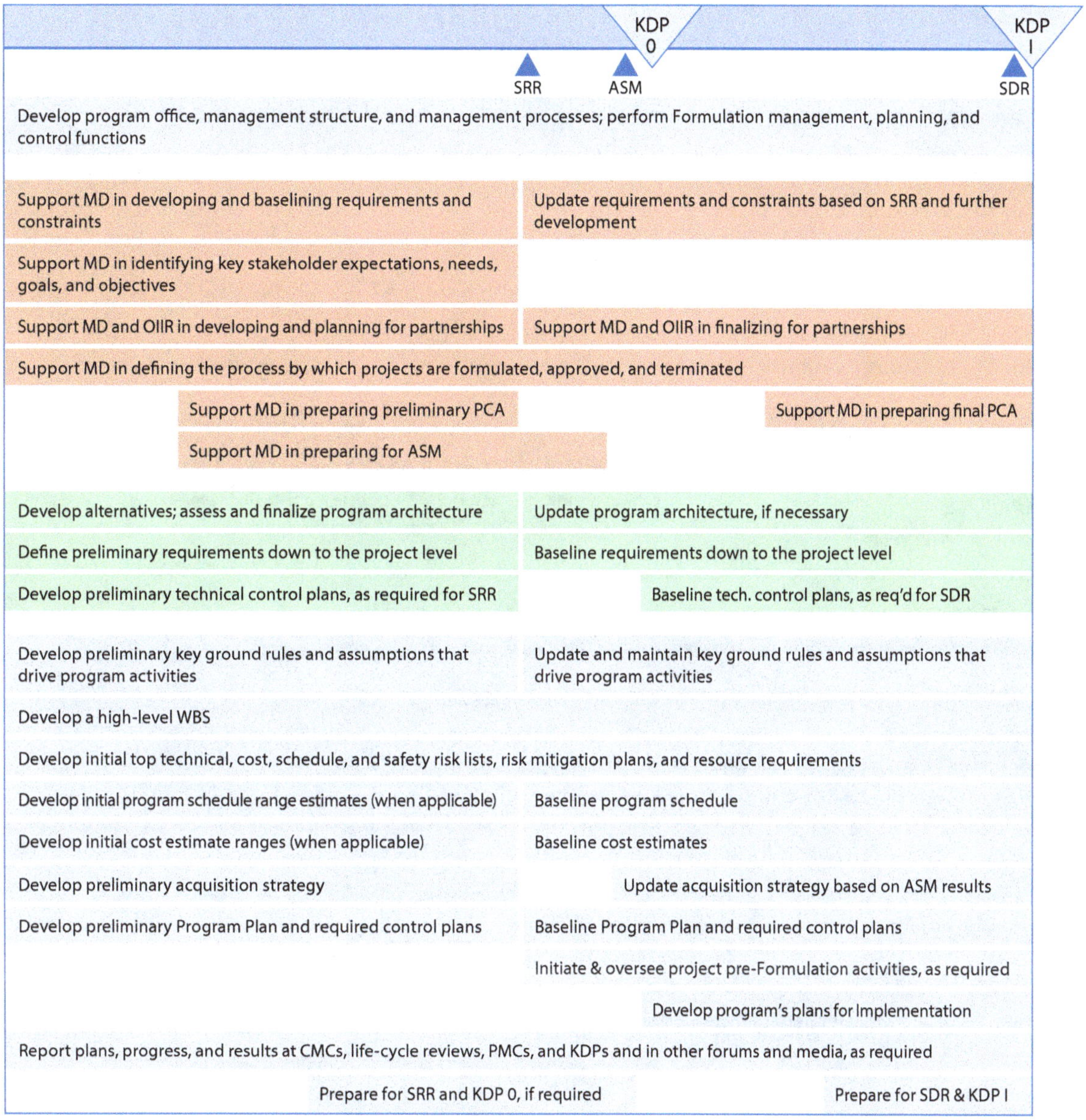
KDP 0
KDP I
SRR
ASM
SDR
Develop program office, management structure, and management processes; perform Formulation management, planning, and control functions
Support MD in developing and baselining requirements and constraints
Update requirements and constraints based on SRR and further development
Support MD in identifying key stakeholder expectations, needs, goals, and objectives
Support MD and OIIR in developing and planning for partnerships
Support MD and OIIR in finalizing for partnerships
Support MD in defining the process by which projects are formulated, approved, and terminated
Support MD in preparing preliminary PCA
Support MD in preparing final PCA
Support MD in preparing for ASM
Develop alternatives; assess and finalize program architecture
Update program architecture, if necessary
Define preliminary requirements down to the project level
Baseline requirements down to the project level
Develop preliminary technical control plans, as required for SRR
Baseline tech. control plans, as req'd for SDR
Develop preliminary key ground rules and assumptions that drive program activities
Update and maintain key ground rules and assumptions that drive program activities
Develop a high-level WBS
Develop initial top technical, cost, schedule, and safety risk lists, risk mitigation plans, and resource requirements
Develop initial program schedule range estimates (when applicable)
Baseline program schedule
Develop initial cost estimate ranges (when applicable)
Baseline cost estimates
Develop preliminary acquisition strategy
Update acquisition strategy based on ASM results
Develop preliminary Program Plan and required control plans
Baseline Program Plan and required control plans
Initiate & oversee project pre-Formulation activities, as required
Develop program's plans for Implementation
Report plans, progress, and results at CMCs, life-cycle reviews, PMCs, and KDPs and in other forums and media, as required
Prepare for SRR and KDP 0, if required
Prepare for SDR & KDP I

Legend: ☐ Program management, planning, and control tasks

☐ Work for which Headquarters is responsible but the program helps accomplish (e.g., international partnerships are a Headquarters responsibility, but the programs help develop and finalize those partnerships)

☐ Technical work the program is doing

Acronyms: MD = Mission Directorate; OIIR = Office of International and Interagency Relations.

Note: These are typical high-level activities that occur during this program phase. Placement of reviews is notional.

Figure 3-8 Uncoupled and Loosely Coupled Program Formulation Flow of Activities

	SRR	ASM	KDP 0 / SDR	KDP I / PDR

Develop program office, mgmt structure, & mgmt processes; perform Formulation management, planning, & control functions

Sup. MD in dev. & baselining req'ts & constr.	Update req'ts & constraints based on life-cycle review/KDP results & further development		
Support MD in identifying key stakeholder expectations, needs, goals, & objectives			
Support MD & OIIR in developing & planning for int'l & interagency partnerships	Support MD and OIIR in finalizing partnerships		

Support MD in defining the process by which projects are formulated, approved, and terminated

	Support MD in preparing for ASM	Support MD in prep. prelim. PCA	Support MD in prep. final PCA

Develop alternatives; assess & finalize program concept, architecture, & concepts of ops		Update concept & architecture, if necessary

Conduct initial assessment of tech. dev. requirements; continue to assess tech. needs as concept(s) evolves; conduct tech. dev, if required

Define prel. req'ts down to project level	Baseline req'ts down to project level	Update req'ts down to project level

Develop preliminary designs

Conduct program system engineering and integrate project technical activities, as required

Assess need for program-level technical products; initiate and develop products as required

Dev. tech. control plans as required for SRR	Dev. tech. control plans as required for SDR	Dev. tech. control plans as required for PDR

Develop and maintain key ground rules and assumptions that drive program activities

Develop a high-level WBS

Develop/update staffing and infrastructure requirements and plans as program Formulation evolves

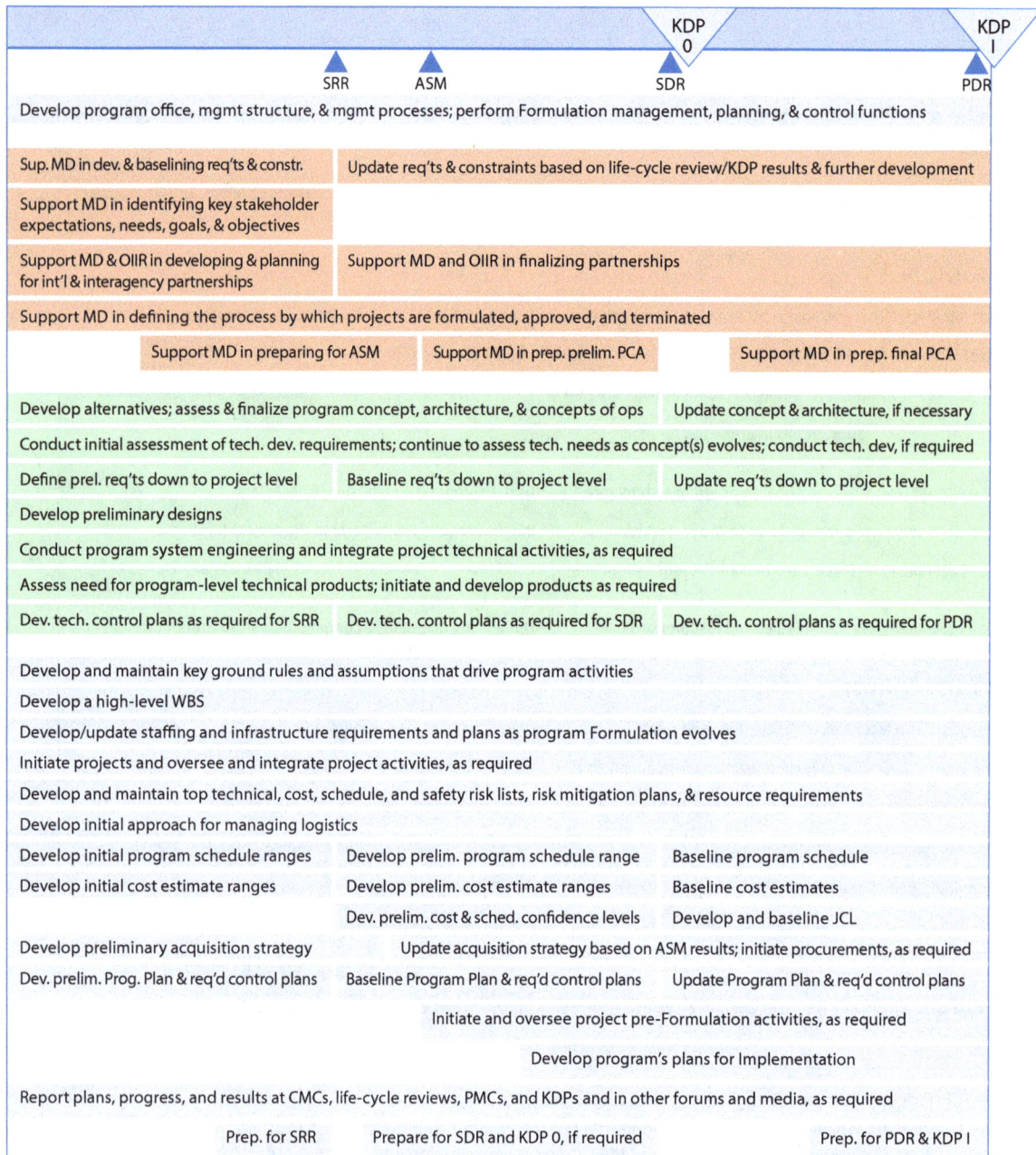

Initiate projects and oversee and integrate project activities, as required

Develop and maintain top technical, cost, schedule, and safety risk lists, risk mitigation plans, & resource requirements

Develop initial approach for managing logistics

Develop initial program schedule ranges	Develop prelim. program schedule range	Baseline program schedule
Develop initial cost estimate ranges	Develop prelim. cost estimate ranges	Baseline cost estimates
	Dev. prelim. cost & sched. confidence levels	Develop and baseline JCL
Develop preliminary acquisition strategy	Update acquisition strategy based on ASM results; initiate procurements, as required	
Dev. prelim. Prog. Plan & req'd control plans	Baseline Program Plan & req'd control plans	Update Program Plan & req'd control plans

	Initiate and oversee project pre-Formulation activities, as required	
	Develop program's plans for Implementation	

Report plans, progress, and results at CMCs, life-cycle reviews, PMCs, and KDPs and in other forums and media, as required

Prep. for SRR	Prepare for SDR and KDP 0, if required	Prep. for PDR & KDP I

Legend: ☐ Program management, planning, and control tasks

☐ Work for which Headquarters is responsible but the program helps accomplish (e.g., international partnerships are a Headquarters responsibility, but the programs help develop and finalize those partnerships)

☐ Technical work the program is doing

Acronyms: MD = Mission Directorate; OIIR = Office of International and Interagency Relations.

Note: These are typical high-level activities that occur during this program phase. Placement of reviews is notional.

Figure 3-9 Tightly Coupled Program Formulation Flow of Activities

KDP A — SRR — ASM — KDP B (MDR/SDR) — PDR — KDP C

KDP A	SRR	ASM	MDR/SDR (KDP B)	PDR (KDP C)

Develop program office, mgmt structure, & mgmt processes; perform Formulation management, planning, & control functions

Support MD in dev. & baselining requirements, constraints, ground rules, etc.	Update requirements, constraints, ground rules, assumptions, etc., based on life-cycle review/KDP results and further development

Sup. MD in identifying key stakeholder expectations, needs, goals, & objectives

Sup. MD & OIIR in developing & planning for int'l & interagency partnerships	Support MD and OIIR in finalizing partnerships

Sup. MD in prep. for ASM	Sup. MD in prep. prelim. PCA	Support MD in prep. final PCA

Develop alternatives; assess & finalize mission concept, architecture, & concepts of ops	Update concept & architecture, if nec.

Conduct initial assessment of tech. dev. req'ts; continue to assess tech. needs as concept(s) evolves; conduct tech. dev., as req'd

Conduct initial assess. of eng. dev. req'ts; continue to assess eng. risk red. needs as concept(s) evolves; conduct eng. dev., as req'd

Assess evolving concepts to ensure heritage is applied properly; identify approp. risk red. activities; conduct heritage assess., as req'd

Baseline req'ts down to proj. & system level	Update req'ts down to system level	Baseline req'ts down to subsystem level

Develop preliminary designs

Conduct program system eng. & integrate project tech. activities, as req'd (for SPP w/ separate program and project structures)

	Develop safety data products

Develop, baseline, and maintain technical control plans, as required

Develop and maintain key ground rules and assumptions that drive program activities

Develop a high-level WBS consistent with NASA standard WBS and the program architecture

Develop/update staffing and infrastructure requirements and plans as program Formulation evolves

Develop and maintain top technical, cost, schedule, and safety risk lists, risk mitigation plans, and resource requirements

Initiate and oversee project Formulation activities, if required (for SPP with separate program and project structures)

Develop initial approach for managing logistics

Develop initial program schedule ranges	Develop prel. program schedule range	Baseline program schedule
Develop initial cost estimate ranges	Develop prelim. cost estimate ranges	Baseline cost estimates
	Dev. prelim. cost & sched. conf. levels	Develop and baseline JCL
Develop preliminary acquisition strategy	Update acq. strategy based on ASM results; initiate procurements, as req'd	
Develop required control plans	Dev. prelim. Prog. Plan & req'd control plans	Baseline Prog. Plan & req'd control plans
	Update Formulation Agreement for Phase B	Dev. program's plans for Implementation

Report plans, progress, and results at CMCs, life-cycle reviews, PMCs, and KDPs and in other forums and media, as required

Prepare for SRR	Prepare for SDR and KDP B	Prep. for PDR & KDP C

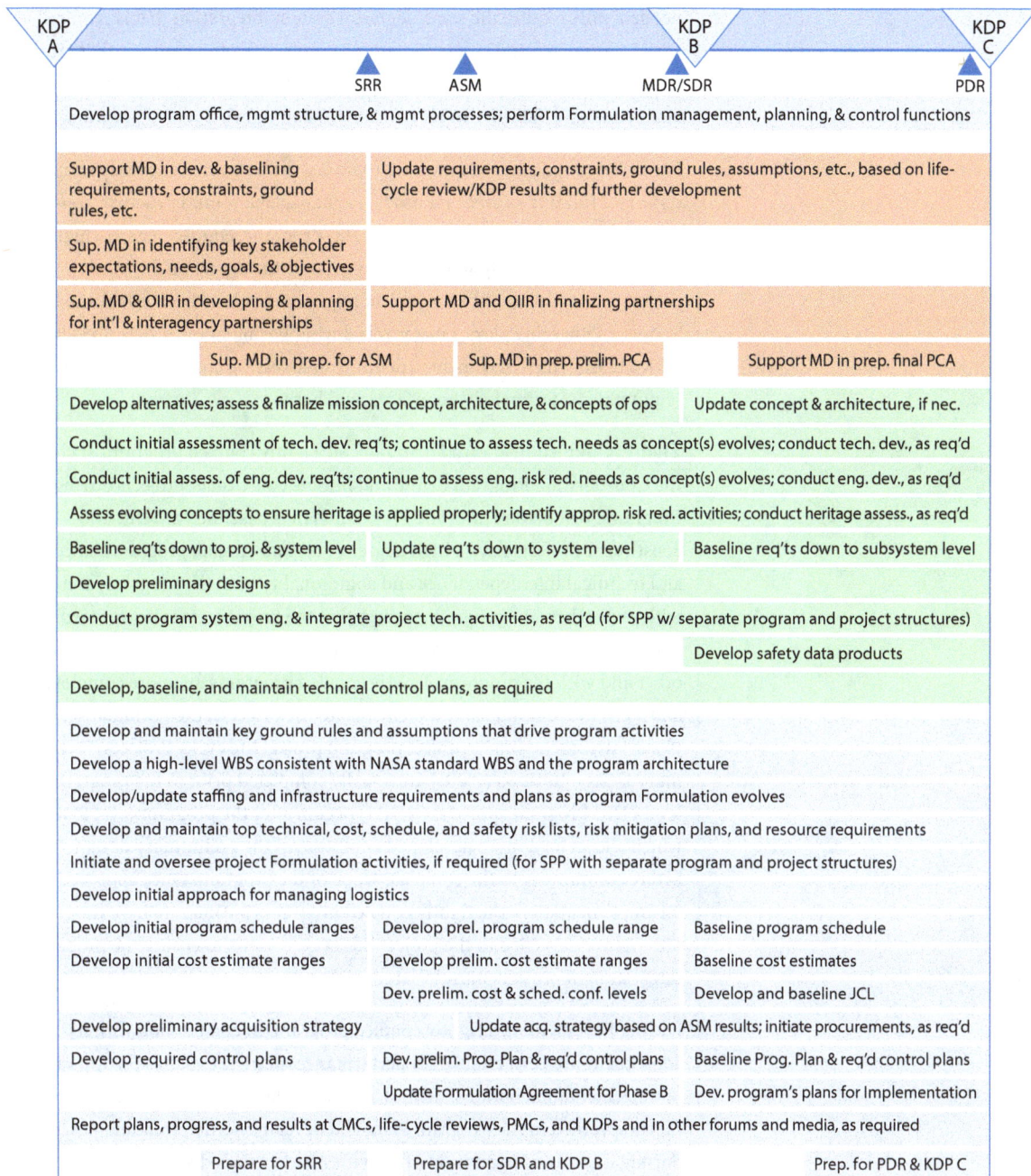

Legend: ☐ Program management, planning, and control tasks

☐ Work for which Headquarters is responsible but the program helps accomplish (e.g., international partnerships are a Headquarters responsibility, but the programs help develop and finalize those partnerships)

☐ Technical work the program is doing

Acronyms: MD = Mission Directorate; OIIR = Office of International and Interagency Relations; SPP = Single-Project Program.

Note: These are typical high-level activities that occur during this program phase. Placement of reviews is notional.

Figure 3-10 Single-Project Program Formulation Flow of Activities

not generally require the same degree of system integration that is required of tightly coupled and single-project programs. Thus the products that are required for these programs are substantially fewer (see Table 3-2 in this Handbook and Table I-1 in NPR 7120.5E) than tightly coupled programs.

For loosely coupled or uncoupled programs, the program office may simply serve as a funding source and provide a management infrastructure, top-level requirements, and project oversight. Program requirements are high level. They are typically stable and have very little impact on day-to-day project management once the project requirements have been established. System engineering plays a major role during Formulation as described in NPR 7123.1, which may include defining or assessing concepts, architecture, requirements, technology, interfaces, and heritage.

Tightly Coupled Program Formulation. Tightly coupled programs such as the Constellation Program define and initiate constituent projects during Program Formulation after the Program Plan is baselined at SDR. The constituent projects have a high degree of organizational, programmatic, and technical interdependence and commonality with the program, and with each other. The program ensures that the projects are synchronized and well integrated throughout their respective life cycles, both with each other and with the program. Tightly coupled programs are more complex, and since the program is intimately tied to the projects, the Formulation Phase mirrors the single-project program project life cycle. Projects' Preliminary Design Reviews (PDRs) are usually completed prior to the program-level PDR. Program approval (KDP I) occurs after the program-level PDR, which allows for a more developed definition of the preliminary design before committing to the complete scope of the program. Once approved for Implementation, the tightly coupled program continues to have program life-cycle reviews tied to the project life-cycle reviews to ensure the proper integration of projects into the larger system.

During Formulation, a tightly coupled program, in conjunction with its constituent projects, establishes performance metrics, explores the full range of implementation options, defines an affordable concept to meet requirements specified in the Program Plan, and develops needed technologies. Formulation is an iterative set of activities, rather than discrete linear steps. System engineering plays a major role during Formulation as described in NPR 7123.1. The primary activities, which, in some cases, may be performed in conjunction with, or by, the constituent projects, include the following:

- Developing and defining the program requirements;
- Assessing the technology requirements, developing the plans to achieve them, and developing the technology;

- Developing the program's knowledge management strategy and processes;

- Examining the Lessons Learned database for lessons that might apply to the current program's planning;

- Developing the program architecture down to the project level;

- Flowing down requirements to the project level;

- Planning acquisitions, including an analysis of the industrial base capability to design, develop, produce, support, and even possibly restart an acquisition program or project;

- Evaluating and refining project to project interfaces;

- Assessing heritage using the *Systems Engineering Handbook, NASA/ SP-2007-6105 Rev 1*, Appendix G (the applicability of designs, hardware, and software in past projects to the present one);

- Conducting safety, performance, technical, cost, and schedule risk trades;

- Identifying and mitigating development and programmatic risks, including supply chain risks;

- Conducting engineering development activities, including developing engineering prototypes and models for the higher risk components and assemblies that have not been previously built or flown in the planned environment and testing them to demonstrate adequate performance;

- Developing time-phased cost and schedule estimates and documenting the basis of these estimates; and

- Preparing the Program Plan for Implementation.

Tightly coupled programs typically have greater integration functions at the program level, such as systems engineering, risk management, and requirements management. The program manager has a significant role and influence over the management and execution of the projects. In the case of a tightly coupled program, major project decisions frequently require the approval of the program manager. Decisions to change elements, such as reduce scope or extend schedule, for one project may affect all other projects within that program. The project manager provides frequent briefings and regular progress status to the program manager. Certain project risks may be integrated into a list of top program risks. Change in program requirements may have a direct impact on project requirements.

Formulation activities continue until Formulation output products (see Tables 3-3 and 3-4) have matured and are acceptable to the program manager, Center Director, MDAA, and Decision Authority. These activities allow the Agency to present to external stakeholders high-confidence cost and schedule commitments at KDP I.

Single-Project Program Formulation. MDAAs may initiate single-project program pre-Formulation activities. In that case, a Mission Directorate provides resources for Pre–Phase A concept studies along with the mission objectives, ground rules and assumptions to be used by the study team. While not formally part of Formulation, concept studies might involve pre-Formulation activities such as Design Reference Mission (DRM) analysis, feasibility studies, technology needs analyses, engineering systems assessments, human systems assessments, logistics support, and analyses of alternatives that typically are performed before a specific single-project program concept emerges. These trade studies are not considered part of formal planning since there is no certainty that a specific proposal will emerge. Pre-Formulation activities also involve identification of risks that are likely to drive the single-project program's cost and schedule range estimates at KDP B and cost and schedule commitments at KDP C, and include development of mitigation plans for those risks. During Pre–Phase A, the program initiates development of a Formulation Agreement (see "Formulation Agreement" box for more information) to document the plans and resources

Formulation Agreement

The Formulation Agreement serves as a tool for communicating and negotiating the single-project program's schedule and funding requirements during Phase A and Phase B with the Mission Directorate. It identifies and prioritizes the technical and acquisition activities that will have the most value during Formulation and inform follow-on plans. The Formulation Agreement focuses on the work necessary to accurately characterize the complexity and scope of the single-project program; increase understanding of requirements; and identify and mitigate safety, technical, cost, and schedule risks. This work enables the single-project program to develop high-fidelity cost and schedule range estimates and associated confidence levels at KDP B, and high-fidelity cost and schedule commitments and associated JCL at KDP C, and to commit to a successful plan for Implementation at KDP C. These activities include establishing the internal management control functions that will be used throughout the life of the single-project program. The Agreement is approved and signed at KDP A (baselined for Phase A and preliminary for Phase B). The Agreement is updated in preparation for SDR/MDR and resubmitted for signature at KDP B (baselined for Phase B). The Formulation Agreement for KDP A includes detailed Phase A information, preliminary Phase B information, and the Formulation Cost, which is based on the estimated costs for Phase A and Phase B. The Formulation Agreement for KDP B identifies the progress made during Phase A, updates and details Phase B information, and updates the Formulation Cost, which is based on the actual cost for Phase A and an updated cost for Phase B. The Formulation Cost at KDP B is the total authorized cost for Formulation activities required to get to KDP C. In practice, the FAD and the Formulation Agreement are developed concurrently so that both documents can be approved at KDP A. Documentation products developed as part of, or as a result of, the Formulation Agreement may be incorporated into the Single-Project Program Plan, if appropriate, as the Single-Project Program Plan is developed during Formulation.

required for Formulation (see NPR 7120.5E, Appendix F for the Formulation Agreement template). Assessments and products developed during Pre–Phase A may be documented in the Formulation Agreement, as opposed to developing separate plans. The Mission Concept Review (MCR) is held at the end of Pre-Phase A. The MCR is the first major life-cycle review in the single-project program life cycle. The purpose of the MCR is to evaluate the feasibility of the proposed mission concept(s) and how well the concept(s) fulfill the program's needs and objectives. After the MCR, the program proceeds to KDP A where the program demonstrates that it has addressed critical NASA needs; the proposed mission concept(s) is feasible; the associated planning is sufficiently mature to begin Phase A; and the mission can probably be achieved as conceived. At the conclusion of Pre–Phase A, the FAD is issued (see NPR 7120.5E, Appendix E) authorizing Formulation to begin.

Single-project program Formulation consists of two sequential phases, i.e., Phase A (Concept and Technology Development) and Phase B (Preliminary Design and Technology Completion). During Formulation, the single-project program establishes performance metrics, explores the full range of implementation options, defines an affordable concept to meet requirements specified in the Program Plan, and develops needed technologies. Formulation is an iterative set of activities rather than discrete linear steps. Systems engineering plays a major role during Formulation as described in NPR 7123.1. The primary activities in these phases include the following:

- Developing and defining the single-project program requirements;
- Assessing the technology requirements, developing the plans to achieve them, and developing the technology;
- Developing the program's knowledge management strategy and processes;
- Examining the Lessons Learned database for lessons that might apply to the current program's planning;
- Developing the system architecture;
- Completing mission and preliminary system designs;
- Flowing down requirements to the system/subsystem level;
- Planning acquisitions, including an analysis of the industrial base capability to design, develop, produce, support, and even possibly restart an acquisition program or project;
- Evaluating and refining subsystem interfaces;
- Assessing heritage using the *Systems Engineering Handbook, NASA/SP-2007-6105 Rev 1*, Appendix G (the applicability of designs, hardware, and software in past projects to the present one);

- Conducting safety, performance, technical, cost, and schedule risk trades;
- Identifying and mitigating development and programmatic risks, including supply chain risks;
- Conducting engineering development activities, including developing engineering prototypes and models for the higher risk components and assemblies that have not been previously built or flown in the planned environment, and testing them to demonstrate adequate performance;
- Developing time-phased cost and schedule estimates and documenting the basis of these estimates; and
- Preparing the Program or Project Plan for Implementation.

During Phase B, there is an overlap between the Formulation Agreement and the preliminary Program Plan. The Formulation Agreement is the agreement between the Mission Directorate and the single-project program that governs the work during Phase B, but the baselined Program Plan control plans govern the management and technical control processes used during this phase.

Formulation activities continue until Formulation output products (i.e., the products listed in Tables 3-5 and 3-6) have matured and are acceptable to the program manager, Center Director, MDAA, and Decision Authority. These activities allow the Agency to present to external stakeholders time-phased cost plans and schedule range estimates at KDP B and high-confidence cost and schedule commitments at KDP C.

Single-project programs follow steps in Formulation and Implementation that are similar to projects. However, because of their importance to the Agency, single-project programs are required to develop and have approved a Program Commitment Agreement (PCA) to move from Formulation to Implementation. A Program Plan is also required, but this document may be combined with the Project Plan if approved by the MDAA and OCE. However, if the Program and Project Plans are combined, the unique parts of the Program and Project Plans still need to be developed. These include the Product Data and Life-Cycle Management (PDLM) control plan, IT control plan, and the Threat Summary. A draft version of the Program Plan is due at KDP B, with final versions baselined by KDP C.

3.3.3 Program Management, Planning, and Control Activities

3.3.3.1 Supporting Headquarters Planning

During Formulation (and possibly pre-Formulation), the program manager and program team support the Mission Directorate in developing the program. When requested, the team helps identify the main stakeholders of the program (e.g., Principal Investigator (PI), science community, technology community, public, education community, and Mission Directorate sponsor) and gather and document key external stakeholder expectations, needs, goals, and objectives. The program also develops the process to be used within the program to ensure stakeholder advocacy. The team supports alignment of the program-level requirements with Agency strategic goals and Mission Directorate requirements and constraints. The MDAA uses this information in developing and obtaining approval of the FAD.

One of the first activities is to select the management team. The program manager is recommended by the Center Director with approval for appointment by the MDAA.

3.3.3.2 Program Structure and Management Framework

The program team, regardless of program type, develops and implements the management framework—including the program team, organizational structure, and management processes—consistent with the program authority, management approach, and Governance structure specified in the FAD. The team identifies the responsibilities related to the respective roles of each involved organization (e.g., Headquarters, Centers, other government agencies, academia, industry, and international partners). The team identifies the chain of accountability along with the frequency of reporting and the decision path outlining the roles and responsibilities of the Mission Directorate sponsor(s), program manager, Center Director, and other authorities (including the Technical Authorities (TAs)), as required. This will delineate clear lines of authority from projects and Centers to the program and to the Mission Directorate. The team also integrates knowledge from applicable lessons learned into the planning and determines how participating Centers' implementation policies and practices will be applied in the execution of the program. The management approach also includes the process by which projects are formulated, approved, and ended.

The program team supports the MDAA and the NASA Headquarters Office of International and Interagency Relations (OIIR) in identifying, planning for, and obtaining approved **interagency and international agreements,**[5] including the planning and negotiation of agreements and recommendations on joint participation in reviews, integration and test, and risk management. To the degree known, these partnership agreements typically are baselined by the SDR/MDR except for single-project programs, where the international partnerships are baselined at PDR.

3.3.3.3 Program Requirements, Ground Rules and Assumptions

The program team, regardless of program type, conducts planning that enables formulation and implementation of program and project concepts, architectures, scenarios/DRMs, and requirements. The team documents the traceability of preliminary **program-level requirements** on both the program and the known individual projects to Agency strategic goals and outcomes as described in *NPD 1001.0, NASA Strategic Plan*. The team selects technical standards in accordance with *NPR 7120.10, Technical Standards for NASA Programs and Projects.*[6]

At the Program/System Requirements Review (SRR), the team baselines the initial program-level requirements and **driving ground rules and assumptions** on the program. After the SRR, the team updates, as required, the baselined program-level requirements and the driving ground rules and assumptions on the program. Specifically:

- The program team identifies and documents the key requirements derived by the program (as opposed to those derived by the Mission Directorate), ground rules, and assumptions that drive development of the program and initial projects. Once the program team has defined the ground rules and assumptions, it tracks them through Formulation to determine if they are being realized (i.e., remain valid) or if they need to be modified.

- When establishing the requirements for the program, there are additional high-level requirements levied on the program from the Agency, Center, and Mission Directorate levels as well as requirements that come from support offices like SMA. The traceability of requirements that flow down from Agency- and Center-level policy to the program and from the program to projects should be documented.

- For all programs, these high-level requirements typically are decomposed into requirements on constituent projects or systems. The requirements are typically specified in the Program Plan or in a separate, configuration-controlled program requirements document prepared by the program team

For each known project, the program team develops an appendix to the Program Plan or separate document that includes a top-level description of the project, including the mission's science or exploration objectives; the project's category, governing PMC, and risk classification; and the project's mission, performance, and safety requirements. For science missions, it includes both baseline and threshold science requirements (see Appendix A) and identifies the mission success criteria for each project based on the threshold science requirements.

---

[5] Bolding indicates required products.

[6] *NASA-STD-8709.20, Management of Safety and Mission Assurance Technical Authority (SMA TA) Requirements* provides further information on selecting SMA standards.

and approved by the MDAA. This documentation is typically controlled by the Mission Directorate. Requirements thus documented, and any subsequent changes, require approval of the program manager and the MDAA.

- Each requirement is stated in objective, quantifiable, and verifiable terms. Requirements can identify the program's principal schedule milestones, including PDR, Critical Design Review (CDR), launch, mission operations critical milestones, and the planned decommissioning date. They can state the development and/or total life-cycle cost constraints on the program and set forth any budget constraints by fiscal year. They can state the specific conditions under which a project Termination Review would be triggered. They can also describe any additional requirements on the project; e.g., international partners. If the mission characteristics indicate a greater emphasis is necessary on maintaining technical, cost, or schedule, then the requirements can identify which is most important (e.g., state if the mission is cost-capped; or if schedule is paramount, as for a planetary mission; or if it is critical to accomplish the technical objectives, as for a technology demonstration mission).

In cases where the program interfaces with other programs, the Mission Directorate may determine that an Architectural Control Document (ACD) needs to be developed to define how the programs will interface. The program team supports the MDAA in developing and obtaining approval of any necessary ACD.

3.3.3.4 Program Activities for Project Initiation

Program offices support the MDAA in beginning project pre-Formulation activities and approving project entry into Formulation. Projects can be initiated in two basic ways: a direct assignment of a project to a Center(s) or a competitive process, typically through a BAA such as an Announcement of Opportunity (AO).[7]

For projects that are not competed, prior to initiating the new project, a Mission Directorate and the program office typically provide resources for concept studies (i.e., Pre–Phase A (Concept Studies)). These pre-Formulation activities involve DRM analysis, feasibility studies, technology needs

The Architectural Control Document (ACD) is a configuration-controlled document or series of documents that embodies cross-Agency mission architecture(s), including the structure, relationships, interfaces, principles, assumptions, and results of the analysis of alternatives that govern the design and implementation of the enabling mission systems. As an example, the Space Communications and Navigation program may use an ACD to describe, manage, and control how science and exploration programs use their communications and navigation services.

---

[7] NASA uses Broad Agency Announcements (BAAs) to solicit bids for work, a form of public/private competition. One form of BAA applicable to space flight programs and projects is Announcements of Opportunity (AOs). Another type is NASA Research Announcements (NRAs). AOs are used to acquire investigations, which may involve complete missions or special instruments to be flown aboard NASA aircraft or spacecraft, and invite investigator-initiated research proposals. NASA solicits, accepts, and evaluates proposals submitted by all categories of proposers in response to an AO, including academia, industry, not-for-profits, Government laboratories, Federally Funded Research and Development Centers (FFRDC), NASA Centers, and the Jet Propulsion Laboratory (JPL).

analyses, engineering systems assessments, and analyses of alternatives. These are performed before a specific project concept emerges. At the conclusion of pre-Formulation with a decision to proceed with the project, the Mission Directorate, supported by the program office, issues a project FAD (see NPR 7120.5E, Appendix E) authorizing project Formulation to begin. The Mission Directorate also agrees to a project Formulation Agreement (see NPR 7120.5E, Appendix F) developed by the project to document the plans and resources required for Formulation.

For competed or "AO-driven" missions, some Mission Directorates have chosen to use one or two steps to initiate projects within a space flight program. In a one-step AO process, projects are competed and selected for Formulation in a single step. In two-step competitions, several projects may be selected in Step 1 and given time to mature their concepts in a funded concept study before the Step 2 down-selection. Program resources are invested (following Step 1 selections) to bring these projects to a state in which their science content, cost, schedule, technical performance, project implementation strategies, SMA strategies, heritage, technology requirements and plans, partnerships, and management approach can be better judged. Programs are not typically involved in the proposal evaluation process or the selection. They generally provide input into the BAA in the form of requirements to ensure that the BAA is consistent with the program's requirements. Once the project is selected, the program assumes management responsibility for the project's development and implementation.

From the point of view of the selected AO-driven project, the proposing teams are clearly doing preparatory work and formal project Formulation (e.g., typical Pre-Phase A and Phase A tasks, such as putting together a detailed WBS, schedules, cost estimates, and implementation plan) during the concept study and the preparation of the Step 2 concept study report. From the point of view of the program, no specific project has been chosen, the total cost is not yet known, and project requirements are not yet finalized, yet Formulation has begun. Therefore, for competed missions, the selection of a proposal for concept development is the equivalent of KDP A. In a one-step AO process, projects enter Phase A after selection (KDP A) and the process becomes the conventional process for directed missions. In a two-step AO process, projects perform concept development in the equivalent of Phase A and go through evaluation for down-selection at the equivalent of KDP B. Following this selection, the process becomes conventional—with the exception that KDP B products requiring Mission Directorate input are finished as early in Phase B as feasible. All NASA space flight programs and projects are subject to NPR 7120.5 requirements. NPR 7120.5 requirements apply to contractors, grant recipients, or parties to agreements only to the extent specified or referenced in the appropriate contracts, grants, or agreements.

3.3.3.5 Management Control Processes and Products

As the program team develops its planning, management processes are documented in control plans, which are designed to keep the program activities aligned, on track, and accounted for as the program moves forward. (See Appendix F for a description of control plans required by NPR 7120.5E.) These control plans are described in this and subsequent sections of this handbook, in conjunction with the phase where they are required. Many control plans are incorporated into the central planning document, which is the Program Plan. NPR 7120.5E, Appendix G, identifies when a control plan may be included in the Program Plan, and when a control plan is required to be a stand-alone document. NPR 7120.5E, Appendix I, Tables I-1, I-3, I-7, and Tables 3-2, 3-4, and 3-6 of this handbook identify when these control plans are required. Centers may have existing plans, which programs may use to satisfy requirements for some of the control plans.

All programs prepare a Program Plan that follows the template in NPR 7120.5E, Appendix G. For uncoupled, loosely coupled, and tightly coupled programs, a preliminary version of the Program Plan is prepared prior to the SRR, and the Program Plan is finalized and baselined by the System Definition Review (SDR). For single-project programs, a preliminary version of the Program Plan is developed prior to the SDR/Mission Definition Review (MDR), and the Program Plan is finalized and baselined by the PDR. Some control plans incorporated into the Program Plan are required to be baselined before the Program Plan is fully finished and baselined. These early control plans are required to assist the program in managing its early work and become part of the preliminary Program Plan.

During early Formulation, the program team (all program types) begins to develop the Technical, Schedule, and Cost Control Plan. A preliminary version of this plan is expected at SRR with the final plan baselined at SDR (SDR/MDR for single-project programs). This plan is required early so that the program team has the tools and processes necessary to manage and control their work during Formulation and the team is prepared to baseline all program content by program approval at KDP I (KDP C for single-project programs). This plan documents how the program plans to control program requirements, technical design, schedule, and cost to achieve its high-level requirements. This control plan includes the program's performance measures in objective, quantifiable, and measurable terms and documents how the measures are traced from the program high-level requirements. The plan establishes baseline and threshold values for the performance metrics to be achieved at each KDP, as appropriate.

Margins are the allowances carried in budget, projected schedules, and technical performance parameters (e.g., weight, power, or memory) to account for uncertainties and risks. Margins are allocated in the formulation process, based on assessments of risks, and are typically consumed as the program or project proceeds through the life cycle.

Tightly coupled and single-project programs also develop and maintain the status of a set of programmatic and technical leading indicators that are defined in the Program Plan to ensure proper progress and management of the program.[8] (See "Required and Recommended Programmatic and Technical Leading Indicators" box.) Per NPR 7123.1B, three indicators are required: Mass Margins, Power Margins, and Request For Action (RFA)/ Review Item Discrepancy (RID)/Action Item burndown. In addition to these required indicators, NASA highly recommends the use of a common set of programmatic and technical indicators to support trending analysis throughout the life cycle. Programs may also identify unique programmatic and technical leading indicators.

The Technical, Schedule, and Cost Control Plan also describes the following:

- How tightly coupled and single-project programs monitor and control the program's Management Agreement and ABC;

Required and Recommended Programmatic and Technical Leading Indicators

Required (per *NPR 7123.1B*)

1. Technical Performance Measures (mass margin, power margin)
2. Review Trends (RID/RFA/action item burndown per review)

Recommended

1. Requirement Trends (percentage growth, TBD/TBR closures, number of requirement changes)
2. Interface Trends (percentage ICD approval, TBD/TBR burn down, # interface requirement changes)
3. Verification Trends (closure burn down, # deviations/waivers approved/open)
4. Software Unique Trends (# software requirements verified and validated per build/release versus plan)[1]
5. Problem Report/Discrepancy Report Trends (# open, # closed)
6. Manufacturing Trends (# nonconformance/corrective actions)
7. Cost Trends (plan, actual, UFE, EVM, new obligation authority)
8. Schedule Trends (critical path slack/float, critical milestones, EVM schedule metrics, etc.)
9. Staffing Trends (FTE, work-year equivalent)
10. Additional project-specific indicators as needed (e.g., human systems integration compliance)

---

[1] Please note that there are non-Technical Leading Indicators software measurement requirements in *NPR 7150.2, NASA Software Engineering Requirements* (e.g., SWE-091) which have implementation guidance in *NASA-HDBK-2203* (http://swehb.nasa.gov).

---

[8] See Section 5.13 and the program and project management community of practice on the NASA Engineering Network (NEN) for a white paper explaining leading indicators and more information on leading indicators.

- The mitigation approach for tightly coupled and single-project programs if the program is exceeding the development cost documented in the ABC to take corrective action prior to triggering the 30 percent breach threshold;

- How tightly coupled and single-project programs will support a rebaseline review in the event the Decision Authority directs one;

- Description of systems engineering organization and structure and how these functions are executed;

- The use of systems of measurement and the identification of units of measure in all product documentation. (See Section 3.3.4, Formulation Technical Activities and Products for more information on the use of Système Internationale (SI) or metrics system.);

- How the program is implementing Technical Authority (Engineering, Safety and Mission Assurance, and Health and Medical), including how the program will address technical waivers and deviations and how Dissenting Opinions will be handled;

- How tightly coupled and single-project programs will use an Earned Value Management System (EVMS); or how loosely coupled or uncoupled programs flow EVM requirements down to the projects, including the reporting of project EVM (see Section 4.3.4.2.2 and Section 5.14 for details on Earned Value Management); and

- Descope plans including key decision dates, savings in cost and schedule, and how the descopes are related to the program's threshold requirements.

The plan describes any additional specific tools the program will use to implement the program control processes, e.g., systems for requirements management; program scheduling; program information management; and how the program will monitor and control the Integrated Master Schedule (IMS), including utilization of its technical and schedule margins and UFE to stay within the terms of the Management Agreement and ABC (if applicable). Finally, the plan documents how the program plans to report technical, schedule, and cost status to the MDAA, including frequency and the level of detail.

All program teams develop a program Work Breakdown Structure (WBS). The NASA standard WBS template is intended to apply to projects, not programs. There is no standard program WBS due to the variance in structure of the Mission Directorates. Tightly coupled and single-project programs generally have a WBS like the standard WBS for space flight projects illustrated in Figure 4-10 in this handbook and in NPR 7120.5E Figure H-2. The WBS for uncoupled and loosely coupled programs will probably be more focused at the project level than the system level shown in the figure. All programs develop a WBS dictionary down to at least the project level. The WBS supports cost and schedule allocation down to a project level that allows for unambiguous cost reporting. (See Section 5.9.1 and Section 5.9.7) for additional guidance on developing a program WBS.)

After developing the WBS and the initial program architecture, the program team develops the cost and schedule estimate and appropriate annual budget submissions. Cost and schedule typically are informed by technology, engineering development and heritage assessments using the *Systems Engineering Handbook, NASA/SP-2007-6105 Rev 1*, Appendix G, acquisition strategies, infrastructure and workforce requirements, and identified risks. Infrastructure requirements include the acquisition, renovation, and/or use of real property/facilities, aircraft, personal property, and information technology. The program identifies the means of meeting infrastructure requirements through synergy with other existing and planned programs and projects to avoid duplication of facilities and capabilities. The program also identifies necessary infrastructure upgrades or new developments, including those needed for environmental compliance.

The program develops the life-cycle cost and schedule estimates consistent with driving assumptions, risks, requirements, and available funding and schedule constraints:

- The program team develops its cost estimates using many different techniques. These include, but are not limited to, bottoms-up estimates where specific work items are estimated by the performing organization using historical data or engineering estimates; vendor quotes; analogies; and parametric cost models. (See Section 5.6 for a discussion of probabilistic cost estimating.)

- The program team develops its resource baseline, which includes funding requirements by fiscal year and the new obligation authority in real-year dollars for all years—prior, current, and remaining. The funding requirements are consistent with the program's WBS and include funding for all cost elements required by the Agency's full-cost accounting procedures. Funding requirements are consistent with the budget. The resource baseline provides a breakdown of the program's funding requirements to at least the WBS Level 2 elements. The resource baseline provides the workforce requirements specific to the program (i.e., not project workforce) by fiscal year, consistent with the program's funding requirements and WBS. The resource baseline identifies the driving ground rules, assumptions and constraints that affect it. Throughout the Implementation Phase, tightly coupled and single-project program baselines are based on the approved JCL in accordance with NPD 1000.5 and NPR 7120.5. (The resource baseline also includes the infrastructure requirements, discussed elsewhere in this section.)

- The program team develops a summary of its IMS, including all critical milestones, major events, life-cycle reviews, and KDPs throughout the program life cycle. The summary of the IMS includes the logical relationships (interdependencies) for the various program elements and projects and critical paths, as appropriate, and identifies the driving ground rules, assumptions, and constraints affecting the schedule baseline. The summary of the IMS is included in the Program Plan.

- In doing these estimates, the program team documents their BoE and the rationales and assumptions that went into their estimate.

- Finally, prior to their PDR, tightly coupled and single-project program teams develop a JCL.

All program types plan, prepare for, and support the ASM, if required, as part of developing their acquisition strategy, generally prior to SRR. The results of this meeting are documented in the ASM minutes and used to finalize the Acquisition Plan, which is baselined at SDR for uncoupled, loosely coupled, and tightly coupled programs. Single-project programs baseline their acquisition plan earlier, at SRR, to allow procurement actions earlier in Formulation. The program Acquisition Plan is developed by the program manager with support by the Office of Procurement. The plan needs to be consistent with the results of the acquisition planning process, which includes such things as assignment of lead Center, considerations for partnering, and decisions made at the ASM.

The Acquisition Plan documents an integrated acquisition strategy that enables the program to meet its mission objectives, provides the best value to NASA, and identifies all major proposed acquisitions (such as engineering design study, hardware and software development, mission and data operations support, and sustainment) in relation to the program WBS. It also describes completed or planned studies supporting make-or-buy decisions, considering NASA's in-house capabilities and the maintenance of NASA's core competencies, as well as cost and best overall value to NASA. The plan describes the state of the industrial base capability and identifies potential critical and single-source suppliers needed to design, develop, produce, support, and, if appropriate, restart an acquisition program or project. The Acquisition Plan for single-project programs, and if applicable, for tightly coupled programs, describes the Integrated Baseline Reviews (IBR) and schedules required for contracts requiring EVM (refer to the NFS), how the program needs to conduct any required IBRs, and how to maintain the contract documentation. (Additional guidance for all programs is provided in the Program Plan template in NPR 7120.5. In addition, further detail for single-project programs is provided in Section 4.3.4.2.2 of this handbook.) The program supports Procurement Strategy Meetings (PSM) for individual

The Basis of Estimate (BoE) documents the ground rules, assumptions, and drivers used in developing the cost and schedule estimates, including applicable model inputs, rationale or justification for analogies, and details supporting cost and schedule estimates. The BoE is contained in material available to the Standing Review Board (SRB) and management as part of the life-cycle review and Key Decision Point (KDP) process. Good BoEs are well-documented, comprehensive, accurate, credible, traceable, and executable. Sufficient information on how the estimate was developed needs to be included to allow review team members, including independent cost analysts, to reproduce the estimate if required. Types of information can include estimating techniques (e.g., bottoms-up, vendor quotes, analogies, parametric cost models), data sources, inflation, labor rates, new facilities costs, operations costs, sunk costs, etc.

The Procurement Strategy Meeting (PSM) provides the basis for approval of the approach for major procurements for programs and projects and ensures they are following the law and the Federal Acquisition Regulations (FAR). Detailed PSM requirements and processes, prescribed by the FAR and NFS and formulated by the Office of Procurement, ensure the alignment of portfolio, mission acquisition, and subsequent procurement decisions. The contents of written acquisition plans and PSMs are delineated in the FAR in Subpart 7.1—Acquisition Plans, the NFS in Subpart 1807.1—Acquisition Plans, and in the Guide for Successful Headquarters Procurement Strategy Meetings at http://prod.nais.nasa.gov/portals/pl/documents/PSMs_091611.html.

procurements that require PSMs. These PSMs and their procurements typically are based on the Acquisition Plan.

All acquisitions over $10 million are required by the NASA FAR Supplement (NFS) to conduct a PSM. The Office of Procurement at Headquarters determines which PSMs require a Headquarters review and which can be delegated to the Centers by reviewing the procurements on the Master Buy List, which is updated periodically by the Centers. The PSM is chaired by the Assistant Administrator for Procurement at Headquarters. Each Center has its own tailored procedure for Center-level PSMs and may specify who chairs their PSMs. (It is usually the Center Procurement Officer.) The PSM covers subjects such as how the acquisition fulfills mission need, budget and funding profile, small business opportunities, contract type, EVM requirements, and length of contract. It implements the decisions that flow from the higher-level meetings.

All program types identify and assess risks that threaten program requirements and development. Uncoupled, loosely coupled, and tightly coupled programs develop a preliminary Risk Management Plan by SRR and baseline the plan by SDR, whereas the single-project program baselines its plan by SRR since system hardware design is being conducted prior to SDR/MDR. This plan summarizes how the program implements the NASA risk management process (including Risk-Informed Decision-Making (RIDM) and Continuous Risk Management (CRM)) in accordance with NPR 8000.4 and NASA/SP-2011-3422. It includes the initial risk list; appropriate actions to mitigate each risk; and the resources needed for managing and mitigating these risks. Programs with international or other U.S. Government agency contributions need to plan for, assess, and report on risks due to international or other government partners and plan for contingencies. For tightly coupled programs, the Risk Management Plan is a stand-alone document.

All programs develop an initial PDLM Plan by SDR (SDR/MDR for single-project programs). (The PDLM is the set of processes and associated information used to manage the entire life cycle of product data from its conception through design, test, and manufacturing to service and disposal.) The PDLM plan is not baselined but is maintained and updated as necessary. The plan documents agreement among the program manager and various providers of PDLM services on how the identified PDLM capabilities will be provided and how authoritative data will be managed effectively in compliance with *NPR 7120.9, NASA Product Data and Life-Cycle Management (PDLM) for Flight Programs and Projects.*

Uncoupled and loosely coupled programs develop an Education Plan and baseline the plan by SDR. Tightly coupled and single-project programs develop a preliminary Education Plan by SDR (SDR/MDR for single-project programs) and baseline the plan by PDR. This plan describes planned efforts and activities to enhance Science, Technology, Engineering, and Math (STEM) education using the program's science and technical content. It describes the plan for coordinating with a Mission Directorate Education Coordinating Council (ECC) member to ensure program education activities are aligned with NASA education portfolio offerings and requirements. It defines goals and outcomes for each activity and addresses how activities will advance NASA strategic goals for education. It also identifies the target audience for each activity and discusses how the activity reaches and engages groups traditionally underrepresented and/or underserved in STEM disciplines. The plan describes how each activity will be evaluated; defines specific metrics and how they will be collected; and includes a timeline with relevant milestones for achieving goals and outcomes for each activity. Finally, the plan describes the relationship between the program and project(s) education plans.

Loosely coupled and uncoupled programs develop a Communications Plan and baseline the plan by SDR. Tightly coupled and single-project programs develop a preliminary plan by SDR (SDR/MDR for single-project programs) and baseline the plan by PDR. This plan describes how the program implements a diverse, broad, and integrated set of efforts and activities to communicate with and engage target audiences, the public, and other stakeholders in understanding the program, its objectives, elements, and benefits. It describes how the plan relates to the larger NASA vision and mission. Focus is placed on activities and campaigns that are relevant, compelling, accessible, and, where appropriate, participatory. The plan describes how these efforts and activities will promote interest and foster participation in NASA's endeavors and develop exposure to and appreciation for STEM education. The program Communications Plan:

- Defines goals and outcomes, as well as key overarching messages and themes;
- Identifies target audiences, stakeholders, and partnerships;
- Summarizes and describes products to be developed and the tools, infrastructure, and methods used to communicate, deploy, and disseminate those products;
- Describes the use of various media, e.g., multimedia, web, social media, and publications for nontechnical audiences, excluding those developed in the context of the Education Plan;
- Describes events, activities, and initiatives focused on public engagement and how they link with planned products and infrastructure;
- Identifies milestones and resources required for implementation;
- Defines metrics to measure success; and

The Operations Concept is a description of how the flight system and the ground system are used together to ensure that the mission operations can be accomplished reasonably. This might include how mission data of interest, such as engineering or scientific data, are captured, returned to Earth, processed, made available to users, and archived for future reference. The Operations Concept typically describes how the flight system and ground system work together across mission phases for launch, cruise, critical activities, science observations, and the end of the mission to achieve the mission. The Operations Concept is baselined at PDR with the initial preliminary operations concept required at MCR.

The term "concept documentation" used in *NPR 7120.5* is the documentation that captures and communicates a feasible concept at MCR that meets the goals and objectives of the mission, including results of analyses of alternative concepts, the concept of operations (baselined at MCR per *NPR 7123.1*), preliminary risks, and potential descopes. (Descope is a particular kind of risk mitigation that addresses risks early in the program Formulation Phase.)

- Describes the relationship between the program and project Communications Plans and the coordination between program and projects regarding communications activities.

Uncoupled, loosely coupled, and tightly coupled programs develop a preliminary Knowledge Management Plan by SRR and baseline the plan by SDR; the single-project program develops a preliminary plan by SDR/MDR and baselines the plan at PDR. This plan contains three elements: creating the program's knowledge management strategy and processes, including practices and approaches for identifying, capturing and transferring knowledge; examining the lessons learned database for relevant lessons that can be reflected into the program early in the planning process to avoid known issues; and creating the plan for how the program continuously captures and documents lessons learned throughout the program life cycle in accordance with *NPD 7120.4, NASA Engineering and Program/Project Management Policy* and as described in *NPD 7120.6, Knowledge Policy on Programs and Projects* and other appropriate requirements and standards documentation.

3.3.4 Technical Activities and Products

The program team for all program types continues to develop the architecture of the program and document its major structural elements, including functional elements and projects, required to make the program work. The architecture includes how the major program components (hardware, software, human systems) will be integrated and are intended to operate together and with legacy systems, as applicable, to achieve program goals and objectives. By implication, the architecture defines the system-level processes necessary for development, production, human systems integration, verification, deployment, operations, support, disposal, and training. The architecture also includes facilities, logistics concepts, and planned mission results and data analysis, archiving, and reporting. The architecture development process usually considers a number of alternative approaches to both the architecture and the program's operations concept.

Tightly coupled and single-project programs develop their Operations Concept and candidate (preliminary) mission, spacecraft, and ground systems architectures. The Operations Concept includes all activities such as integration and test, launch integration, launch, deployment and on-orbit checkout (robotic programs) or initial operations (human space flight programs), in-space operations, landing and recovery (as applicable), and decommissioning and disposal.

In analyzing the operations concept, the tightly coupled and single-project programs develop the preliminary approach to verification and validation;

system integration; and human rating, if applicable. Identifying these at this point enables the tightly coupled and single-project programs to assess unique workforce and infrastructure needs early enough to include the requirements for these in the initial concept(s).

As the single-project program approaches the MCR, it develops and documents at least one feasible preliminary concept (included as part of concept documentation in NPR 7120.5E, Table I-6 and Table 3-5 at the end of this chapter), including the key preliminary ground rules and assumptions that drive the concept(s) and the operations concept. A feasible concept is one that is probably achievable technically within the cost and schedule resources allocated by the Mission Directorate. This preliminary concept includes key drivers, preliminary estimates of technical margins for candidate architectures, and a preliminary Master Equipment List (MEL). (Tightly coupled programs develop the preliminary MEL no later than SRR.) This concept is sometimes referred to as the mission concept, particularly in the robotic community. As a minimum, the principal concept will be approved following the MCR and KDP A. Future changes to this concept (and others, if approved for further study) will be identified at each follow-on life-cycle review and KDP so that management understands how the concept is evolving as formulation progresses.

Based on the leading concept, the tightly coupled and single-project programs develop and mature the initial mission objectives and requirements and develop a mission or science traceability matrix that shows how the requirements flow from the objectives of the mission through the operational requirements (such as science measurement requirements) to the top-level infrastructure implementation requirements (such as orbit characteristics and pointing stability). At this point, tightly coupled and single-project programs, with guidance from their stakeholders, begin to select technical standards for use as tightly coupled and single-project program requirements in accordance with NPR 7120.10. Based on currency and applicability, technical standards required by law and those designated as mandatory by NPDs and NPRs are selected first. When all other factors are the same, NASA promotes the use of voluntary consensus standards when they meet or can be tailored to meet the needs of NASA and other Government agency technical standards.

In addition, the tightly coupled and single-project programs develop an initial assessment of engineering development needs, including defining the need for engineering prototypes and models for the higher risk components and assemblies that have not been previously built or flown in the planned environment and testing them to demonstrate adequate performance. As with technology development, identification at this point will enable tightly

The Master Equipment List (MEL) summarizes all major components of each flight element subsystem and each instrument element component. Description for each major component includes current best estimates and contingency allocation for mass and power (including for individual components), number of flight units required, and some description of the heritage basis. Power values generally represent nominal steady-state operational power requirements. Information includes identification of planned spares and prototypes, required deliveries/exchanges of simulators for testing, and other component description/characteristics. Certain items (like electronic boxes and solar arrays) usually include additional details to identify and separate individual elements. The MEL is useful to tightly coupled and single-project program managers for understanding where the design is, where the mass is being carried, what the power needs are, what the margins are, and other parameters as the tightly coupled and single-project programs progress in development.

coupled and single-project programs to plan and initiate engineering development activities early in Formulation knowing that the funding has been planned for these activities.

For concepts and architectures that plan to use heritage systems, tightly coupled and single-project programs develop an initial assessment, using the *Systems Engineering Handbook, NASA/SP-2007-6105 Rev 1*, Appendix G, of heritage hardware and software systems that may be utilized outside of environments and configurations for which they were originally designed and used.

All of these activities help tightly coupled and single-project programs develop an initial assessment of preliminary technical risks for candidate architectures, including engineering development risks.

If not already defined, tightly coupled and single-project programs identify their payload risk classification in accordance with *NPR 8705.4, Risk Classification for NASA Payloads*.

Following the SRR, tightly coupled and single-project programs update the concept documentation, architectures, and operations plans based on the results of the SRR and continue to perform analyses and trades in support of concept/design refinement. They prepare the preliminary design documentation for use during the peer reviews, subsystem reviews, and system reviews leading to tightly coupled and single-project programs' SDR/Mission Definition Review (MDR). Tightly coupled and single-project programs update the design documentation as changes are made during this process and finalize them at PDR.

Tightly coupled and single-project programs implement engineering development plans, heritage hardware and software assessments, and risk mitigation plans identified in tightly coupled programs' formulation plans and a single-project's Formulation Agreement for Phase A. As these risk reduction plans are executed, tightly coupled and single-project programs monitor, assess, and report the status of engineering development results and heritage assessments.

To provide additional options in the event that development begins to exceed the resources allocated, tightly coupled and single-project programs typically begin to develop an initial list of descope options. Descope is a particular kind of risk mitigation that addresses risks early in the program Formulation Phase. Documentation of tightly coupled and single-project programs' descope plans typically includes a detailed description of the potential descope, the effect of the descope on tightly coupled and single-project programs' success criteria, the cost and schedule savings resulting from the descope, and key decision dates by when the descope needs to be exercised to realize these savings.

Tightly coupled and single-project programs develop preliminary Safety Data Packages and other safety process deliverables as required by NPR 7120.5 and the NPRs and NPDs identified below:

Currently, requirements for the safety data packages can be found in the following documents:

- Safety per *NPR 8715.3, NASA General Safety Program Requirements*, *NPR 8715.5, Range Flight Safety Program, NPR 8715.7, Expendable Launch Vehicle Payload Safety Program*, and local range requirements; *NPR 8621.1, NASA Procedural Requirements for Mishap and Close Call Reporting, Investigating, and Recordkeeping*;

- Human rating requirements per *NPR 8705.2, Human-Rating Requirements for Space Systems*;

- Quality assurance per *NPD 8730.5, NASA Quality Assurance Program Policy, NPD 8730.2, NASA Parts Policy*, and *NPR 8735.1, Procedures for Exchanging Parts, Materials, Software, and Safety Problem Data Utilizing the Government-Industry Data Exchange Program (GIDEP) and NASA Advisories*; and

- Limiting the generation of orbital debris per *NPR 8715.6, NASA Procedural Requirements for Limiting Orbital Debris*.

For all program types, the program team assesses the ability of the program and its component project(s) and all contributors to the program and its projects (including contractors, industrial partners, and other partners) to use the International System of Units (the Système Internationale (SI), commonly known as the metric system of measurement). This assessment determines an approach that maximizes the use of SI while minimizing short- and long-term risk to the extent practical and economically feasible or to the extent that the supply chain can support utilization without loss of markets to U.S. firms. Use of the SI or metric system of measurement is especially encouraged in cooperative efforts with international partners. This assessment documents an integration strategy if both SI and U.S. customary units are used in the program or its projects. The assessment is completed and documented in the Program Plan no later than the SDR/MDR. To the degree possible, programs need to use consistent measurement units throughout all documentation to minimize risk of errors.

All programs that plan to develop technologies develop a Technology Development Plan. Generally, technologies developed at the program level cut across projects within the program. Uncoupled, loosely coupled, and tightly coupled programs develop a preliminary Technology Development Plan prior to SRR and baseline the plan at SDR. The single-project program baselines its plan by SRR so that technology requirements can be implemented early in Formulation. The Technology Development Plan describes the technology assessment, development, management, and acquisition strategies needed to achieve the program's mission objectives. It describes how the program will assess its technology development requirements, including how the program will evaluate the feasibility, availability, readiness, cost, risk, and benefit of the new technologies. It describes how the program will identify opportunities for leveraging ongoing technology efforts, including technology developed on other NASA programs or projects, at other governmental agencies, or in industry. The Technology Development Plan:

- Identifies the supply chain needed to manufacture the technology and any costs and risks associated with the transition from development to the manufacturing and production phases;

- Documents appropriate mitigation plans for the identified risks;

- Describes the program's strategy for ensuring that there are alternative development paths available if/when technologies do not mature as expected;

- Describes how the program will remove technology gaps, including:
  - Maturation, validation, and insertion plans;
  - Performance measurement at quantifiable milestones;
  - Off-ramp decision gates (i.e., the point during development where the program assesses whether or not the technology is maturing adequately and, if not, decides to terminate continued technology development); and
  - Resources required;

- Describes briefly how the program will ensure that all planned technology exchanges, contracts, and partnership agreements comply with all laws and regulations regarding export control and the transfer of sensitive and proprietary information;

- Describes the program's technology utilization and commercialization plan in accordance with the requirements of *NPD 7500.2, NASA Innovative Partnerships Program* and *NPR 7500.1, NASA Technology Commercialization Process*;

- Describes how the program will transition technologies from the development stage to manufacturing, production, and insertion into the end system; and

- Identifies any potential costs and risks associated with the transition to manufacturing, production, and insertion; and documents appropriate mitigation plans for the identified risks.

Loosely coupled, uncoupled, and tightly coupled programs develop a preliminary **Safety and Mission Assurance (SMA) Plan** by SRR and baseline the plan by SDR. Single-project programs baseline their plan at SRR to ensure that proper SMA procedures are in place for the system design activities. The SMA Plan addresses life-cycle SMA functions and activities. The plan identifies and documents program-specific SMA roles, responsibilities, and relationships. This is accomplished through a program-unique mission assurance process map and matrix developed and maintained by the program with the appropriate support and guidance from the Headquarters and/or Center SMA organizations. The plan reflects the program life cycle from the SMA process perspective addressing areas, including procurement, management, design and engineering, design verification and test, software design, software verification and test, manufacturing, manufacturing verification and test, operations, preflight verification and test, maintenance, and retirement.

The plan also addresses specific critical SMA disciplines, including the following at a minimum:

- Safety per NPR 8715.3, *NPR 8715.5, Range Flight Safety Program*, *NPR 8715.7, Expendable Launch Vehicle Payload Safety Program*, and local range requirements; *NPR 8621.1, NASA Procedural Requirements for Mishap and Close Call Reporting, Investigating, and Recordkeeping*;

- Human rating requirements per NPR 8705.2;

- Quality assurance per *NPD 8730.5, NASA Quality Assurance Program Policy*, *NPD 8730.2, NASA Parts Policy*, and *NPR 8735.1, Procedures for Exchanging Parts, Materials, Software, and Safety Problem Data Utilizing the Government-Industry Data Exchange Program (GIDEP) and NASA Advisories*;

- Limiting the generation of orbital debris per NPR 8715.6;

- Compliance verification, audit, safety and mission assurance reviews, and safety and mission assurance process maps per *NPR 8705.6, Safety and Mission Assurance (SMA) Audits, Reviews, and Assessments*;

- Reliability and maintainability per *NPD 8720.1, NASA Reliability and Maintainability (R&M) Program Policy*;

- Software safety and assurance per *NASA-STD-8719.13, NASA Software Safety Standard* and *NASA-STD-8739.8, Software Assurance Standard*;

- Software engineering requirements for safety critical software *NPR 7150.2, NASA Software Engineering Requirements*;

- Quality assurance functions per *NPR 8735.2, Management of Government Quality Assurance Functions for NASA Contracts*; and

- Other applicable NASA procedural safety and mission success requirements.

The Mission Assurance Process Matrix is constructed to identify program life-cycle assurance agents and specific assurance activities, processes, responsibilities, accountability, depth of penetration, and independence. The matrix includes key assurance personnel in engineering, manufacturing, program management, operations, and SMA.

The Mission Assurance Process Map is a high-level graphic representation of governing SMA policy and requirements, processes, and key participant roles, responsibilities, and interactions. It also includes the reporting structure that constitutes a program's/project's SMA functional flow.

The plan describes how the program will track and resolve problems, including developing and managing a Closed Loop Problem Reporting and Resolution System. The process typically includes a well-defined data collection system and process to record hardware and software problems and anomaly reports, problem analysis, and corrective action.

Loosely coupled, uncoupled, and tightly coupled programs develop a preliminary Systems Engineering Management Plan (SEMP) that includes the content required by NPR 7123.1 by SRR and baseline the plan by SDR. Single-project programs baseline their plan at SRR to ensure that proper system engineering procedures are in place for the system design activities. Single-project programs update their SEMP at SDR/MDR and PDR. The plan summarizes the key elements of the program systems engineering and includes descriptions of the program's overall approach for systems engineering. The systems engineering process typically includes system design and product realization processes (implementation and/or integration, verification and validation, and transition), as well as the technical management processes. For tightly coupled programs, the SEMP is a stand-alone document.

If applicable, in accordance with NPR 7123.1B, programs develop a Human Systems Integration (HSI) Plan. (More information may be available in the second revision of the Systems Engineering Handbook.) The HSI Plan is baselined at SRR, and updated at SDR (SDR/MDR for single-project programs), and PDR. Human systems integration is an interdisciplinary and comprehensive management and technical process that focuses on the integration of human considerations into the system acquisition and development processes to enhance human system design, reduce life-cycle ownership cost, and optimize total system performance. Human system domain design activities associated with manpower, personnel, training, human factors engineering, safety, health, habitability, and survivability are considered concurrently and integrated with all other systems engineering design activities.

Tightly coupled and single-project programs develop a preliminary Verification and Validation Plan by SDR (SDR/MDR for single-project programs) and baseline the plan by PDR. This plan summarizes the approach for performing verification and validation of the program products. It indicates the methodology to be used in the verification/validation (test, analysis, inspection, or demonstration) as defined in NPR 7123.1.

Tightly coupled and single-project programs develop a preliminary Information Technology (IT) Plan by SRR and baseline the plan by SDR/MDR. The plan describes how the program will acquire and use information technology and addresses the following:

- The program's approach to knowledge capture, including methods for contributing knowledge to other entities and systems in compliance with *NPD 2200.1, Management of NASA Scientific and Technical Information* and *NPR 2200.2, Requirements for Documentation, Approval, and Dissemination of NASA Scientific and Technical Information*;

- How the program will manage information throughout its life cycle, including the development and maintenance of an electronic program library and how the program will ensure identification, control, and disposition of program records in accordance with *NPD 1440.6, NASA Records Management*, and *NPR 1441.1, NASA Records Retention Schedules*;

- The program's approach to implementing IT security requirements in accordance with *NPR 2810.1, Security of Information Technology*; and

- The steps the program will take to ensure that the information technology it acquires and/or uses complies with *NPR 2830.1, NASA Enterprise Architecture Procedures*.

All programs develop and baseline a Review Plan by SRR in time to establish the independent SRB and permit adequate planning and definition of the program's approach for conducting the series of reviews. The reviews include internal reviews and program life-cycle reviews in accordance with Center best practices, Mission Directorate review requirements, and the requirements in NPR 7123.1 and *NPR 7120.5, NASA Space Flight Program and Project Management Requirements*. The Review Plan identifies the life-cycle reviews the program plans to conduct and the purpose, content, and timing of those life-cycle reviews, and documents any planned deviations or waivers granted from the requirements in NPR 7123.1 and NPR 7120.5E. It also provides the technical, scientific, schedule, cost, and other criteria that will be used in the consideration of a Termination Review.

For tightly coupled programs the Review Plan documents the program life-cycle review requirements on the supporting projects that represent an integrated review process for the various projects. When multiple Centers are involved, review plans take into consideration the participating Centers' review process best practices. For each program life-cycle review and KDP, the Review Plan documents the sequencing of the associated project life-cycle reviews and KDPs, i.e., whether the associated project life-cycle reviews and KDPs precede or follow the program life-cycle review and KDP. In addition, the plan documents which projects need to proceed to their KDPs together, which projects need to proceed to their KDPs simultaneously with the program KDP, and which projects may proceed to their KDPs as individual projects. The sequencing of project life-cycle reviews and KDPs with respect to program life-cycle reviews and KDPs is especially important for project PDR life-cycle reviews that precede the KDPs to transition to Implementation (KDP I (KDP C for single-project programs). Since changes to one project can easily impact other projects' technical, cost, and schedule baselines, and potentially impact other projects' risk assessments and mitigation plans, projects and their program generally need to proceed to KDP C (KDP I for single-project programs) together.

All programs develop an **Environmental Management Plan**. Uncoupled, loosely coupled, and single-project program baseline the plan by SDR. This plan describes the activities to be conducted to comply with *NPR 8580.1, NASA National Environmental Policy Act Management Requirements* and Executive Order 12114. Based on consultation with the NASA Headquarters National Environmental Protection Act (NEPA) coordinator, the plan describes the program's NEPA strategy at all affected Centers, including decisions regarding NEPA documents. This consultation enables the program to determine the most effective and least resource-intensive strategy to meet NEPA requirements across the program and its constituent projects. Any critical milestones associated with complying with these regulations are inserted into the program's IMS.

> Environmental management is the activity of ensuring that program and project actions and decisions that may potentially affect or damage the environment are assessed during the Formulation Phase and reevaluated throughout Implementation. This activity is performed according to all NASA policy and Federal, State, Tribal Government, and local environmental laws and regulations. Additionally, this activity identifies constraints and impediments from external requirements on the program so alternatives can be identified to minimize impacts to cost, schedule, and performance.

Early in Formulation tightly coupled and single-project programs develop a logistics support concept that supports the overall mission concept and that accommodates the specific characteristics of the program's component projects, including identifying the infrastructure and procurement strategies necessary to support the program. This concept typically includes expected levels of contractor effort for life-cycle logistics support functions through all life-cycle phases. These logistics support concepts are integrated into the system design process. Tightly coupled and single-project programs finalize a preliminary **Integrated Logistics Support (ILS) Plan** by SDR/MDR and baseline the document by PDR. The Integrated Logistics Support Plan describes how the program will implement *NPD 7500.1, Program and Project Life Cycle Logistics Support Policy*, including a maintenance and support concept; participation in the design process to enhance supportability; supply support, including spares, procurement and replenishment, resupply and return, and supply chain management related to logistics support functions; maintenance and maintenance planning; packaging, handling, and transportation of deliverable products; technical data and documentation; support and test equipment; training; manpower and personnel for ILS functions; facilities required for ILS functions; and logistics information systems for the life of the program.

Tightly coupled and single-project programs develop a preliminary **Science Data Management Plan** by PDR that describes how the program will manage the scientific data generated and captured by the operational mission(s) and any samples collected and returned for analysis. (For uncoupled and loosely coupled programs, this plan is developed at the project level.) The plan includes descriptions of how data will be generated, processed, distributed, analyzed, and archived. It also describes how any samples will be collected and stored during the mission and managed when returned to Earth, including any planetary protection measures. The Plan typically includes definitions of data rights and services and access to samples, as appropriate. It explains how the program will accomplish the knowledge capture and information management and disposition requirements in *NPD 2200.1, Management of NASA Scientific and Technical Information, NPR 2200.2, Requirements for Documentation, Approval, and Dissemination of NASA Scientific and Technical Information*, and *NPR 1441.1, NASA Records Retention Schedules* as applicable to program science data. The plan further describes how the program will adhere to all NASA sample handling, curation, and planetary protection directives and rules, including *NPR 8020.12, Planetary Protection Provisions for Robotic Extraterrestrial Missions.*

Uncoupled, loosely coupled, and single-project programs develop a **Configuration Management Plan** and baseline the plan by SDR/MDR. Tightly coupled programs develop a preliminary plan by SDR and baseline the plan by PDR. This plan describes the configuration management approach that the program team will implement, consistent with NPR 7123.1 and *NASA-STD-0005, NASA Configuration Management (CM) Standard*. It describes the structure of the configuration management organization and tools to be used; the methods and procedures to be used for configuration identification, configuration control, interface management, configuration traceability, and configuration status accounting and communications; how configuration management will be audited; and how contractor configuration management processes will be integrated with the program. This plan may be a separate stand-alone plan or integrated into the Program Plan. The plan is developed early in Formulation to assist the program in managing requirements and the control plans that are needed before the Program Plan is finalized.

Uncoupled and loosely coupled programs develop a **Security Plan** and baseline the plan by SDR. Tightly coupled and single-project programs develop a preliminary plan by SDR/MDR and baseline the plan by PDR. This plan describes the program's plans for ensuring security and technology protection. It includes three types of requirements: security, IT security, and emergency response. Security requirements include the program's approach for planning and implementing the requirements for information, physical, personnel, industrial, and counterintelligence/counterterrorism security; provisions to protect personnel, facilities, mission-essential infrastructure, and critical program information from potential threats and other vulnerabilities that may be identified during the threat and vulnerability assessment process; and the program's approach to security awareness/education requirements in accordance with *NPR 1600.1, NASA Security Program Procedural Requirements* and *NPD 1600.2, NASA Security Policy.* IT security requirements document the program's approach to implementing requirements in accordance with *NPR 2810.1, Security of Information Technology.* Emergency response requirements describe the program's emergency response plan in accordance with *NPR 1040.1, NASA Continuity of Operations (COOP) Planning Procedural Requirements* and define the range and scope of potential crises and specific response actions, timing of notifications and actions, and responsibilities of key individuals.

Uncoupled and loosely coupled programs baseline a preliminary Threat Summary by SDR. Tightly coupled programs prepare a preliminary Threat Summary by SDR and baseline the document by PDR. Single-project programs prepare a preliminary Project Protection Plan by SDR/MDR and baseline the document by PDR. The Threat Summary attempts to document the threat environment that a NASA space system is most likely to encounter as it reaches operational capability. For more information on Threat Summary and Project Protection Plan specifics, go to the Community of Practice for Space Asset Protection at https://nen.nasa.gov/web/sap.

Uncoupled and loosely coupled programs develop a Technology Transfer Control Plan and baseline the plan by SDR. Tightly coupled and single-project programs develop a preliminary plan by SDR (SDR/MDR for single-project programs) and baseline the plan by PDR. This plan describes how the program will implement the export control requirements specified in *NPR 2190.1, NASA Export Control Program.*

Tightly coupled and single-project programs identify potential nonconformance to orbital debris requirements in their Orbital Debris Assessment per *NPR 8715.6, NASA Procedural Requirements for Limiting Orbital Debris* and *NASA-STD-8719.14, Process for Limiting Orbital Debris* for planned breakups, reentry of major components that potentially could reach the surface, the planned orbital lifetime, and the use of tethers. Deviations typically are submitted to the Chief, SMA for approval prior to the ASM. For single-project programs, the initial assessment is due at MCR with the assessment of the preliminary design due at PDR. For tightly coupled programs, the Orbital Debris Assessment can be done at either the program or project level as appropriate. For uncoupled and loosely coupled programs, these assessments are performed at the project level.

If a program includes human space flight systems, the program develops a Human Rating Certification Package (HRCP) per NPR 8705.2. The initial

HRCP is delivered at SRR and updated at SDR/MDR, PDR, CDR, and ORR. Human rating certification focuses on the integration of the human into the system, preventing catastrophic events during the mission and protecting the health and safety of humans involved in or exposed to space activities, specifically the public, crew, passengers, and ground personnel.

Single-project programs without a project (for example, the Space Launch System (SLS)) are required to prepare the following control plans in accordance with Table I-7 in NPR 7120.5E. These control plans are described in the appropriate paragraphs in Chapter 4:

- Software Management Plan
- Integration Plan
- Planetary Protection Plan
- Nuclear Safety Launch Approval Plan
- Range Safety Risk Management Process Documentation.

3.3.5 Completing Formulation Activities and Preparing for Implementation

3.3.5.1 Establishing the Program's Baseline

As the program approaches its program milestone for approval to enter Implementation, KDP I (KDP C for single-project programs), the program team finalizes the program baselines: technical (including requirements), resource (including funding, NOA, infrastructure and staffing), and cost and schedule. Once approved and documented in the Decision Memorandum, these baselines are maintained under configuration control as part of the program plan. Section 5.5 provides additional detail on maturing, approving and maintaining cost and schedule baselines.

For single-project programs and tightly coupled programs with EVM requirements, the program works with the Mission Directorate to conduct a program-level pre-approval IBR as part of the preparations for KDP I/ KDP C to ensure that the program's work is properly linked with its cost, schedule, and risk and that the systems are in place to conduct EVM. Section 5.14 provides additional details on this review.

The program documents the driving ground rules, assumptions, and constraints affecting the resource baseline. (See Section 3.3.3.5 for details on the resource baseline.)

When the project resource baselines are approved, the Program Plan is updated with the approved project baselines.

All programs are required to have a **Program Commitment Agreement (PCA)** approved to proceed into Implementation. (For a definition of the PCA, see Section 3.3.2.1.) Programs support the MDAA in developing the preliminary PCA when required. Uncoupled and loosely coupled programs prepare a preliminary PCA by SRR. Tightly coupled and single-project programs prepare their preliminary PCAs as part of their SDR/MDR preparations. All programs support the MDAA in finalizing and obtaining approval of the PCA in preparation for their KDP I (KDP C for single-project programs). The PCA is finalized at SDR for uncoupled and loosely coupled and at PDR for tightly coupled and single-project programs.

Uncoupled, loosely coupled, and tightly coupled programs support the MDAA in the selection of projects, either directly assigned or through a competitive process.

All programs develop the program's plans for work to be performed during the Implementation Phase.

All programs document the results of Formulation activities. As part of this activity, the programs generate the appropriate documentation per Appendix G of NPR 7123.1, Tables I-1, I-2, I-3, and I-6 and I-7 of NPR 7120.5E, and Tables 3-2, 3-3, 3-4, 3-5 and 3-6 at the end of this chapter. These documentation requirements may be satisfied, in whole or in part, by the FAD, the basis of cost and schedule estimates, draft and preliminary versions of program documents and plans, and the final life-cycle review briefing packages.

3.3.5.2 Program Reporting Activities and Preparing for Major Milestones

3.3.5.2.1 Program Reporting

The program reports to the Center, as requested by the Center, on whether Center engineering, Safety and Mission Assurance (SMA), health and medical, and management best practices (e.g., program and project management, resource management, procurement, institutional best practices) are being followed, and whether Center resources support program or project requirements. The program also provides program and project risks and the status and progress of activities so the Center can identify and report trends and provide guidance to the Agency and affected programs and projects. The CMC (or equivalent) provides its findings and recommendations to program managers and to the appropriate PMCs regarding the performance and technical and management viability of the program prior to the KDPs.

Aside from the Center and Agency reporting already mentioned, many stakeholders will be interested in the status of the program from Congress on down. The program manager will probably be required to report status and performance in many forums, including Mission Directorate monthly meetings and the Agency's monthly BPRs. See Section 5.12 for further information regarding potential program external reporting.

3.3.5.2.2 Program Internal Reviews

Prior to the program Formulation life-cycle reviews, programs conduct internal reviews in accordance with NPR 7123.1, Center practices, and NPR 7120.5. These internal reviews are the decisional meetings where the programs solidify their plans, technical approaches, and programmatic commitments. This is accomplished as part of the normal systems engineering work processes defined in NPR 7123.1 where major technical and programmatic requirements are assessed along with the system design and other implementation plans. Major technical and programmatic performance metrics are reported and assessed against predictions.

For tightly coupled and single-project programs:

- Non-SRB program technical reviews are divided into several categories: major systems reviews (one or two levels down from the program), Engineering Peer Reviews (EPRs), internal reviews, and tabletop reviews. Program systems reviews are major technical milestones of the program that typically precede the life-cycle review, covering major systems milestones such as the completion of a spacecraft, instrument, or ground system design. The technical progress of the program is assessed at key milestones such as these systems reviews to ensure that the program's maturity is progressing as required. In many cases these reviews are conducted by the program in coordination with a Center-sponsored independent review panel if the Center is using these reviews as one means to oversee the program's work. In these cases, the program manager works with the Center to ensure that there is a suitable independent review panel in place for each such review and works with systems engineering to ensure that clear technical criteria and an agreed agenda have been established well in advance of each such review.

- System engineering collects and reviews the documentation that demonstrates the technical progress planned for the major systems review and submits the materials as a data package to the review team prior to the review. This allows adequate review by the selected technical representatives to identify problems and issues that can be discussed at the review. Systems engineering is responsible for the agenda, organization, and conduct of the systems review as well as obtaining closure on any action items and corrective actions. Systems engineering acts as recorder, noting all comments and questions that are not adequately addressed during the presentations. At the conclusion of a major systems review, the independent review panel, if in place, makes a determination as to whether or not the predetermined criteria for a successful review have been met and makes a recommendation on whether or not the system is ready to proceed into the next phase of its development.

- An EPR can address an entire system or subsystem, but more typically addresses a lower level of assembly or component. An EPR is a focused, in-depth technical review of a subsystem, lower level of assembly, or a component, which adds value and reduces risk through expert knowledge infusion, confirmation of approach, and specific recommendations. The mission systems engineer works with the respective product manager (program manager, program formulation manager, instrument manager, or principal investigator) to ensure that the EPR review panel is comprised of technical experts with significant practical experience relevant to the technology and requirements of the subsystem, lower level of assembly, or a component to be reviewed. The key distinction between an EPR and a major subsystem review is that the review panel is selected by personnel supporting the program and not by the Center. An EPR plan is produced that lists the subsystems, lower levels of assembly and components to be reviewed, and the associated life-cycle milestones for the reviews. A summary of results of the EPRs is presented at each major subsystem review and/or at each life-cycle review.

- Additional program technical reviews sometimes called "internal reviews" or "tabletop reviews" are conducted by program team members as necessary and are one of their primary mechanisms for internal technical program control. These reviews follow the general protocols described above for subsystem reviews and EPRs.

3.3.5.2.3 Preparing for Approval for Program Transition

Programs support the program Formulation life-cycle reviews (SRR, SDR/MDR, and PDR) in accordance with NPR 7123.1, Center practices, and NPR 7120.5, including the life-cycle review objectives and expected maturity states defined in Appendix E of this handbook and in NPR 7120.5E Appendix I. Life-cycle review entrance and success criteria in NPR 7123.1 and the life-cycle phase and KDP information in the maturity states tables in Appendix E of this handbook provide specifics for addressing the six criteria required to demonstrate the program has met the expected maturity state. MCRs are generally conducted by the Center, but the Decision Authority may request an SRB to perform this review. If this is the case, Section 5.10 and the *NASA Standing Review Board Handbook* provide guidance.

Program teams plan, prepare for, and support the governing PMC review prior to KDP 0 (if required by the Decision Authority) and KDP I (KDP C for single-project programs) and provide or obtain the KDP readiness products listed in Section 3.2.3.

Once the KDP has been completed and the Decision Memorandum signed, the program team updates its documents as required and plans to reflect the decisions made and actions assigned at the KDP.

3.4 Program Implementation

Program Implementation begins when the program receives approval to proceed to Implementation with the successful completion of KDP I (KDP C for single-project programs) and a fully executed Decision Memorandum. Implementation encompasses program acquisition, operations and sustainment. If constituent projects have not already been initiated, or if new projects are identified, projects may be initiated during program Implementation. Constituent projects' formulation, approval, implementation, integration, operation, and ultimate decommissioning are constantly monitored. The program is adjusted to respond as needs, risks, opportunities, constraints, resources, and requirements change, managing technical and programmatic margins and resources to ensure successful completion of Implementation. The program develops products required during Implementation in accordance with the applicable program Product Maturity tables (Tables 3-2, 3-3, 3-4, 3-5, 3-6) at the end of this chapter.

Single-project programs and tightly coupled programs have the characteristics of very large projects and are run with similar requirements to the project requirements in NPR 7120.5. For tightly coupled programs, the Implementation Phase synchronization with the projects' life cycles continues to ensure that the program and all its projects are properly integrated, including proper interface definition and resource allocation across all internal projects and with external programs and organizations.

Uncoupled and loosely coupled programs oversee the implementation of the projects in the program, helping with funding, assisting the MDAA in such activities as selecting projects, performing systems engineering between projects, and potentially developing and ensuring technology insertion at appropriate points of the program.

As the program evolves and matures, the program manager ensures that the Program Plan and the attendant program resources remain aligned. Program life-cycle reviews for uncoupled or loosely coupled programs ensure that the program continues to contribute to Agency and Mission

Directorate goals and objectives within funding constraints. Program life-cycle reviews for tightly coupled programs ensure that the program's projects are properly integrated as development and operations activities are implemented. In some cases, programs may recycle through Formulation when program changes are sufficient to warrant such action.

The general flow of activities for the various program types in Implementation are shown in Figures 3-11, 3-12, 3-13.

Once in Implementation, the program manager works with the program team, the program's constituent projects, and with the MDAA to execute the Program Plan. As the program conducts its activities, it continues to support the MDAA in ensuring continuing alignment of the program and projects with applicable Agency strategic goals, and Mission Directorate requirements and constraints. When changes occur to the program requirements or resource levels, the program manager works with the MDAA to update the PCA and Program Plan, as appropriate.

All program teams also continue to support the MDAA and the Office of International and Interagency Relations (OIIR) in obtaining updated interagency and international agreements (including the planning and negotiation of updated agreements and recommendations on joint participation in reviews, integration and test, and risk management), as appropriate.

All programs continue management, planning, and control activities. They ensure appropriate infrastructure and in coordination with the Centers engaged in the program, ensure trained/certified staff that cut across multiple projects within the program are available and ready when needed to support Implementation activities.

The program updates life-cycle cost and schedule baselines, as needed, for any changes in the program during Implementation. It documents the BoE for the cost and schedule baselines, as needed. It reviews and approves annual project budget submissions and prepares annual program budget submissions.

The program confirms key ground rules and assumptions that drive development of the program and projects. Once the program has defined the ground rules and assumptions, it tracks them to determine if they are being realized (i.e., remain valid) or if they need to be modified. The program continues to track, manage, and mitigate risks.

The program executes procurement activities in accordance with the Acquisition Plan. In doing so, it maintains programmatic oversight of industrial base and supply chain issues that might pose a risk to the program or projects and provides timely notification of supply chain disruptions to the

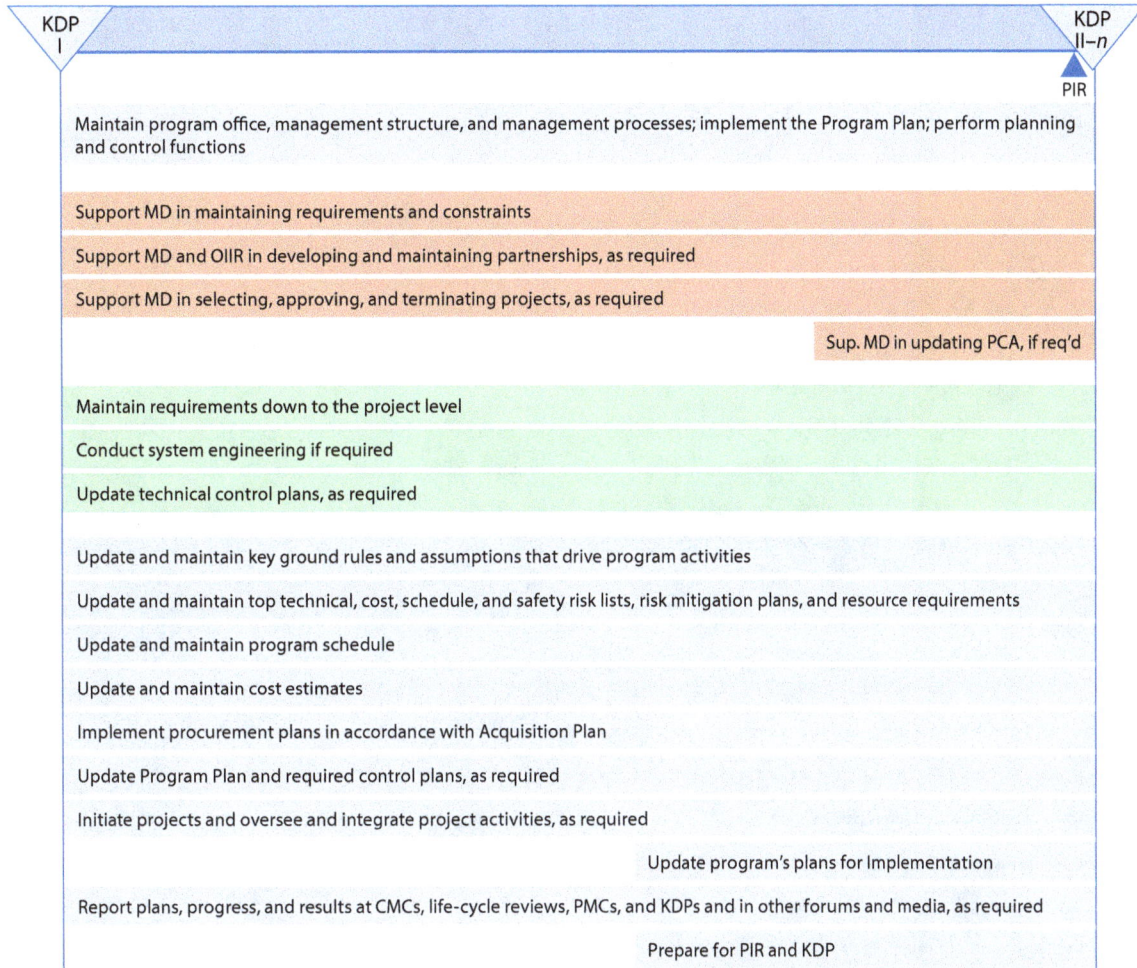

KDP I

KDP II–*n*

PIR

Maintain program office, management structure, and management processes; implement the Program Plan; perform planning and control functions

Support MD in maintaining requirements and constraints

Support MD and OIIR in developing and maintaining partnerships, as required

Support MD in selecting, approving, and terminating projects, as required

Sup. MD in updating PCA, if req'd

Maintain requirements down to the project level

Conduct system engineering if required

Update technical control plans, as required

Update and maintain key ground rules and assumptions that drive program activities

Update and maintain top technical, cost, schedule, and safety risk lists, risk mitigation plans, and resource requirements

Update and maintain program schedule

Update and maintain cost estimates

Implement procurement plans in accordance with Acquisition Plan

Update Program Plan and required control plans, as required

Initiate projects and oversee and integrate project activities, as required

Update program's plans for Implementation

Report plans, progress, and results at CMCs, life-cycle reviews, PMCs, and KDPs and in other forums and media, as required

Prepare for PIR and KDP

Legend: ☐ Program management, planning, and control tasks

☐ Work for which Headquarters is responsible but the program helps accomplish (e.g., international partnerships are a Headquarters responsibility, but the programs help develop and finalize those partnerships)

☐ Technical work the program is doing

Acronyms: MD = Mission Directorate; OIIR = Office of International and Interagency Relations.

Note: These are typical high-level activities that occur during this program phase. Placement of reviews is notional.

Figure 3-11   Uncoupled and Loosely Coupled Program Implementation Flow of Activities

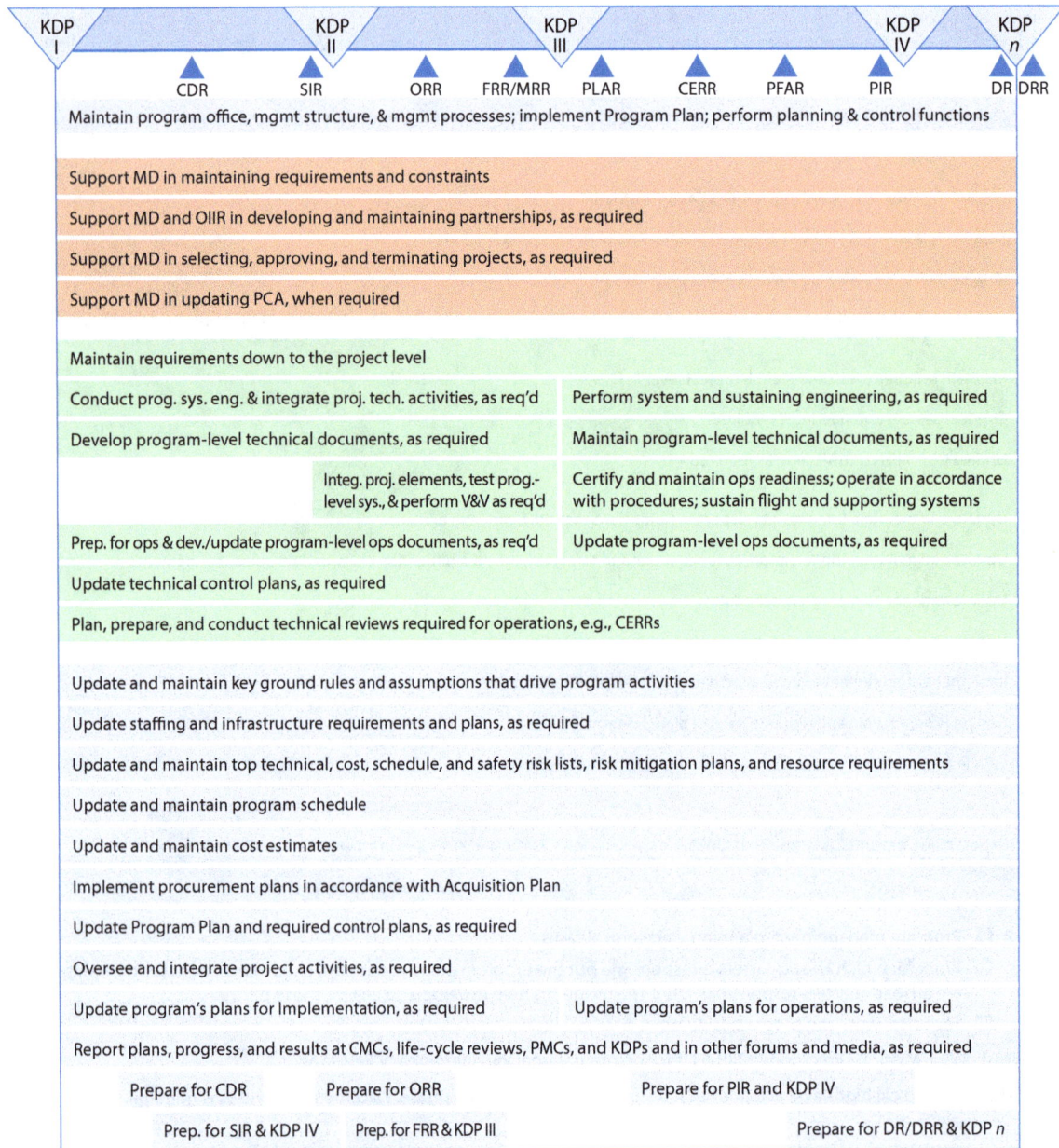

Figure 3-12 Tightly Coupled Program Implementation Flow of Activities

Legend: ☐ Program management, planning, and control tasks

☐ Work for which Headquarters is responsible but the program helps accomplish (e.g., international partnerships are a Headquarters responsibility, but the programs help develop and finalize those partnerships)

☐ Technical work the program is doing

Acronyms: MD = Mission Directorate; OIIR = Office of International and Interagency Relations.

Note: These are typical high-level activities that occur during this program phase. Placement of reviews is notional.

KDP C — CDR — KDP D — SIR — ORR — KDP E — FRR/MRR — PLAR — CERR — PFAR — KDP E–n — PIR — KDP F — DR DRR

Maintain program office, mgmt structure, & mgmt processes; implement the Program Plan; perform planning and control functions

Support MD in maintaining requirements and constraints

Support MD and OIIR in developing and maintaining partnerships, as required

Support MD in selecting, approving, and terminating projects, as required

Support MD in updating PCA, when required

Conduct program system engineering and integrate project technical activities, as required	Perform system and sustaining engineering, as required
Develop program-level technical documents, as required	Maintain program-level technical documents, as required
Complete design	Start fabrication	Integrate project elements, test program-level systems, & perform V&V, as required	Certify and maintain ops readiness; operate in accordance with procedures; sustain flight and supporting systems
Prep. for ops & dev./update program-level ops documents, as req'd	Update program-level ops documents, as required

Update technical control plans, as required

Plan, prepare, and conduct technical reviews, e.g., SAR, TRRs, CERRs, etc.

Update and maintain key ground rules and assumptions that drive program activities

Update staffing and infrastructure requirements and plans, as required

Update and maintain top technical, cost, schedule, and safety risk lists, risk mitigation plans, and resource requirements

Update and maintain program schedule

Update and maintain cost estimates

Implement procurement plans in accordance with Acquisition Plan

Update Program Plan and required control plans, as required

Oversee and integrate project activities, as required

Update program's plans for Implementation, as required | Update program's plans for operations, as required

Report plans, progress, and results at CMCs, life-cycle reviews, PMCs, and KDPs and in other forums and media, as required

Prepare for CDR — Prep. for ORR — Prep. for PIR & KDP E–n

Prep. for SIR & KDP D — Prep. for FRR & KDP E — Prepare for DR/DRR and KDP F

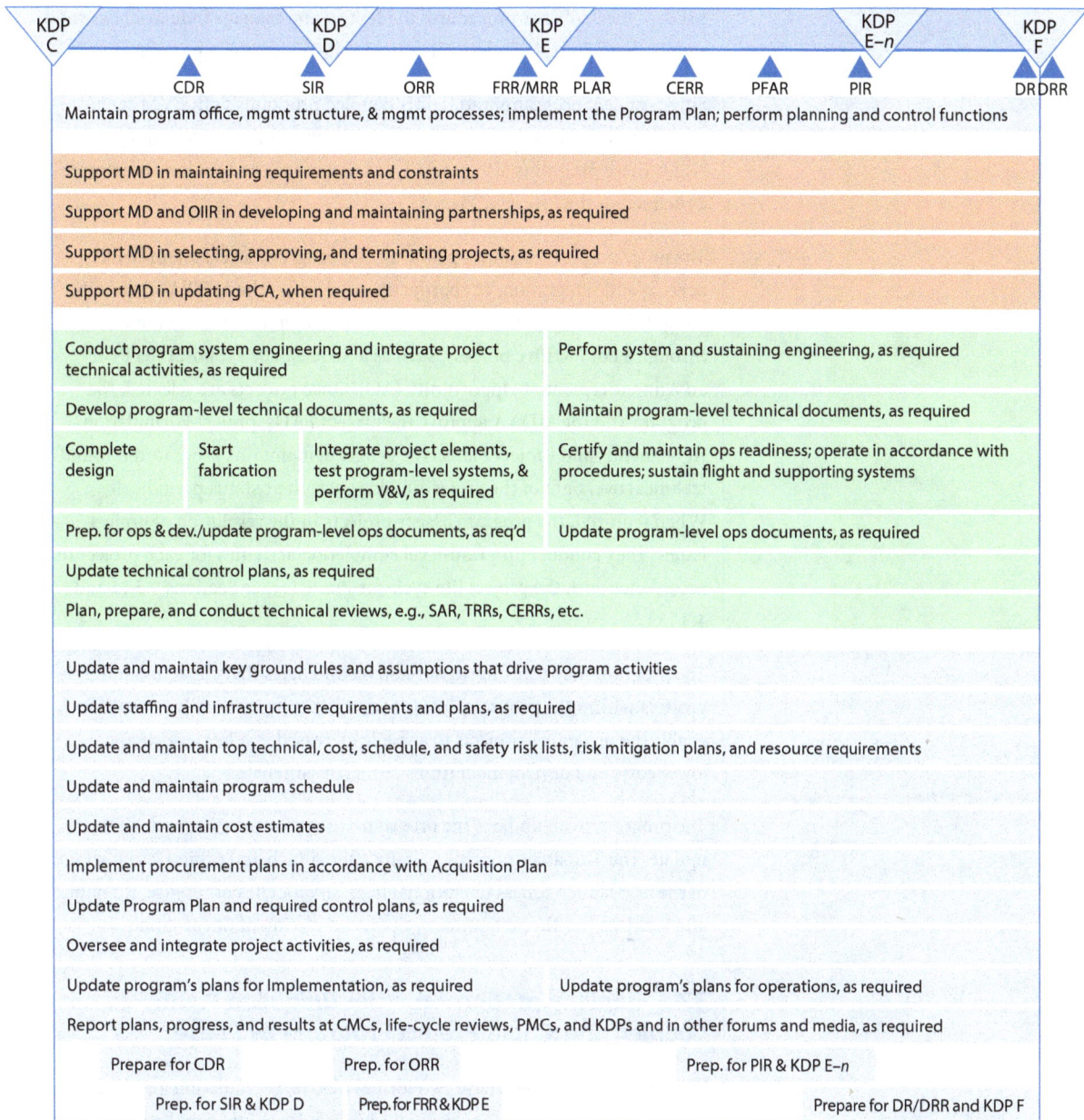

Legend: ☐ Program management, planning, and control tasks

☐ Work for which Headquarters is responsible but the program helps accomplish (e.g., international partnerships are a Headquarters responsibility, but the programs help develop and finalize those partnerships)

☐ Technical work the program is doing

Acronyms: MD = Mission Directorate; OIIR = Office of International and Interagency Relations.

Note: These are typical high-level activities that occur during this program phase. Placement of reviews is notional.

Figure 3-13 Single-Project Program Implementation Flow of Activities

MDAA. It establishes procedures to identify and manage industrial base and supply chain risks, including all critical and single-source partners.

Single-project programs and tightly coupled programs with EVM requirements conduct any required IBRs for contracts requiring EVM (refer to the NFS). These programs also report EVM metrics to the Mission Directorate as defined in the Program Plan.

The program team conducts planning and program-level systems engineering and integration, as appropriate, to support the MDAA when initiating the project selection process, either through direct assignment or through a competitive process, such as a Request For Proposal (RFP) or an Announcement of Opportunity (AO). Once projects are selected, the program and the MDAA approve the project FADs, project Formulation Agreements, and Project Plans. The program maintains programmatic and technical oversight of the projects and reports their status periodically. When required, the program assists projects in the resolution of project issues. They conduct program-level completion activities for each project in accordance with the project life cycle for Phase F. (See Sections 4.3.14 and 4.3.15.)

The program may continue to develop technologies that cut across multiple projects within the program. These technologies are generally pursued to enable the program's projects to achieve increased results/performance, lower costs and development times or increased reliability.

The program team updates the program Threat Summary at each KDP. It updates the Threat Summary annually after launch and orbital verification of the first launch unless operational necessity dictates otherwise. It finalizes and archives the Threat Summary upon completion of the program.

3.4.1 Implementation Activities Unique to Tightly Coupled and Single-Project Programs by Phase

Whereas programs only have two formal phases, Formulation and Implementation, the project life cycle is also broken down into subphases. For single-project programs and tightly coupled programs, the activities of the two formal program phases also break down into roughly the equivalent of the project subphases. For uncoupled and loosely coupled programs, these activities are carried out at the project level.

3.4.2 Final Design and Fabrication

The purpose of this phase for tightly coupled and single-project programs is to complete and document the final design that meets the detailed requirements and synchronize with the program's project(s) as the program team implements the program in accordance with the Program Plan. During Final Design and Fabrication, the program, in conjunction with its project(s):

- Ensures that the systems engineering activities are performed to determine if the design is mature enough to proceed with full-scale implementation within the constraints of the Management Agreement and the ABC;
- Performs qualification testing;
- Develops product specifications and begins fabrication of test and flight architecture (e.g., flight article components, assemblies, subsystems, and associated software); and
- Develops integration plans and procedures and ensures that all integration facilities and personnel are ready and available.

Final Design and Fabrication is a long phase and these activities will overlap during the phase.

For programs that develop or acquire multiple copies of a product or system(s), the program ensures that the system developers include a production process for multiple copies. When this occurs, the program holds a Production Readiness Review (PRR). The objectives of the PRR are to evaluate the readiness of system developer(s) to produce the required number of systems within defined program constraints for programs developing multiple similar flight or ground support systems and to evaluate the degree to which the production plans meet the system's operational support requirements. (See Table G-8 in Appendix G of the NPR 7123.1 for entrance and success criteria of the PRR.)

Final Design and Fabrication activities are focused toward the CDR and the System Integration Review (SIR), the life-cycle review preceding KDP II/ KDP D.

The objectives of the CDR are to evaluate (1) the integrity of the program integrated design, including its projects and supporting infrastructure; (2) the program's ability to meet mission requirements with appropriate margins and acceptable risk within cost and schedule constraints; and (3) whether the integrated design is appropriately mature to continue with the final design and fabrication phase.

The objective of the SIR is to evaluate the readiness of the program, including its projects and supporting infrastructure, to begin the System Assembly, Integration, and Test (AI&T) part of Implementation with acceptable risk and within cost and schedule constraints.

At KDP II (KDP D for single-project programs), the program demonstrates that it is still on plan; the risk is commensurate with the projects' payload classifications (or Mission Directorate's risk definition if not a payload in accordance with NPR 8705.4); and the program is ready for AI&T with acceptable risk within its ABC.

The program team continues to perform the technical activities required in NPR 7123.1 for this phase. It completes the engineering design and development activities (e.g., qualification and life tests) and incorporates the results into the final design. It completes and documents final flight and ground designs by CDR and updates them, as necessary, at SIR. It begins to implement the defined validation and verification program on flight and/or ground products. It updates the technology readiness assessment documentation by CDR if any technology development activities were performed after PDR. Finally, it develops system integration plans and procedures.

The program documents lessons learned in accordance with NPD 7120.4 and per NPD 7120.6 and the program's Knowledge Management Plan.

Tightly coupled and single-project programs develop a preliminary **Mission Operations Plan** by SIR and baseline the plan by ORR. Uncoupled and loosely coupled programs do not have these plans since they are only necessary for their projects. This plan is required at this point in development to document the activities required to transition to operations and operate the mission. It describes how the program will implement the associated facilities, hardware, software, and procedures required to complete the mission. It includes mission operations plans, rules, and constraints and describes the Mission Operations System (MOS) and Ground Data System (GDS) in the following terms: MOS and GDS human resources and training requirements; procedures to ensure that operations are conducted in a reliable, consistent, and controlled manner using lessons learned during the program and from previous programs; facilities requirements (offices, conference rooms, operations areas, simulators, and test beds); hardware (ground-based communications and computing hardware and associated documentation); and software (ground-based software and associated documentation). Single-project programs develop the Operations Handbook by SIR (see "Operations Handbook" box for more information) and baseline the plan by ORR.

The handbook identifies the commands for the spacecraft, defines the functions of these commands, and provides supplemental reference material for use by the operations personnel. The main emphasis is placed on command types, command definitions, command sequences, and operational constraints. Additional document sections may describe uploadable operating parameters, the telemetry stream data contents (for both the science

---

Operations Handbook

The Operations Handbook provides information essential to the operation of the spacecraft. It generally includes the following:

1. A description of the spacecraft and the operational support infrastructure;

2. Operational procedures, including step-by-step operational procedures for activation and deactivation;

3. Malfunction detection procedures; and

4. Emergency procedures.

---

and the engineering data), the Mission Operations System displays, and the spacecraft health monitors.

The program updates the following control plans at CDR: Safety and Mission Assurance (SMA) Plan, Verification and Validation Plan, Environmental Management Plan, Integrated Logistics Support (ILS) Plan, Integration Plan, Threat Summary, Technology Transfer Control Plan, Education Plan, and Communications Plan. Programs also update the Human Rating Certification Package and the Human Systems Integration Plan, if applicable. It is expected that these plans will be updated at this point, but other plans need to be updated as necessary. Single-project programs also update the Integration Plan and the Range Safety Risk Management Process Documentation.

The program updates the following control plans at SIR: Verification and Validation Plan and Threat Summary.

3.4.2.1 System Assembly, Integration and Test, Launch and Checkout

Program Implementation continues with System Assembly, Integration and Test, Launch and Checkout as the program team implements the program in accordance with the Program Plan. During this part of Implementation, the program with its constituent projects(s):

- Performs system AI&T;
- Completes validation testing, finalizes operations preparations, and completes operational training;
- Resolves failures, anomalies, and issues;
- Conducts various internal reviews such as Test Readiness Reviews (TRRs), the SAR, and pre-ship reviews;
- Certifies the system for launch;
- Launches the system; and

The decision to launch and conduct early operations is a critical decision for the Agency. The KDP III (KDP E for projects and single-project programs) decision occurs before launch to provide Decision Authority approval for this decision. The KDP III/KDP E decision includes approval for the transition to the operations phase of the life cycle, however, unlike other life-cycle phase transitions, the transition to operations does not occur immediately after the KDP III/KDP E. This transition occurs after launch and checkout. The timing for this transition stems from the historical practice of funding missions through on-orbit checkout, transitioning from the development team to the operations team following on-orbit checkout, and funding mission operations separately.

- Completes on-orbit system checkout (robotic space flight programs) or initial operations (human space flight programs).

The transition from this subphase to the next, Operations and Sustainment, is different than other transitions in that the transition does not occur immediately after the KDP. KDP III (KDP E for single-project programs) marks the decision to conduct launch and early operations. However, the transition to operations occurs after on-orbit checkout (robotic space flight programs) or initial operations (human space flight programs) at the conclusion of the Post-Launch Assessment Review (PLAR) or, for certain human space flight programs, the Post-Flight Assessment Review (PFAR).

The flow of activities in preparation for launch is very formal and involves important reviews by the Agency's stakeholders. Section 4.3.11 provides a detailed description of the flow of the review process in preparation for launch for human and robotic space flight programs and projects. This process is the same for both single-project programs and tightly coupled programs.

The phase activities focus on preparing for the Operational Readiness Review (ORR), Flight Readiness Review (FRR) (for human space flight programs) or the Mission Readiness Review (MRR) (for robotic space flight programs), KDP III (KDP E for single-project programs), launch, the Post-Launch Assessment Review (PLAR), and for certain human space flight programs the Post-Flight Assessment Review (PFAR).

The objectives of the ORR are to evaluate the readiness of the program, including its projects, ground systems, personnel, procedures, and user documentation, to operate the flight system and associated ground systems in compliance with program requirements and constraints during the operations phase.

The objectives of the FRR/MRR are to evaluate the readiness of the program and its projects, ground systems, personnel, and procedures for a safe and successful launch and flight/mission.

At KDP III (KDP E for single-project programs) the program is expected to demonstrate that it is ready for launch and early operations with acceptable risk within Agency commitments.

The PLAR is a non-KDP-affiliated review that is conducted after the mission has launched and on-orbit checkout has completed. The objectives of the PLAR are to evaluate the in-flight performance of the program and its projects and to determine the program's readiness to begin the operations phase of the life cycle and transfer responsibility to the operations organization. At the PLAR, the program is expected to demonstrate that it is ready to conduct mission operations with acceptable risk within Agency commitments. (For human space flight programs that develop flight systems that return to Earth, the PLAR may be combined with the Post-Flight Assessment Review (PFAR), which is conducted after landing and recovery. See Section 4.3.10 for a detailed discussion of this topic.)

The program continues to perform the technical activities required in NPR 7123.1. As the various project assemblies arrive at the integration facility, the program team begins to assemble, integrate, and test the various system pieces and complete verification and validation on the products as they are integrated. It prepares the preliminary verification and validation report before the ORR and then baselines the report by FRR/MRR. Once the hardware is shipped to the launch site, the program with its constituent projects and with launch site support begins the process of receiving and inspecting the hardware, reassembling the spacecraft as required, integrating spacecraft and vehicles produced by constituent projects (tightly coupled programs), completing final spacecraft testing, completing integrated spacecraft/vehicle testing (tightly coupled programs), and resolving any open issues that remain. The program transitions or delivers the final products and baselines the as-built hardware and software documentation. It supports launch rehearsals, participates in press conferences, and supports the launch approval process. It prepares for operations and updates the Operations Concept and the Mission Operations Plan.

The program team updates the following control plans at ORR if necessary: Mission Operations Plan, Science Data Management Plan, Threat Summary, Education Plan, and Communications Plan. Programs also obtain the certification of the Human Rating Certification Package. Single-project programs without a project also baseline the range safety risk management process.

The program updates the following control plans at FRR/MRR if necessary: Safety and Mission Assurance (SMA) Plan (by the Safety and Mission Success Review (SMSR)) and Threat Summary. (See Section 4.3.11 for a detailed description of the review process in preparation for launch.) The single-project program updates the Operations Handbook at FRR/MRR.

3.4.2.2 Operations and Sustainment

During Operations and Sustainment, the program implements the Missions Operations Plan. For human space flight programs, this phase begins after initial operations have been successfully completed and all flight test objectives have been met. For robotic space flight programs, the phase begins following a successful launch and on-orbit checkout. (See Section 4.3.11 for robotic and human space flight programs.) Mission operations may be periodically punctuated with Critical Event Readiness Reviews (CERR). Human

Safety and Mission Success Review (SMSR) is held to prepare Agency safety and engineering management to participate in program final readiness reviews preceding flights or launches, including experimental/ test launch vehicles, or other reviews as determined by the Chief, Safety and Mission Assurance. The SMSR provides the knowledge, visibility, and understanding necessary for senior safety and engineering management to either concur or nonconcur in program decisions to proceed with a launch or significant flight activity.

space flight missions may conduct Post-Flight Assessment Reviews (PFARs) specific to their needs. These are non-KDP-affiliated reviews. The program periodically has PIRs followed by a KDP as determined by the Decision Authority. The Decision Authority makes an annual assessment of whether the program needs to have a PIR in the coming year to assess the program's performance, evaluate its continuing relevance to the Agency's Strategic Plan, and authorize its continuation.

The Operations and Sustainment subphase ends with the Decommissioning Review (DR) and KDP n (tightly coupled programs) or KDP F (single-project programs), at which time the end of the program is approved. (See "Sustainment and Sustainment Engineering" box.)

(After KDP F, single-project programs are also required to conduct the project-level Disposal Readiness Review (DRR). (See Section 3.4.1.4.) The DR and DRR may be combined if the disposal of the spacecraft will be done immediately after the DR.)

The objectives of the PIR are to evaluate the program's continuing relevance to the Agency's Strategic Plan, assess performance with respect to expectations, and determine the program's ability to execute its Program Plan with acceptable risk within cost and schedule constraints. The program is expected to demonstrate that it still meets Agency needs and is continuing to meet Agency commitments as planned. (See Section 5.11.3 in this handbook for guidance on PIRs.)

Sustainment and Sustainment Engineering

Sustainment generally refers to supply, maintenance, transportation, sustaining engineering, data management, configuration management, manpower, personnel, training, habitability, survivability, environment, safety, supportability, and interoperability functions.

The term "sustaining engineering" refers to technical activities that can include, for example, updating designs (e.g., geometric configuration), introducing new materials, and revising product, process, and test specifications. These activities typically involve first reengineering items to solve known problems, and then qualifying the items and sources of supply. The problems that most often require sustaining engineering are lack of a source (e.g., vendor going out of business), component that keeps failing at a high rate, and long production lead time for replacing items.

As parts age, the need and opportunity for sustaining engineering increase. The practice of sustaining engineering includes not only the technical activity of updating designs, but also the business judgment of determining how often and on what basis the designs need to be reviewed.

The objectives of the CERR are to evaluate the readiness of the program and its projects to execute a critical event during the flight operations phase of the life cycle. CERRs are established at the discretion of the program office.

The objectives of the PFAR are to evaluate how well mission objectives were met during a human space flight mission and the status of the flight and ground systems, including the identification of any anomalies and their resolution.

The objectives of the Decommissioning Review are to evaluate the readiness of the program and its projects to conduct closeout activities, including final delivery of all remaining program or project deliverables and safe decommissioning/disposal of space flight systems and other program or project assets.

The objective of the Disposal Readiness Review is to evaluate the readiness of the project and the flight system for execution of the spacecraft disposal event.

Tightly coupled and single-project programs and their projects eventually cease as a natural evolution of completing their mission objectives. When this occurs, the Mission Directorate, program, and project(s) need to be sure that all the products or systems produced by the program (e.g., spacecraft, ground systems, test beds, spares, science data, operational data, returned samples) are properly dispositioned and that all program and project activities (e.g., contracts, financial obligations) are properly closed out. The program develops a Decommissioning/Disposal Plan (which includes the project Decommissioning/Disposal Plans) in preparation for the Decommissioning Review to cover all activities necessary to close the program and its projects out. It conducts a Decommissioning Review in preparation for final approval to decommission by the Decision Authority at the final program KDP. This process is the same for both programs and projects and is described in Section 4.3.14. This section provides an overview of the disposal of a spacecraft, the various documents that are produced as part of this, and the order and timing of major activities and document deliveries.

At KDP n/KDP F following the Decommissioning Review, the program is expected to demonstrate that decommissioning is consistent with program objectives and that the program is ready for final analysis and archival of mission and science data and safe disposal of its assets.

3.4.2.3 Closeout

During Closeout, the program and its projects perform the technical activities required in NPR 7123.1. They perform spacecraft and other in-space asset disposal and closeout and disposition ground systems, test beds, and spares. They monitor decommissioning and disposal risks, actively assess open risks, and develop and implement mitigation plans.

They complete archiving of mission/operational and science data and document the results of all activities. They complete storage and cataloging of returned samples and archive project engineering and technical management data. They close out contracts, as appropriate. They develop mission reports and document lessons learned in accordance with NPD 7120.4 and per NPD 7120.6 and the program's Knowledge Management Plan.

After KDP F, single-project programs are also required to conduct the project-level Disposal Readiness Reviews (DRRs). The objective of the Disposal Readiness Review is to evaluate the readiness of the project and the flight system for execution of the spacecraft disposal event.

3.4.3 Preparing for Program Decommissioning and Closing Out

Program teams plan, prepare for, and support the governing PMC review prior to the Decommissioning KDP *n*/KDP F and provide or obtain the KDP readiness products listed in Section 3.2.3. Once the Implementation KDPs have been completed and the Decision Memoranda signed, the program updates its documents and plans as needed to reflect the decisions made and actions assigned.

3.5 Program Products by Phase

3.5.1 Non-Configuration-Controlled Documents

For non-configuration-controlled documents, the following terms and definitions are used in Tables 3-2 through 3-6:

- "Initial" is applied to products that are continuously developed and updated as the program or project matures.
- "Final" is applied to products that are expected to exist in this final form, e.g., minutes and final reports.
- "Summary" is applied to products that synthesize the results of work accomplished.

- "Plan" is applied to products that capture work that is planned to be performed in the following phases.

- "Update" is applied to products that are expected to evolve as the formulation and implementation processes evolve. Only expected updates are indicated. However, any document may be updated, as needed.

3.5.2 Configuration-Controlled Documents

For configuration-controlled documents, the following terms and definitions are used in Tables 3-2 through 3-6:

- "Preliminary" is the documentation of information as it stabilizes but before it goes under configuration control. It is the initial development leading to a baseline. Some products will remain in a preliminary state for multiple life-cycle reviews. The initial preliminary version is likely to be updated at subsequent life-cycle reviews but remains preliminary until baselined.

- "Baseline" indicates putting the product under configuration control so that changes can be tracked, approved, and communicated to the team and any relevant stakeholders. The expectation on products labeled "baseline" is that they will be at least final drafts going into the designated life-cycle review and baselined coming out of the life-cycle review. Baselining of products that will eventually become part of the Program or Project Plan indicates that the product has the concurrence of stakeholders and is under configuration control. Updates to baselined documents require the same formal approval process as the original baseline. "Baseline" indicates that the product needs to be baselined no later than this point, however, there is no penalty for baselining earlier at the program or project's discretion, no waiver required.

- "Approve" is used for a product, such as Concept Documentation, that is not expected to be put under classic configuration control but still requires that changes from the "Approved" version are documented at each subsequent "Update."

- "Update" is applied to products that are expected to evolve as the formulation and implementation processes evolve. Only expected updates are indicated. However, any document may be updated, as needed. Updates to baselined documents require the same formal approval process as the original baseline.

Table 3-2 Uncoupled and Loosely Coupled Program Milestone Products and Control Plans Maturity Matrix

Products	Formulation KDP I[1] SRR	Formulation KDP I[1] SDR	Implementation KDP II–n PIR
1. FAD	Baseline		
2. PCA	Preliminary	Baseline	
3. Program Plan	Preliminary	Baseline	Update
3.a. Mission Directorate requirements and constraints	Baseline	Update	
3.b. Traceability of program-level requirements on projects to the Agency strategic goals and Mission Directorate requirements and constraints	Preliminary	Baseline	
3.c. Documentation of driving ground rules and assumptions on the program	Preliminary	Baseline	
4. Interagency and international agreements	Preliminary	Baseline	
5. ASM minutes		Final	
6. Risk mitigation plans and resources for significant risks	Initial	Update	Update
7. Documented Cost and Schedule Baselines	Preliminary	Baseline	Update
8. Documentation of Basis of Estimate (cost and schedule)	Preliminary	Baseline	Update
9. Documentation of performance against plan/baseline, including status/closure of formal actions from previous KDP	Summary	Summary	Summary
10. Plans for work to be accomplished during Implementation		Plan	Plan
Program Plan Control Plans[2]			
1. Technical, Schedule, and Cost Control Plan	Preliminary	Baseline	
2. Safety and Mission Assurance Plan	Preliminary	Baseline	
3. Risk Management Plan	Preliminary	Baseline	
4. Acquisition Plan	Preliminary	Baseline	
5. Technology Development Plan	Preliminary	Baseline	
6. Systems Engineering Management Plan	Preliminary	Baseline	
7. Product Data and Life-Cycle Management Plan		Initial	Update annually
8. Review Plan	Baseline	Update	
9. Environmental Management Plan		Baseline	
10. Configuration Management Plan		Baseline	
11. Security Plan		Baseline	
12. Threat Summary		Baseline	Update annually
13. Technology Transfer Control Plan		Baseline	
14. Education Plan		Baseline	
15. Communications Plan		Baseline	
16. Knowledge Management Plan	Preliminary	Baseline	

[1] If desired, the Decision Authority may request a KDP 0 be performed generally following SRR.

[2] Requirements for and scope of control plans will depend on scope of program. As noted in NPR 7120.5, Appendix G, control plans may be a part of the basic Program Plan.

Table 3-3 Tightly Coupled Program Milestone Products Maturity Matrix

Products	Formulation			Implementation				
	KDP 0		KDP I	KDP II		KDP III		KDP n
	SRR	SDR	PDR	CDR	SIR	ORR	MRR/FRR	DR
1. FAD	Baseline							
2. PCA		Prelim.	Baseline					
3. Program Plan	Prelim.	Baseline	Update	Update	Update	Update	Update	Update
3.a. Mission Directorate requirements and constraints	Baseline	Update	Update					
3.b. Traceability of program-level requirements on projects to the Agency strategic goals and Mission Directorate requirements and constraints	Prelim.	Baseline	Update					
3.c. Documentation of driving ground rules and assumptions on the program	Prelim.	Baseline	Update	Update	Update			
4. Interagency and international agreements	Prelim.	Baseline	Update					
5. ASM minutes		Final						
6. Risk mitigation plans and resources for significant risks	Initial	Update	Update	Update	Update	Update	Update	Update
7. Documented Cost and Schedule Baselines	Prelim.	Prelim.	Baseline	Update	Update	Update	Update	Update
8. Documentation of Basis of Estimate (cost and schedule)	Prelim.	Prelim.	Baseline	Update	Update	Update	Update	Update
9. Confidence Level(s) and supporting documentation		Prelim. cost confidence level & prelim. schedule confidence level	Baseline joint cost & schedule confidence level					
10. Shared Infrastructure,[1] Staffing, and Scarce Material Requirements and Plans	Initial	Update	Update	Update				
11. Documentation of performance against plan/baseline, including status/closure of formal actions from previous KDP		Summary	Summary	Summary	Summary	Summary	Summary	Summary
12. Plans for work to be accomplished during next life-cycle phase	Plan		Plan		Plan		Plan	Plan

[1] Shared infrastructure includes facilities that are required by more than one of the program's projects.

Table 3-4 Tightly Coupled Program Plan Control Plans Maturity Matrix

NPR 7120.5 Program Plan—Control Plans[1]	Formulation			Implementation				
	KDP 0		KDP I	KDP II		KDP III		KDP *n*
	SRR	SDR	PDR	CDR	SIR	ORR	MRR/FRR	DR
1. Technical, Schedule, and Cost Control Plan	Preliminary	Baseline	Update					
2. Safety and Mission Assurance Plan	Preliminary	Baseline	Update	Update			Update (SMSR)	
3. Risk Management Plan	Preliminary	Baseline	Update					
4. Acquisition Plan	Preliminary strategy	Baseline	Update					
5. Technology Development Plan	Preliminary	Baseline	Update					
6. Systems Engineering Management Plan	Preliminary	Baseline						
7. Product Data and Life-Cycle Management Plan		Initial	Update annually thereafter					
8. Verification and Validation Plan		Preliminary	Baseline	Update	Update			
9. Information Technology Plan	Preliminary	Baseline	Update					
10. Review Plan[2]	Baseline	Update	Update					
11. Missions Operations Plan					Preliminary	Baseline	Update	
12. Environmental Management Plan		Preliminary	Baseline	Update				
13. Integrated Logistics Support Plan		Preliminary	Baseline	Update				
14. Science Data Management Plan			Preliminary			Baseline	Update	
15. Configuration Management Plan[3]	Preliminary	Baseline	Update					
16. Security Plan		Preliminary	Baseline					
17. Threat Summary		Preliminary	Baseline	Update	Update	Update	Update annually	
18. Technology Transfer Control Plan		Preliminary	Baseline	Update				
19. Education Plan		Preliminary	Baseline	Update		Update		
20. Communications Plan		Preliminary	Baseline	Update		Update		
21. Knowledge Management Plan	Preliminary	Baseline	Update					

[1] See template in NPR 7120.5E, Appendix G, for control plan details.

[2] Review Plan needs to be baselined before the first review.

[3] Software and hardware configuration management may be preliminary at SRR and updated at SDR.

Table 3-5 Single-Project Program Milestone Products Maturity Matrix

Products	Pre–Phase A KDP A	Phase A KDP B		Phase B KDP C	Phase C KDP D		Phase D KDP E		Phase E KDP F	Phase F
	MCR	SRR	SDR/MDR	PDR	CDR	SIR	ORR	MRR/FRR	DR	DRR
Headquarters Products[1]										
1. FAD	Baseline									
2. PCA			Preliminary	Baseline						
3. Traceability of Agency strategic goals and Mission Directorate requirements and constraints to pro-gram/project-level requirements and constraints	Preliminary	Baseline	Update	Update						
4. Documentation of driving mission, tech-nical, and program-matic ground rules and assumptions	Preliminary	Preliminary	Baseline	Update	Update	Update				
5. Partnerships and interagency and international agree-ments	Preliminary	Update	Baseline U.S. partnerships and agree-ments	Baseline international agreements						
6. ASM minutes		Final								
7. NEPA compliance documentation per NPR 8580.1				Final documen-tation per NPR 8580.1						
8. Mishap Preparedness and Contingency Plan				Preliminary		Update		Baseline (SMSR)	Update	Update

Table 3-5 Single-Project Program Milestone Products Maturity Matrix

Products	Pre–Phase A KDP A — MCR	Phase A KDP B — SRR	Phase A KDP B — SDR/MDR	Phase B KDP C — PDR	Phase C KDP D — CDR	Phase C KDP D — SIR	Phase D KDP E — ORR	Phase D KDP E — MRR/FRR	Phase E KDP F — DR	Phase F — DRR
Single-Project Program Technical Products[2]										
1. Concept Documentation	Approve	Update	Update	Update						
2. Mission, Spacecraft, Ground, and Payload Architectures	Preliminary mission and spacecraft architecture(s) with key drivers	Baseline mission and spacecraft architecture, preliminary ground and payload architectures. Classify payload(s) by risk per NPR 8705.4	Update mission and spacecraft architecture, baseline ground and payload architectures	Update mission, spacecraft, ground, and payload architectures						
3. Project-Level, System, and Subsystem Requirements	Preliminary project-level requirements	Baseline project-level and system-level requirements	Update Project-level and system-level requirements, Preliminary subsystem requirements	Update project-level and system-level requirements. Baseline subsystem requirements						
4. Design Documentation			Preliminary	Baseline Preliminary Design	Baseline Detailed Design	Update		Baseline as-built hardware and software		
5. Operations Concept	Preliminary	Preliminary	Preliminary	Baseline						
6. Technology Readiness Assessment Documentation	Initial	Update	Update	Update	Update					
7. Engineering Development Assessment Documentation	Initial	Update	Update	Update						
8. Heritage Assessment Documentation	Initial	Update	Update	Update						

Table 3-5 Single-Project Program Milestone Products Maturity Matrix (continued)

Products	Pre–Phase A KDP A — MCR	Phase A KDP B — SRR	Phase A KDP B — SDR/MDR	Phase B KDP C — PDR	Phase C KDP D — CDR	Phase C KDP D — SIR	Phase D KDP E — ORR	Phase D KDP E — MRR/FRR	Phase E KDP F — DR	Phase F — DRR
9. Safety Data Packages				Preliminary	Baseline	Update	Update	Update		
10. ELV Payload Safety Process Deliverables				Preliminary	Preliminary	Baseline				
11. Verification and Validation Report							Preliminary	Baseline		
12. Operations Handbook						Preliminary	Baseline	Update		
13. Orbital Debris Assessment per NPR 8715.6	Preliminary Assessment			Preliminary design ODAR	Detailed design ODAR			Final ODAR (SMSR)		
14. End of Mission Plans per NPR 8715.6/ NASA-STD 8719.14, App B								Baseline End of Mission Plan (SMSR)	Update EOMP annually	Update EOMP
15. Mission Report										Final
Single-Project Program Management, Planning, and Control Products										
1. Formulation Agreement	Baseline for Phase A; Preliminary for Phase B		Baseline for Phase B							
2. Program Plan[3]			Preliminary	Baseline						
3. Project Plan[3]			Preliminary	Baseline						
4. Plans for work to be accomplished during next Implementation life-cycle phase				Baseline for Phase C		Baseline for Phase D		Baseline for Phase E	Baseline for Phase F	

Table 3-5 Single-Project Program Milestone Products Maturity Matrix

Products	Pre–Phase A KDP A	Phase A KDP B		Phase B KDP C	Phase C KDP D		Phase D KDP E		Phase E KDP F	Phase F
	MCR	SRR	SDR/MDR	PDR	CDR	SIR	ORR	MRR/FRR	DR	DRR
5. Documentation of performance against Formulation Agreement (see #1 above) or against plans for work to be accomplished during Implementation life-cycle phase (see #4 above), including performance against baselines and status/closure of formal actions from previous KDP		Summary	Summary		Summary	Summary	Summary	Summary	Summary	
6. Project Baselines			Preliminary	Baseline	Update	Update	Update	Update		
6.a. Top technical, cost, schedule and safety risks, risk mitigation plans, and associated resources	Initial	Update	Update	Update	Update	Update	Update	Update	Update	Update
6.b. Staffing requirements and plans	Initial	Update	Update	Update	Update		Update			
6.c. Infrastructure requirements and plans, business case analysis for infrastructure Alternative Future Use Questionnaire (NASA Form 1739), per NPR 9250.1	Initial	Update	Update base-line for NF 1739 Section A	Update base-line for NF 1739 Section B	Update					

Table 3-5 Single-Project Program Milestone Products Maturity Matrix

Products	Pre–Phase A / KDP A	Phase A / KDP B	Phase A	Phase B KDP C	Phase C KDP D	Phase C	Phase D KDP E	Phase D	Phase E KDP F	Phase F
	MCR	SRR	SDR/MDR	PDR	CDR	SIR	ORR	MRR/FRR	DR	DRR
6.d. Schedule	Risk informed at project level with preliminary Phase D completion ranges	Risk informed at system level with preliminary Phase D completion ranges	Risk informed at subsystem level with preliminary Phase D completion ranges. Preliminary Integrated Master Schedule	Risk informed and cost- or resource-loaded. Baseline Integrated Master Schedule	Update IMS	Update IMS	Update IMS			
6.e. Cost Estimate (Risk-Informed or Schedule-Adjusted Depending on Phase)	Preliminary range estimate	Update	Risk-informed schedule-adjusted range estimate	Risk-informed and schedule-adjusted baseline	Update	Update	Update	Update	Update	Update
6.f. Basis of Estimate (cost and schedule)	Initial (for range)	Update (for range)	Update (for range)	Update for cost and schedule estimate	Update	Update	Update	Update	Update	Update
6.g. Confidence Level(s) and supporting documentation			Preliminary cost confidence level and preliminary schedule confidence level	Baseline joint cost and schedule confidence level						
6.h. External Cost and Schedule Commitments			Preliminary for ranges	Baseline						
6.i. CADRe			Baseline	Update	Update	Update		Update[4]	Update	
7. Decommissioning/Disposal Plan									Baseline	Update disposal portions

[1] These products are developed by the Mission Directorate.

[2] These document the work of the key technical activities performed in the associated phases.

[3] The program and project plans may be combined with the approval of the MDAA.

[4] The CADRe for MRR/FRR is considered the "Launch CADRe" to be completed after the launch.

Table 3-6 Single-Project Program Plan Control Plans Maturity Matrix

NPR 7120.5 Program and Project Plan—Control Plans[1]	Pre–Phase A	Phase A KDP B		Phase B KDP C	Phase C KDP D		Phase D KDP E		Phase E KDP F
	MCR	SRR	SDR/MDR	PDR	CDR	SIR	ORR	MRR/FRR	DR
1. Technical, Schedule, and Cost Control Plan	Approach for managing schedule and cost during Phase A[2]	Preliminary	Baseline	Update					
2. Safety and Mission Assurance Plan		Baseline	Update	Update	Update			Update (SMSR)	Update
3. Risk Management Plan	Approach for managing risks during Phase A[2]	Baseline	Update	Update					
4. Acquisition Plan	Preliminary Strategy	Baseline	Update	Update					
5. Technology Development Plan (may be part of Formulation Agreement)	Baseline	Update	Update	Update					
6. Systems Engineering Management Plan	Preliminary	Baseline	Update	Update					
7. Product Data and Life-Cycle Management Plan			Initial	Update annually thereafter					
8. Information Technology Plan		Preliminary	Baseline	Update					
9. Software Management Plan(s)		Preliminary	Baseline	Update					
10. Verification and Validation Plan	Preliminary Approach[3]		Preliminary	Baseline	Update	Update			
11. Review Plan	Preliminary	Baseline	Update	Update					
12. Mission Operations Plan						Preliminary	Baseline	Update	
13. Environmental Management Plan			Baseline						
14. Integrated Logistics Support Plan	Approach for managing logistics[2]	Preliminary	Preliminary	Baseline	Update				
15. Science Data Management Plan				Preliminary			Baseline	Update	
16. Integration Plan	Preliminary approach[2]		Preliminary	Baseline	Update				
17. Threat Summary			Preliminary	Baseline	Update	Update	Update	Update annually	
18. Configuration Management Plan		Baseline	Update	Update					

Table 3-6 Single-Project Program Plan Control Plans Maturity Matrix   (continued)

NPR 7120.5 Program and Project Plan— Control Plans[1]	Pre– Phase A	Phase A KDP B		Phase B KDP C	Phase C KDP D		Phase D KDP E		Phase E KDP F
	MCR	SRR	SDR/ MDR	PDR	CDR	SIR	ORR	MRR/ FRR	DR
19. Security Plan			Preliminary	Baseline					Update annually
20. Project Protection Plan			Preliminary	Baseline	Update	Update	Update	Update	Update annually
21. Technology Transfer Control Plan			Preliminary	Baseline	Update				
22. Knowledge Management Plan	Approach for managing during Phase A[2]		Preliminary	Baseline	Update				
23. Human Rating Certification Package	Preliminary approach[3]	Initial	Update	Update	Update		Update	Approve Certification	
24. Planetary Protection Plan			Planetary Protection Certification (if required)	Baseline					
25. Nuclear Safety Launch Approval Plan			Baseline (mission has nuclear materials)						
26. Range Safety Risk Management Process Documentation				Preliminary	Preliminary	Baseline			
27. Education Plan			Preliminary	Baseline	Update		Update		
28. Communications Plan			Preliminary	Baseline	Update		Update		

[1] See template in NPR 7120.5E, Appendix G, for control plan details.

[2] Not the Plan, but documentation of high-level process. May be documented in MCR briefing package.

[3] Not the Plan, but documentation of considerations that might impact the cost and schedule baselines. May be documented in MCR briefing package.

4 Project Life Cycle, Oversight, and Activities by Phase

4.1 NASA Projects

Projects[1] are the means by which NASA accomplishes the work needed to explore space, expand scientific knowledge, and perform aeronautics research on behalf of the Nation. They develop the hardware and software required to deliver NASA's missions. NASA's technologically challenging projects regularly extend the Nation's scientific and technological boundaries. These complex endeavors require a disciplined approach framed by a management structure and institutional processes essential to mission success.

As with space flight programs, projects vary in scope and complexity and, thus, require varying levels of management requirements and Agency attention and oversight. These differing Agency expectations are defined by different categories of projects, which determine both the project's oversight council and the specific approval requirements. Projects are assigned Category 1, 2, or 3 based initially on:

- The project life-cycle cost estimate (LCCE),[2]
- The inclusion of significant radioactive material,[3] and
- Whether the system being developed is for human space flight.

> A project is a specific investment identified in a Program Plan having defined requirements, a life-cycle cost, a beginning, and an end. A project also has a management structure and may have interfaces to other projects, agencies, and international partners. A project yields new or revised products that directly address NASA's strategic goals.

---

[1] Some single-project programs may integrate their project structures directly into the program and therefore not have a specific project. See Chapter 3 for guidance on implementing single-project programs.

[2] The project LCCE includes Phases A through F and all Work Breakdown Structure (WBS) Level 2 elements (see Section 5.9), and is measured in real year (nominal) dollars.

[3] Nuclear safety launch approval is required by the Administrator or Executive Office of the President when significant radioactive materials are included onboard the spacecraft and/or launch vehicle. (Levels are defined in *NPR 8715.3, NASA General Safety Program Requirements*.)

Secondarily, projects are assigned a category based on a priority level related to the importance of the activity to NASA, as determined by:

- The extent of international participation (or joint effort with other government agencies),
- The degree of uncertainty surrounding the application of new or untested technologies, and
- Spacecraft/payload development risk classification. (See *NPR 8705.4, Risk Classification for NASA Payloads*.)

The determination of the priority level is subjective based on how the Agency's senior management assesses the risk of the project to NASA's overall mission success, including the project's importance to its external stakeholders.

Guidelines for determining project categorization are shown in Table 4-1, but the Mission Directorate Associate Administrator (MDAA) may recommend different categorization that considers additional risk factors facing the project. The NASA Associate Administrator (AA) approves the final project categorization. The Office of the Chief Engineer (OCE) is responsible for the official listing of all NASA projects in accordance with *NPD 7120.4, NASA Engineering and Program/Project Management Policy*.[4] This listing in the Metadata Manager (MdM) database provides the basis for the Agency Work Breakdown Structure (WBS). See Section 5.9 for an explanation of how projects are documented in the MdM and how the MdM, WBS, and the financial system interrelate.

Table 4-1 Project Categorization Guidelines

Priority Level	LCC < $250 million	LCC ≥ $250 million and ≤ $1 billion	LCC > $1 billion, significant radioactive material, or human space flight
High	Category 2	Category 2	Category 1
Medium	Category 3	Category 2	Category 1
Low	Category 3	Category 2	Category 1

Projects can be initiated in a variety of ways. Generally, a program initiates a project, with support and guidance from the Mission Directorate, as part of the program's overall strategy and consistent with the program's objectives and requirements. These program-initiated projects are usually either "directed" or "competed" by the Mission Directorate with support from the program:

[4] These data are maintained by the Office of Chief Financial Officer (OCFO) in a database called the Metadata Manager (MdM). This database is the basis for the Agency's work breakdown and forms the structure for program and project status reporting across all Mission Directorates and Mission Support Offices.

- A "directed" mission is generated in a top-down process from the Agency strategic goals and through the strategic acquisition planning process. It is defined and directed by the Agency, assigned to a Center[5] or implementing organization by the MDAA[6] consistent with direction and guidance from the strategic acquisition planning process, and implemented through a program or project management structure. Direction may also come from outside NASA and implementing organizations may include other Government agencies.

- A "competed" mission is opened up to a larger community for conceptualization and definition through a Request For Proposal (RFP) or competitive selection process, such as an Announcement of Opportunity (AO), before entering the conventional life-cycle process. (See Section 4.3.3.) In a competed mission, a Center is generally part of the proposal.

Projects can also be initiated in many other ways. In some cases, other Federal agencies ask NASA to design and develop projects. As part of the agreement with that agency, these projects are usually funded by the sponsoring agency and are known as "reimbursable" projects. As an example, NASA has been supporting the National Oceanographic and Atmospheric Administration (NOAA) by developing spacecraft for them and has turned the operation of those spacecraft over to NOAA after launch and on-orbit checkout. The Geostationary Operational Environmental Satellite–R Series (GOES-R) is a good example of this type of project. The requirements of NPR 7120.5, including doing an ABC and Management Agreement, apply to reimbursable projects unless waived, as well as any additional requirements the sponsoring partner adds, as negotiated.

Projects can also come from other types of acquisition authorities. These authorities include, but are not limited to, grants, cooperative agreements, and Space Act Agreements (SAA). As an example, the Commercial Crew & Cargo Program (C3PO) is using SAAs for initiating and managing NASA's Commercial Orbital Transportation Services (COTS) projects. NPR 7120.5 requirements apply to contractors, grant recipients, or parties to agreements only to the extent specified or referenced in the appropriate contracts, grants, or agreements.

---

[5] For Category 1 projects, the assignment to a Center or other implementing organization is with the concurrence of the NASA Associate Administrator.

[6] As part of the process of assigning projects to NASA Centers, the affected program manager may recommend project assignments to the MDAA.

The National Aeronautics and Space Act of 1958, as amended (51 U.S.C. 20113(e)), authorizes NASA "to enter into and perform such…other transactions as may be necessary in the conduct of its work and on such terms as it may deem appropriate…" This authority enables NASA to enter into "Space Act Agreements (SAAs)" with organizations in the public and private sector. SAA partners can be a U.S. or foreign person or entity, an academic institution, a Federal, state, or local governmental unit, a foreign government, or an international organization, for profit, or not for profit.

SAAs establish a set of legally enforceable terms between NASA and the other party to the agreement, and constitute Agency commitments of resources such as personnel, funding, services, equipment, expertise, information, or facilities. SAAs can be reimbursable, non-reimbursable, and funded agreements. Under reimbursable agreements, NASA's costs are reimbursed by the agreement partner, either in full or in part. Non-reimbursable agreements are those in which NASA is involved in a mutually beneficial activity that furthers the Agency's missions, with each party bearing its own costs, and no exchange of funds between the parties. Funded agreements are those under which NASA transfers appropriated funds to an agreement partner to accomplish an Agency mission. (See *NPD 1000.5* and http://www.nasa.gov/open/plan/space-act.html for additional information on SAAs.)

4.1.1 Project Life Cycle

Figure 4-1 illustrates the project life-cycle phases, gates, and major events, including Key Decision Points (KDPs) (see paragraph below), life-cycle reviews, and principal documents that govern the conduct of each phase. It also shows how projects recycle through Formulation when changes warrant such action.

The Standing Review Board is a group of independent experts who assess and evaluate project activities, advise projects and convening authorities, and report their evaluations to the responsible organizations, as identified in Figure 4-5. They are responsible for conducting independent reviews (life cycle and special) of a project and providing objective, expert judgments to the convening authorities. The reviews are conducted in accordance with approved terms of reference and life-cycle requirements per *NPR 7120.5* and *NPR 7123.1*. For more detail see Section 5.10 of this handbook and the *NASA Standing Review Board Handbook*.

Each project life-cycle phase includes one or more life-cycle reviews. A life-cycle review is designed to provide a periodic assessment of a project's technical and programmatic status and health at a key point in the life cycle and enables the project to assure itself that it has completed the work required for this point in the life cycle. Thus, life-cycle reviews are essential elements of conducting, managing, evaluating, and approving space flight projects and are an important part of NASA's system of checks and balances. Most life-cycle reviews are conducted by the project and an independent Standing Review Board (SRB).[7] NASA accords special importance to maintaining the integrity of its independent review process to gain the value of independent technical and programmatic perspectives.

Life-cycle reviews provide the project and NASA senior management with a credible, objective assessment of how the project is progressing. The final life-cycle review in a given project life-cycle phase provides essential information for the KDP, which marks the end of that life-cycle phase. A KDP is the point at which a Decision Authority determines whether and how a project proceeds through the life cycle, and authorizes key project cost, schedule, and content parameters that govern the remaining life-cycle activities. A KDP serves as a mandatory gate through which a project must pass to proceed to the next life-cycle phase. During the period between the life-cycle review and the KDP, the project continues its planned activities unless otherwise directed by the Decision Authority.

For Category 1 projects, the Decision Authority is the NASA Associate Administrator. For Category 2 and 3 projects, the Decision Authority is the MDAA. KDPs for projects are labeled with capital letters, e.g., KDP A. The letter corresponds to the project phase that will be entered after successfully passing through the gate.

Figure 4-1 shows two separate life-cycle lines: one for human space flight, and one for robotic space flight. The reason for this is to acknowledge that these two communities have developed slightly different terms and launch approval processes over the years. Despite these subtle differences, the project management life cycles are essentially the same.

---

[7] Life-cycle reviews required to be performed by the SRB are depicted by red triangles in Figure 4-1.

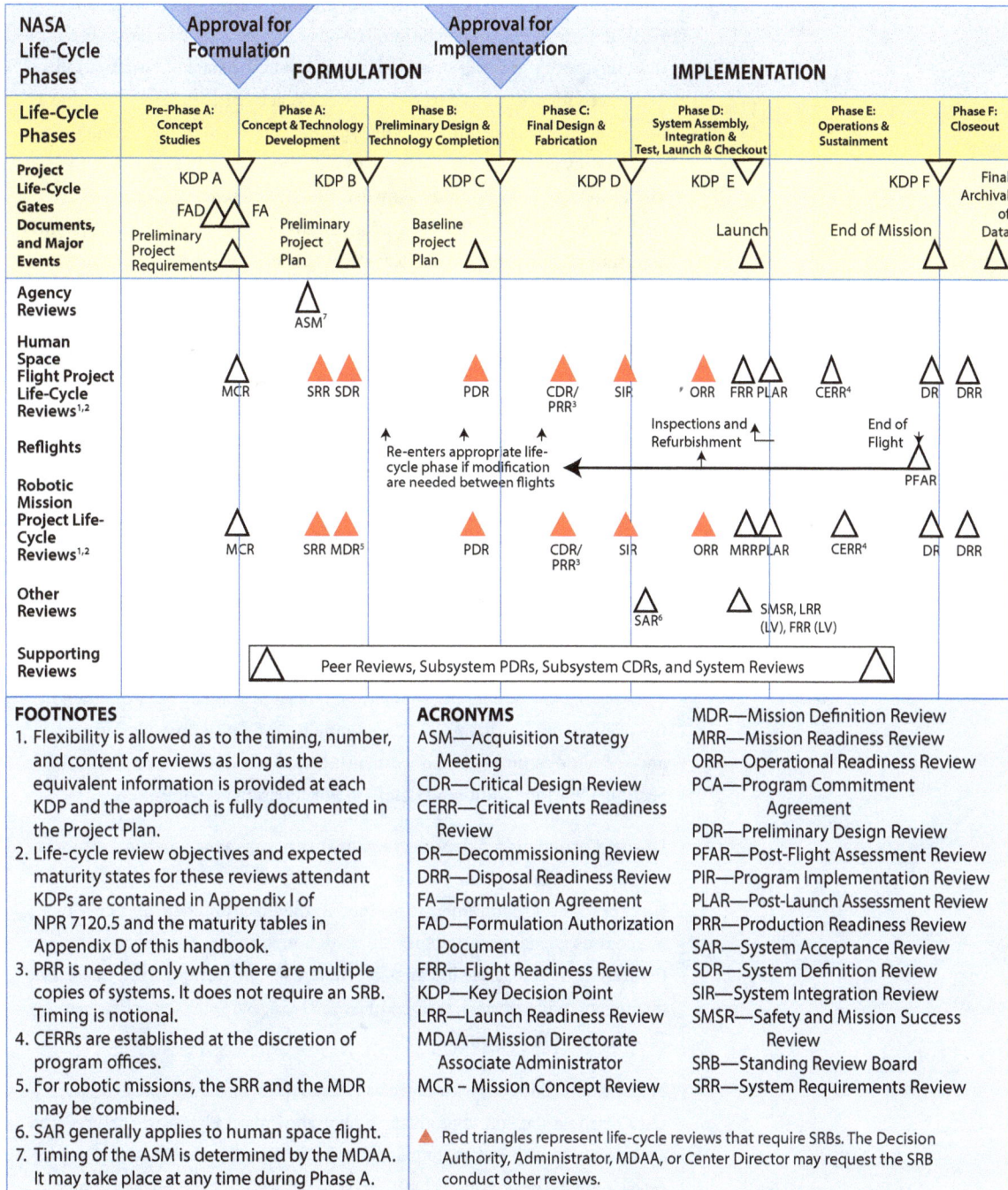

Figure 4-1 NASA Project Life Cycle

FOOTNOTES

1. Flexibility is allowed as to the timing, number, and content of reviews as long as the equivalent information is provided at each KDP and the approach is fully documented in the Project Plan.
2. Life-cycle review objectives and expected maturity states for these reviews attendant KDPs are contained in Appendix I of NPR 7120.5 and the maturity tables in Appendix D of this handbook.
3. PRR is needed only when there are multiple copies of systems. It does not require an SRB. Timing is notional.
4. CERRs are established at the discretion of program offices.
5. For robotic missions, the SRR and the MDR may be combined.
6. SAR generally applies to human space flight.
7. Timing of the ASM is determined by the MDAA. It may take place at any time during Phase A.

ACRONYMS

ASM—Acquisition Strategy Meeting
CDR—Critical Design Review
CERR—Critical Events Readiness Review
DR—Decommissioning Review
DRR—Disposal Readiness Review
FA—Formulation Agreement
FAD—Formulation Authorization Document
FRR—Flight Readiness Review
KDP—Key Decision Point
LRR—Launch Readiness Review
MDAA—Mission Directorate Associate Administrator
MCR – Mission Concept Review

MDR—Mission Definition Review
MRR—Mission Readiness Review
ORR—Operational Readiness Review
PCA—Program Commitment Agreement
PDR—Preliminary Design Review
PFAR—Post-Flight Assessment Review
PIR—Program Implementation Review
PLAR—Post-Launch Assessment Review
PRR—Production Readiness Review
SAR—System Acceptance Review
SDR—System Definition Review
SIR—System Integration Review
SMSR—Safety and Mission Success Review
SRB—Standing Review Board
SRR—System Requirements Review

▲ Red triangles represent life-cycle reviews that require SRBs. The Decision Authority, Administrator, MDAA, or Center Director may request the SRB conduct other reviews.

As noted earlier, project life cycles are fundamentally divided between Formulation and Implementation. However, projects also undergo activities preparatory to being stood up as a project at the start of Formulation. These activities occur as part of Pre–Phase A activities. Prior to initiating a new project, a Mission Directorate, typically supported by a program office, provides resources for concept studies (i.e., Pre–Phase A (Concept Studies)). These Concept Study activities involve Design Reference Mission (DRM) analysis, feasibility studies, technology needs analyses, engineering systems assessments, and analyses of alternatives that need to be performed before a specific project concept emerges.

Project Formulation consists of two sequential phases, Phase A (Concept and Technology Development) and Phase B (Preliminary Design and Technology Completion). Formulation activities include developing project requirements; assessing technology requirements; developing the system architecture; completing mission and preliminary system designs; flowing down requirements to the system/subsystem level; planning acquisitions; assessing heritage (the applicability of designs, hardware, and software from past projects to the present one); conducting safety, performance, cost, and risk trades; identifying and mitigating development and programmatic risks; conducting engineering development activities, including developing and testing engineering prototypes and models for the higher risk components and assemblies that have not been previously built or flown in the planned environment; and developing high-fidelity time-phased cost and schedule estimates and documenting the basis of these estimates. (See Section 4.3.4.1 for additional detail on Formulation activities.)

During Formulation, the project establishes performance metrics, explores the full range of implementation options, defines an affordable project concept to meet requirements specified in the Program Plan, and develops or acquires needed technologies. Formulation is an iterative set of activities, rather than discrete linear steps. Systems engineering plays a major role during Formulation as described in *NPR 7123.1, NASA Systems Engineering Processes and Requirements*.

Formulation continues with execution of activities, normally concurrently, until Formulation output products, such as the Project Plan, have matured and are acceptable to the program manager, Center Director, and MDAA. For projects with LCC greater than $250M, these activities allow the Agency to present to external stakeholders time-phased cost plans and schedule range estimates at KDP B and high-confidence cost and schedule commitments at KDP C.

Project Implementation consists of Phases C, D, E, and F. Decision Authority approval at KDP C marks the transition from Phase B of Formulation to Phase C of Implementation:

- Phase C (Final Design and Fabrication) includes completion of final system design and the fabrication, assembly, and test of components, assemblies, and subsystems.

- Phase D (System Assembly, Integration and Test, and Launch and Checkout) includes system assembly, integration, and test (AI&T); verification/certification; prelaunch activities; launch; and checkout. Completing KDP E and authorizing launch is complex and unique because completing the KDP does not lead immediately to transition to Phase E. Transition to Phase E occurs after successful checkout of the flight system. Section 4.4.4 provides details on the launch review and approval process and the transition to Phase E for human and robotic space flight projects.

- The start of Phase E (Operations and Sustainment) marks the transition from system development and acquisition activities to primarily system operations and sustainment activities (see "Sustainment and Sustainment Engineering" box for an explanation of sustainment activities) in Section 4.4.6.1.

- In Phase F (Closeout), project space flight and associated ground systems are taken out of service and safely disposed of or reused for other activities, although scientific and other analyses might still continue under project funding.

Independent evaluation activities occur throughout all phases.

4.1.2 Project Life-Cycle Reviews

The project life-cycle reviews identified in the project life cycle are essential elements of conducting, managing, evaluating, and approving space flight projects. The project manager is responsible for planning for and supporting the life-cycle reviews. These life-cycle reviews assess the following six assessment criteria identified in NPR 7120.5:

- **Alignment with and contribution to Agency strategic goals and the adequacy of requirements that flow down from those.** The scope of this criterion includes, but is not limited to, alignment of project requirements/designs with Agency strategic goals, project requirements and constraints, mission needs and success criteria; allocation of program requirements to projects; and proactive management of changes in project scope and shortfalls.

- **Adequacy of management approach.** The scope of this criterion includes, but is not limited to, project authorization, management framework and plans, acquisition strategies, and internal and external agreements.

- **Adequacy of technical approach,** as defined by NPR 7123.1 entrance and success criteria. The scope of this criterion includes, but is not limited to, flow down of project requirements to systems/subsystems; architecture and design; and operations concepts that respond to and satisfy the requirements and mission needs.

The joint cost and schedule confidence level is the product of a probabilistic analysis of the coupled cost and schedule to measure the likelihood of completing all remaining work at or below the budgeted levels and on or before the planned completion of the development phase. A JCL is required for all tightly coupled and single-project programs, and for all projects with an LCC greater than $250 million. Small Category 3/Class D projects with a life-cycle cost estimate less than $150 million are not required to do a JCL, but are required to develop a NASA internal cost and schedule commitment (ABC internal commitment) per October 2014 guidance from the NASA AA on tailoring NPR 7120.5 requirements, available on the OCE tab in NODIS under "Other Policy Documents" at http/nodis3.gsfc.nasa.gov/OCE_docs/OCE_25.pdf. The JCL calculation includes consideration of the risk associated with all elements, regardless of whether or not they are funded from appropriations or managed outside of the program or project. JCL calculations include the period from approval for Implementation (KDP I for tightly coupled programs, KDP C for projects and single-project programs) through the handover to operations. Per *NPR 7120.5*, Mission Directorates plan and budget tightly coupled and single-project programs (regardless of life-cycle cost) and projects with an estimated life-cycle cost greater than $250 million based on a 70 percent JCL or as approved by the Decision Authority. Mission Directorates ensure funding for these projects is consistent with the Management Agreement and in no case less than the equivalent of a 50 percent JCL.

- **Adequacy of the integrated cost and schedule estimate and funding strategy** in accordance with *NPD 1000.5, Policy for NASA Acquisition.* The scope of this criterion includes, but is not limited to, cost and schedule control plans; cost and schedule estimates (prior to KDP C) and baselines (at KDP C) that are consistent with the project requirements, assumptions, risks, and margins; Basis of Estimate (BoE); Joint Cost and Schedule Confidence Level (JCL) (when required); and alignment with planned budgets.

- **Adequacy and availability of resources other than budget.** The scope of this criterion includes, but is not limited to, planning, availability, competency and stability of staffing, infrastructure, and the industrial base/supplier chain requirements.

- **Adequacy of the risk management approach and risk identification and mitigation** per *NPR 8000.4, Agency Risk Management Procedural Requirements* and *NASA/SP-2011-3422, NASA Risk Management Handbook.* The scope of this criterion includes, but is not limited to risk-management plans, processes (e.g., Risk-Informed Decision Making (RIDM) and Continuous Risk Management (CRM)), open and accepted risks, risk assessments, risk mitigation plans, and resources for managing/mitigating risks.

Life-cycle reviews are designed to provide the project with an opportunity to ensure that it has completed the work of that phase and an independent assessment of the project's technical and programmatic status and health. Life-cycle reviews are conducted under documented Agency and Center review processes. (See Section 5.10 in this handbook and the *NASA Standing Review Board Handbook.*)

The life-cycle review process provides:

- The project with a credible, objective independent assessment of how it is progressing.
- NASA senior management with an understanding of whether
  - The project is on track to meet objectives,
  - The project is performing according to plan, and
  - Impediments to project success are addressed.
- For a life-cycle review that immediately precedes a KDP, a credible basis for the Decision Authority to approve or disapprove the transition of the project at a KDP to the next life-cycle phase.

The independent review also provides vital assurance to external stakeholders that NASA's basis for proceeding is sound.

The project finalizes its work for the current phase during the life-cycle review. In some cases, the project uses the life-cycle review meeting(s) to make formal programmatic and technical decisions necessary to complete its work. In all cases, the project utilizes the results of the independent assessment and the resulting management decisions to finalize its work.

In addition, the independent assessment serves as a basis for the project and management to determine if the project's work has been satisfactorily completed, and if the plans for the following life-cycle phases are acceptable. If the project's work has not been satisfactorily completed, or its plans are not acceptable, the project addresses the issues identified during the life-cycle review, or puts in place the action plans necessary to resolve the issues.

Prior to the project life-cycle reviews, projects conduct internal reviews in accordance with NPR 7123.1, Center practices, and NPR 7120.5. These internal reviews are key components of the process used by projects to solidify their plans, technical approaches, and programmatic commitments and are part of the normal systems engineering work processes as defined in NPR 7123.1, where major technical and programmatic requirements are assessed along with the system design and other implementation plans. For both robotic and human space flight projects, these internal reviews are typically lower level system and subsystem reviews that lead to and precede the life-cycle review. Major technical and programmatic performance metrics are reported and assessed against predictions. Figure 4-2 shows how these

> A life-cycle review is complete when the governing PMC and Decision Authority complete their assessment and sign the Decision Memorandum.

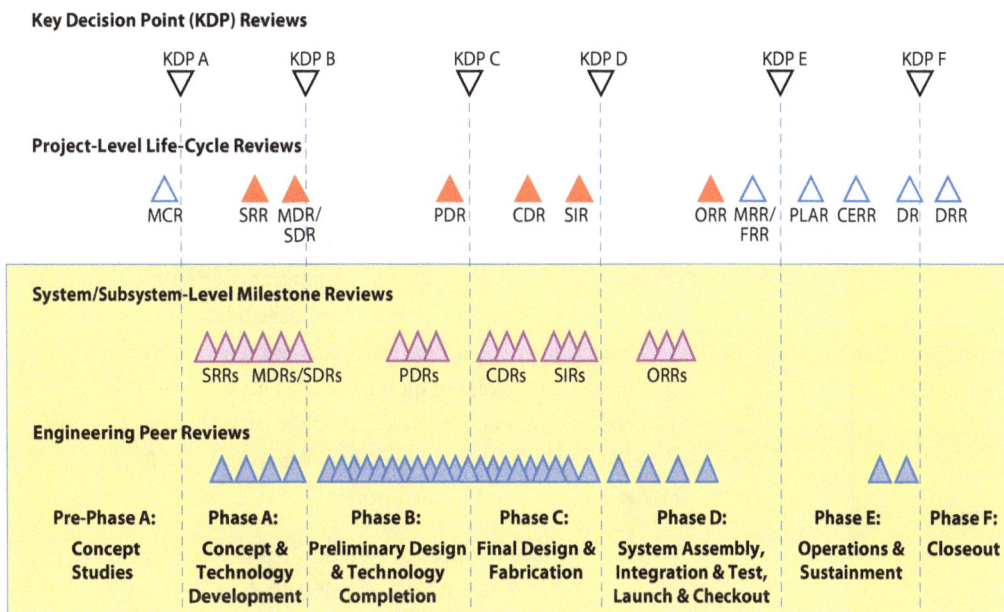

Figure 4-2 Work Led by the Project Throughout the Life Cycle

internal reviews relate to life-cycle reviews. (This graphic is an example based on Goddard Space Flight Center practices. Each Center may have a different approach.)

The project manager has the authority to determine whether to hold a one-step review or a two-step review. This determination usually depends on the state of the project's cost and schedule maturity as described below. Any life-cycle review can be either a one-step review or a two-step review. The project manager documents the project's review approach in the project review plan.

Descriptions of the one-step and two-step life-cycle review processes are provided in Figures 4-3 and 4-4. (These descriptions are written from the perspective of life-cycle reviews conducted by a project and an SRB. For life-cycle reviews that do not require an Agency-led SRB, the project manager will work with the Center Director or designee to prepare for and conduct the life-cycle review in accordance with Center practices and a Center-assigned independent review team. Small Category 3, Class D projects with a life-cycle cost of under $150 million should refer to guidance on using an independent review team to perform independent assessments of the project in place of an SRB. Guidance can be found on the OCE tab in NODIS under "Other Policy Documents" at http/nodis3.gsfc.nasa.gov/OCE_docs/OCE_25.pdf. When the life-cycle review is conducted by the project and a Center independent review team rather than an Agency-led SRB, the remaining references to SRB are replaced with Center independent review team.)

In a one-step review, the project's technical maturity and programmatic posture are assessed together against the six assessment criteria. In this case the project has typically completed all of its required technical work as defined in NPR 7123.1 life-cycle review entrance criteria and has aligned the scope of this work with its cost estimate, schedule, and risk posture before the life-cycle review. The life-cycle review is then focused on presenting this work to the SRB. Except in special cases, a one-step review is chaired by the SRB. The SRB assesses the work against the six assessment criteria and then provides an independent assessment of whether or not the project has met these criteria. Figure 4.3 illustrates the one-step life-cycle review process.

In a two-step review, the project typically has not fully integrated the project's cost and schedule with the technical work. In this case, the first step of the life-cycle review is focused on finalizing and assessing the technical work described in NPR 7123.1. However as noted in Figure 4-4, which illustrates the two-step life-cycle review process, the first step does consider the preliminary cost, schedule, and risk as known at the time of the review. This first step is only one half of the life-cycle review. At the end of the first step, the SRB will have fully assessed the technical approach criteria but will only be able to determine preliminary findings on the remaining criteria since

There are special cases, particularly for human space flight projects, where the project uses the life-cycle review to make formal decisions to complete the project's technical work and align it with the cost and schedule. In these cases, the project manager may co-chair the life-cycle review since the project manager is using this forum to make project decisions, and the SRB will conduct the independent assessment concurrently. The project manager will need to work with the SRB chair to develop the life-cycle review agenda and agree on how the life-cycle review will be conducted to ensure that it enables the SRB to fully accomplish the independent assessment. The project manager and the SRB chair work together to ensure that the life-cycle review Terms of Reference (ToR) reflect their agreement and the convening authorities approve the approach.

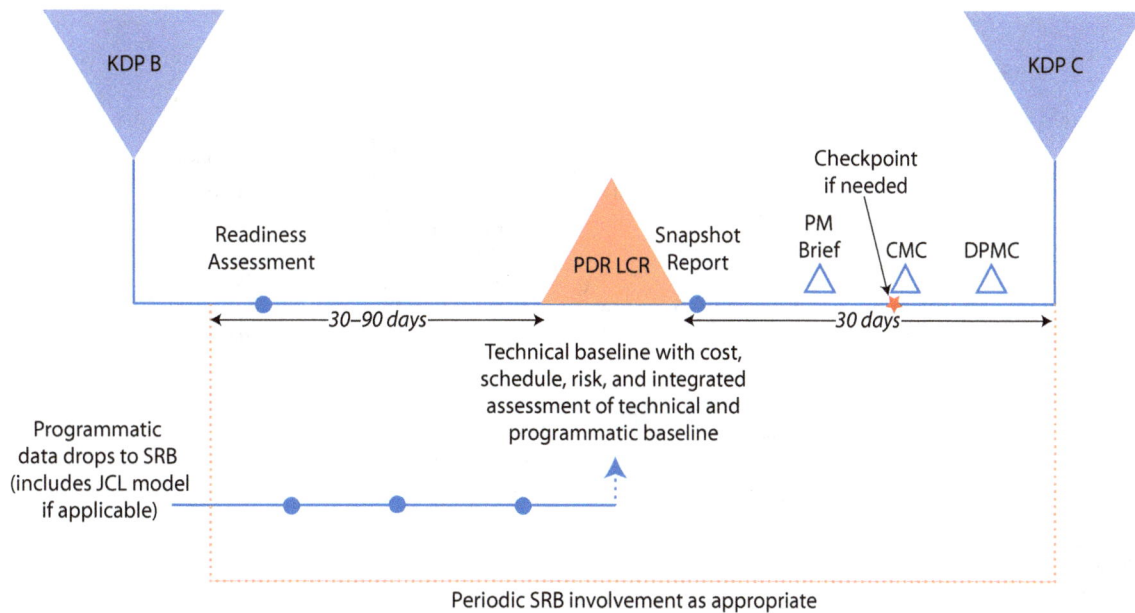

Notes: A one-or two-step review may be used for any life-cycle review. Section 5.10 and the *NASA Standing Review Board Handbook* provide information on the readiness assessment, snapshot reports, and checkpoints associated with life-cycle reviews. Time is not to scale.

Figure 4-3 One-Step PDR Life-Cycle Review Overview

the project has not yet finalized its work. Thus, the second step is conducted after the project has taken the results of the first step and fully integrated the technical scope with the cost, schedule, and risk, and has resolved any issues that may have arisen as a result of this integration. The period between steps may take up to six months depending on the complexity of the project. In the second step, which may be referred to as the Independent Integrated Life-Cycle Review Assessment, the project typically presents the integrated technical, cost, schedule, and risk, just as is done for a one-step review, but the technical presentations may simply update information provided during the first step. The SRB then completes its assessment of whether or not the project has met the six assessment criteria. In a two-step life-cycle review, both steps are necessary to fulfill the life-cycle review requirements. Except in special cases, the SRB chairs both steps of the life-cycle review.

Details on project review activities by life-cycle phase are provided in the sections below. The *NASA Standing Review Board Handbook* and Section 5.10 in this handbook also contain more detailed information on conducting life-cycle reviews. NPR 7123.1 provides life-cycle review entrance and success criteria, and Appendix I in NPR 7120.5E and

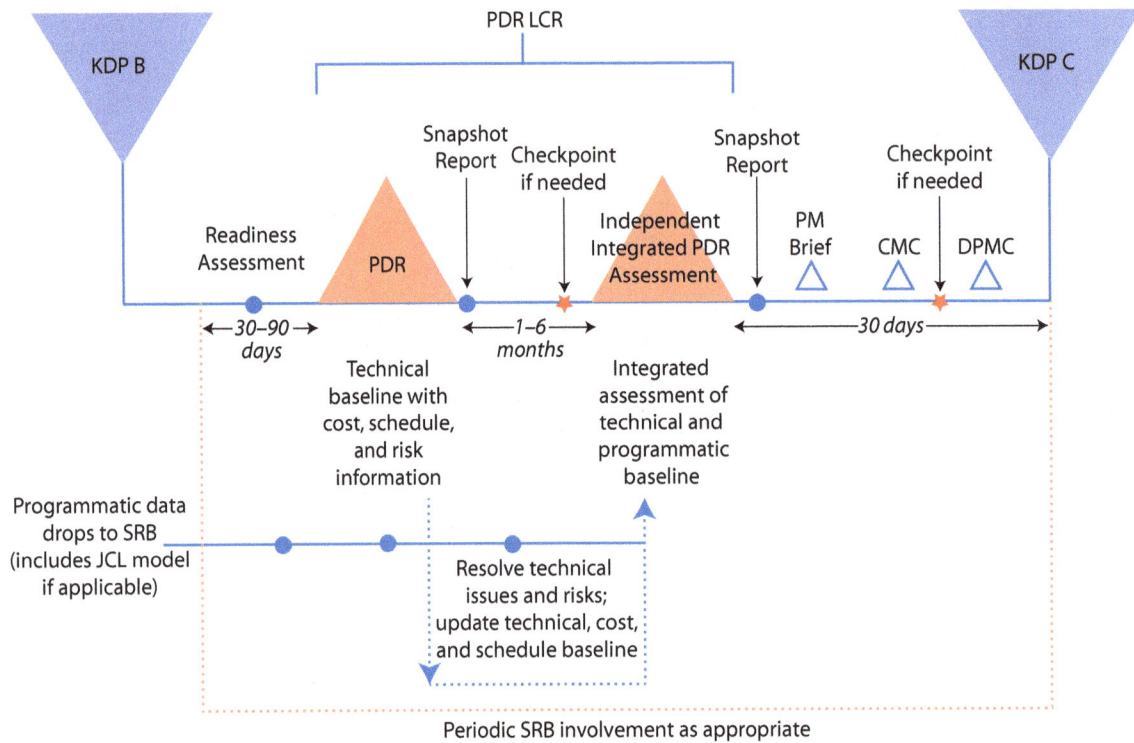

Notes: A one-or two-step review may be used for any life-cycle review. The *NASA Standing Review Board Handbook* provides information on the readiness assessment, snapshot reports, and checkpoints associated with life-cycle reviews. Time is not to scale.

Figure 4-4 Two-Step PDR Life-Cycle Review Overview

Appendix E in this handbook provides specifics for addressing the six assessment criteria required to demonstrate the project has met the expected maturity state to transition to the next phase.

4.1.3 Other Reviews and Resources

Special reviews may be convened by the Office of the Administrator, MDAA, Center Director, the Technical Authority (TA),[8] or other convening authority. Special reviews may be warranted for projects not meeting expectations for achieving safety, technical, cost, or schedule requirements; not being able to develop an enabling technology; or experiencing some unanticipated change to the project baseline. Special reviews include a Rebaseline Review and Termination Review. Rebaseline reviews are conducted

---

[8] That is, individuals with specifically delegated authority in Engineering (ETA), Safety and Mission Assurance (SMA TA), and Health and Medical (HMTA). See Section 5.2 for more information on Technical Authorities.

when the Decision Authority determines the Agency Baseline Commitment (ABC) needs to be changed. (For more detail on Rebaseline Reviews, see Section 5.5.4.1. For more detail on the ABC, see Section 4.2.4 and Section 5.5.) A Termination Review may be recommended by a Decision Authority, MDAA, or program executive if he or she believes it may not be in the Government's best interest to continue funding a project. Other reviews, such as Safety and Mission Assurance (SMA) reviews, are part of the regular management process. For example, SMA Compliance/Verification reviews are spot reviews that occur on a regular basis to ensure projects are complying with NASA safety principles and requirements. For more detail on Termination Reviews and SMA reviews, see Section 5.11.

Other resources are also available to help a project manager evaluate and improve project performance. These resources include the following:

- The NASA Engineering and Safety Center (NESC), an independently funded organization with a dedicated team of technical experts, provides objective engineering and safety assessments of critical, high-risk projects. The NESC is a resource to benefit projects and organizations within the Agency, the Centers, and the people who work there by promoting safety through engineering excellence, unaffected and unbiased by the projects it is evaluating. The NESC mission is to proactively perform value-added independent testing, analysis, and assessments to ensure safety and mission success and help NASA avoid future problems. Projects seeking an independent assessment or expert advice on a particular technical problem can contact the NESC at http://www.nasa.gov/offices/nesc/contacts/index.html or the NESC chief engineer at their Center.

- The NASA Independent Verification and Validation (IV&V) Facility strives to improve the software safety, reliability, and quality of NASA projects and missions through effective applications of systems and software IV&V methods, practices, and techniques. The NASA IV&V Facility applies software engineering best practices to evaluate the correctness and quality of critical and complex software systems. When applying systems and software IV&V, the NASA IV&V Facility seeks to ensure that the software exhibits behaviors exactly as intended, does not exhibit behaviors that were not intended, and exhibits expected behaviors under adverse conditions. Software IV&V has been demonstrated to be an effective technique on large, complex software systems to increase the probability that software is delivered within cost and schedule, and that software meets requirements and is safe. When performed in parallel with systems development, software IV&V provides for the early detection and identification of risk elements, enabling early mitigation of the risk elements. For projects that are required or desire to do software IV&V, go to the "Contact Us" link at http://www.nasa.gov/centers/ivv/home/index.

html. (All Category 1 projects; all Category 2 projects that have Class A or Class B payload risk classification per *NPR 8705.4, Risk Classification for NASA Payloads*; and projects specifically selected by the NASA Chief, Safety and Mission Assurance are required to do software IV&V. See NPR 7120.5E and Section 4.1 in this handbook for project categorization guidelines.)

4.1.4 Project Evolution and Recycle

A project may evolve over time in ways that require it to go back and restart parts of its life cycle. A project may evolve as a result of a planned series of upgrades, when the need for new capabilities is identified, or when the project includes reflights.

When the requirements imposed on a project significantly change, the project needs to evaluate whether the changes impact its current approved approach and/or system design and performance. In these cases, the project may be asked by the Decision Authority to go back through the necessary life-cycle phases and reviews and to update project documentation to ensure that the changes have been properly considered in light of the overall project/system performance. Each case is likely to be different and thus may not require completely restarting the process at the beginning. The decision on when and where to recycle through the life-cycle reviews will be based on a discussion between the project, the program, the Mission Directorate, and the Decision Authority. For example, a project may need to refurbish operational reusable systems after each flight, or a project may be required to make modifications between flights. A project going back through a part of its life cycle is depicted in the project life-cycle figure on the "Reflights" line (Figure 4-1). "Reflight" may involve updates of the Project Plan and other documentation.

4.1.5 Project Tailoring

Project teams are expected to tailor the requirements of NPR 7120.5 to meet the specific needs of the project. In general, all the requirements would be expected to be applicable to Category 1 projects, while Category 3 projects, for example, may only need some of the more significant requirements for success. Small Category 3, Class D projects with a life-cycle cost of under $150 million should refer to guidance on tailoring NPR 7120.5 require-ments from the NASA AA, which can be found on the OCE tab in NODIS under "Other Policy Documents" at http/nodis3.gsfc.nasa.gov/OCE_docs/OCE_25.pdf. When a project team and its management determine that a requirement is not needed, the process for tailoring that requirement requires getting permission from the requirement owner to waive the requirement as described in Section 5.4. This can be done using the required Compliance Matrix in the Project Plan, which also shows the requirement owner. Tailoring allows projects to perform only those activities that are needed for mission success while still meeting Agency external requirements and receiving the benefits of NASA policy, reflecting lessons learned and best practice. Project managers and their management are encouraged to thoughtfully examine and tailor the requirements so projects perform only those requirements that contribute to achieving mission success. Requirements imposed by Federal law or external entities generally cannot be waived.

Management tools to guide project managers in tailoring the requirements for their project category can be found on the Engineering Program and Project Management Division (EPPMD) community of practice site on the NASA Engineering Network (NEN). Four areas that often need tailoring are areas where (1) requirements do not apply, such as requirements for nuclear materials if the spacecraft does not use nuclear materials, requirements for a different category of project, or requirements for projects in a different cost category; (2) documents can be combined, such as smaller projects, including all their control plans within their Project Plan; (3) type and timing of reviews are adjusted, such as combining reviews; (4) projects do not need to satisfy the requirement at the same level as a Category 1 project would; and (5) the intent of the requirement is met by other means, given the solution that has been decided upon.

Tailoring can also be applied at a more detailed level for both programmatic and technical areas. As an example, a Category 3 project may not need or wish to:

- Develop a WBS structure and attendant schedules to the fourth or fifth level due to the increased time needed to manage to this level of detail;

- Develop requirements for the fourth or fifth levels if sufficient definition exists to satisfactorily describe what needs to be developed at the third or fourth level; or

- Conduct Verification and Validation (V&V) on heritage systems if the heritage systems are sufficiently understood and the changes sufficiently minor that the performance of the heritage systems will not be affected.

Changes such as these typically are documented in the project's Formulation Agreement and the Project Plan so that the project's management can assess the rationale and agree to the tailoring. (See Section 4.3.2.1 for a detailed description of the Formulation Agreement.) Tailoring documented in the Formulation Agreement or Project Plan is approved when the proper authorities for those documents and the requirement holders have signed off on the tailoring. The Agency's requirements and handbooks have been developed to assist the project managers in achieving project mission success by establishing requirements and best practices. It is not possible to generate the proper requirements and guidelines for every possible scenario. Project managers and their teams need to use good common sense when developing their plans, processes, and tools so that they can be effective, efficient, and successful with acceptable risk. Managers work with their Center and the Mission Directorate when tailoring to ensure that all parties are in agreement with the proposed approach. Table 4-2 illustrates tailoring for a small-scale, low risk, Category 3, Class D project with a life-cycle cost estimate of less than $250 million for a technology demonstration program being planned and implemented under NPR 7120.5. The example shows a summarized Compliance Matrix (see complete Compliance Matrix template in Appendix C of 7120.5E) that reflects coordination with the requirements owners.

Table 4-2 Example of Tailoring for Small Projects

Requirement/Paragraph	Comply	Justification	Approval
Table I-4: 10. ELV Payload Safety	FC	Projects that fall under the applicability of NPR 8715.7 will produce the Safety Process Deliverables as defined. Projects that do not fall under the applicability of NPR 8715.7 will comply with NPR 8715.3 to ensure adherence to appropriate local requirements.	
Table I-4: 11. V&V Report	FC		
Table I-4: 12. Operations Handbook	T	List of Operations Procedures for launch site, on-orbit verification and checkout, and demonstration operations to be provided as part of review briefing package	OCE
Table I-4: 13. Orbital debris, 14. End of Mission Plan, 15. Mission Report, 1. Formulation Agreement, 2. Project Plan	FC		
Table I-4: 3. Plans for work	T	Sufficient detail to be provided in the project IMS tasks/milestones to define plans for work to be accomplished in the next phase.	OCE
Table I-4: 4. Performance against plan	FC		
Table I-4: 5.a. techn, cost, schedule, and safety risks	T	To be provided in review briefing package.	OCE
Table I-4: 5.b Staffing requirements and plans	T	To be provided as part of the basis of estimate that goes along with the cost and schedule package generated for each life-cycle review.	OCE
Table I-4: 5.c Infrastructure plans and business case	FC / NA	(FC) Infrastructure requirements/plans—To be provided as part of the basis of estimate that goes along with the cost and schedule package generated for each life-cycle review. Project to coordinate infrastructure plan content with affected Center(s). (NA) Business Case Analysis & AFUQ–Deemed not necessary for Class D missions due to low dollar value.	OSI (EMD) Approved / OCFO Approved
Table I-4: 5.d Schedule, 5.e. Cost estimate, 5.f. BoE	FC	Basis of Estimate provided as part of the cost and schedule packages that are generated for each life-cycle review.	

Table 4-2 Example of Tailoring for Small Projects (continued)

Requirement/Paragraph	Comply	Justification	Approval
Table I-4: 5.g. JCL	NA	JCL not required for projects with an LCCE less than $250 million	CAD approved
Table I-4: 5.h. External Cost and Schedule commitments	NA	This requirement is Not Applicable since this product is applicable only to projects with LCCs of $250 million or greater (which would have an externally reported baseline). The projects covered under this compliance matrix will establish an LCC baseline at KDP C in accordance with requirements 2.4.1, 2.4.1.1, 2.4.1.2, 2.4.1.3, and 2.4.1.5 of this compliance matrix.	OCE OCFO Concurred with N/A and rationale
Table I-4: 5.i. CADRe	T	Program office is working with HQ/CAD to produce a CADRe for projects where it is feasible as agreed to by HQ/CAD and the program office. A tailored CADRe format for technology missions, as defined by HQ/CAD, is used.	CAD approved
Table I-5: Project Plan Control Plans	FC	Projects were compliant with all control plans except for the three exceptions below.	
Table I-5: 6. SEMP	T	Plan included in project plan section. Approach summarized in review briefing package.	OCE
Table I-5: 12. Environmental Management Plan	T	Plan included in project plan section. Approach summarized in review briefing package. Project to coordinate plan content with affected Center Environmental Manager. Plan content to be consistent with the form provided by OSI.	OSI (EMD) Approved
Table I-5: 13. Integrated Logistics Support Plan	T	Primarily concerning Packaging, Handling Storage, Transportation, and GSE. Plan included in project plan section. Approach summarized in review briefing package. Project to coordinate plan content with affected Center logistics manager.	OSI (LMD) Approved
2.2.8–2.3.1	FC		
2.3.1.1 ABC	NA	Projects have an LCCE less than $250 million.	OCE concurred OCFO Approved
2.3.2–2.4.1.5	FC		
2.4.1.6 5 Tightly coupled programs shall document their life-cycle cost estimate…	NA	Not applicable to projects. This requirement is for programs.	OCE concurred
2.4.1.7, 2.4.2	FC		
2.4.3–2.4.4.2	NA	Not applicable for projects with an LCCE less than $250 million.	CAD approved
3.3.1–3.7.1	FC		

Note: The tables and sections referenced here are in NPR 7120.5E.

The tailoring for these small projects shows requirements that are tailored to combine some reviews and products, requirements that are not applicable to projects or to projects with a life-cycle cost estimate under $250 million and places where the intent of the requirement was met in a different way. (See Table 4-2.) Requirements in Tables 4-6 and 4-7 in this handbook and Tables I-4 and I-5 in NPR 7120.5E apply for small projects.

4.2 Project Oversight and Approval

NASA has established a project management oversight process to ensure that experience, diverse perspectives, and thoughtful programmatic and technical judgment at all levels is available and applied to project activities. The Agency employs management councils and management forums, such as the Baseline Performance Review (BPR), to provide insight to upper management on the status and progress of projects and their alignment with Agency goals. (See Section 4.2.5.) This section describes NASA's oversight approach and the process by which a project is approved to move forward through its life cycle. It defines and describes NASA's Decision Authority, Key Decision Points (KDPs), management councils, and the BPR.

The general flows of the project oversight and approval process for life-cycle reviews that require SRBs and of the periodic reporting activity for projects are shown in Figure 4-5. Prior to the life-cycle review, the project conducts its internal reviews. Then the project and the SRB conduct the life-cycle review. Finally, the results are reported to senior management via the management councils.

Additional insight is provided by the independent perspective of SRBs at life-cycle reviews identified in Figure 4-1. Following each life-cycle review, the independent SRB chair and the project manager brief the applicable management councils on the results of the life-cycle review to support the councils' assessments. These briefings are completed within 30 days of the life-cycle review. The 30 days ensures that the Decision Authority is informed in a timely manner as the project moves forward to preclude the project from taking action that the Decision Authority does not approve. These briefings cover the objectives of the review; the maturity expected at that point in the life cycle; findings and recommendations to rectify issues or improve mission success; the project's response to these findings; and the project's proposed cost, schedule, safety, and technical plans for the follow-on life-cycle phases. This process enables a disciplined approach for developing the Agency's assessment, which informs the Decision Authority's KDP determination of project readiness to proceed to the next life-cycle phase. Life-cycle reviews are conducted under documented Agency and Center review processes.

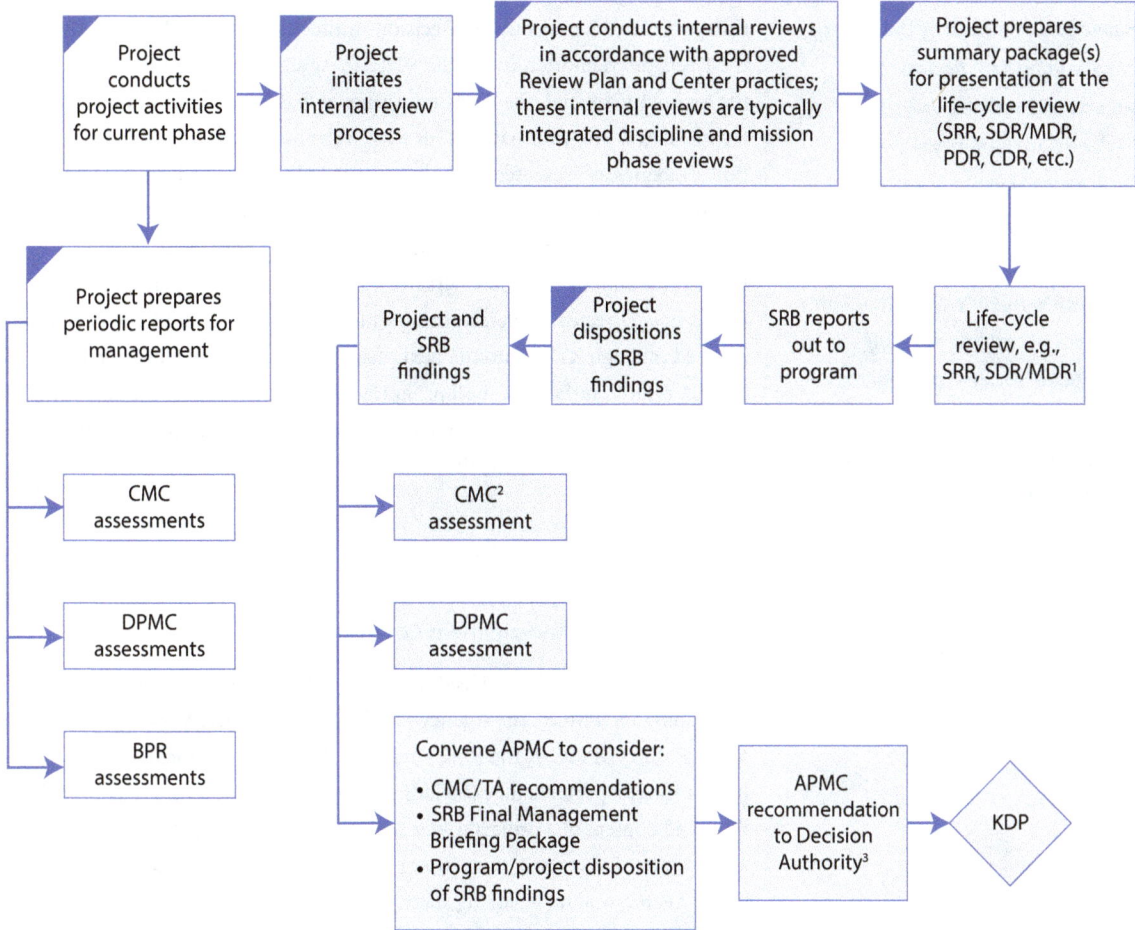

Legend: ▼ Project activity ☐ Periodic reporting activity ☐ Life-cycle review activity

[1] See Section 5.10 and the *NASA Standing Review Board Handbook* for details.

[2] May be an Integrated Center Management Council when multiple Centers are involved.

[3] Life-cycle review is complete when the governing PMC and Decision Authority complete their assessment.

Figure 4-5 Project Life-Cycle Review Process and Periodic Reporting Activity

4.2.1 Decision Authority

The Decision Authority is an Agency individual who is responsible for making the KDP determination on whether and how the project proceeds through the life cycle and for authorizing the key project cost, schedule, and content parameters that govern the remaining life-cycle activities.

For Category 1 projects, the Decision Authority is the NASA Associate Administrator, who signs the Decision Memorandum at the KDP. The

The Decision Authority is the individual authorized by the Agency to make important decisions on projects under their purview. The Decision Authority makes the KDP decision by considering a number of factors, including technical maturity; continued relevance to Agency strategic goals; adequacy of cost and schedule estimates; associated probabilities of meeting those estimates (confidence levels); continued affordability with respect to the Agency's resources; maturity and the readiness to proceed to the next phase; and remaining project risk (safety, cost, schedule, technical, management, and programmatic).

AA may delegate this authority to the MDAA for Category 1 projects. For Category 2 and 3 projects, the Decision Authority is the MDAA, who signs the Decision Memorandum at the KDP. These signatures signify that, as the approving official, the Decision Authority has been made aware of the technical and programmatic issues within the project, approves the mitigation strategies as presented or with noted changes requested, and accepts technical and programmatic risk on behalf of the Agency. The MDAA may delegate some of their Programmatic Authority to appropriate Mission Directorate staff or to Center Directors. Decision authority may be delegated to a Center Director for determining whether Category 2 and 3 projects may proceed through KDPs into the next phase of the life cycle. However, the MDAA retains authority for all program-level requirements, funding limits, launch dates, and any external commitments. All delegations are documented and approved in the Program Plan.

4.2.2 Management Councils

4.2.2.1 Program Management Councils

At the Agency level, NASA Headquarters has two levels of program management councils (PMCs): the Agency PMC (APMC) and the Mission Directorate PMCs (DPMCs). The PMCs evaluate the safety, technical, and programmatic performance (including cost, schedule, risk, and risk mitigation) and content of a project under their purview for the entire life cycle. These evaluations focus on whether the project is meeting its commitments to the Agency and on ensuring successful achievement of NASA strategic goals. Table 4-3 shows the governing management councils for projects (by category).

Table 4-3 Relationship Between Projects and PMCs

	Agency PMC	Mission Directorate PMC
Category 1 Projects	●	■
Category 2 Projects		●
Category 3 Projects		●

Legend: ● Governing PMC; ■ PMC evaluation

For all Category 1 projects, the governing PMC is the APMC. The APMC is chaired by the NASA Associate Administrator and consists of Headquarters senior managers and Center Directors. The council members are advisors to the AA in his or her capacity as the PMC Chair and Decision Authority. The APMC is responsible for the following:

- Ensuring that NASA is meeting the commitments specified in the relevant management documents for project performance and mission assurance;

- Ensuring implementation and compliance with NASA program and project management processes and requirements;

- Reviewing projects routinely, including institutional ability to support project commitments;

- Reviewing special and out-of-cycle assessments; and

- Approving the Mission Directorate strategic portfolio and its associated risk.

As the governing PMC for Category 1 projects, the APMC evaluates projects in support of KDPs. For these projects, the KDP normally occurs at the conclusion of an APMC review as depicted in Figure 4-5. The APMC makes a recommendation to the NASA AA (or delegated Decision Authority) on a Category 1 project's readiness to progress in its life cycle and provides an assessment of the project's proposed cost, schedule, and content parameters. The NASA AA (or delegate), as the Decision Authority for Category 1 projects, makes the KDP determination on whether and how the project progresses in its life cycle, and authorizes the key project cost, schedule, and content parameters that govern the remaining life-cycle activities. Decisions are documented in a formal Decision Memorandum, and actions are tracked in a Headquarters tracking system (e.g., the Headquarters Action Tracking System (HATS)). See Sections 4.2.4 and 5.5 for a description of the Decision Memorandum.

A DPMC provides oversight for the MDAA and evaluates all projects executed within that Mission Directorate. For all Category 2 and 3 projects, the DPMC is the governing PMC. The DPMC is usually chaired by the MDAA and is composed of senior Headquarters executives from that Mission Directorate. The MDAA may delegate the chairmanship to one of his or her senior executives. The activities of the DPMC are directed toward periodically (usually monthly) assessing projects' performance and conducting in-depth assessments of projects at critical milestones. The DPMC makes recommendations regarding the following:

- Initiation of new projects based on the results from advanced studies;

- Action on the results of periodic or special reviews, including rebaselining or terminating projects; and

- Transition of ongoing projects from one phase of the project life cycle to the next.

As the governing PMC for Category 2 and 3 projects, the DPMC evaluates projects in support of KDPs. The KDP normally occurs at the conclusion of the DPMC, as depicted in Figure 4-5. The DPMC makes a recommendation to the MDAA (or delegated Decision Authority) on a Category 2 or 3 project's readiness to progress in its life cycle and provides an assessment of the project's proposed cost, schedule, and content parameters. The MDAA (or delegate), as the Decision Authority for Category 2 and 3 projects, makes the KDP determination on whether and how the project progresses in its life cycle, and authorizes the key project cost, schedule, and content parameters that govern the remaining life-cycle activities. The results of the DPMC are documented in a formal Decision Memorandum and include decisions made and actions to be addressed.

The DPMC also evaluates Category 1 projects in support of the review by the APMC and the KDP. For Category 1 projects the MDAA carries forward the DPMC findings and recommendations to the APMC. However, the MDAA may determine in some cases that a Category 1 project is not ready to proceed to the APMC and may direct corrective action.

4.2.2.2 Center Management Council

In addition to the APMC and DPMCs, Centers have a Center Management Council (CMC) that provides oversight and insight for the Center Director (or designee) for all project work executed at that Center. The CMC evaluation focuses on whether Center engineering, Safety and Mission Assurance (SMA), health and medical, and management best practices (e.g., project management, resource management, procurement, institutional) are being followed by the project under review; whether Center resources support project requirements; and whether the project is meeting its approved plans successfully. Centers typically conduct CMCs monthly. The Center Director, as chair of the CMC, or his/her designated chair, may provide direction to the project manager to correct project deficiencies with respect to these areas. However, the Center Director does not provide direction, but only recommendations with respect to programmatic requirements, budgets, and schedules to the project manager, Mission Directorate, or Agency leadership. The CMC also assesses project risk and evaluates the status and progress of activities to identify and report trends and provide guidance to the Agency and affected projects. For example, the CMC may note a trend of increasing risk that potentially indicates a bow wave of accumulating work or may communicate industrial base issues to other programs or projects that might be affected. The Center Director/CMC chair provides the Center's findings and recommendations to project managers, program managers, the DPMC, and the APMC (if applicable), regarding the performance, technical, and management viability of the project prior to KDPs. This includes making

In accordance with *NPR 7120.5*: "Center Directors are responsible and accountable for all activities assigned to their Center. They are responsible for the institutional activities and for ensuring the proper planning for and assuring the proper execution of programs and projects assigned to the Center." This means that the Center Director is responsible for ensuring that projects develop plans that are executable within the guidelines from the Mission Directorate and for assuring that these projects are executed within the approved plans. In cases where the Center Director believes a project cannot be executed within approved guidelines and plans, the Center Director will work with the project and Mission Directorate to resolve the problem. (See Section 5.1.2 for additional information on the Center Directors' responsibilities.)

recommendations to the Decision Authority at KDPs regarding the ability of the project to execute successfully. (Figure 4-5 shows this process.) These recommendations consider all aspects, including safety, technical, programmatic, and major risks and strategy for their mitigation and are supported by independent analyses, when appropriate.

The relationship of the various management councils to each other is shown in Figure 4-6.

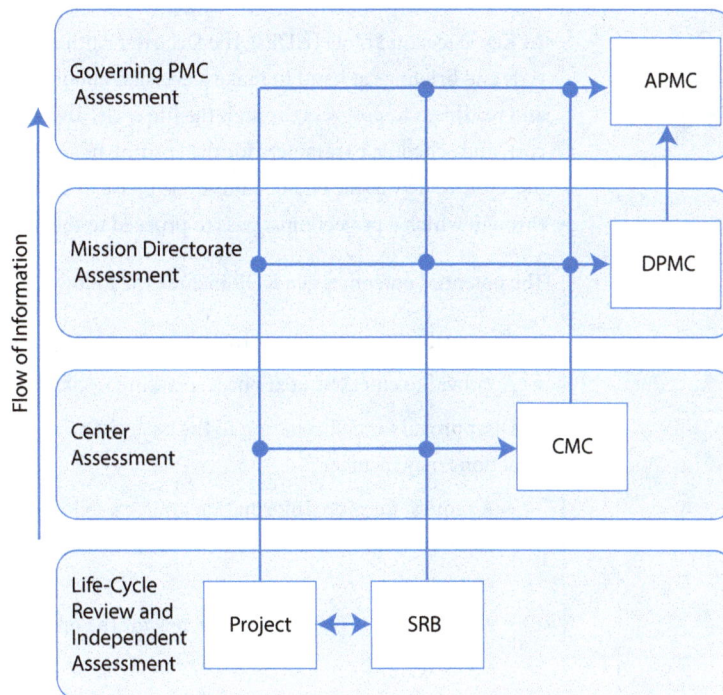

Figure 4-6 Management Council Reviews in Support of KDPs

4.2.2.3 Integrated Center Management Councils

An Integrated Center Management Council (ICMC) may be used for any project conducted by multiple Centers. The ICMC performs the same functions as the CMC but generally includes the Center Director (or representative) from each Center with a substantial project development role. The ICMC is chaired by the Center Director (or representative) of the Center responsible for the project management.

When an ICMC is used to oversee the project, the participating Centers work together to define how the ICMC will operate, when it will meet, who will participate, how decisions will be made, and how Dissenting Opinions will be resolved. (See Section 5.3 on Dissenting Opinion.) In general, final decisions are made by the chair of the ICMC. When a participating Center Director disagrees with a decision made at the ICMC, the standard Dissenting Opinion process is used. As an example, this would generally require that the NASA Chief Engineer resolve disagreements for engineering or project management policy issues.

4.2.3 Key Decision Points

At Key Decision Points (KDPs), the Decision Authority reviews all the materials and briefings at hand to make a decision about the project's maturity and readiness to progress through the life cycle, and authorizes the content, cost, and schedule parameters for the ensuing phase(s). KDPs conclude the life-cycle review at the end of a life-cycle phase. A KDP is a mandatory gate through which a project must pass to proceed to the next life-cycle phase.

The potential outcomes at a KDP include the following:

- Approval to enter the next project phase, with or without actions.
- Approval to enter the next phase, pending resolution of actions.
- Disapproval for continuation to the next phase. In such cases, follow-up actions may include:
  - A request for more information and/or a follow-up review that addresses significant deficiencies identified as part of the life-cycle review;
  - A request for a Termination Review for the project (Phases B, C, D, and E only);
  - Direction to continue in the current phase; or
  - Redirection of the project.

To support a KDP decision process, appropriate KDP readiness products are submitted to the Decision Authority and members of the governing PMC. These materials include the following:

- The project's proposed cost, schedule, safety, and technical plans for their follow-on phases. This includes the proposed preliminary and final project baselines at KDPs B and C, respectively.
- Summary of accepted risks and waivers.
- Project documents or updates signed or ready for signature; for example, the project Formulation Authorization Document (FAD) (see Section 4.3.1.3.1), project Formulation Agreement, Project Plan, Memoranda of Understanding (MOUs), and Memoranda of Agreement (MOAs).

> A life-cycle review is complete when the governing PMC and Decision Authority complete their assessment and sign the Decision Memorandum.

- Summary status of action items from previous KDPs (with the exception of KDP A).

- Draft Decision Memorandum and supporting data. (See Section 4.2.4.)

- The program manager recommendation.

- The project manager recommendation.

- The SRB Final Management Briefing Package.

- The CMC/ICMC recommendation.

- The MDAA recommendation (for Category 1 projects).

- The governing PMC review recommendation.

After reviewing the supporting material and completing discussions with all parties, the Decision Authority determines whether and how the project proceeds and approves any additional actions. These decisions are summarized and recorded in the Decision Memorandum. The Decision Authority completes the KDP process by signing the Decision Memorandum. The expectation is to have the Decision Memorandum signed by concurring members as well as the Decision Authority at the conclusion of the governing PMC KDP meeting. (See more information on the Decision Memorandum, including signatories and their respective responsibilities, in Section 5.5.6.) The Decision Authority archives the KDP documents with the Agency Chief Financial Officer, and the project manager attaches the approved KDP Decision Memorandum to the Formulation Agreement or Project Plan. Any appeals of the Decision Authority's decisions go to the next higher Decision Authority. (See Section 4.3.2.1 for a detailed description of the Formulation Agreement.)

4.2.4 Decision Memorandum, Management Agreement, and Agency Baseline Commitment

The Decision Memorandum is a summary of key decisions made by the Decision Authority at a KDP, or as necessary, in between KDPs. Its purpose is to ensure that major project decisions and their basis are clearly documented and become part of the retrievable records. The Decision Memorandum supports clearly defined roles and responsibilities and a clear line of decision making and reporting documented in the official project documentation.

When the Decision Authority approves the project's entry into the next phase of its life cycle at a KDP, the Decision Memorandum describes this approval, and the key project cost, schedule, and content parameters authorized by the Decision Authority that govern the remaining life-cycle activities. The Decision Memorandum also describes the constraints and

The Management Agreement contained within the Decision Memorandum defines the parameters and authorities over which the project manager has management control. A project manager has the authority to manage within the Management Agreement and is accountable for compliance with the terms of the agreement. The Management Agreement, which is documented at every KDP, may be changed between KDPs as the project matures, with approval from the Decision Authority. The Management Agreement typically is viewed as a contract between the Agency and the project manager and requires renegotiation and acceptance if it changes.

parameters within which the Agency and the project manager operate, i.e., the Management Agreement; the extent to which changes in plans may be made without additional approval, and any additional actions that came out of the KDP.

During Formulation, the Decision Memorandum documents the key parameters, including LCC and schedule, related to work to be accomplished during each phase of Formulation. For projects with a LCC greater than $250 million, this includes a target Life-Cycle Cost (LCC) range and schedule range that the Decision Authority determines is reasonable to accomplish the project. Given the project's lack of maturity during Formulation, this range reflects the broad uncertainties regarding the project's scope, technical approach, safety objectives, acquisition strategy, implementation schedule, and associated costs. The range is also the basis for coordination with the Agency's stakeholders, including the White House and Congress. At KDP B, a more refined LCC range is developed.

During Implementation, the Decision Memorandum documents the parameters for the entire life cycle of the project. Projects transition from Formulation to Implementation at KDP C. At this point, the approved Life-Cycle Cost Estimate (LCCE) of the project is no longer documented as a range but instead as a single number. The LCCE includes all costs, including all Unallocated Future Expenses (UFE) and funded schedule margins for development through prime mission operations (the mission operations as defined to accomplish the prime mission objectives) to disposal, excluding extended operations.[9]

The prime mission is approved for operations at KDP E. This mission has a defined operations span, but in many cases, the mission can be extended beyond the currently approved operational span. During the prime mission phase, the Mission Directorate may initiate consideration for approval for an extended mission:

● Generally for science missions, the Mission Directorate solicits a proposal from the project and establishes a process for proposal evaluation. This process usually includes submitting the proposal to a science theme-specific Senior Review, a peer review panel, for evaluation of the merits of the proposal. The Mission Directorate can accept, modify, or reject the proposal, and can establish new budget authority for operating in the extended phase.

Unallocated Future Expenses (UFE) are the portion of estimated cost required to meet the specified confidence level that cannot yet be allocated to the specific WBS sub-elements because the estimate includes probabilistic risks and specific needs that are not known until these risks are realized. (For projects that are not required to perform probabilistic analysis, the UFE should be informed by the projects unique risk posture in accordance with Mission Directorate and Center guidance and requirements. The rationale for the UFE, if not conducted via a probabilistic analysis, should be appropriately documented and be traceable, repeatable, and defendable.) UFE may be held at the project level, program level, and Mission Directorate level.

---

[9] Projects that are part of tightly coupled and single-project programs document their life-cycle cost estimate in accordance with the life-cycle scope defined in their program's Program Plan, PCA or FAD, or the project's FAD and other parameters in their Decision Memorandum and ABC at KDP C.

- For human space flight (HSF) missions, the Mission Directorate asks the program office to develop a proposal for extending the mission. The Mission Directorate evaluates the proposal and work with Agency senior management to determine the viability and cost of the extension. Extending HSF missions generally requires close coordination with the Agency stakeholders and approval of funding by Congress.

The project baseline, called the Agency Baseline Commitment (ABC), is established at approval for Implementation, KDP C. The ABC and other key parameters are documented in the Decision Memorandum.

4.2.5 Management Forum—Baseline Performance Review

NASA's Baseline Performance Review (BPR) serves as NASA's monthly, internal senior performance management review, integrating Agency-wide communication of performance metrics, analysis, and independent assessment for both mission and mission support projects and activities. While not a council, the Baseline Performance Review (BPR) is closely linked with the councils and integral to council operations. As an integrated review of institutional and project activities, the BPR highlights interrelated issues that impact performance and project risk enabling senior management to quickly address issues, including referral to the governing councils for decision, if needed. The BPR forum fosters communication across organizational boundaries to identify systemic issues and address mutual concerns and risks. The BPR is the culmination of all of the Agency's regular business rhythm performance monitoring activities, providing ongoing performance assessment between KDPs. The BPR is also used to meet requirements for quarterly progress reviews contained in the Government Performance Reporting and Accountability Modernization Act of 2010 (GPRAMA) and OMB Circular A-11 Section 6.[10]

The NASA Associate Administrator and Associate Deputy Administrator co-chair the BPR. Membership includes Agency senior management and Center Directors. The Office of the Chief Engineer (OCE) leads the project performance assessment process conducted by a team of independent assessors drawn from OCE, the Office of the Chief Financial Officer (OCFO), and the Office of Safety and Mission Assurance (OSMA).

A typical BPR agenda includes an assessment of each Mission Directorates' project performance against Management Agreements and ABCs, with

The Agency Baseline Commitment (ABC) is an integrated set of project requirements, cost, schedule, technical content, and JCL when applicable. The ABC cost is equal to the project LCC approved by the Agency at approval for Implementation. The ABC is the baseline against which the Agency's performance is measured during the Implementation Phase of a project. Only one official baseline exists for a project, and it is the ABC. The ABC for projects with a LCC greater than $250 million forms the basis for the Agency's external commitment to the Office of Management and Budget (OMB) and Congress and serves as the basis by which external stakeholders measure NASA's performance for these projects. Changes to the ABC are controlled through a formal approval process. See Section 5.5 for a detailed description of maturing, approving and maintaining project plans, LCCs, baselines, and commitments.

---

[10] Additional information on GPRAMA can be found at http://www.gpo.gov/fdsys/pkg/PLAW-111publ352/pdf/PLAW-111publ352.pdf. Additional information on A-11 Section 6 can be found at http://www.whitehouse.gov/sites/default/files/omb/assets/a11_current_year/s200.pdf).

rotating in-depth reviews of specific mission areas. The schedule ensures that each mission area is reviewed on a quarterly basis. Mission support functions are included in the BPR. Assessors use existing materials when possible. Table 4-4 shows typical information sources that may be used by the BPR assessors. Different emphasis may be placed on different sources depending on which mission is being assessed.

Table 4-4 Typical Information Sources Used for BPR Assessment

	Info Sources for BPR Assessments
Program/Project Documents	FAD, Formulation Agreement, and Project Plans
Reviews	Life-cycle reviews
	Monthly, quarterly, midyear, and end-of-year Mission Directorate reviews
	Other special reviews (see Section 4.1.3)
	Monthly Center status reviews
Meetings	APMC (presentations and decision memorandums)
	DPMC (presentations and decision memorandums)
	Recurring staff/status meetings including project monthly status
	Project Control Board (meetings and weekly status reports)
	Biweekly tag-ups with the SMA TAs supporting and overseeing the project.
Reports	Reports from Agency assessment studies (CAD, IPAO, etc.)
	PPBE presentations
	Quarterly cost and schedule reports on major projects delivered to OCFO
	Center summaries presentations at BPR
	Weekly Mission Directorate report
	Weekly project reports
	Weekly reports from the NESC
	Monthly EVM data
	Project anomaly reports
	Center SMA reports
	Technical Authority reports
Databases	N2 budget database
	SAP and Business Warehouse financial databases
	OMB/Congressional cost/schedule data

4.3 Project Formulation

The following paragraphs explain the project activities chronologically by phase.

NASA places significant emphasis on project Formulation (including activities leading to the start of Formulation) to ensure adequate preparation of project concepts and plans and mitigation of high-risk aspects of the project essential to position the project for the highest probability of mission success.

In practice, the activities described for each phase below are not always carried out exclusively in that phase; their timing depends on the particular schedule requirements of the project. For example, some projects procure long-lead flight hardware in Phase B to enable them to achieve their launch dates.

4.3.1 Concept Studies (Pre–Phase A) Activities

4.3.1.1 Project Activities Leading to the Start of Formulation (Pre–Phase A)

The process for initiating projects begins at the senior NASA management level with the strategic acquisition process. This process enables NASA management to consider the full spectrum of acquisition approaches for its projects from commercial off-the-shelf buys to total in-house design and build efforts where NASA has a unique capability and capacity or the need to maintain or develop such capability and capacity. The Agency goes through this "make or buy" decision on whether to acquire the capability in-house, acquire it from outside the Agency, or acquire it by a combination of the two. Strategic acquisition is used to promote best-value approaches (taking into account the Agency as a whole), encourage innovation and efficiency, and take advantage of state-of-the-art solutions available within NASA and from industry, academia, other Federal agencies, and international partners.

Many processes support acquisition, including the program and project management system, the budget process, and the procurement system. The NASA Planning, Programming, Budgeting, and Execution (PPBE) process supports allocating the resources of programs to projects through the Agency's annual budgeting process. (See *NPR 9420.1, Budget Formulation* and *NPR 9470.1, Budget Execution*.) The NASA procurement system supports the acquisition of assets and services from external sources. (See NPD 1000.5, the Federal Acquisition Regulation (FAR), and the NASA FAR Supplement (NFS) for NASA's specific implementation of the FAR.)

The strategic acquisition process is the Agency process for ensuring that NASA's strategic vision, programs, projects, and resources are properly developed and aligned throughout the mission and life cycle. (See *NPD 1000.0, NASA Governance and Strategic Management Handbook* and *NPD 1000.5, Policy for NASA Acquisition* for additional information on the strategic acquisition process.)

4.3.1.2 Project Pre–Phase A Life-Cycle Activities

An MDAA has the authority to begin project pre-Formulation activities. Prior to initiating a new project, a Mission Directorate, typically supported by a program office, provides resources for concept studies (i.e., Pre–Phase A Concept Studies) along with the mission objectives, ground rules, and assumptions to be used by the study team. While not formally a part of Formulation, some formulation-type activities naturally occur as part of earlier advanced studies. These pre-Formulation activities involve Design Reference Mission (DRM) analysis, feasibility studies, technology needs analyses, engineering systems assessments, human systems assessments, and analyses of alternatives that need to be performed before a specific project concept emerges. These trade studies are not considered part of formal project planning since there is no certainty that a specific project proposal will emerge. Pre-Formulation activities also involve identification of risks that are likely to drive the project's cost and schedule estimates, or cost and schedule range estimates (projects with an LCC greater than $250 million), at KDP B and cost and schedule commitments at KDP C and include development of mitigation plans for those risks.

During Pre–Phase A, a pre-project team studies a broad range of mission concepts that contribute to program and Mission Directorate goals and objectives. These advance studies, along with interactions with customers and other potential stakeholders, help the team to identify promising mission concept(s) and to draft project-level requirements. The Mission Directorate uses the results of this work to determine if the mission concepts warrant continued development. A major focus of Pre–Phase A is to conduct technology and engineering system assessments to identify risks that are likely to drive the project's cost and schedule estimates, or cost and schedule range estimates (projects with an LCC greater than $250 million), at KDP B. The team identifies potential technology needs (based on the best mission concepts) and assesses the gaps between the needed technology and current or planned technology, the technology readiness levels (TRLs) (see NPR 7123.1B, Appendix E for TRL definitions) and the technology risks. The team also identifies risks in engineering development, payload, supply chain, and heritage hardware and software. The team defines risk mitigation plans and resource requirements for the top risks. These activities are focused toward the Mission Concept Review (MCR) and KDP A. These activities also inform development of the Formulation Agreement in response to the Formulation Authorization Document (FAD) (see Section 4.3.1.3) generated by the Mission Directorate to authorize formulation of the mission. (See Section 4.3.2.1 for a detailed description of the Formulation Agreement.) At the conclusion of pre-Formulation, a FAD is issued (see NPR 7120.5, Appendix E authorizing Formulation to begin, and a Formulation Agreement is developed and approved to document the plans and resources required for Formulation.

The following paragraphs describe the activities a project needs to accomplish to develop one or more sound concepts, conduct a successful Mission Concept Review (MCR), and get approval at KDP A to enter project Formulation. The MCR is the first major life-cycle review in a project life cycle. The **purpose of the MCR** is to evaluate the feasibility of the proposed mission concept(s) and how well the concept(s) fulfill the project's needs and objectives. After the MCR, the project proceeds to KDP A where the project demonstrates that it has addressed critical NASA needs; the proposed mission concept(s) is feasible; the associated planning is sufficiently mature to begin Phase A; and the mission can probably be achieved as conceived.

The general flow of activities for a project in pre-Formulation is shown in Figure 4-7.

4.3.1.3 Project Pre–Phase A Management, Planning, and Control Activities

4.3.1.3.1 Supporting Headquarters Planning

Once the Mission Directorate decides to begin pre-Formulation, the project manager and project team (designated as the pre-project manager and pre-project team until the project is formalized) support the Mission Directorate in developing the concept for the project. When requested, the team helps identify the main stakeholders of the project (e.g., Principal Investigator, science community, technology community, public, education community, Mission Directorate sponsor) and gather and document key external stakeholder expectations, needs, goals, and objectives. The project team supports the program manager and the MDAA in the development of the preliminary program requirements, constraints, ground rules and assumptions on the project and stakeholder expectations, including preliminary mission objectives/goals and mission success criteria. The project also supports the program manager and the MDAA in ensuring alignment of the project requirements with the Program Plan and applicable Agency strategic goals. These requirements are eventually put into an appendix of the Program Plan. The MDAA uses this information in developing and obtaining approval of the FAD. The project also develops the process to be used within the project to ensure stakeholder advocacy.

One of the first activities is to select the management team. The project managers for Category 1 projects are recommended by the Center Director with approval for appointment by the MDAA. For Category 2 and 3 projects, the Center Director appoints the project manager with concurrence

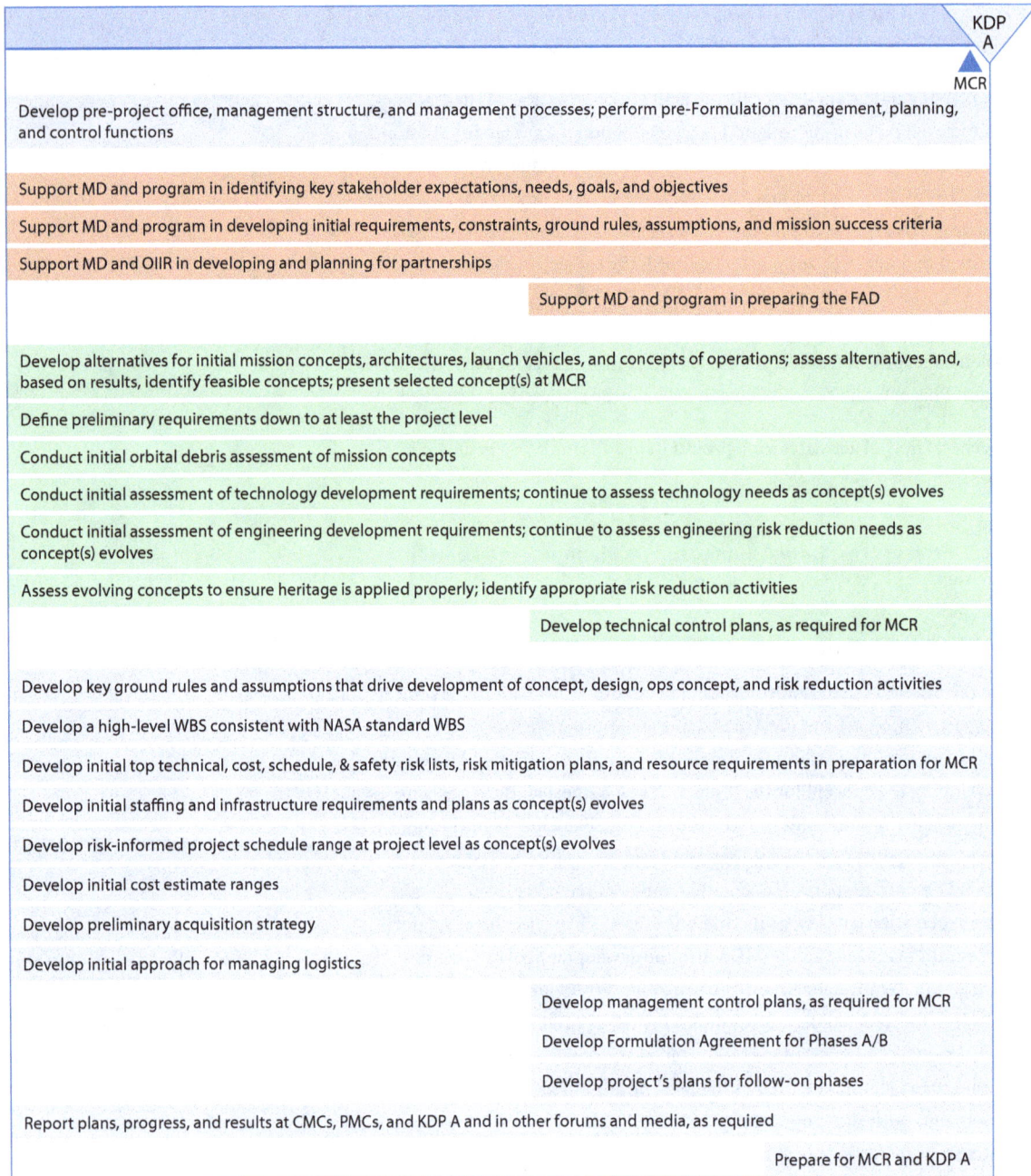

KDP A

MCR

Develop pre-project office, management structure, and management processes; perform pre-Formulation management, planning, and control functions

Support MD and program in identifying key stakeholder expectations, needs, goals, and objectives

Support MD and program in developing initial requirements, constraints, ground rules, assumptions, and mission success criteria

Support MD and OIIR in developing and planning for partnerships

Support MD and program in preparing the FAD

Develop alternatives for initial mission concepts, architectures, launch vehicles, and concepts of operations; assess alternatives and, based on results, identify feasible concepts; present selected concept(s) at MCR

Define preliminary requirements down to at least the project level

Conduct initial orbital debris assessment of mission concepts

Conduct initial assessment of technology development requirements; continue to assess technology needs as concept(s) evolves

Conduct initial assessment of engineering development requirements; continue to assess engineering risk reduction needs as concept(s) evolves

Assess evolving concepts to ensure heritage is applied properly; identify appropriate risk reduction activities

Develop technical control plans, as required for MCR

Develop key ground rules and assumptions that drive development of concept, design, ops concept, and risk reduction activities

Develop a high-level WBS consistent with NASA standard WBS

Develop initial top technical, cost, schedule, & safety risk lists, risk mitigation plans, and resource requirements in preparation for MCR

Develop initial staffing and infrastructure requirements and plans as concept(s) evolves

Develop risk-informed project schedule range at project level as concept(s) evolves

Develop initial cost estimate ranges

Develop preliminary acquisition strategy

Develop initial approach for managing logistics

Develop management control plans, as required for MCR

Develop Formulation Agreement for Phases A/B

Develop project's plans for follow-on phases

Report plans, progress, and results at CMCs, PMCs, and KDP A and in other forums and media, as required

Prepare for MCR and KDP A

Legend: ☐ Project management, planning, and control tasks

☐ Work for which Headquarters is responsible but the project helps accomplish (e.g., international partnerships are a Headquarters responsibility, but the projects help develop and finalize those partnerships)

☐ Technical work the project is doing

Acronyms: MD = Mission Directorate; OIIR = Office of International and Interagency Relations.

Note: These are typical high-level activities that occur during this project phase. Placement of reviews is notional.

Figure 4-7 Project Pre–Phase A Flow of Activities

from the program manager. The MDAA issues the Formulation Authorization Document (FAD) to authorize the formulation of a project whose goals fulfill part of the Agency's Strategic Plan and Mission Directorate strategies. The FAD describes the purpose of the project, including a clear traceability from the goals and objectives in the Mission Directorate strategies and/or Program Plan. It describes the level or scope of work, and the goals and objectives to be accomplished in the Formulation Phase. It also describes the structure for managing the Formulation process from the MDAA to the NASA Center program or project managers, as applicable, and includes lines of authority, coordination, and reporting. It identifies Mission Directorates, Mission Support Offices, and Centers to be involved in the activity, their scope of work, and any known constraints related to their efforts (e.g., the project is cofunded by a different Mission Directorate). It identifies any known participation by organizations external to NASA, their scope of work, and any known constraints related to their efforts. It identifies the funding to be committed to the project during each year of Formulation. Finally, it specifies the project life-cycle reviews planned during the Formulation Phase.

4.3.1.3.2 Initial Project Structure and Management Framework

The project team works with the Center to develop and implement an initial management framework, including the project team, organizational structure, and initial management processes consistent with the direction from the MDAA and program identifying the roles and responsibilities of each organization (e.g., Headquarters, Centers, other Government agencies, academia, industry, and international partners).

The project team supports the MDAA and the NASA Headquarters Office of International and Interagency Relations (OIIR) in identifying and planning for any preliminary **partnerships and interagency and international agreements**[11] as they are known at the time.

4.3.1.3.3 Management Control Processes and Products

The project team conducts planning that enables formulation and implementation of the mission concept(s), architectures, scenarios/DRMs and requirements. The results of this planning, much of which is described below, supports the MCR and KDP A by demonstrating how the project plans to implement the mission concept(s) being proposed.

As the project team develops its planning, management processes are documented in control plans, which are designed to keep the project activities

---

[11] Bolding indicates a required product.

aligned, on track, and accounted for as the project moves forward. These control plans are described in this and subsequent sections of this handbook, in conjunction with the phase where they are required. Many control plans are incorporated into the central planning document, which is the Project Plan. NPR 7120.5E, Appendix H, identifies when a control plan may be included in the Project Plan, and when a control plan is required to be a stand-alone document. NPR 7120.5E, Appendix I, Table I-5 and Table 4-7 at the end of this chapter identify when these control plans are required. Centers may have existing plans, which projects may use to satisfy requirements for some of the control plans.

The project supports the MDAA and the program in the development of driving mission, technical, and programmatic ground rules and assumptions. The project also responds to the FAD and assists the program manager as necessary to prepare the FAD for baselining at the MCR/KDP A.

The project team develops a high-level WBS that is consistent with the NASA standard space flight project WBS. (See Section 5.9.)

As the concepts mature and for each concept being considered, the team iteratively performs an assessment of potential infrastructure and workforce needs, as well as opportunities to use that infrastructure and workforce in other Government agencies, industry, academia, and international organizations. Based on this assessment, the project team develops the initial **requirements and plans for staffing and infrastructure**.

Additionally, the team develops the preliminary **strategy for acquisition**, including:

- A preliminary assessment of supply chain risks, including potential critical or single-source suppliers needed to design, develop, produce, and support required capabilities at planned cost and schedule;
- An approach for managing logistics;
- Plans for in-house work versus procurements, including major proposed procurements, types of procurements, and "no later than" procurement schedules; and
- Preliminary plans for partners (i.e., other Government agencies, domestic and international), their roles and anticipated contributions, and plans for obtaining commitments for these contributions.

Consistent with the technical team's work, the project develops the initial **top safety, technical, cost, and schedule risks**, including technology development, engineering development, payload (robotic spaceflight), and procurement risks; risks associated with the use of heritage hardware and software; and risks that are likely to drive the project's cost and schedule estimates, or cost and schedule range estimates (projects with an LCC greater than $250 million), at KDP B. The project identifies the initial risk mitigation plans and associated resources and the approach for managing risks during Phase A. This activity forms the foundation for the Risk Management Plan.

Based on the concepts that are to be carried forward, the project team develops a risk-informed schedule at the project level (as a minimum) with a preliminary date, or a preliminary range for Phase D completion. In addition, the team develops project cost and schedule estimates or cost and schedule range estimates covering Phase A (excluding Pre–Phase A) through completion of Phase D. These cost and schedule estimates typically are informed by technology needs; engineering development and heritage assessments using the *Systems Engineering Handbook, NASA/SP-2007-6105 Rev 1*, Appendix G; acquisition strategies; infrastructure and workforce requirements; and need to accommodate resolution of identified risks. The project typically also identifies the initial phased life-cycle cost and schedule estimates, or cost and schedule range estimates (Phase A through Phase F, excluding any extended operations). These estimates need to be consistent with the preliminary Phase D completion estimate. The project documents the basis for initial cost and schedule estimates and develops the initial approach for managing schedule and cost during Phase A. This is the first effort in developing the Technical, Schedule, and Cost Control Plan, which eventually becomes part of the Project Plan.

The project develops an approach for knowledge management and managing the identification and documentation of lessons learned during Phase A. This includes the project's knowledge management strategy; how the project will take advantage of lessons learned identified by others; and how the project will continuously capture lessons learned during Formulation and Implementation. This approach evolves to a formal Knowledge Management Plan that is one of the Control Plans in the Project Plan.

4.3.1.4 Project Pre–Phase A Technical Activities and Products

The project team performs the technical activities required in NPR 7123.1 for this phase, starting with gathering key internal stakeholder expectations, needs, goals, and objectives. Based on these and the program-level requirements, constraints, ground rules, and assumptions, the project begins to develop concepts and architectures that satisfy these expectations. This process usually considers a number of alternative approaches to both the architecture and the Operations Concept, and the project develops candidate (preliminary) mission, spacecraft, and ground systems architectures. The architecture includes how the major project components (hardware,

The Operations Concept is a description of how the flight system and the ground system are used together to ensure that the mission operations can be accomplished reasonably. This might include how mission data of interest, such as engineering or scientific data, are captured, returned to Earth, processed, made available to users, and archived for future reference. The Operations Concept typically describes how the flight system and ground system work together across mission phases for launch, cruise, critical activities, science observations, and the end of the mission to achieve the mission. The Operations Concept is baselined at PDR with the initial preliminary operations concept required at MCR.

software, human systems) will be integrated and are intended to operate together and with heritage systems, as applicable, to achieve project goals and objectives. By implication, the architecture defines the system-level processes necessary for development, production, human systems integration, verification, deployment, operations, support, disposal, and training. The architecture also includes facilities, logistics concepts, and planned mission results and data analysis, archiving, and reporting. The operations concept includes all activities such as integration and test, launch integration, launch, deployment and on-orbit checkout (robotic projects) or initial operations (human space flight projects), in-space operations, landing and recovery (if applicable), and decommissioning and disposal.

If the architecture and operations concept require a launch service, the project will begin to work with the NASA Launch Services Program (LSP) at KSC to develop and assess the mission's launch options. (Launch options can include any methods specified in *NPD 8610.12, Human Exploration and Operation Mission Directorate (HEOMD) Space Transportation Services for NASA and NASA-Sponsored Payloads*; however, most missions use a launch service procured and managed by the LSP to facilitate the application of the launch services risk mitigation and technical oversight policies as described in *NPD 8610.7, Launch Services Risk Mitigation Policy for NASA-Owned and/or NASA-Sponsored Payloads/Missions* and *NPD 8610.23, Launch Vehicle Technical Oversight Policy*.) LSP evaluates the project's spacecraft needs and pairs the requirements of the project with an appropriate launch service. Early interaction and involvement helps to ensure that the potential viable launch options are encompassed and accommodated in the spacecraft design and test plans. LSP acquires the launch service through a competitive process whenever possible, awarding based on best value to the government. The project is typically part of the proposal evaluation team. The project funds LSP's acquisition efforts required to perform preliminary studies (if necessary) and ultimately to procure the launch service. LSP provides the launch service management, as well as mission assurance activities, payload launch site processing services, payload integration activities and launch phase telemetry and command services. LSP works diligently to ensure mission success, providing technical guidance through the entire process from the pre-mission planning to the post-launch phase of the project's spacecraft. The interaction with LSP will also include coordination with the project's Mission Directorate, e.g., SMD, and the Human Exploration and Operations Mission Directorate (HEOMD), which oversees the LSP. Figure 4-8 shows the interaction of the project and the LSP throughout the project's life cycle and illustrates the end-to-end support that LSP provides, beginning years before the spacecraft is created, until well after the spacecraft is launched.

Spacecraft Life-Cycle Phases	**Pre– Phase A** Concept Studies	**Phase A** Concept & Technology Development	**Phase B** Preliminary Design & Technology Completion	**Phase C** Final Design & Fabrication	**Phase D** System Assembly, Integration & Test, Launch & Checkout	**Phase E** Operations & Sustainment
LSP Mission Life-Cycle Phases	Pre-Mission Planning	Mission Planning	Baseline Mission & Procure Launch Services	Launch Vehicle & Spacecraft Engineering & Manufacturing	Launch Site Operations / Launch	Post-Launch
Activity Timeframe	Launch minus (L–) 4 to 10 yrs	L – 3 to 4 yrs	L – 2 to 3 yrs	L – 3 yrs to 3 mos	L – 3 mos to 10 days / L – 10 days	L + 3 mos

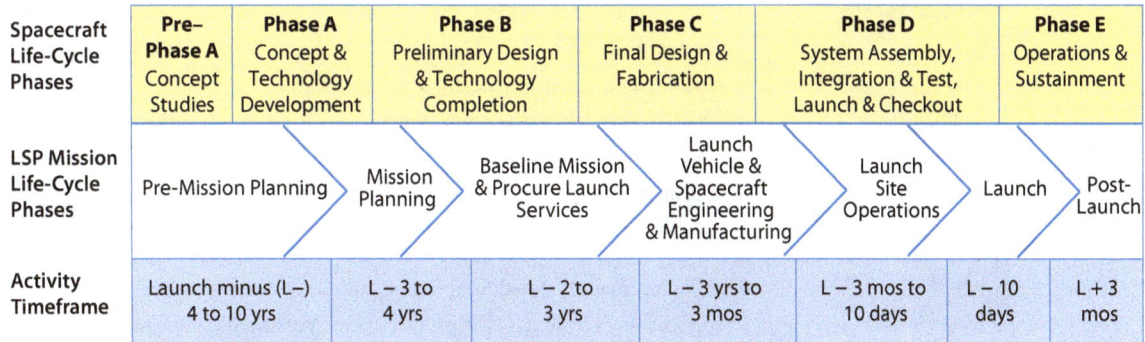

Figure 4-8 Mission Life Cycle for Project/LSP Interaction

In addition, the project develops a preliminary assessment of orbital debris per *NPR 8715.6, NASA Procedural Requirements for Limiting Orbital Debris* and identifies the planned orbital lifetime, any potential nonconformance to orbital debris requirements for planned intentional breakups, reentry of major components that potentially could reach the surface, and the use of tethers. Any deviations are submitted to the Chief SMA for approval prior to the Acquisition Strategy Meeting (ASM).

In analyzing the Operations Concept, the project develops the preliminary approach to V&V; system integration; and human rating, if applicable. Identifying these at this point enables the project to assess unique workforce and infrastructure needs early enough to include the requirements for these in the initial concept(s).

As the Pre–Phase A work approaches the MCR, the project develops and documents at least one feasible preliminary concept (included as part of concept documentation in NPR 7120.5, Table I-4, and Table 4-6 at the end of this chapter), including the key preliminary ground rules and assumptions that drive the concept(s) and the operations concept. A feasible concept is one that is probably achievable technically within the cost and schedule resources allocated by the program in the project's FAD. This preliminary concept includes key drivers, preliminary estimates of technical margins for candidate architectures, and a preliminary Master Equipment List (MEL). This concept is sometimes referred to as the mission concept, particularly in the robotic community. As a minimum, the principal concept will be approved following the MCR and KDP A. Future changes to this concept (and others, if approved for further study) will be identified at each follow-on life-cycle review and KDP so that management understands how the concept is evolving as the formulation process progresses.

The term "concept documentation" used in NPR 7120.5 is the documentation that captures and communicates a feasible concept at MCR that meets the goals and objectives of the mission, including results of analyses of alternative concepts, the concept of operations (baselined at MCR per NPR 7123.1), preliminary risks, and potential descopes. (Descope is a particular kind of risk mitigation that addresses risks early in the project Formulation Phase.)

The Master Equipment List (MEL) summarizes all major components of each flight element subsystem and each instrument element component. For each major component, current best estimates and contingency allocation for mass and power (including for individual components), number of flight units required, and some description of the heritage basis is included. Power values generally represent nominal steady-state operational power requirements. Information includes identification of planned spares and prototypes, required deliveries/exchanges of simulators for testing, and other component description/characteristics. Certain items (like electronic boxes and solar arrays) usually include additional details, as applicable, to identify and separate individual elements. The MEL is useful to program and project managers for understanding where the design is, where the mass is being carried, what the power needs are, what the margins are, and other parameters as the project progresses in development.

Based on the leading concept, the project develops the initial recommendations for mission objectives and requirements and preliminary project-level requirements and typically develops a mission or science traceability matrix that shows how the requirements flow from the objectives of the mission through the operational requirements (such as science measurement requirements) to the top-level infrastructure implementation requirements (such as orbit characteristics and pointing stability).

Each requirement is stated in objective, quantifiable, and verifiable terms. Requirements can identify the project's principal schedule milestones, including Preliminary Design Review (PDR), Critical Design Review (CDR), launch, mission operations critical milestones, and the planned decommissioning date. They can state the development and/or total Life-Cycle Cost (LCC) constraints on the project and set forth any budget constraints by fiscal year. They can state the specific conditions under which a Termination Review would be triggered. They can also describe any additional requirements on the project; e.g., international partners. If the mission characteristics indicate a greater emphasis is necessary on maintaining technical, cost, or schedule, then the requirements can identify which is most important; e.g., state if the mission is cost-capped; or if schedule is paramount, as for a planetary mission; or if it is critical to accomplish the technical objectives, as for a technology demonstration mission.

For each known project, the program team develops an appendix to the Program Plan or a separate document that includes a top-level description of the project, including the mission's science or exploration objectives; the project's category, governing PMC, and risk classification; and the project's mission, performance, and safety requirements. For science missions, it includes both baseline and threshold science requirements (see Appendix A for definitions) and identifies the mission success criteria for each project based on the threshold science requirements.

At this point, with guidance from its stakeholders, the project begins to select technical standards for use as project requirements in accordance with *NPR 7120.10, Technical Standards for NASA Programs and Projects*.[12] Based on currency and applicability, technical standards required by law and those mandated by NPDs and NPRs are selected first. When all other factors are the same, NASA promotes the use of voluntary consensus standards over NASA and other Government agency technical standards when they meet or can be tailored to meet NASA's needs.

---

[12] *NASA STD 8709.20, Management of Safety and Mission Assurance Technical Authority (SMA TA) Requirements* provides further information on selecting SMA standards.

During Pre–Phase A, the project develops multiple assessments and products, described below, that may be documented in the project's Formulation Agreement, as opposed to developing separate plans. See Section 4.3.2.1 for a detailed description of the Formulation Agreement.

For each of the candidate concepts that will be carried forward into Phase A, the project develops an initial assessment of potential technology needs and their current technology readiness level, as well as potential opportunities to use commercial, academic, and other Government agency sources of technology. The project team develops and baselines the Technology Development Plan[13] so that the needed technology development can be initiated once formal Formulation starts after KDP A. This plan describes the technology assessment, development, management, and acquisition strategies needed to achieve the project's mission objectives; describes how the project will transition technologies from the development stage to the manufacturing and production phases; identifies the supply chain needed to manufacture the technology and any costs and risks associated with the transition to the manufacturing and production phases; develops and documents appropriate mitigation plans for the identified risks; and describes the project's strategy for ensuring that there are alternative development paths available if and when technologies do not mature as expected.

In addition, the project develops an initial assessment of engineering development needs, including defining the need for engineering prototypes and models for the higher risk components and assemblies that have not been previously built or flown in the planned environment and testing them to demonstrate adequate performance. As with technology development, identification at this point will enable the project to plan and initiate engineering development activities early in Formulation knowing that the funding has been planned for these activities.

For concepts and architectures that plan to use heritage systems, using the *Systems Engineering Handbook, NASA/SP-2007-6105 Rev 1*, Appendix G, the project develops an initial assessment of heritage hardware and software systems that may be utilized outside of environments and configurations for which they were originally designed and used.

All of these activities help the project develop an initial assessment of preliminary technical risks for candidate architectures, including engineering development risks.

---

[13] At this point in its development, the Technology Development Plan may be part of the Formulation Agreement.

The project team develops a preliminary Systems Engineering Management Plan (SEMP) prior to MCR. The SEMP summarizes the key systems engineering elements and enables the project to initiate system engineering activities once formulation has been started following KDP A. It includes descriptions of the project's overall approach for systems engineering to include system design and product realization processes (implementation and/or integration, V&V, and transition), as well as the technical management processes.

If applicable in accordance with NPR 7123.1B, the project develops a preliminary Human Systems Integration (HSI) Plan. Human systems integration is an interdisciplinary and comprehensive management and technical process that focuses on the integration of human considerations into the system acquisition and development processes to enhance human system design, reduce life-cycle ownership cost, and optimize total system performance. Human system domain design activities associated with manpower, personnel, training, human factors engineering, safety, health, habitability, and survivability are considered concurrently and integrated with all other systems engineering design activities.

The project also develops the preliminary review plan and identifies preliminary plans, if any, for combining life-cycle reviews in future life-cycle phases.

4.3.2 Completing Pre–Phase A (Concept Studies) Activities and Preparing for Phase A (Concept and Technology Development)

4.3.2.1 Finalizing Plans for Phase A

As the project FAD is being developed at Headquarters, the project concurrently begins to develop its project Formulation Agreement. (See "Formulation Agreement" box for additional information.)

In preparation for completing the Pre–Phase A activities, the project documents the results of its efforts in this period. The project team generates the documentation specified in NPR 7123.1 and the product Tables I-4 and I-5 in NPR 7120.5E and Tables 4-6 and 4-7 at the end of this chapter. Most of these documents have been described above. Inclusion of information in the Formulation Agreement, the basis of cost and schedule estimates, draft and preliminary versions of project documents and plans, and/or the Mission Concept Review (MCR) briefing package may satisfy some of the documentation.

Formulation Agreement

The Formulation Agreement serves as a tool for communicating and negotiating the project's schedule and funding requirements during Phase A and Phase B with the Mission Directorate. It identifies and prioritizes the technical and acquisition activities that will have the most value during Formulation and informs follow-on plans. The Formulation Agreement focuses on the work necessary to accurately characterize the complexity and scope of the project; increase understanding of requirements; identify and mitigate safety, technical, cost, and schedule risks, and develop high quality cost and schedule estimates. (For projects with a LCC greater than $250 million, this work enables the project to develop high-fidelity cost and schedule range estimates and associated confidence levels at KDP B, and high-fidelity cost and schedule commitments and associated JCL at KDP C, and to commit to a successful plan for Implementation at KDP C.) These activities include establishing the internal management control functions that will be used throughout the life of the project. The Agreement is approved and signed at KDP A (baselined for Phase A and preliminary for Phase B). The Agreement is updated in preparation for the System Definition Review (SDR)/Mission Definition Review (MDR) and resubmitted for signature at KDP B (baselined for Phase B). The Formulation Agreement for KDP A includes detailed Phase A information, preliminary Phase B information, and the Formulation Cost, which is based on the estimated costs for Phase A and Phase B. The Formulation Agreement for KDP B identifies the progress made during Phase A, updates and details Phase B information, and updates the Formulation Cost, which is based on the actual cost for Phase A and an updated cost for Phase B. The Formulation Cost at KDP B is the total authorized cost for Formulation activities required to get to KDP C. In practice, the FAD and the Formulation Agreement are developed concurrently so that both documents can be approved at KDP A. Documentation products developed as part of, or as a result of, the Formulation Agreement may be incorporated into the Project Plan, if appropriate, as the Project Plan is developed during Formulation.

4.3.2.2 Project Pre–Phase A Reporting Activities and Preparing for Major Milestones

4.3.2.2.1 Project Reporting

The project reports to the Center, as requested by the Center, to enable the Center Director to evaluate whether engineering, SMA, health and medical, and management best practices (e.g., project management, resource management, procurement, and institutional best practices) are being followed, and whether Center resources support project requirements. The project also provides project risks and the status and progress of activities so the Center can identify and report trends and provide guidance to the Agency and affected programs and projects. The CMC (or equivalent) provides its findings and recommendations to project managers and to the appropriate PMCs regarding the performance and technical and management viability of the project prior to KDPs.

Aside from the Center and Agency reporting already mentioned, many stakeholders are interested in the status of the project from Congress on down. The project manager supports the program executive in reporting the status of project Formulation at many other forums, including Mission Directorate monthly status meetings and the Agency's monthly BPR. See Section 4.2.5 for more information on BPRs and Section 5.12 for more information on external reporting.

4.3.2.2.2 Project Internal Reviews

Prior to life-cycle reviews, projects conduct internal reviews in accordance with NPR 7123.1, Center practices, and NPR 7120.5. These internal reviews are the decisional meetings wherein the projects solidify their plans, technical approaches, and programmatic commitments. This is accomplished as part of the normal systems engineering work processes as defined in NPR 7123.1 wherein major technical and programmatic requirements are assessed along with the system design and other implementation plans. For both robotic and human space flight projects, these internal reviews are typically lower level system and subsystem reviews that lead to and precede the life-cycle review. Major technical and programmatic performance metrics are reported and assessed against predictions.

Non-SRB project technical reviews are divided into several categories: major systems reviews (one or two levels down from the project), Engineering Peer Reviews (EPRs), internal reviews, and tabletop reviews. Project systems reviews are major technical milestones of the project that typically precede the life-cycle review, covering major systems milestones. The technical progress of the project is assessed at key milestones such as these systems reviews to ensure that the project's maturity is progressing as required. In many cases, these reviews are conducted by the project in coordination with a Center-sponsored independent review panel if the Center is using these reviews as one means to oversee the project's work. In these cases, the project manager works with the Center to ensure that there is a suitable independent review panel in place for each such review and works with systems engineering to ensure that clear technical criteria and an agreed agenda have been established well in advance of each such review.

System engineering collects and reviews the documentation that demonstrates the technical progress planned for the major systems review and submits the materials as a data package to the review team prior to the review. This allows the selected technical representatives to identify problems and issues that can be discussed at the review. Systems engineering is responsible for the agenda, organization, and conduct of the systems review as well as obtaining closure on any action items and corrective actions.

Systems engineering acts as recorder, noting all comments and questions that are not adequately addressed during the presentations. At the conclusion of a major systems review, the independent review panel, if in place, makes a determination as to whether or not the predetermined criteria for a successful review have been met and makes a recommendation on whether or not the system is ready to proceed into the next phase of its development.

An EPR is a focused, in-depth technical review of a subsystem, lower level of assembly, or a component. An EPR can address an entire system or subsystem, but more typically addresses a lower level. The EPR adds value and reduces risk through expert knowledge infusion, confirmation of approach, and specific recommendations. The key distinction between an EPR and a major subsystem review is that the review panel is selected by personnel supporting the project, and not by the Center. The mission systems engineer works with the respective product manager (project manager, project formulation manager, instrument manager, or Principal Investigator) to ensure that the EPR review panel is comprised of technical experts with significant practical experience relevant to the technology and requirements of the subsystem, lower level of assembly, or component to be reviewed. They also work together to produce an EPR plan, which lists the subsystems, lower levels of assembly, and components to be reviewed and the associated life-cycle milestones for the reviews. A summary of results of the EPRs is presented at each major subsystem review and/or at each life-cycle review.

Additional informal project technical reviews, sometimes called "table top reviews," are conducted by project team members as necessary and are one of their primary mechanisms for internal technical project control. These reviews follow the general protocols described above for subsystem reviews and EPRs.

4.3.2.3 Preparing for Approval to Enter Formulation (Phase A)

Projects support the Mission Concept Review (MCR) life-cycle review in accordance with NPR 7123.1, Center practices, and NPR 7120.5, including ensuring that the life-cycle review objectives and expected maturity states defined in NPR 7120.5 have been satisfactorily met. Life-cycle review entrance and success criteria in Appendix G of NPR 7123.1 and the expected maturity states in Appendix E of this handbook provide specifics for addressing the six assessment criteria required to demonstrate that the project has met its expected maturity state. MCRs are generally conducted by the Center, but the Decision Authority may request an SRB to perform this review. If this is the case, Section 5.10 of this handbook and the *NASA Standing Review Board Handbook* provide guidance.

Projects plan prepare for and support the governing PMC review prior to KDP A and provide or obtain the KDP readiness products listed in Section 4.2.3.

Once the KDP has been completed and the Decision Memorandum signed, the project updates its documents and plans as required to reflect the decisions made and actions assigned at the KDP.

4.3.3 Initiation of Competed Mission Projects

For competed or "Announcement of Opportunity (AO)-driven" missions, some Mission Directorates, primarily the Science Mission Directorate (SMD), have chosen to use one or two steps to initiate projects within a space flight program:

- In a one-step AO process, projects are competed and selected for Formulation in a single step.

- In two-step competitions, several projects may be selected in Step 1 and given time to mature their concepts in a funded concept study before the Step 2 down-selection. Program resources are invested (following Step 1 selections) to bring these projects to a state in which their science content, cost, schedule, technical performance, project implementation strategies, SMA strategies, heritage, technology requirements and plans, partnerships, and management approach can be better judged.

From the point of view of the selected AO-driven project, the proposing teams are clearly doing preparatory work and formal project Formulation (e.g., typical Pre-Phase A and Phase A tasks, such as putting together a detailed WBS, schedules, cost estimates, and implementation plan) during the concept study and the preparation of the Step 2 concept study report. From the point of view of the program, no specific project has been chosen, the total cost is not yet known, and project requirements are not yet finalized, yet Formulation has begun. Therefore, for competed missions, the selection of a proposal for concept development is the equivalent of KDP A. In a one-step AO process, projects enter Phase A after selection (KDP A) and the process becomes the conventional process for directed missions. In a two-step AO process, projects perform concept development in the equivalent of Phase A and go through evaluation for down-selection at the equivalent of KDP B. Following this selection, the process becomes conventional—with the exception that KDP B products requiring Mission Directorate input are finished as early in Phase B as feasible.

4.3.4 Project Phase A, Concept and Technology Development Activities

4.3.4.1 Project Phase A Life-Cycle Activities

Project Formulation consists of two sequential phases, Phase A (Concept and Technology Development) and Phase B (Preliminary Design and Technology Completion). Formulation is an iterative set of activities, rather than discrete linear steps. The purpose of Phase A is to develop a proposed mission/system architecture that is credible and responsive to program requirements and constraints on the project, including resources. The Phase A work products need to demonstrate that the maturity of the project's mission/system definition and associated plans are sufficient to begin Phase B, and the mission can probably be achieved within available resources with acceptable risk.

During Phase A, a project team is formed or expanded (if already formed in Pre–Phase A) to update and fully develop the mission concept and begin or assume responsibility for the technology development; engineering prototyping; heritage hardware and software assessments using the *Systems Engineering Handbook, NASA/SP-2007-6105 Rev 1*, Appendix G; and other risk-mitigation activities identified in the Project Formulation Agreement. The project establishes performance metrics, explores the full range of implementation options, defines an affordable project concept to meet requirements specified in the Program Plan, and develops needed technologies. The primary activities in these phases include:

- Developing and defining the project requirements down to at least the system level;
- Flowing down requirements to the system and preliminary requirements to the subsystem level;
- Assessing the technology requirements, developing the plans to achieve them, and initiating development of the technology;
- Developing the project's knowledge management strategy and processes;
- Examining the Lessons Learned database for lessons that might apply to the current project's planning;
- Developing the system architecture;
- Conducting acquisition planning, including an analysis of the industrial base capability to design, develop, produce, support, and—if appropriate— restart an acquisition project;
- Assessing heritage using the Systems Engineering Handbook, NASA/SP-2007-6105 Rev 1, Appendix G (the applicability of designs, hardware, and software in past projects to the present one);

- Conducting safety, performance, cost, and risk trades;

- Identifying and mitigating development and programmatic risks, including supply chain risks;

- Conducting engineering development activities, including initiating development of engineering prototypes and models for the higher risk components and assemblies that have not been previously built or flown in the planned environment and initiating testing of them to demonstrate adequate performance;

- Completing mission and preliminary system-level designs;

- Evaluating and refining subsystem interfaces; and

- Developing time-phased cost and schedule estimates and documenting the basis of these estimates.

Finally, the project team develops the preliminary Project Plan and the preliminary project technical baselines (preliminary design documentation at the system level), and cost and schedule estimates. Projects with an LCC greater than $250 million develop cost range estimates with confidence levels and schedule range estimates with confidence levels. Formulation activities continue, normally concurrently, until Formulation output products, such as the Project Plan have matured and are acceptable to the program manager, Center Director, MDAA, and AA (if the AA is the Decision Authority). When applicable, these activities allow the Agency to present to external stakeholders time-phased cost plans and schedule range estimates at KDP B and high-confidence cost and schedule commitments at KDP C.

Phase A completes when the Decision Authority approves transition from Phase A to Phase B at KDP B. Major project and life-cycle reviews leading to approval at KDP B are the ASM, the System Requirements Review (SRR), and the System Definition Review (SDR)/Mission Definition Review (MDR),[14] and the governing PMC review.

The MDAA and Associate Administrator determine when and whether an ASM is required. The purpose of the ASM is for senior Agency management to review and agree upon the acquisition strategy before authorizing resource expenditures for major acquisitions. This includes implementation of the decisions and guidance that flowed out of the Agency strategic planning and considerations such as resource availability, impact on the Agency workforce, maintaining core capabilities, make-or-buy planning, supporting Center assignments, and the potential for partnerships. The development of an acquisition strategy also includes an analysis of the industrial base

The Acquisition Strategy Meeting (ASM) is a decision-making forum where senior Agency management reviews and approves project acquisition strategies. The ASM considers impacts to Agency workforce and maintaining core capabilities, make-or-buy decisions, Center assignments and potential partners, risk, and other planning decisions from an Agency perspective. The ASM is held at the Agency level, implementing the decisions that flow out of the earlier Agency Strategic Implementation Planning (SIP) process. (See Section 5.8.3.1 for information on the SIP process.)

---

[14] The SDR and MDR are the same review: robotic programs tend to use the terminology MDR and human programs tend to use SDR.

capability as well as the mechanisms used to identify, monitor, and mitigate industrial base and supply chain risks. The ASM review is based on information provided by the associated Mission Directorate or Mission Support Office, and results in approval of plans for Formulation and Implementation. Decisions are documented in the ASM meeting minutes. The results of the ASM are used to finalize the Acquisition Plan.

The purpose of the SRR is to evaluate whether the functional and performance requirements defined for the system are responsive to the program's requirements on the project and represent achievable capabilities.

The purpose of the SDR/MDR is to evaluate the credibility and responsiveness of the proposed system/mission architecture to the program requirements and constraints on the project, including available resources, and to determine whether the maturity of the project's system/mission definition and associated plans are sufficient to begin Phase B.

At KDP B, the project is expected to demonstrate its credibility and maturity to begin Phase B and to have shown that the mission can probably be achieved within available resources with acceptable risk.

The general flow of activities for a project in Phase A is shown in Figure 4-9.

4.3.4.2 Project Phase A Management, Planning, and Control Activities

4.3.4.2.1 Supporting Headquarters Planning

During Phase A, the project manager and project team support the program manager and the MDAA in developing the baseline program requirements (on the project), selection and use of technical standards products, and constraints on the project, including mission objectives, goals, and success criteria.[15] The program and the project also document any important program/Mission Directorate-imposed driving mission, technical, and programmatic ground rules and assumptions. In doing this, the project supports the program manager and the MDAA in ensuring continuing alignment of the project requirements with applicable Agency strategic goals.

Early in Phase A, the Mission Directorate, with support from both the program and the project, begins to plan and prepare for the ASM if the ASM is required. This is done prior to partnership agreements so the Agency can ensure that all elements are engaged in the project in accordance with the Agency's strategic planning. The project obtains the ASM minutes after the

[15] Program requirements on the project are contained in the Program Plan.

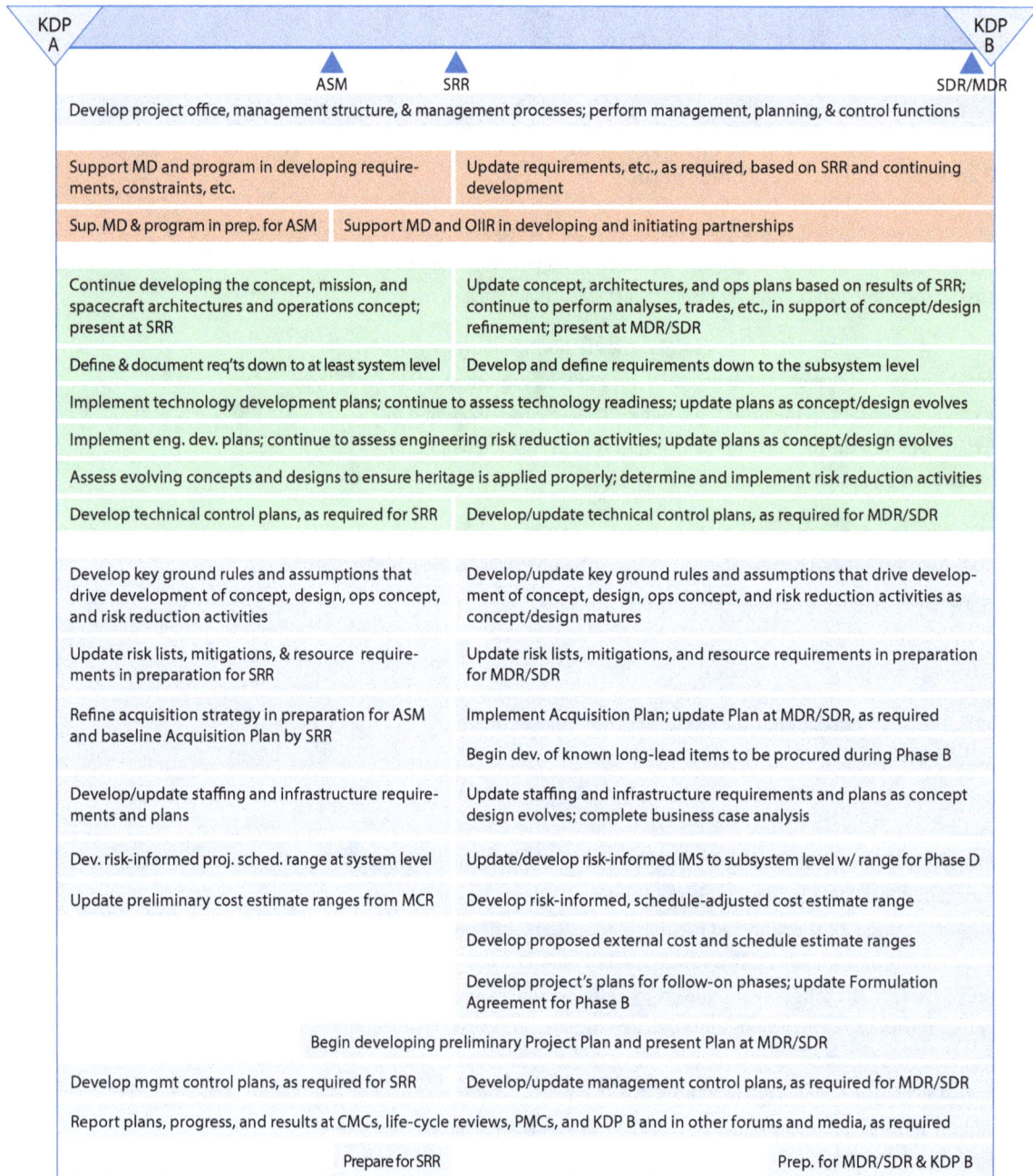

Figure 4-9 Project Phase A Flow of Activities

Legend: ☐ Project management, planning, and control tasks

☐ Work for which Headquarters is responsible but the project helps accomplish (e.g., international partnerships are a Headquarters responsibility, but the projects help develop and finalize those partnerships)

☐ Technical work the project is doing

Acronyms: MD = Mission Directorate; OIIR = Office of International and Interagency Relations.

Note: These are typical high-level activities that occur during this project phase. Placement of reviews is notional.

meeting and uses them as guidance/direction to finalize the project's acquisition strategy.

Once the Agency has completed its strategic planning and has held the ASM, the project supports the program manager, the MDAA, and the NASA Headquarters Office of International and Interagency Relations (OIIR) in initiating **interagency and international agreements**, including planning and negotiating agreements and making recommendations on joint participation in reviews, integration and test, and risk management if applicable. The project works with the appropriate NASA Headquarters offices to initiate the development of Memoranda of Understanding (MOUs)/Memoranda of Agreement (MOAs) with external partners as needed.

4.3.4.2.2 Management Control Processes and Products

The project team conducts planning that enables formulation and implementation of the mission concept(s), architectures, scenarios/DRMs and requirements and implements the Formulation Agreement. The results of this planning, much of which is described below, support the System Requirements Review (SRR), the SDR/MDR, and KDP B by demonstrating how the project plans to implement the mission concept(s) being proposed.

The project team continues to work with the Center to further develop and implement the management framework, fill out the project team and organizational structure, and define the initial management processes consistent with the direction from the MDAA and the program.

The project develops a preliminary **Technical, Schedule, and Cost Control Plan** by SRR and baselines the plan by SDR/MDR. This plan documents how the project plans to control project requirements, technical design, schedule, and cost to achieve the program requirements on the project. The plan describes how the project monitors and controls the project requirements, technical design, schedule, and cost to ensure that the high-level requirements levied on the project are met. It describes the project's performance measures in objective, quantifiable, and measurable terms and documents how the measures are traced from the program requirements on the project. In addition, it documents the minimum mission success criteria associated with the program requirements on the project that, if not met, trigger consideration of a Termination Review. The minimum success criteria are generally defined by the project's threshold science requirements. The project also develops and maintains the status of a set of programmatic and technical leading indicators. (See "Required and Recommended Programmatic and Technical Leading Indicators" box.) Per NPR 7123.1B, three indicators are required: mass margins, power margins, and Request For Action (RFA)/Review Item Discrepancy (RID)/action item burn down.

Margins are the allowances carried in budget, projected schedules, and technical performance parameters (e.g., weight, power, or memory) to account for uncertainties and risks. Margins are allocated in the formulation process, based on assessments of risks, and are typically consumed as the program or project proceeds through the life cycle.

Required and Recommended Programmatic and Technical Leading Indicators

Required (per *NPR 7123.1B*)

1. Technical Performance Measures (mass margin, power margin)
2. Review Trends (RID/RFA/action item burndown per review)

Recommended

1. Requirement Trends (percentage growth, TBD/TBR closures, number of requirement changes)
2. Interface Trends (percentage ICD approval, TBD/TBR burn down, # interface requirement changes)
3. Verification Trends (closure burn down, # deviations/waivers approved/open)
4. Software Unique Trends (# software requirements verified and validated per build/release versus plan)[1]
5. Problem Report/Discrepancy Report Trends (# open, # closed)
6. Manufacturing Trends (# nonconformance/corrective actions)
7. Cost Trends (plan, actual, UFE, EVM, new obligation authority)
8. Schedule Trends (critical path slack/float, critical milestones, EVM schedule metrics, etc.)
9. Staffing Trends (FTE, work-year equivalent)
10. Additional project-specific indicators as needed (e.g., human systems integration compliance)

---

[1] Please note that there are non-Technical Leading Indicators software measurement requirements in *NPR 7150.2, NASA Software Engineering Requirements* (e.g., SWE-091) which have implementation guidance in *NASA-HDBK-2203* (http://swehb.nasa.gov).

In addition to these required indicators, NASA highly recommends the use of a common set of programmatic and technical indicators to support trend analysis throughout the life cycle. Projects may also identify unique programmatic and technical leading indicators. (See Section 5.13 and the program and project management community of practice on the NASA Engineering Network (NEN) for a white paper explaining leading indicators and more information on leading indicators.

The plan describes the following:

- The approach to monitor and control the project's ABC and how the project will periodically report performance.

- The mitigation approach if the project is exceeding the development cost documented in the ABC to take corrective action prior to triggering the 30 percent breach threshold.

- How the project will support a rebaseline review in the event the Decision Authority directs one.

- The project's implementation of Technical Authority (Engineering, Health and Medical, and Safety and Mission Assurance).

- How the project will implement metric (Système Internationale (SI)) and nonmetric systems of measurement and the identification of units of measure in all product documentation. (See Section 4.3.4.3 for more details.)

- The project's implementation of EVM,[16] including:

  - How the Performance Measurement Baseline (PMB) will be developed and maintained;

  - The methods the project will use to authorize work and to communicate changes to the scope, schedule, and budget of all suppliers;

  - The process the project team will use to communicate the time-phased levels of funding that have been forecast to be made available to each supplier;

  - For the class of suppliers not required to use EVM, the schedule and resource information that will be required of the suppliers to establish and maintain a baseline and to quantify schedule and cost variances;

  - What contractor performance reports will be required; and

  - How the cost and schedule data from all partners/suppliers will be integrated to form a total project-level assessment of cost and schedule performance.

- Any additional specific tools necessary to implement the project's control processes (e.g., the requirements management system, project scheduling system, project information management systems, budgeting, and cost accounting system);

- The process for monitoring and controlling the Integrated Master Schedule (IMS). The project develops a summary of its IMS, including all critical milestones, major events, life-cycle reviews, and KDPs throughout the project life cycle. The summary of the IMS includes the logical relationships (interdependencies) for the various project elements and projects and critical paths as appropriate, and identifies driving ground rules, assumptions, and constraints affecting the schedule. The summary of the IMS is included in the Project Plan;

- The process for utilizing the project's technical and schedule margins and UFE to stay within the terms of the Management Agreement and ABC;

The Performance Measurement Baseline is a time-phased cost plan for accomplishing all authorized work scope in a project's life cycle, which includes both NASA internal costs and supplier costs. The project's performance against the PMB is measured using EVM if EVM is required, or other performance measurement techniques if EVM is not required. The PMB does not include UFE.

---

[16] Small Category 3/Class D projects with development costs greater than $20 million and a life-cycle cost estimate less than $150 million should reference the EVM guide for applying EVM principles to small projects. The guidance for tailoring 7120.5 requirements for small Cat 3/Class D projects can be found on the OCE tab in NODIS under "Other Policy Documents" at http/nodis3.gsfc.nasa.gov/OCE_docs/OCE_25.pdf.

- How the project plans to report technical, schedule, and cost status to the program manager, including the frequency and level of detail of reporting;

- The project's internal processes for requesting technical waivers and deviations and handling Dissenting Opinions;

- The project's descope plans, including key decision dates, savings in cost and schedule, and how the descopes are related to the project's threshold performance requirements; and

- A description of the systems engineering organization and structure and how the project executes the overall systems engineering functions.

The project team expands the WBS, consistent with the NASA standard space flight project WBS (see Section 5.9) and provides the project's WBS and WBS dictionary to the Level 2 elements in accordance with the standard template in Figure 4-10. The WBS supports cost and schedule allocation down to a work package level; integrates both Government and contracted work; integrates with the Earned Value Management System (EVMS) approach; allows for unambiguous cost reporting; and is designed to allow project managers to monitor and control work package/product deliverable costs and schedule.

> A work package is a defined group or set of work performed by an organization that can be tracked by cost and performance milestones.

The project team develops its resource baseline, which includes funding requirements by fiscal year and the new obligation authority in real-year dollars for all years—prior, current, and remaining. The funding requirements are consistent with the project's WBS and include funding for all cost elements required by the Agency's full-cost accounting procedures. Funding requirements are consistent with the budget. The resource baseline

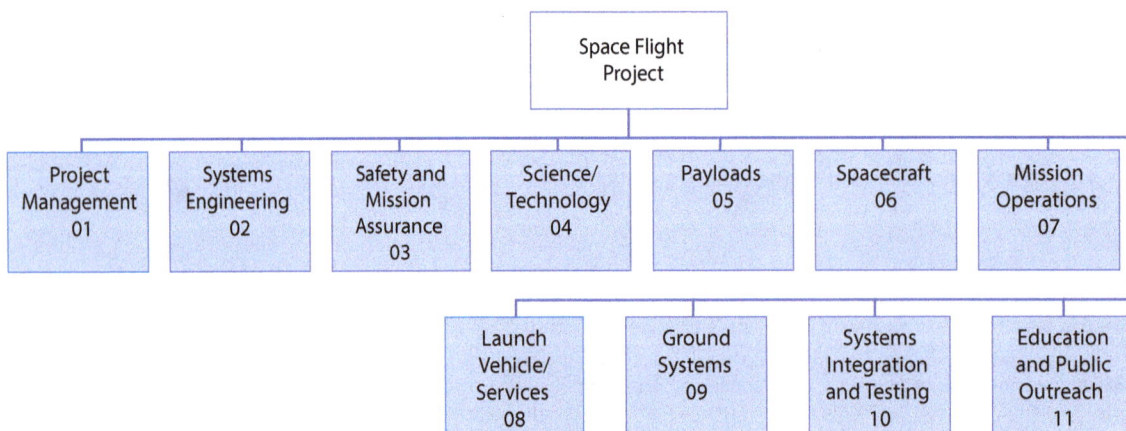

Figure 4-10 Standard Level 2 WBS Elements for Space Flight Projects

provides a breakdown of the project's funding requirements to the WBS Level 2 elements. The resource baseline provides the workforce requirements by fiscal year, consistent with the project's funding requirements and WBS. Throughout the Implementation Phase, for projects with a LCC greater than $250M, baselines are based on and maintained consistent with the approved JCL in accordance with NPD 1000.5 and NPR 7120.5. (The resource baseline also includes the infrastructure requirements, discussed elsewhere in this section.)

The project further develops and baselines the project's key ground rules and assumptions that drive development of the mission concept, engineering prototyping plans/status, required funding profiles and schedules for Phases A and B, results of technology heritage assessments and key subsystem trade studies, technical requirements, and programmatic preliminary baseline. Once the project has defined the ground rules and assumptions, it tracks them through Formulation to determine if they are being realized (i.e., remain valid) or if they need to be modified.

As the concepts mature, the project team updates its assessment of potential infrastructure and workforce needs versus current plans, as well as opportunities to use infrastructure and workforce in other Government agencies, industry, academia, and international organizations for each concept being considered. Based on this assessment, the project team updates the initial requirements and plans for staffing and infrastructure at both the SRR and the SDR/MDR. As part of this activity, the project completes a preliminary business case analysis[17] for infrastructure for each proposed project real property infrastructure investment consistent with *NPD 8820.2, Design and Construction of Facilities* and *NPR 8820.2, Facility Project Requirements* and for the acquisition of new aircraft consistent with *NPR 7900.3, Aircraft Operations Management Manual*. The business case analysis needs to be initiated in sufficient time to allow the analysis, documentation, review, approval and funding of the infrastructure to support the mission requirements. Also in coordination with the OCFO and in accordance with *NPR 9250.1, Property, Plant, and Equipment and Operating Materials and Supplies*, the project team completes the **Alternative Future Use Questionnaire** (Form NF 1739),[18] Section A, to determine the appropriate accounting treatment of capital assets. Once it has completed the questionnaire, the project team forwards it to the OCFO, Property Branch.

---

[17] See the NASA Business Case Guide for Facilities Projects at http://www.hq.nasa.gov/office/codej/codejx/Assets/Docs/Case_Guide_4-20-06.pdf.

[18] The questionnaire can be found in NASA's Electronics Forms Database website: http://itcd.hq.nasa.gov/eforms.html.

The project team expands the preliminary strategy for acquisition developed in Pre–Phase A and develops its Acquisition Plan. The acquisition strategy is the plan or approach for using NASA's acquisition authorities to achieve the project's mission. The strategy includes recommendations from make/buy analyses, the recommendations from competed/directed analyses, proposed partnerships and contributions, proposed infrastructure use and needs, budget, and any other applicable considerations. This strategy addresses the project's initial plans for obtaining the systems, research, services, construction, and supplies that it needs to fulfill its mission, including any known procurement(s); the availability of the industrial base capability and supply chain needed to design, develop, produce, and support the project; identifying risks associated with single source or critical suppliers; and attendant mitigation plans.

The project team works with the Mission Directorate and the program to prepare for the ASM if one is required. Once the ASM is completed, the project team obtains a copy of the ASM minutes and finalizes the Acquisition Plan based on the ASM direction. The **Acquisition Plan** is baselined by SRR and updated at SDR/MDR. The project Acquisition Plan:

- Is developed by the project manager, supported by the Office of Procurement.

- Documents an integrated acquisition strategy that enables the project to meet its mission objectives and provides the best value to NASA.

- Identifies all major proposed acquisitions (such as engineering design study, hardware and software development, mission and data operations support, and sustainment) in relation to the project WBS and provides summary information on each proposed acquisition, including a contract WBS; major deliverable items; recommended type of procurement (e.g., competitive, Announcement of Opportunity for instruments); type of contract (e.g., cost-reimbursable, fixed-price); source (e.g., institutional, contractor, other Government agency, or international organization); procuring activity; and surveillance approach.

- Identifies the major procurements that require a Procurement Strategy Meeting (PSM).

- Describes completed or planned studies supporting make-or-buy decisions, considering NASA's in-house capabilities and the maintenance of NASA's core competencies, as well as cost and best overall value to NASA.

- Describes the supply chain and identifies potential critical and single-source suppliers needed to design; develop; produce; support; and, if appropriate, restart an acquisition project.

- Promotes sufficient project stability to encourage industry to invest in, plan for, and bear their share of risk.

The Procurement Strategy Meeting (PSM) provides the basis for approval of the approach for major procurements for programs and projects and ensures they are following the law and the Federal Acquisition Regulations (FAR). Detailed PSM requirements and processes, prescribed by the FAR and NFS and formulated by the Office of Procurement, ensure the alignment of portfolio, mission acquisition, and subsequent procurement decisions. The contents of written acquisition plans and PSMs are delineated in the FAR in Subpart 7.1—Acquisition Plans, the NFS in Subpart 1807.1—Acquisition Plans, and in the Guide for Successful Headquarters Procurement Strategy Meetings at http://prod.nais.nasa.gov/portals/pl/documents/PSMs_091611.html.

- Describes the internal and external mechanisms and procedures used to identify, monitor, and mitigate supply chain risks and includes data reporting relationships to allow continuous surveillance of the supply chain that provides for timely notification and mitigation of potential risks.

- Describes the process for reporting supply chain risks to the program.

- Identifies the project's approach to strengthening SMA in contracts and describes how the project will establish and implement a risk management process per NPR 8000.4 and NASA/SP-2011-3422.

- Describes all agreements, MOUs, barters, in-kind contributions, and other arrangements for collaborative and/or cooperative relationships, including partnerships created through mechanisms other than those prescribed in the FAR and NFS. It lists all such agreements (the configuration control numbers, the date signed or projected dates of approval, and associated record requirements) necessary for project success. It includes or references all agreements concluded with the authority of the project manager and references agreements concluded with the authority of the program manager and above. These include NASA agreements, e.g., space communications, launch services, inter-Center MOAs; Government agencies, and international agreements.

- Lists long-lead procurements that will need to be procured in Phase B, which will need to be approved by the program manager.

During this period, projects with contracts requiring EVM (refer to the NFS) will conduct the required Integrated Baseline Reviews (IBRs) focusing on EVM system planning.

By SRR, the project team baselines a Risk Management Plan that includes the content required by NPR 8000.4 and NASA/SP-2011-3422. The plan summarizes how the project will implement a risk management process (including Risk-Informed Decision Making (RIDM) and Continuous Risk Management (CRM)) in accordance with NPR 8000.4 and NASA/SP-2011-3422. It includes the initial Risk List and appropriate actions to mitigate each risk. Projects with international or other Government agency contributions need to assess and report on risks due to international or other Government partners and plan for contingencies. The Risk Management Plan is required to be a stand-alone plan unless an alternate approach is approved. Consistent with the technical team's work, the project continues to identify, assess, and update the technical, cost, schedule and safety risks that threaten the system requirements, mission concept, operations concept, and technology development. Risks include, but are not limited to, technology development, engineering development, payload (robotic space flight), and procurement risks; risks associated with use of heritage hardware and software; and risks that are likely to drive the project's cost and schedule, or cost and schedule ranges (projects with an LCC greater than $250 million) at KDP B. The project team updates, identifies, assesses, and mitigates (if feasible) supply chain risks, including potential critical or single-source suppliers needed to design, develop, produce, and support required capabilities at planned cost and schedule. The project team reports risks to the program in accordance with the approved Acquisition Plan. The project team identifies risk mitigation plans and associated resources for managing and mitigating risks in accordance with the Risk Management Plan.

Projects develop a preliminary Education Plan by SDR/MDR (the plan is baselined at PDR). This plan describes planned efforts and activities to enhance Science, Technology, Engineering, and Math (STEM) education using the project's science and technical content. It describes the plan for coordinating with the Mission Directorate Education Coordinating Council (ECC) member to ensure project education activities are aligned with NASA education portfolio offerings and requirements. It defines goals and outcomes for each activity and addresses how activities will advance NASA strategic goals for education. It also identifies the target audience for each activity and discusses how it reaches and engages groups traditionally underrepresented and/or underserved in STEM disciplines. The plan describes how each activity will be evaluated; defines specific metrics and describes how they will be collected; and includes a timeline with relevant milestones for achieving goals and outcomes for each activity. Finally, the plan describes the relationship between the program and project(s) education plans.

Projects develop a preliminary Communications Plan by SDR/MDR (the plan is baselined at PDR). This plan describes how the project team will implement a diverse, broad, and integrated set of efforts and activities to communicate with and engage target audiences, the public, and other stakeholders in understanding the project, its objectives, elements, and benefits. It describes how the project team's efforts and activities relate to the larger NASA vision and mission. Focus typically is placed on activities and campaigns that are relevant; compelling; accessible; and, where appropriate, participatory. The plan describes how these efforts and activities will promote interest and foster participation in NASA's endeavors and address how these efforts and activities will develop exposure to and appreciation for the STEM disciplines. The plan:

- Defines goals and outcomes, as well as key overarching messages and themes;
- Identifies target audiences, stakeholders, and partnerships;

- Summarizes and describes products to be developed and the tools, infrastructure, and methods that will be used to communicate, deploy, and disseminate those products, including media, multimedia, Web, social media, and publications for nontechnical audiences, excluding those developed in the context of the Education Plan;

- Describes events, activities, and initiatives focused on public engagement and how they link with planned products and infrastructure; and

- Identifies milestones and resources required for implementation, and defines metrics to measure success.

Finally, the plan describes the relationship between the program and project Communications Plans and the coordination between a program and its projects regarding communications activities.

All projects prepare a preliminary Project Plan that follows the template in NPR 7120.5E, Appendix H, by the SDR/MDR. The Project Plan contains a number of required Project Plan control plans. NPR 7120.5E, Appendix I, Table I-5, and Table 4-7 at the end of this chapter, show which of the control plans are required during this phase and also describes when the control plans are required to be developed. Each of these control plans is described in this chapter, and some of them are required to be baselined before the Project Plan is fully finished and baselined at PDR. These early control plans help the project team manage its early work and become part of the preliminary Project Plan. During Phase B, there is an overlap between the Formulation Agreement and the preliminary Project Plan. The Formulation Agreement is the agreement between the Mission Directorate and the project that governs the work during Phase B; however, the Project Plan control plans that are baselined govern the management and technical control processes used during this phase.

All project teams prepare cost and schedule estimates for both SRR and SDR/MDR consistent with driving assumptions, risks, requirements, and available funding and schedule constraints:

- Based on the refined concept/design and its risks at SRR, the project team develops a risk-informed schedule at the system level (as a minimum) with a preliminary Phase D completion date estimate by SRR. In addition, the project team updates the initial project cost estimate, prepared for the MCR/KDP A, by SRR. For projects with a LCC greater than $250 million, these cost and schedule estimates are ranges that represent optimistic outcomes and pessimistic outcomes if all risks and unknown-unknowns materialize. In other words, the ranges ensure the upper limits will not be exceeded by the final cost and schedule commitments made at KDP C. The costs need to include institutional funding requirements, technology investments, and multi-Center operations; costs associated with Agency constraints (e.g., workforce allocations at Centers); and costs associated with efficient use of Agency capital investments, facilities, and workforce.

- As the project approaches SDR/MDR and KDP B, the project team prepares its **project preliminary baselines**. The project develops and documents preliminary project baselines and a proposed Management Agreement for all work to be performed by the project. All preliminary baselines are consistent with the program requirements and constraints levied on the project, key assumptions, workforce estimates, key acquisitions, and significant risks. The preliminary project baselines support the Decision Authority in establishing cost and schedule estimates, or cost and schedule range estimates that can be provided to external stakeholders, if applicable. The preliminary project baseline cost and schedule estimates include:

  - A risk-informed and schedule-adjusted life-cycle cost estimate or cost range estimate based on the project's preliminary baselines and mission concept (this product includes phased life-cycle costs and is developed using the latest accounting guidance and practices). The project team develops its cost estimates using many different techniques. These include, but are not limited to, bottoms-up estimates where specific work items are estimated by the performing organization using historical data or engineering estimates; vendor quotes; analogies; and parametric cost models. (See Section 5.6 for a discussion of probabilistic cost estimating.);

  - Proposed annual budgeted costs, or range of annual budgeted costs by Government fiscal year and by the project's WBS;

  - Proposed annual UFE, or range of annual UFE; and

  - A risk-informed, preliminary **Integrated Master Schedule (IMS)** that contains the following key data elements: all task/milestone sequence interdependency assignments, WBS code assignment on all tasks/milestones, current task/milestone progress, and clearly identifiable schedule margin.

- Projects with an LCCE greater than $250 million develop their range of cost and range for schedule estimates with confidence levels identified for the low and high values of the range. These confidence levels are established by a probabilistic analysis and are based on identified resources and associated uncertainties by fiscal year. These analyses can be separate analyses of cost and schedule; a JCL is not required at this point but may be used. These cost and schedule range estimates typically are informed by technology needs, engineering development and heritage assessments, acquisition strategies, infrastructure and workforce requirements, and identified risks.

Projects document the Basis of Estimate (BoE) for initial cost and schedule estimates at both SRR and SDR/MDR.

Flight projects provide a preliminary Cost Analysis Data Requirement (CADRe) (parts A, B, and C) consistent with the *NASA Cost Estimating Handbook* 60 days prior to the KDP B milestone with a final version 30 to 45 days after the KDP event to reflect any decisions from the KDP. This CADRe is based on the project's preliminary baselines presented at the SDR/MDR. (A CADRe is not mandatory for small Category 3/Class D projects, but data collection for smaller projects is critical for future estimating capabilities and is strongly encouraged. See guidance for tailoring NPR 7120.5 requirements on the OCE tab in NODIS under "Other Policy Documents" at http/nodis3.gsfc.nasa.gov/OCE_docs/OCE_25.pdf.)

4.3.4.3 Project Phase A Technical Activities and Products

The project team continues developing the concept and architecture of the project, its major components, and the way they will be integrated, including its operations concepts. This includes continuing to work with the Launch Services Program at KSC to refine the viable launch service options, if applicable. In this phase, the LSP begins to refine spacecraft customer requirements, prepare the acquisition strategy for the launch service, identify support services and estimated costs, establish dates for spacecraft delivery and complete a launch service assessment. System engineering plays a major role during Formulation as described in NPR 7123.1. The project performs the iterative and recursive process of functional analysis, requirements allocation, trade studies, preliminary synthesis, evaluation, and requirements analysis. As the project approaches the SRR, the project documents the updated concept and mission and spacecraft architecture and defines and documents the preliminary ground and payload architectures and preliminary operations concept.

Based on the leading concept, the project finalizes the initial mission objectives and project-level requirements, including allocated and derived requirements down to at least the system level. If not already defined, the project team identifies the payload risk classification as described in NPR 8705.4. The project needs to continue to update and maintain the requirements traceability matrix initially developed in Pre–Phase A.

The project team assesses the ability of the project and all contributors to the project, including contractors, industrial partners, and other partners to use the International System of Units (the Système Internationale (SI),

The Basis of Estimate (BoE) documents the ground rules, assumptions, and drivers used in developing the cost and schedule estimates, including applicable model inputs, rationale or justification for analogies, and details supporting cost and schedule estimates. The BoE is contained in material available to the Standing Review Board (SRB) and management as part of the life-cycle review and Key Decision Point (KDP) process. Good BoEs are well-documented, comprehensive, accurate, credible, traceable, and executable. Sufficient information on how the estimate was developed needs to be included to allow review team members, including independent cost analysts, to reproduce the estimate if required. Types of information can include estimating techniques (e.g., bottoms-up, vendor quotes, analogies, parametric cost models), data sources, inflation, labor rates, new facilities costs, operations costs, sunk costs, etc.

commonly known as the metric system of measurement). This assessment determines an approach that maximizes the use of SI while minimizing short- and long-term risk to the extent practical and economically feasible or to the extent that the supply chain can support utilization without loss of markets to U.S. firms. Use of the SI or metric system of measurement is especially encouraged in cooperative efforts with international partners. This assessment documents an integration strategy if both SI and U.S. customary units are used in a project. The assessment is completed and documented in the preliminary Project Plan no later than the SDR/MDR. To the degree possible, projects need to use consistent measurement units throughout all documentation to minimize the risk of errors. Where full implementation of the metric system of measurement is not practical, hybrid configurations (i.e., a controlled mix of metric/nonmetric system elements) may be used to support maximum practical use of metric units for design, development, and operations. Where hybrid configurations are used, the project describes the specific requirements established to control interfaces between elements using different measurement systems.

Following the SRR, the project updates the concept documentation, architectures, and operations plans based on the results of the SRR and continues to perform analyses and trades in support of concept/design refinement. It prepares the preliminary design documentation for use during the peer reviews, subsystem reviews, and system reviews leading to the project's SDR/MDR. The project updates the design documentation as changes are made during this process in accordance with the project team's and Center's standard practices.

Projects that plan to develop technologies initiate the development of technologies as agreed to in the Formulation Agreement. As the technologies develop, the project monitors, assesses, and reports the status of technology readiness advancement. Projects update their technology development plans, including assessment points to terminate development of technologies that are not maturing adequately with corresponding alternate approaches.

Projects implement engineering development plans, heritage hardware and software assessments (using the *Systems Engineering Handbook, NASA/ SP-2007-6105 Rev 1*, Appendix G), and risk mitigation plans identified in the project Formulation Agreement for Phase A. As these risk reduction plans are executed, the project monitors, assesses, and reports the status of engineering development results and heritage assessments. Projects update their plans when needed.

In accordance with *NPR 2190.1, NASA Export Control Program*, the project supports the appropriate NASA export control officials to identify and assess export-controlled technical data that potentially will be provided to

International Partners and the approval requirements for release of that data as a part of developing the project's preliminary Technology Transfer Control Plan.

To provide additional options in the event that development begins to exceed the resources allocated, the project typically begins to refine the list of descope options. Documentation of the project's descope plans typically includes a detailed description of the potential descope, the effect of the descope on the project's success criteria, the cost and schedule savings resulting from the descope, and key decision dates by when the descope needs to be exercised to realize these savings.

The Project Plan contains a number of required control plans, many of which are technical. NPR 7120.5E, Appendix I, Table I-5, and Table 4-7 at the end of this chapter, show the control plans that are required during Phase A and when they need to be developed. Each of the technical control plans is described in this handbook section.

The project team baselines at SRR the preliminary SEMP developed at MCR and updates the SEMP at SDR/MDR. The SEMP is required to be a stand-alone plan unless an alternate approach is approved.

If applicable, in accordance with NPR 7123.1B, the project baselines the Human Systems Integration (HSI) Plan at SRR. The HSI Plan is updated at SDR/MDR.

The project baselines a SMA Plan by SRR and updates the plan at SDR/MDR. The SMA Plan addresses life-cycle SMA functions and activities. The plan identifies and documents project-specific SMA roles, responsibilities, and relationships. This is accomplished through a project-unique mission assurance process map and matrix developed and maintained by the project with appropriate support and guidance from the Headquarters and/or Center-level SMA organizations. The plan addresses areas that include procurement, management, design and engineering, design verification and test, software design, software verification and test, manufacturing, manufacturing verification and test, operations, and preflight verification and test. The plan also addresses specific critical SMA disciplines, including the following as a minimum:

- Safety per NPR 8715.3; *NPR 8715.5, Range Flight Safety Program*; *NPR 8715.7, Expendable Launch Vehicle Payload Safety Program*; local range requirements; and *NPR 8621.1, NASA Procedural Requirements for Mishap and Close Call Reporting, Investigating, and Recordkeeping*.
- Human rating requirements per *NPR 8705.2, Human-Rating Requirements for Space Systems*.

The Mission Assurance Process Map is a high-level graphic representation of governing SMA policy and requirements, processes, and key participant roles, responsibilities, and interactions. It also includes the reporting structure that constitutes a project's SMA functional flow.

The Mission Assurance Process Matrix is constructed to identify project life-cycle assurance agents and specific assurance activities, processes, responsibilities, accountability, depth of penetration, and independence. The matrix includes key assurance personnel in engineering, manufacturing, project management, operations, and SMA.

- Quality assurance per *NPD 8730.5, NASA Quality Assurance Program Policy*; *NPD 8730.2, NASA Parts Policy*; and *NPR 8735.1, Procedures for Exchanging Parts, Materials, Software, and Safety Problem Data Utilizing the Government-Industry Data Exchange Program (GIDEP) and NASA Advisories.*

- Limiting the generation of orbital debris per *NPR 8715.6*;

- Compliance verification, audit, SMA reviews and SMA process maps per *NPR 8705.6, Safety and Mission Assurance (SMA) Audits, Reviews, and Assessments.*

- Reliability and maintainability per *NPD 8720.1, NASA Reliability and Maintainability (R&M) Program Policy.*

- Software safety and assurance per *NASA-STD-8719.13, NASA Software Safety Standard* and *NASA-STD-8739.8, Software Assurance Standard.*

- Software engineering requirements for safety critical software per *NPR 7150.1, NASA Software Engineering Requirements.*

- Quality Assurance functions per *NPR 8735.2, Management of Government Quality Assurance Functions for NASA Contracts.*

- Other applicable NASA procedural safety and mission success requirements.

In the SMA Plan, the project describes how it will develop and manage a closed-loop corrective action system/problem reporting and resolution system and how it develops, tracks, and resolves problems. The process needs to include a well-defined process for data collection and a data collection system for hardware and software problem and anomaly reports, problem analysis, and corrective action. The SMA Plan is required to be a stand-alone plan unless an alternate approach is approved.

At SDR/MDR, the project updates the Technology Development Plan, baselined at MCR. This plan may be part of the Formulation Agreement rather than a separate plan.

The project develops the preliminary Information Technology Plan by SRR and baselines the plan by SDR/MDR. This plan describes how the project will acquire and use IT. It documents the project's approach to implementing IT security requirements in accordance with *NPR 2810.1, Security of Information Technology* with emphasis on conducting the Information/System Security Categorization required by NPR 2810.1 for IT systems during Phase A of the project. The plan describes:

- The steps the project will take to ensure that the information technology it acquires and/or uses complies with *NPR 2830.1, NASA Enterprise Architecture Procedures*;

- How the project will manage information throughout its life cycle, including the development and maintenance of an electronic project library;

- How the project will ensure identification, control, and disposition of project records in accordance with *NPD 1440.6, NASA Records Management* and *NPR 1441.1, NASA Records Retention Schedules*;

- The project's approach to knowledge capture, as well as the methods for contributing knowledge to other entities and systems, including compliance with *NPD 2200.1, Management of NASA Scientific and Technical Information* and *NPR 2200.2, Requirements for Documentation, Approval, and Dissemination of NASA Scientific and Technical Information.*

The project develops one or more preliminary **Software Management Plan(s)** by SRR and baselines them by SDR/MDR. This plan summarizes how the project will develop and/or manage the acquisition of software required to achieve project and mission objectives. It is developed as a stand-alone Software Management Plan that includes the content required by *NPR 7150.2, NASA Software Engineering Requirements* and *NASA-STD-8739.8, Software Assurance Standard*, unless approved otherwise. The plan needs to be aligned with the SEMP.

The project develops a preliminary **Verification and Validation Plan by SDR/MDR**. (This plan is baselined at PDR.) This plan summarizes the project team's approach for performing V&V of the project products. It indicates the methodology to be used in the V&V (test, analysis, inspection or demonstration) as defined in NPR 7123.1. At this point in time, the level of detail is consistent with the level of detail of the concept/design.

The project updates the preliminary **Review Plan presented at MCR**, baselines the plan at SRR, and updates it at SDR/MDR. This plan summarizes the project's approach for conducting a series of reviews, including internal reviews and project life-cycle reviews in accordance with Center best practices, program review requirements, and the requirements in NPR 7123.1 and NPR 7120.5. The Review Plan identifies the life-cycle reviews the project plans to conduct and the purpose, content, and timing of those life-cycle reviews, and documents any planned deviations or waivers granted from the requirements in NPR 7123.1 and NPR 7120.5E. It also provides the technical, scientific, schedule, cost, and other criteria that will be utilized in the consideration of a Termination Review.

The project baselines the **Environmental Management Plan** by SDR/MDR. This plan describes the activities to be conducted at all project locations with support from the responsible Environmental Management Office to comply

For projects that are part of tightly coupled programs, project life-cycle reviews and Key Decision Points (KDPs) are planned in accordance with the project life cycle and KDP sequencing guidelines in the Program Plan. The Review Plan documents the sequencing of each project life-cycle review and KDP with respect to the associated program life-cycle review and KDP. In addition, the Review Plan documents which project KDPs are conducted simultaneously with other projects' KDPs and which project KDPs are conducted simultaneously with the associated program KDPs. The sequencing of project life-cycle reviews and KDPs with respect to program life-cycle reviews and KDPs is especially important for project Preliminary Design Review (PDR) life-cycle reviews that precede KDP Cs. Since changes to one project can easily impact other projects' technical, cost, and schedule baselines, and potentially impact other projects' risk assessments and mitigation plans, projects and their program generally need to proceed to KDP C/KDP I together.

with *NPR 8580.1, NASA National Environmental Policy Act Management Requirements* and Executive Order 12114. Specifically, the plan:

- Identifies all required permits, waivers, documents, approvals, or concurrences required for compliance with applicable Federal, State, and Tribal Government, and local environmental laws and regulations;

- Plans the level of National Environmental Policy Act (NEPA) documentation required to satisfy NEPA requirements prior to KDP C and Project Implementation; e.g., the Environmental Checklist and Record of Environmental Consideration (REC), Environmental Assessment (EA), and Environmental Impact Statement (EIS). The project's plans are based on the program's NEPA strategy at all affected Centers, which is developed in consultation with the NASA Headquarters NEPA coordinator to ensure the most effective, least resource-intensive strategy to meet NEPA requirements across the program and its constituent projects.

- Describes the documentation and schedule of events for complying with these regulations, including identifying any modifications to the Center's Environmental Management System (EMS) that would be required for compliance; and

- Defines the critical milestones associated with complying with these regulations that need to be inserted into the project schedule.

Smaller projects and projects with limited environmental impacts may be able to satisfy this requirement, with approval from the requirements holder, using the "Project Environmental Management Planning Checklist," which can be found on the EPPMD Community of Practice. A checklist is completed for each NASA Center or facility that the project will use.

The project develops a preliminary **Integrated Logistics Support (ILS) Plan** by SRR and updates it at SDR/MDR (the plan is baselined at PDR). This plan describes how the project will implement *NPD 7500.1, Program and Project Life Cycle Logistics Support Policy*, including a maintenance and support concept; participation in the design process to enhance supportability; supply support; maintenance and maintenance planning; packaging, handling, and transportation; technical data and documentation; support and test equipment; training; manpower and personnel for ILS functions; facilities required for ILS functions; and logistics information systems for the life of the project.

The project develops a preliminary **Integration Plan** by SDR/MDR. This plan defines the integration and verification strategies and is structured to show how components come together to assemble each subsystem and how the subsystems are assembled into the system/product. The primary purposes of the Integration Plan are to: (1) describe this coordinated integration effort that supports the implementation strategy, (2) describe for the participants what needs to be done in each integration step, and (3) identify the required resources and when and where they will be needed.

The project baselines the Configuration Management Plan by SRR and updates it at SDR/MDR. This plan describes the configuration management approach that the project team will implement, consistent with NPR 7123.1 and *NASA-STD-0005, NASA Configuration Management (CM) Standard*. It describes:

- The structure of the configuration management organization and tools to be used;
- The methods and procedures to be used for configuration identification, configuration control, interface management, configuration traceability, and configuration status accounting and communications;
- How configuration management will be audited; and
- How contractor configuration management processes will be integrated with the project.

The project develops a preliminary Security Plan by SDR/MDR. This plan describes the project's plans for ensuring security and technology protection. It includes three types of requirements: security, IT security, and emergency response. It describes the project's approach for planning and implementing the requirements for information, physical, personnel, industrial, and counterintelligence/counterterrorism security and for security awareness/education requirements in accordance with *NPR 1600.1, NASA Security Program Procedural Requirements* and *NPD 1600.2, NASA Security Policy*. It includes provisions to protect personnel, facilities, mission-essential infrastructure, and critical project information from potential threats and other vulnerabilities that may be identified during the threat and vulnerability process. IT security requirements document the project's approach to implementing requirements in accordance with *NPR 2810.1, Security of Information Technology*. The plan also describes the project's emergency response plan to meet the emergency response requirements in *NPR 1040.1, NASA Continuity of Operations (COOP) Planning Procedural Requirements* and defines the range and scope of potential crises and specific response actions, the timing of notifications and actions, and the responsibilities of key individuals.

The project manager ensures development of a preliminary Project Protection Plan by SDR/MDR. For more information on about Project Protection Plan specifics, go to the Community of Practice for Space Asset Protection at https://nen.nasa.gov/web/sap. The project develops a preliminary Technology Transfer Control Plan by SDR/MDR. It describes how the project will control its technology to implement the export control requirements specified in *NPR 2190.1, NASA Export Control Program.*

The project develops a preliminary **Knowledge Management Plan** by SDR/MDR. This plan describes the project's approach to creating the knowledge management strategy and processes, including practices and approaches for identifying, capturing and transferring knowledge; and capturing and documenting lessons learned throughout the project life cycle in accordance with NPD 7120.4 and as described in *NPD 7120.6, Knowledge Policy on Programs and Projects* and other appropriate requirements and standards documentation.

The **Human Rating Certification Package (HRCP)** is required for human space flight missions. If the program has done an HRCP, the project may refer to the program HRCP. The initial HRCP is developed by SRR and updated at SDR/MDR. The HRCP is developed per NPR 8705.2. Human rating certification focuses on the integration of the human into the system, preventing catastrophic events during the mission, and protecting the health and safety of humans involved in or exposed to space activities, specifically the public, crew, passengers, and ground personnel.

The project prepares a **Planetary Protection Plan** by SDR/MDR that specifies management aspects of the planetary protection activities of the project. Planetary protection encompasses: (1) the control of terrestrial microbial contamination associated with space vehicles intended to land, orbit, flyby, or otherwise encounter extraterrestrial solar system bodies and (2) the control of contamination of the Earth by extraterrestrial material collected and returned by missions. The scope of the plan contents and level of detail will vary with each project based upon the requirements in NASA policies *NPR 8020.12, Planetary Protection Provisions for Robotic Extraterrestrial Missions* and *NPD 8020.7, Biological Contamination Control for Outbound and Inbound Planetary Spacecraft.* The project also obtains a **planetary protection certification** for the mission (if required) in accordance with these two policy documents.

The project baselines a **Nuclear Safety Launch Approval Plan** for missions with nuclear materials. Planning begins in Formulation, and the Plan is baselined at SDR/MDR. This plan documents the project's approach for meeting the nuclear safety requirements in *NPR 8715.3, General Safety Program Requirements.* NPR 8715.3, Chapter 6 specifies the internal NASA procedural requirements for characterizing and reporting potential risks associated with a planned launch of radioactive materials into space, on launch vehicles and spacecraft, and during normal or abnormal flight conditions. Procedures and levels of review and analysis required for nuclear launch safety approval vary with the quantity of radioactive material planned for use and the potential risk to the general public and the environment. NPR 8715.3 requirements include identification of the amount of radioactive material; developing a schedule for nuclear launch safety approval activities; identifying the process for documenting the risk represented by the use of radioactive materials; identifying and developing required analyses and reports, which may include the Radioactive Materials Onboard Report, the Safety Analysis Summary (SAS), and the Safety Analysis Report; conducting the nuclear safety analysis; developing the nuclear safety review; obtaining nuclear safety concurrence; and obtaining approval for launch of nuclear materials.

The launch of any radioactive material requires some level of analysis, review, reporting, notification, and approval. The requirements for the level of analysis, review, reporting, notification, and approval of missions involving radioactive material is dependent on the A_2 mission multiple, which is an International Atomic Energy Agency (IAEA) measure of radioactive material. (Specific details for calculating the A_2 mission multiple are provided in NPR 8715.3, Appendix D.) Table 4-5 depicts these requirements including levels of approval and timeframes for the notification and approval process. The launch approval process can take more than 3 years.

The NASA Nuclear Flight Safety Assurance Manager (NFSAM) is the person appointed by the Chief, Safety and Mission Assurance to help projects meet the required nuclear launch safety requirements. The project works with and

Table 4-5 Approval of Missions Involving Radioactive Material

A_2 Mission Multiple	Launch Reported to NFSAM	Launch Reported to OSTP	Submit Request for Launch Concurrence or SAR (Months Before Launch)[1]	Required Level of Review and Reports	Approval/ Concurrence
$A_2 < 0.001$	YES	NO	>4 months	Radioactive Material On-Board Report (RMOR) and Launch Request	Concurrence Letter from NFSAM
$0.001 < A_2 < 10$	YES	YES	>4 months	RMOR and Launch Request	Concurrence Letter from NFSAM
$10 < A_2 < 500$	YES	YES	>5 months	Nuclear Safety Review of Radiological Risk	Approval Letter from Chief, OSMA
$500 < A_2 < 1000$	YES	YES	>6 months	Safety Analysis Summary (SAS)	Approval Letter from NASA Administrator
$1000 < A_2$	YES	YES	>10 months	INSRP Safety Analysis Report (SAR)	NASA Administrator Requests Approval From Executive Office of the President (via Director, OSTP)

[1] For missions with A_2 less than 1000, the project needs to submit the Request for Launch Concurrence no later than the indicated number of months prior to launch. For missions with A_2 greater than 1000, the project needs to submit the SAR no later than 10 months before launch. The end-to-end launch approval process, including safety analysis/risk assessment and the INSRP process, can take more than three years.

through their MDAA program executive to coordinate with the NFSAM to obtain nuclear launch safety approval or launch concurrence. This includes notifying the NASA NFSAM in writing as soon as radioactive sources are identified for potential use on NASA spacecraft to schedule nuclear launch safety approval activities; identifying the amount of radioactive material and the process for documenting the risk represented by the use of radioactive materials to the NFSAM; and developing and providing required analyses and reports to the NFSAM.

A nuclear safety evaluation may be required by an Interagency Nuclear Safety Review Panel (INSRP)[19] in obtaining nuclear launch safety approval. In such cases, the NASA Administrator requests empanelment of an INSRP, and appoints a NASA member to the panel (with consideration of the recommendation from the Chief, Safety and Mission Assurance). The NASA coordinator is the person appointed by the Chief, Safety and Mission Assurance to coordinate NASA's participation in activities supporting the empanelled INSRP to ensure adequate information is available to the INSRP. The INSRP receives and reviews the Safety Analysis Report (SAR) developed by the project and prepares a Safety Evaluation Report (SER). The project prepares and coordinates the Nuclear Launch Safety Approval Request based on information in the SER. The Request is signed by the NASA Administrator for submittal to the Office of the President. The Office of the President renders a Nuclear Launch Safety Approval decision and notifies NASA in writing of the results. A positive Nuclear Launch Safety Approval decision is mandatory for launch.

In coordination with the program executive, projects involving the launch of radioactive materials also need to:

- Comply with the provisions of the National Environmental Policy Act of 1969, in accordance with the policy and procedures contained in 14 CFR Part 1216, Subpart 1216.3, Procedures for Implementing the National Environmental Policy Act (NEPA), *NPR 8580.1, Implementing the National Environmental Policy Act*, and Executive Order 12114, and

- Develop radiological contingency plans in accordance with *NPD 8710.1, Emergency Preparedness Program*, and *NPR 8715.2, NASA Emergency Preparedness Plan Procedural Requirements*.

---

[19] The Interagency Nuclear Safety Review Panel (INSRP) is an ad-hoc panel assembled for nuclear missions with greater than 1000 A2 quantities of radioactive material. The INSRP consists of members who are subject matter experts from the Department of Defense, Department of Energy, NASA and the Environmental Protection Agency. The INSRP evaluates the risks associated with missions requiring the President's approval and prepares a Nuclear Safety Evaluation Report to the NASA Administrator and the Office of Science and Technology Policy (OSTP).

4.3.5 Completing Concept and Technology Development (Phase A) Activities and Preparing for Preliminary Design and Technology Completion (Phase B)

4.3.5.1 Establishing the Project's Preliminary Baseline

As the project approaches SDR/MDR and KDP B, the project team finalizes the project's preliminary baselines, described in detail in Section 4.3.4.2.2— technical (including requirements), resource (including funding, NOA, infrastructure and staffing), and cost and schedule. The project documents the driving ground rules, assumptions, and constraints affecting the resource baseline. During Phase B, these preliminary baselines continue to be updated in preparation for project approval at KDP C.

4.3.5.2 Finalizing Plans for Phase B

The project develops its plans for work to be performed during the subsequent life-cycle phases, including generation of life-cycle review plans and the project IMS, details on technical work to be accomplished, key acquisition activities planned, and plans for monitoring performance against plan. As the project approaches the SDR/MDR review and KDP B, the project updates its project Formulation Agreement to finalize the plans for Phase B.

The project prepares and finalizes work agreements for Phase B. These work agreements can be between Centers or between organizations within a Center. They are usually used by the project to gain commitments from the performing organizations for the scope of work, the cost to perform that work, and the schedule for delivering the products for the next phase.

The project documents the results of Phase A activities and generates the appropriate documentation per NPR 7123.1 and NPR 7120.5E Tables I-4 and I-5 and Tables 4-6 and 4-7 at the end of this chapter. Documentation requirements may be satisfied by including in the Formulation Agreement the basis of cost and schedule estimates, draft and preliminary versions of project documents and plans, and/or the SDR/MDR briefing package.

4.3.5.3 Project Phase A Reporting Activities and Preparing for Major Milestones

4.3.5.3.1 Project Reporting

The project manager reports to the Center Director or designee (see Section 4.2.2.2) and supports the program executive in reporting the status of project Formulation at many other forums, including Mission Directorate monthly status meetings and the Agency's monthly BPR. Section 5.12 provides further information regarding potential project reporting.

4.3.5.3.2 Project Internal Reviews

Prior to life-cycle reviews, projects conduct internal reviews in accordance with NPR 7123.1, Center practices, and NPR 7120.5. These internal reviews are described in Section 4.3.2.2.2.

4.3.5.3.3 Preparing for Major Milestone Reviews

Projects support the SRR and SDR/MDR life-cycle reviews in accordance with NPR 7123.1, Center practices, and NPR 7120.5, ensuring that the life-cycle review objectives and expected maturity states defined in NPR 7120.5 have been satisfactorily met. Life-cycle review entrance and success criteria in Appendix G of NPR 7123.1 and the life-cycle phase and KDP information in Appendix E of this handbook provide specifics for addressing the six assessment criteria required to demonstrate the project has met its expected maturity state. The *NASA Standing Review Board Handbook* also provides additional detail on this process for those reviews requiring an independent SRB.

Projects plan, prepare for, and support the governing PMC review prior to KDP B and provide or obtain the KDP readiness products listed in Section 4.2.3.

Once the KDP has been completed and the Decision Memorandum signed, the project updates its documents and plans, as required, to reflect the decisions made and actions assigned at the KDP.

For tightly coupled and single-project programs, project(s) transition to KDP B in accordance with the Review Plan documented in the Program or Project Plan(s).

4.3.6 Project Phase B, Preliminary Design and Technology Completion Activities

4.3.6.1 Project Phase B Life-Cycle Activities

Project Formulation completes with the second of two sequential phases, Phase B (Preliminary Design and Technology Completion). The purpose of Phase B is for the project team to complete their technology development; engineering prototyping; heritage hardware and software assessments using the *Systems Engineering Handbook, NASA/SP-2007-6105 Rev 1*, Appendix G; and other risk-mitigation activities identified in the project Formulation

Agreement and to complete the preliminary design. The project demonstrates that its planning, technical, cost, and schedule baselines developed during Formulation are complete and consistent; the preliminary design complies with its requirements; the project is sufficiently mature to begin Phase C; and the cost and schedule are adequate to enable mission success with acceptable risk. It is at the conclusion of this phase that the project and the Agency commit to accomplishing the project's objectives for a given cost and schedule. For projects with an LCC greater than $250 million, this commitment is made with the Congress and the U.S. Office of Management and Budget (OMB). This external commitment is the ABC.

Phase B Formulation continues to be an iterative set of activities, rather than discrete linear steps. These activities are focused toward baselining the Project Plan, completing the preliminary design, and assuring that the systems engineering activities are complete to ensure the design is feasible for proceeding into Implementation. Phase B completes when the Decision Authority approves transition from Phase B to Phase C at KDP C. The major project life-cycle review leading to approval at KDP C is the Preliminary Design Review (PDR).

The **objectives of the PDR** are to evaluate the completeness and consistency of the planning, technical, cost, and schedule baselines developed during Formulation; to assess compliance of the preliminary design with applicable requirements; and to determine if the project is sufficiently mature to begin Phase C.

At the KDP C, the project is expected to demonstrate that the objectives of the PDR have been met and the approved cost and schedule are adequate to enable mission success with acceptable risk.

The general flow of activities for a project in Phase B is shown in Figure 4-11.

4.3.6.2 Project Phase B Management, Planning, and Control Activities

4.3.6.2.1 Supporting Headquarters Planning

During Phase B, the project manager and project team support the program manager and the MDAA in maintaining the baseline program requirements and constraints on the project, including mission objectives/goals; mission success criteria; and driving mission, technical, and programmatic ground rules and assumptions. The project obtains an update to these, if needed, and updates the project's documentation and plans accordingly.[20] In doing

---

[20] Program requirements on the project are contained in the Program Plan.

KDP B | | KDP C
PDR

Finalize project office, mgmt structure, & mgmt processes; continue to perform management, planning, & control functions

Support MD and program in maintaining requirements, etc., and alignment with Agency goals as required, based on MDR/SDR and continuing development

Support MD and OIIR in finalizing and baselining partnerships and agreements

Support MD & program in completing env. planning process & developing preliminary Mishap Preparedness & Contingency Plan

Coordinate with HEOMD for space transportation, space communications, navigation capabilities, and launch services

Continue developing and updating the concept, architectures, and ops plans based on results of SRR; continue to perform analyses, trades, etc., in support of design refinement; finalize, baseline, and present baseline at PDR

Update mission objectives and project-level and system-level requirements; baseline requirements down to the subsystem level

Implement technology development plans; continue to assess technology readiness; update plans as concept/design evolves

Implement eng. dev. plans; continue to assess engineering risk reduction activities; update plans as concept/design evolves

Assess evolving designs to ensure heritage is applied properly; determine and implement risk reduction activities

Develop preliminary safety data packages and expendable launch vehicle payload safety process deliverables (if applicable)

Develop preliminary orbital debris assessment and perform IT system risk assessment

Develop/update technical control plans, as required for PDR

Confirm, refine, and update key ground rules and assumptions that drive development of concept, design, ops concept, and risk reduction activities as concept/design matures

Update risk lists, mitigations, and resource requirements in preparation for PDR

Update staffing and infrastructure requirements and plans as concept design evolves; update business case analysis; complete Alternative Future Use Questionnaire

Implement Acquisition Plan; update Plan at PDR, as required; begin procurement of approved long-lead items during Phase B

Update and baseline risk-informed, cost/resource-loaded IMS to the subsystem level

Update and baseline risk-informed, schedule-adjusted cost estimate; develop and baseline the JCL, if required

Update bases of estimates and baseline cost and schedule commitments

Develop/update management control plans, as required for PDR

Finalize and baseline Project Plan and present Plan at PDR

Develop project's plans for follow-on phases

Report plans, progress, and results at CMCs, life-cycle reviews, PMCs, and KDP C and in other forums and media, as required

Conduct preapproval project-level IBR; prepare for PDR and KDP C

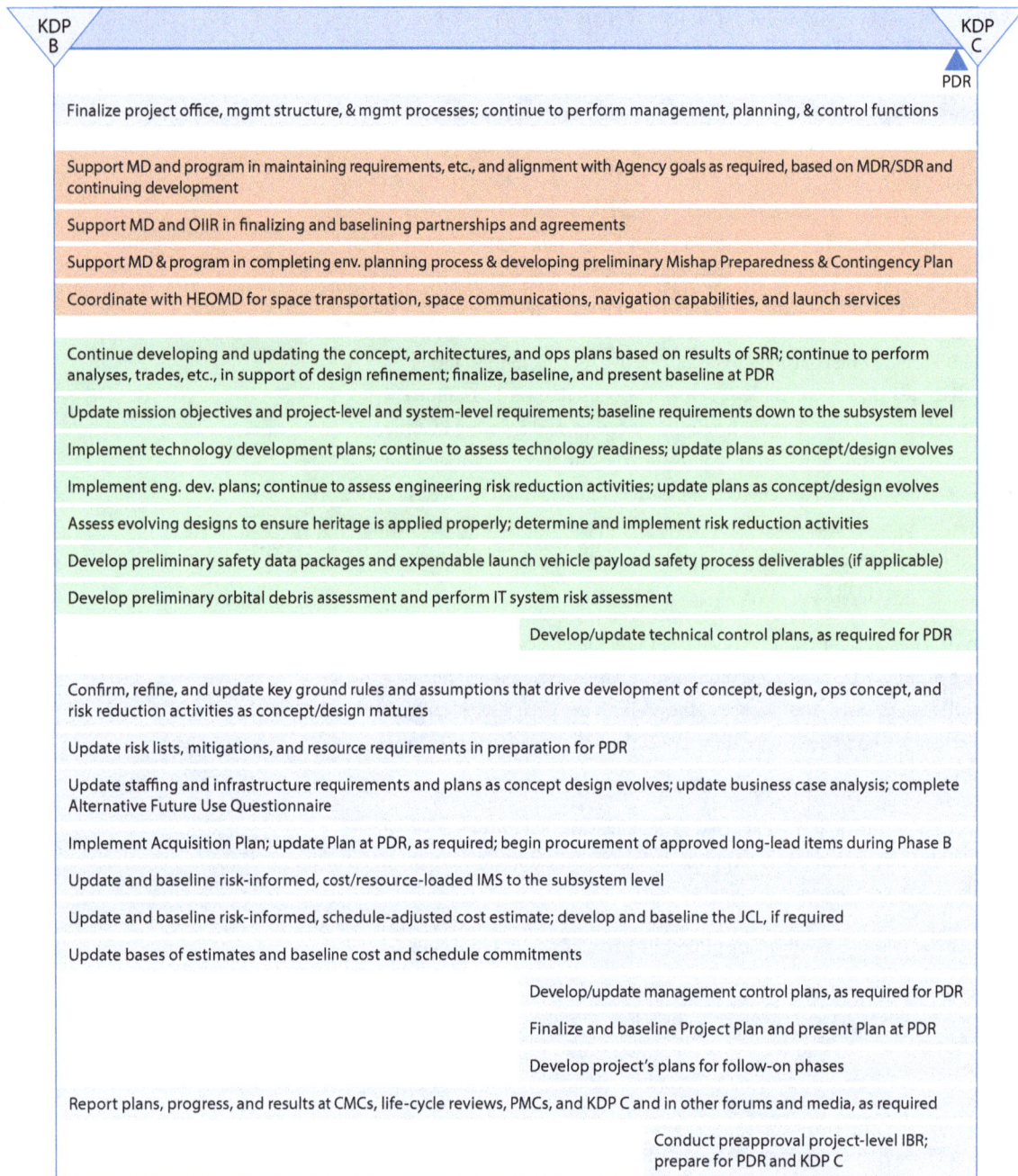

Legend: ☐ Project management, planning, and control tasks
☐ Work for which Headquarters is responsible but the project helps accomplish (e.g., international partnerships are a Headquarters responsibility, but the projects help develop and finalize those partnerships)
☐ Technical work the project is doing

Acronyms: MD = Mission Directorate; OIIR = Office of International and Interagency Relations.

Note: These are typical high-level activities that occur during this project phase. Placement of reviews is notional.

Figure 4-11 Project Phase B Flow of Activities

this, the project supports the program manager and the MDAA in ensuring the continuing alignment of the project requirements, design approaches, and conceptual design with applicable Agency strategic goals.

The project works with the program, Mission Directorate, and other NASA organizations to obtain approval of any necessary modifications to prescribed requirements with waivers or deviations that are updated and reflected in modifications to the Formulation Agreement. Approved waivers and deviations that apply to project activities during Implementation are documented in the Project Plan.

In coordination with the program manager, the MDAA, and the NASA Headquarters Office of International and Interagency Relations (OIIR), the project manager supports the finalization and baselining of external agreements, such as interagency and international agreements (including the planning and negotiation of agreements and recommendations on joint participation in reviews, integration and test, and risk management), if applicable.

The project works with the program and the Mission Directorate to complete the environmental planning process as explained in NPR 8580.1 and planned in the project's Environment Management Plan. This includes preparing the final NEPA documentation. (Note: For certain projects utilizing nuclear power sources, completion of the environmental planning process can be extended, with the approval of the Decision Authority, into Phase C, but need to be completed by the project Critical Design Review (CDR)).

The project works with the program and the Mission Directorate to develop a preliminary Mishap Preparedness and Contingency Plan in accordance with *NPR 8621.1, NASA Procedural Requirements for Mishap and Close Call Reporting, Investigating, and Recordkeeping.*

The project coordinates with HEOMD, including LSP, to schedule space transportation services, space communication and navigation capabilities, or launch services, if applicable, in compliance with NPD 8610.7 and NPD 8610.12.

4.3.6.2.2 Management Control Processes and Products

The project team continues planning that enables formulation and implementation of the mission concept(s), architectures, scenarios/DRMs and requirements, and implement the Formulation Agreement as updated at KDP B. The project team continues to work with the Center to obtain support for the project.

The project updates the Technical, Schedule, and Cost Control Plan as necessary to reflect adjustments to the project management approach. It continues to monitor and control the project requirements, technical design, schedule, and cost of the project to ensure that the high-level requirements levied on the project are met.

The project further confirms, refines, and updates the project's key ground rules and assumptions that drive implementation of the mission design and the funding profiles and schedules necessary for Phases C through F. The project continues to track them through Formulation to determine if they are being realized (i.e., remain valid) or if they need to be modified.

As the design matures, the project team updates their assessment of potential infrastructure and workforce needs versus current plans, as well as any further opportunities to use infrastructure and workforce in other Government agencies, industry, academia, and international organizations. Based on this assessment, the project team updates the requirements and plans for staffing and infrastructure at PDR. As part of this activity, the project updates the business case analysis[21] for infrastructure for each proposed project real property infrastructure investment consistent with *NPD 8820.2, Design and Construction of Facilities* and *NPR 8820.2, Facility Project Requirements* and, for the acquisition of new aircraft, consistent with NPR 7900.3. This analysis needs to be completed in sufficient time to allow the analysis, documentation, review, approval, and funding of the infrastructure in time to support the mission requirements.

Also in coordination with the OCFO and in accordance with *NPR 9250.1, Property, Plant, and Equipment and Operating Materials and Supplies*, projects complete the Alternative Future Use Questionnaire (Form NF 1739),[22] Section B, to identify the acquisition components of the project and to determine the appropriate accounting treatment of the capital acquisitions within the project. Once completed, projects forward the questionnaire to the OCFO, Property Branch.

The project team implements its plans for acquisition in accordance with its approved Acquisition Plan. The project finalizes its plans and executes long-lead procurements. (Long-lead procurements can be initiated in Phase B only when specifically approved by the Mission Directorate and/or program.) In accordance with the approved Acquisition Plan, the project

---

[21] Business case analyses require the approval of the MDAA and the Assistant Administrator for Strategic Infrastructure or designee. See the *NASA Business Case Guide for Facilities Projects* at http://www.hq.nasa.gov/office/codej/codejx/Assets/Docs/Case_Guide_4-20-06.pdf.

[22] The questionnaire can be found in NASA's Electronics Forms Database website: http://itcd.hq.nasa.gov/eforms.html.

also updates, identifies, assesses, and mitigates (if feasible) supply chain risks, including critical or single-source suppliers needed to design, develop, produce, and support required capabilities at planned cost and schedule and report risks to the program. The Acquisition Plan is updated at PDR to reflect any adjustments to procurement plans for the following phases.

The Acquisition Plan describes the IBRs and schedules required for contracts requiring EVM (refer to the NFS), how the project needs to conduct any required IBRs, and how to maintain the contract documentation.

The project updates the Risk Management Plan. As the concept and design evolve, the project continues to identify, assess, and update the technical, cost, schedule, and safety risks that threaten the system development, approved mission concept, operations concept, and technology development. Risks include but are not limited to technology development, engineering development, payload (robotic space flight), and procurement risks; risks associated with use of heritage hardware and software; and risks that are likely to drive the project's cost and schedule estimates at KDP C. The project identifies risk mitigation plans and associated resources for managing and mitigating risks in accordance with the Risk Management Plan.

The project team finalizes and baselines the Education and Communications Plans, which were developed in preliminary form during Phase A.

The project prepares the final Project Plan that follows the template in Appendix G of NPR 7120.5 and has the plan ready for baselining at PDR/ KDP C. (See the product maturity Tables 4-6 and 4-7 at the end of this chapter or NPR 7120.5E Tables I-4 and I-5) for a list of required control plans and their required maturity by phase.)

The project continues to update its cost and schedule estimates as the design matures. As the project approaches PDR, the project finalizes its cost and schedule estimates in preparation for establishing the project's baseline at KDP C.

The results of this work include:

- Risk-informed and a cost-loaded or resource-loaded IMS;
- Risk-informed and schedule-adjusted cost estimate;
- JCL for projects greater than $250 million, consistent with the confidence level approved by the Decision Authority (see Section 5.7 for more information on the JCL); UFE and schedule margins that have been determined by the confidence level provided by the joint cost and schedule

calculations. (For projects that are not required to perform probabilistic analysis, the UFE is informed by the project's unique risk posture in accordance with Mission Directorate and Center guidance and requirements. The rationale for the UFE, if not conducted via a probabilistic analysis, is appropriately documented and is traceable, repeatable, and defendable.);

- Proposed annual estimated costs by Government fiscal year and by the project's WBS;

- Assessment of the consistency of the time-phased Government Fiscal Year (GFY) LCCE with anticipated budget availability.

    - Proposed external cost and schedule commitments, if applicable; Updated basis for cost and schedule estimates at PDR. A BoE documents the ground rules, assumptions, and drivers used in cost and schedule estimate development and includes applicable model inputs, rationale/justification for analogies, and details supporting bottom-up cost and schedule estimates. Good BoEs are well-documented, comprehensive, accurate, credible, traceable, and executable. Sufficient information on how the estimate was developed needs to be included to allow review team members, including independent cost analysts, to reproduce the estimate if required. Types of information can include estimating techniques (e.g., bottoms-up, vendor quotes, analogies, parametric cost models), data sources, inflation, labor rates, new facilities costs, operations costs, sunk costs, etc.

These products provide for adequate technical, schedule, and cost margins and incorporate the impacts of performance to UFE and schedule margin. At any point, a convening authority can request an Independent Cost Assessment (ICA) and/or an Independent Cost Estimate (ICE) from either the internal independent review board (e.g., the SRB) or from external organizations outside of NASA (e.g., Aerospace Corp.). Multiple cost estimates are reconciled by identifying the key differences in underlying assumptions used for the various estimate models, risks, and sensitivities to the project, and briefing the results to the convening authorities to enable the Decision Authority to make an informed decision. For LCCEs, the result of the reconciliation is a recommendation to the Decision Authority on what the LCCE needs to be. For projects with an LCC greater than $250 million, the goal is to provide sufficient understanding of the risks and associated impacts on cost and schedule to allow determination of a cost estimate and its associated confidence levels consistent with the estimate NASA commits to the external stakeholders. The estimates can be reconciled through the independent review process, the management review process (e.g., the DPMC) or at the KDP, which is the last point for reconciliation.

Space flight projects baseline a CADRe (parts A, B, and C) consistent with the *NASA Cost Estimating Handbook* and tailoring guidance for small projects 60 days prior to KDP C with an update 30 to 45 days after the event to reflect any changes from the KDP. This CADRe is based on the project baselines at PDR.

4.3.6.3 Project Phase B Technical Activities and Products

The project team continues developing the concept and architecture of the project, its major components, and the way they will be integrated, including its operations concepts through the system engineering process described in NPR 7123.1. The project continues engineering development activities (e.g., engineering models, brass boards, bread boards, test beds, and full-up models) and incorporates the results into the preliminary design. As the project approaches the PDR, the project finalizes and baselines the concept, mission and spacecraft architectures, launch service requirements (with LSP support as described in Section 4.3.4.3), the ground and payload architectures, and the operations concept. The project updates the mission objectives and project-level and system-level requirements as needed, and baselines the subsystem-level requirements. In support of the launch service procurement, if applicable, the project completes the spacecraft-to-launch vehicle Interface Requirements Document (IRD), which becomes an input to the Request for Launch Services Proposal that is developed by LSP. In addition, the project typically supports the evaluation of such proposals. (The project's level of involvement in evaluating such proposals is per mutual agreement between the project and LSP.) The project ensures that all requirements are traceable back to the program-level requirements on the project and develops an updated list of descope options in case some requirements cannot be met.

The project develops, documents, and baselines the preliminary design documentation as described in NPR 7123.1 as a minimum.

The project completes its risk reduction and mitigation activities and updates its technology, engineering, and heritage assessments. The project completes mission-critical or enabling technology, as needed, to the level of a system/subsystem model or prototype demonstration in a relevant environment (ground or space) (Technology Readiness Level (TRL) 6 by KDP C) unless otherwise documented in the Technology Development Plan. It also finishes its engineering model and prototype developments.

The project develops preliminary Safety Data Packages and other safety process deliverables as required by NPR 7120.5 and the NPRs and NPDs identified below. Currently, the requirements for the safety data packages can be found in the following documents:

- Safety per NPR 8715.3, *NPR 8715.5, Range Flight Safety Program, NPR 8715.7, Expendable Launch Vehicle Payload Safety Program*, and local range requirements; *NPR 8621.1, NASA Procedural Requirements for Mishap and Close Call Reporting, Investigating, and Recordkeeping.*

- Human rating requirements per NPR 8705.2.

- Quality assurance per *NPD 8730.5, NASA Quality Assurance Program Policy, NPD 8730.2, NASA Parts Policy*, and *NPR 8735.1, Procedures for Exchanging Parts, Materials, Software, and Safety Problem Data Utilizing the Government-Industry Data Exchange Program (GIDEP) and NASA Advisories.*

- Limiting the generation of orbital debris per NPR 8715.6.

The project develops a preliminary orbital debris assessment in accordance with *NPR 8715.6, NASA Procedural Requirements for Limiting Orbital Debris* using the format and requirements contained in NASA-STD-8719.14.

The project performs an IT system risk assessment required by NPR 2810.1 during Phase B of the project.

The project documents lessons learned in accordance with the project's Knowledge Management Plan.

Based on the evolving design, the project team updates the following control plans: Safety and Mission Assurance (SMA) Plan, Technology Development Plan, SEMP, Information Technology (IT) Plan, Software Management Plan(s), Review Plan, and Configuration Management Plan.

The project team finalizes the Verification and Validation (V&V) Plan and baselines the plan by PDR. This plan summarizes the approach for performing V&V of the project products.

The project finalizes the Integrated Logistics Support (ILS) Plan in accordance with *NPD 7500.1, Program and Project Life Cycle Logistics Support Policy* and baselines the plan by PDR.

The project develops a preliminary Science Data Management Plan by PDR (the plan is baselined at ORR). This plan describes how the project will manage the scientific data generated and captured by the operational mission(s) and any samples collected and returned for analysis. It includes descriptions of how data will be generated, processed, distributed, analyzed, and archived, as well as how any samples will be collected, stored during the mission, and managed when returned to Earth. The plan typically includes definition of data rights and services and access to samples, as appropriate, and explains how the project will accomplish the knowledge capture and information management and disposition requirements in *NPD 2200.1,*

Management of NASA Scientific and Technical Information; NPR 2200.2, Requirements for Documentation, Approval, and Dissemination of NASA Scientific and Technical Information; and *NPR 1441.1, NASA Records Retention Schedules* as applicable to project science data.

The project finalizes the Integration Plan and baselines the plan by PDR. This plan defines the integration and verification strategies and is structured to show how components come together to assemble each subsystem and how all of the subsystems are assembled into the system/product.

The project finalizes the Security Plan and baselines the plan by PDR. This plan describes the project's plans for ensuring security and technology protection.

The project finalizes the Project Protection Plan, Technology Transfer Control Plan,[23] Knowledge Management Plan, and Planetary Protection Plan, if applicable, and baselines the plan(s) by PDR.

If required, the project updates the Human Rating Certification Package by PDR as described in NPR 8705.2. Per NPR 7123.1B, the project updates the Human System Integration Plan, if required, by PDR.

The project develops preliminary Range Safety Risk Management process documentation, in accordance with *NPR 8715.5, Range Flight Safety Program.* This applies to launch and entry vehicle projects, scientific balloons, sounding rockets, drones, and Unmanned Aircraft Systems. This does not apply to projects developing a payload that will that fly onboard a vehicle. The range flight safety concerns associated with a payload are addressed by the vehicle's range safety process. The focus is on the protection of the public, workforce, and property during range flight operations.

4.3.7 Completing Preliminary Design and Technology Completion (Phase B) Activities and Preparing for Final Design and Fabrication (Phase C)

4.3.7.1 Establishing the Project's Baseline

As the project approaches its project approval milestone, KDP C, the project team finalizes its project baselines. This effort is described in more detail in Section 4.3.6.2.2. All projects finalize their project baselines and the Management Agreement as part of the preparations for the PDR. This includes the project's technical baseline, risk posture, IMS, baseline LCCE,

---

[23] This plan describes how the project will implement the export control requirements specified in *NPR 2190.1, NASA Export Control Program.*

and resource baseline, all consistent with the program requirements and constraints on the project, the key assumptions, workforce estimates, and infrastructure requirements. This typically includes an internal review of the entire scope of work with a series of in-depth assessments of selected critical work elements of the WBS prior to and following the project's PDR life-cycle review preceding KDP C. For projects with EVM requirements, the project works with the program and the Mission Directorate to conduct a project-level preapproval IBR as part of the preparations for KDP C to ensure that the project's work is properly linked with its cost, schedule, and risk and that the systems are in place to conduct EVM. Section 5.14 provides additional details on this review.

Once approved at KDP C and documented in the Decision Memorandum, the project baselines are maintained under configuration control. See Section 5.5 for maintaining baselines.

4.3.7.1.1 Finalizing Plans for Phase C

The project develops and updates its plans for work to be performed during Phase C and the subsequent life-cycle phases, including updates, if needed, to life-cycle review plans, the project IMS, details on technical work to be accomplished, key acquisition activities planned, and plans for monitoring performance against plan. The project incorporates the impact of performance against the plan established at KDP B.

The project prepares and finalizes work agreements for Phase C and D. The work scope and price for Phase C and D contracts may be negotiated but not executed prior to approval to proceed at KDP C unless otherwise approved. Once the project has been approved and funding is available, the negotiated contracts may be executed, assuming no material changes.

The project documents the results of Phase B activities and generates the appropriate documentation as described in NPR 7123.1, NPR 7120.5E Tables I-4 and I-5, and Tables 4-6 and 4-7 at the end of this chapter and captures it in retrievable project records.

4.3.7.2 Project Phase B Reporting Activities and Preparing for Implementation Approval Reviews

4.3.7.2.1 Project Reporting

The project manager reports to the Center Director or designee (see Section 4.2.2.2) and supports the program executive in reporting the status of project Formulation at many other forums, including Mission Directorate monthly status meetings and the Agency's monthly BPR. Section 5.12 provides further information regarding potential project reporting.

4.3.7.2.2 Project Internal Reviews

Prior to the life-cycle reviews, projects conduct internal reviews in accordance with NPR 7123.1, Center practices, and NPR 7120.5. These internal reviews are described in Section 4.3.2.2.2.

4.3.7.2.3 Preparing for Project Implementation Approval

Projects support the PDR life-cycle review in accordance with NPR 7123.1, Center practices, and NPR 7120.5, including ensuring that the life-cycle review objectives and expected maturity states defined in NPR 7120.5 have been satisfactorily met. Life-cycle review entrance and success criteria in Appendix G of NPR 7123.1 and the life-cycle phase and KDP information in Appendix E of this handbook provide specifics for addressing the six assessment criteria required to demonstrate the project has met its expected maturity state. The *NASA Standing Review Board Handbook* provides additional detail on this process for those reviews requiring an independent SRB. Projects plan, prepare for, and support the governing PMC review prior to KDP C and provide or obtain the KDP readiness products listed in Section 4.2.3.

Once the KDP has been completed and the Decision Memorandum signed, the project updates its documents and plans, as required, to reflect the decisions made and actions assigned at the KDP.

In tightly coupled and single-project programs, project(s) transition to KDP C in accordance with the Review Plan documented in the Program or Project Plan.

4.4 Project Implementation

4.4.1 Project Phase C, Final Design and Fabrication Activities

4.4.1.1 Project Phase C Life-Cycle Activities

Project Implementation begins with Phase C as the project team implements the project in accordance with the Project Plan. The purpose of Phase C is to:

- Complete and document the final design that meets the detailed requirements;
- Ensure that the systems engineering activities are performed to determine if the design is mature enough to proceed with full-scale implementation within the constraints of the Management Agreement and the ABC;
- Perform qualification testing
- Develop product specifications and begin fabrication of test and flight architecture (e.g., flight article components, assemblies, subsystems, and associated software);
- Develop detailed integration plans and procedures; and
- Ensure that all integration facilities and personnel are ready and available.

For projects that will develop or acquire multiple copies of systems, the project ensures that the system developers are ready to efficiently produce the required number of systems. The general flow of activities for a project in Phase C is shown in Figure 4-12.

These activities are focused toward the CDR, the Production Readiness Review (PRR) (for four or more copies), and the System Integration Review (SIR). Phase C completes when the Decision Authority approves transition from Phase C to Phase D at KDP D:

- The objectives of the CDR are to evaluate the integrity of the project design and its ability to meet mission requirements with appropriate margins and acceptable risk within defined project constraints, including available resources, and to determine if the design is appropriately mature to continue with the final design and fabrication phase.

- The objectives of the PRR are to evaluate the readiness of system developer(s) to produce the required number of systems within defined project constraints for projects developing multiple similar flight or ground support systems and to evaluate the degree to which the production plans meet the system's operational support requirements. (See Table G-8 in Appendix G of the NPR 7123.1 for entrance and success criteria of the PRR.)

- The objectives of the SIR are to evaluate the readiness of the project and associated supporting infrastructure to begin system AI&T, to evaluate whether the remaining project development can be completed within available resources, and to determine if the project is sufficiently mature to begin Phase D.

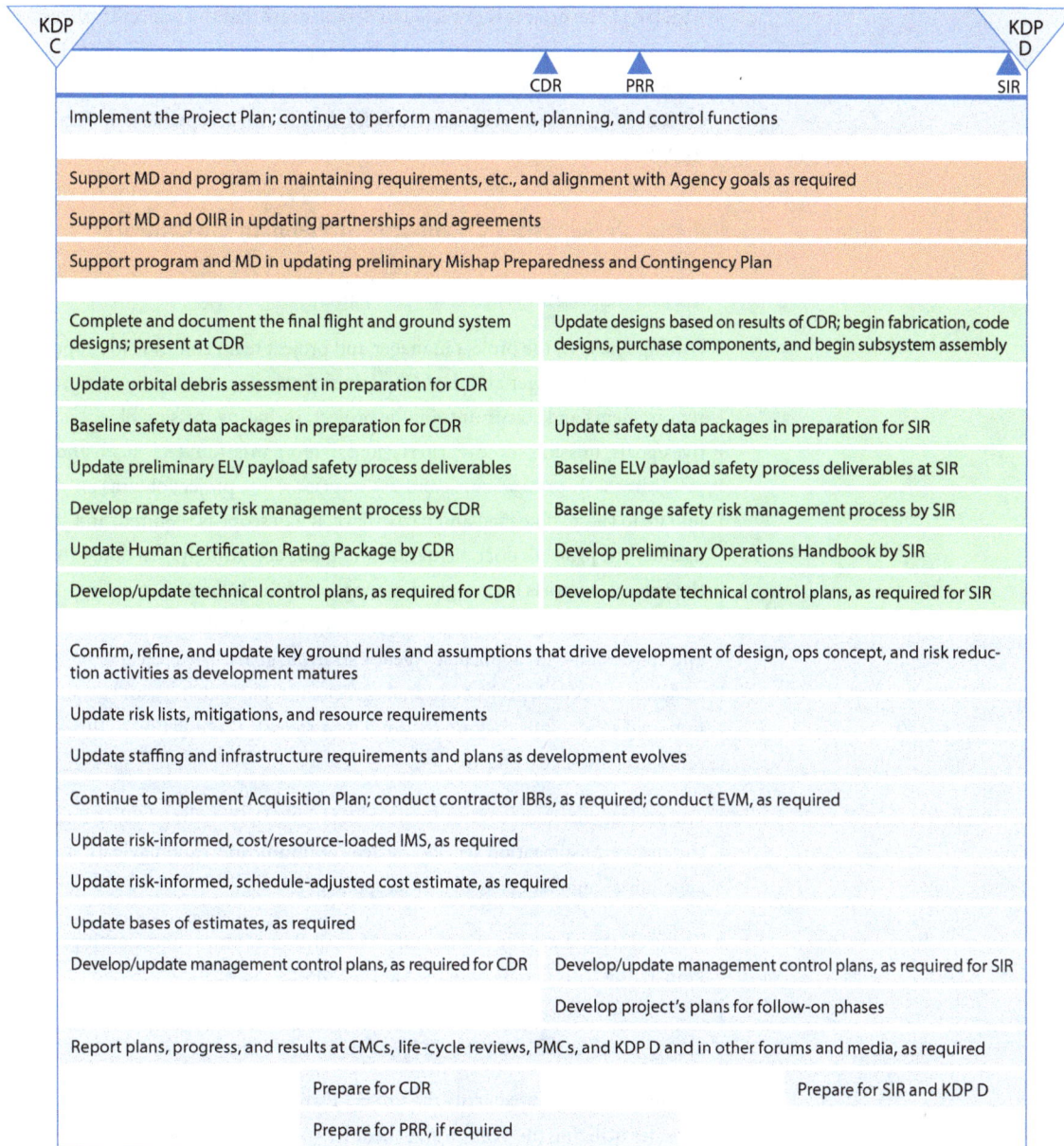

Legend: ☐ Project management, planning, and control tasks
☐ Work for which Headquarters is responsible but the project helps accomplish (e.g., international partnerships are a Headquarters responsibility, but the projects help develop and finalize those partnerships)
☐ Technical work the project is doing

Acronyms: ELV = Expendable Launch Vehicle; MD = Mission Directorate; OIIR = Office of International and Interagency Relations.

Note: These are typical high-level activities that occur during this project phase. Placement of reviews is notional.

Figure 4-12 Project Phase C Flow of Activities

At KDP D, the project is expected to demonstrate that the project is still on plan, the risk is commensurate with the project's payload classification, (or the Mission Directorate's risk definition if not a payload in accordance with NPR 8705.4), and the project is ready for AI&T with acceptable risk within its ABC.

4.4.1.2 Project Phase C Management, Planning and Control Activities

4.4.1.2.1 Supporting Headquarters Planning

During Phase C, the project manager and project team continue to support the program manager and the MDAA in maintaining the baseline program requirements and constraints on the project, including mission objectives/goals; mission success criteria; and driving mission, technical, and programmatic ground rules and assumptions. The project obtains an update to these, if needed and particularly if a descope is required, and updates the project's documentation and plans accordingly.[24] In doing this, the project supports the program manager and the MDAA in ensuring the continuing alignment of the project requirements, design approaches, and the design with applicable Agency strategic goals. The project team updates, as needed, project external agreements, partnerships, and acquisition, and other plans that are required for successful completion of this and remaining life-cycle phases.

The project continues to coordinate with HEOMD, including LSP, to finalize the space transportation services, space communication and navigation capabilities, and launch services, as applicable.

The project updates the preliminary Mishap Preparedness and Contingency Plan at SIR. This plan is baselined at the SMSR.

4.4.1.2.2 Management Control Processes and Products

The project team implements the Project Plan approved at KDP C. This includes utilizing the Technical, Schedule, and Cost Control Plan and management tools to guide monitoring, managing, and controlling the project requirements and technical design, schedule, and cost of the project to ensure that the high-level requirements levied on the project are met. The project further confirms, refines, and updates the project's key ground rules and assumptions that will drive implementation of the design and the funding profiles and schedules necessary for Phases C through F. The project continues to track them through Implementation to determine if they are being realized (i.e., remain valid) or if they need to be modified.

---

[24] Program requirements on the project are contained in the Program Plan.

As the design matures and fabrication begins, the project team updates its assessment of potential infrastructure and workforce needs versus current plans. Based on this assessment, the project team updates the **requirements and plans for staffing and infrastructure** at CDR.

The project team implements its plans for acquisition in accordance with its approved **Acquisition Plan**. The project also updates, identifies, assesses, and mitigates (if feasible) supply chain risks, including critical or single-source suppliers needed to design, develop, produce, and support required capabilities at planned cost and schedule. The project reports risks to the program. For contracts requiring EVM (refer to the NFS), the project conducts any required IBRs.

For projects using EVM, the project reports EVM metrics to the program and the Mission Directorate as defined in the Project Plan.

As the design finalizes and fabrication begins, the project continues to identify, assess, and update the **technical, cost, schedule and safety risks** that threaten the system development and risks that are likely to drive the project's cost and schedule estimates. The project maintains a record of accepted risks and the associated rationale for their acceptance, actively assesses open risks, and develops and implements mitigation plans. It updates resources being applied to manage and mitigate risks, including supply chain risks in accordance with the approved Acquisition Plan.

Projects manage within the approved baselines identified in their **Management Agreement**. This includes the technical baseline, project's risk posture, IMS, and baseline LCCE, all consistent with the program requirements and constraints on the project, the key assumptions, workforce estimates, and infrastructure requirements.

The project maintains and updates, if required, the project baselines and **Management Agreement** under configuration management with traceability to the ABC approved at KDP C. As a minimum, the project:

- Confirms key ground rules and assumptions that drive project requirements, designs, and the programmatic baseline. The project tracks the status of the realization of these, as appropriate, to determine if they are being realized (i.e., remain valid) or if they need to be modified.
- Manages technical and programmatic margins and resources to ensure successful completion of this and remaining life-cycle phases within budget, schedule, and risk constraints.
- Updates the risk-informed, cost-loaded IMS when changes warrant.
- Updates the risk-informed, schedule adjusted cost estimate when internal or external changes warrant.

- Updates and documents the basis for cost and schedule estimates for any tasks or system components added since KDP C.

- Assesses the adequacy of anticipated budget availability against phased life-cycle cost requirements and commitments, incorporating the impact of performance to date.

- Provides the program manager and the MDAA with immediate written notice if the latest estimate for the development cost (Phase C through D) exceeds the ABC cost for Phase C through D by 15 percent or more. Development cost growth of 15 percent or more for projects over $250 million is reported to Congress.

- Provides a written report to the program manager and MDAA explaining the reasons for the change in the cost and a recovery plan within 15 days of the above notification.

- Provides the program manager and the MDAA with immediate notification of a breach if the projected cost estimate for development cost exceeds the ABC cost for Phase C through D by 30 percent or more. Projects with a LCC greater than $250 million prepare to respond to Agency direction and a potential requirement for reauthorization by Congress.

- Provides the program manager and the MDAA with immediate written notice and a recovery plan if a milestone listed for Phase C and D on the project life-cycle chart (see Figures 2-4 and 4-1) is estimated to be delayed in excess of six months from the date scheduled in the ABC.

- If in breach, updates the Project Plan in accordance with direction and written notice.

- See Section 5.5 for more information on maintaining and updating project baselines, and Section 5.12 for more information on external reporting requirements associated with development cost growth of 15 percent or more, development schedule slip of 6 months or more, and breach due to development cost growth of 30 percent or more.

Projects update the CADRe (parts A, B, and C) consistent with the *NASA Cost Estimating Handbook* and tailoring guidance for small projects 30 to 45 days after CDR to reflect any changes from the CDR.

4.4.1.3 Project Phase C Technical Activities and Products

The project continues to perform the technical activities required in NPR 7123.1 for this phase. It completes the engineering design and development activities (e.g., qualification and life tests) and incorporates the results into the final design. It completes and documents final flight and ground designs by CDR and updates them, as necessary, at SIR, performing the systems engineering activities to determine if the design is mature enough to proceed with full-scale implementation. It develops product specifications and fabricates, purchases, and/or codes designs after the appropriate CDR(s) (e.g., flight article components, assemblies, and subsystems), and begins to implement the defined V&V program on flight and/or ground products. It updates the technology readiness assessment documentation by CDR, if required, and develops integration plans and procedures.

The project continues to work with the LSP, if applicable, to refine plans for the integration and test of the spacecraft at the launch site, preparations for launch, launch and post-launch support.

The project updates the orbital debris assessment a minimum of 45 days prior to the project CDR in accordance with NPR 8715.6 using the format and requirements contained in *NASA-STD-8719.14, Process for Limiting Orbital Debris.*

The project updates, documents, and baselines safety data packages by the CDR in accordance with *NPR 8715.7, Expendable Launch Vehicle Payload Safety Program* and local range requirements and NPR 7120.5 (for HSF). These safety data packages are updated at SIR. In addition, the project updates the preliminary ELV payload safety process deliverables at CDR and baselines them at SIR. For launch vehicles, if applicable, the project updates documentation that details the range safety risk management process in accordance with *NPR 8715.5, Range Flight Safety Program* at CDR and baselines it by SIR.

The project develops the preliminary Mission Operations Plan and the Operations Handbook by SIR (see "Operations Handbook" box for more information). The Mission Operations Plan describes the activities required to perform the mission and describes how the project will implement the associated facilities, hardware, software, and procedures required to complete the mission. It describes mission operations plans, rules, and constraints and describes the Mission Operations System (MOS) and Ground Data System (GDS) in the following terms:

- MOS and GDS human resources and training requirements;
- Procedures to ensure that operations are conducted in a reliable, consistent, and controlled manner using lessons learned during the project and from previous programs and projects;
- Facilities requirements (offices, conference rooms, operations areas, simulators, and test beds);
- Hardware (ground-based communications and computing hardware and associated documentation); and
- Software (ground-based software and associated documentation).

Operations Handbook

The Operations Handbook provides information essential to the operation of the spacecraft. It generally includes the following:

1. A description of the spacecraft and the operational support infrastructure;

2. Operational procedures, including step-by-step operational procedures for activation and deactivation;

3. Malfunction detection procedures; and

4. Emergency procedures.

The handbook identifies the commands for the spacecraft, defines the functions of these commands, and provides supplemental reference material for use by the operations personnel. The main emphasis is placed on command types, command definitions, command sequences, and operational constraints. Additional document sections may describe uploadable operating parameters, the telemetry stream data contents (for both the science and the engineering data), the Mission Operations System displays, and the spacecraft health monitors.

For HSF missions, the project updates the Human Rating Certification Package as described in NPR 8705.2 45 days prior to CDR. Per NPR 7123.1B, the project updates the Human Systems Integration Plan, if required, prior to CDR.

The project documents lessons learned in accordance with NPD 7120.4 and NPD 7120.6 and the project's Knowledge Management Plan.

The project updates the following control plans by CDR: SMA Plan, V&V Plan, ILS Plan, Integration Plan, Project Protection Plan, Technology Transfer Control Plan, Knowledge Management Plan, Education Plan, and Communications Plan.

The project updates the following control plans by SIR: V&V Plan and Project Protection Plan.

4.4.2 Completing Final Design and Fabrication (Phase C) Activities and Preparing for System Assembly, Integration and Test, Launch and Checkout (Phase D)

4.4.2.1 Finalize Plans for Phase D

The project develops and updates its plans for work to be performed during Phase D and the subsequent life-cycle phases, including updates, if needed, to life-cycle review plans and the project IMS; details on technical work to be accomplished; key acquisition activities planned; and plans for monitoring performance against plan. The project incorporates the impact of performance against the plan established at KDP C.

The project prepares and finalizes Phase D work agreements.

The project documents the results of Phase C activities and generates the appropriate documentation as described in NPR 7123.1 and NPR 7120.5E Tables I-4 and I-5 and Tables 4-6 and 4-7 at the end of this chapter.

4.4.2.2 Project Phase C Reporting Activities and Preparing for Major Milestones

4.4.2.2.1 Project Reporting

The project manager reports to the Center Director or designee, and supports the program executive in reporting the status of project Implementation at many other forums, including Mission Directorate monthly status meetings and the Agency's monthly BPR. Section 5.12 provides further information regarding potential project reporting.

4.4.2.2.2 Project Internal Reviews

Prior to life-cycle reviews, projects conduct internal reviews in accordance with NPR 7123.1, Center practices, and NPR 7120.5. These internal reviews are described in Section 4.3.2.2.2.

4.4.2.2.3 Preparing for Major Milestones

Projects plan, prepare for, and support the CDR, PRR (if required), and SIR life-cycle reviews in accordance with NPR 7123.1, Center practices, and NPR 7120.5, including ensuring that the life-cycle review objectives and expected maturity states defined in NPR 7120.5 have been satisfactorily met. Life-cycle review entrance and success criteria in Appendix G of NPR 7123.1 and the life-cycle phase and KDP information in Appendix E of this handbook provide specifics for addressing the six assessment criteria required to demonstrate the project has met the expected maturity state. The *NASA Standing Review Board Handbook* and Section 5.10 of this handbook provide additional detail on this process for those reviews requiring an independent SRB.

Projects plan, prepare for, and support the governing PMC review prior to KDP D and provide or obtain the KDP readiness products listed in Section 4.2.3.

Once the KDP has been completed and the Decision Memorandum signed, the project updates its documents and plans as required to reflect the decisions made and actions assigned at the KDP.

In tightly coupled and single-project programs, project(s) transition to KDP D in accordance with the plan for reviews documented in the Program or Project Plan.

4.4.3 Project Phase D, System Assembly, Integration and Test, Launch and Checkout Activities

4.4.3.1 Project Phase D Life-Cycle Activities

Project Implementation continues with Phase D as the project team implements the project in accordance with the Project Plan. The purpose of Phase D is to perform system AI&T; complete validation testing; finalize operations preparations; complete operational training; resolve failures, anomalies, and issues; certify the system for launch; launch the system; and complete on-orbit system checkout (robotic space flight projects) or initial operations (human space flight projects).

The transition from Phase D to Phase E is different from other phase transitions in the life cycle. KDP E marks the decision to conduct launch and early operations. However, the transition from Phase D to Phase E occurs after on-orbit checkout (robotic space flight projects) or initial operations (human space flight projects) at the conclusion of the Post-Launch Assessment Review (PLAR). The flow of activities in preparation for launch is very formal and involves important reviews. Section 4.4.4 provides a detailed description of the flow of the review and approval process in preparation for launch for human and robotic space flight programs and projects. This process is the same for projects and programs.

The phase activities focus on preparing for the ORR, SMSR, FRR, and LRR for HSF projects; or the ORR, LVRR, MRR, SMSR, FRR and LRR for robotic space flight projects; KDP E; launch; PLAR, and for certain HSF projects, PFAR. The objectives of these reviews are described in detail in Section 4.4.4. At KDP E, the project is expected to demonstrate that the project and all supporting systems are ready for safe, successful launch and early operations with acceptable risk within its ABC.

The general flow of activities for a project in Phase D is shown in Figure 4-13.

KDP D			KDP E

ORR   FRR/MRR

Implement the Project Plan; continue to perform management, planning, and control functions

Support MD and program in maintaining requirements, etc., and alignment with Agency goals as required

Support MD and OIIR in updating partnerships and agreements

Support MD and program in baselining Mishap Preparedness and Contingency Plan by SMSR

Assemble, integrate, and test various system pieces and perform verification and validation on products as they are integrated	Resolve open issues; close all waivers and deviations; prepare vehicle for transportation to launch site
Finalize and baseline as-built hardware and software documentation	
Plan, prepare, & conduct TRRs, SAR, & other reviews as needed	Plan, prepare, & conduct reviews required for shipment & launch
Prepare for operations and develop/update operations documentation	Prepare for launch and update operations documentation
Baseline Operations Handbook by ORR	Update Operations Handbook by FRR/MRR
Update safety data packages in preparation for ORR	Update safety data packages in preparation for SMSR
	Update orbital debris assessment in preparation for SMSR
	Develop and baseline the End of Mission Plan by SMSR
Update Human Certification Rating Package by ORR	Approve Human Certification Rating Package by SMSR
Develop/update technical control plans, as required for ORR	Update technical control plans, as required for FRR/MRR

Confirm, refine, and update key ground rules and assumptions that drive development and risk reduction activities as development matures

Update risk lists, mitigations, and resource requirements

Update staffing and infrastructure requirements and plans as development evolves

Continue to implement Acquisition Plan; conduct contractor IBRs, as required; conduct EVM, as required

Update risk-informed, cost/resource-loaded IMS, as required

Update risk-informed, schedule-adjusted cost estimate, as required

Update bases of estimates, as required

Develop/update management control plans, as required for ORR	Develop/update management control plans, as required for FRR/MRR
	Develop project's plans for follow-on phases

Report plans, progress, and results at CMCs, life-cycle reviews, PMCs, and KDP E and in other forums and media, as required

Prepare for ORR   Prepare for FRR/MRR and KDP E

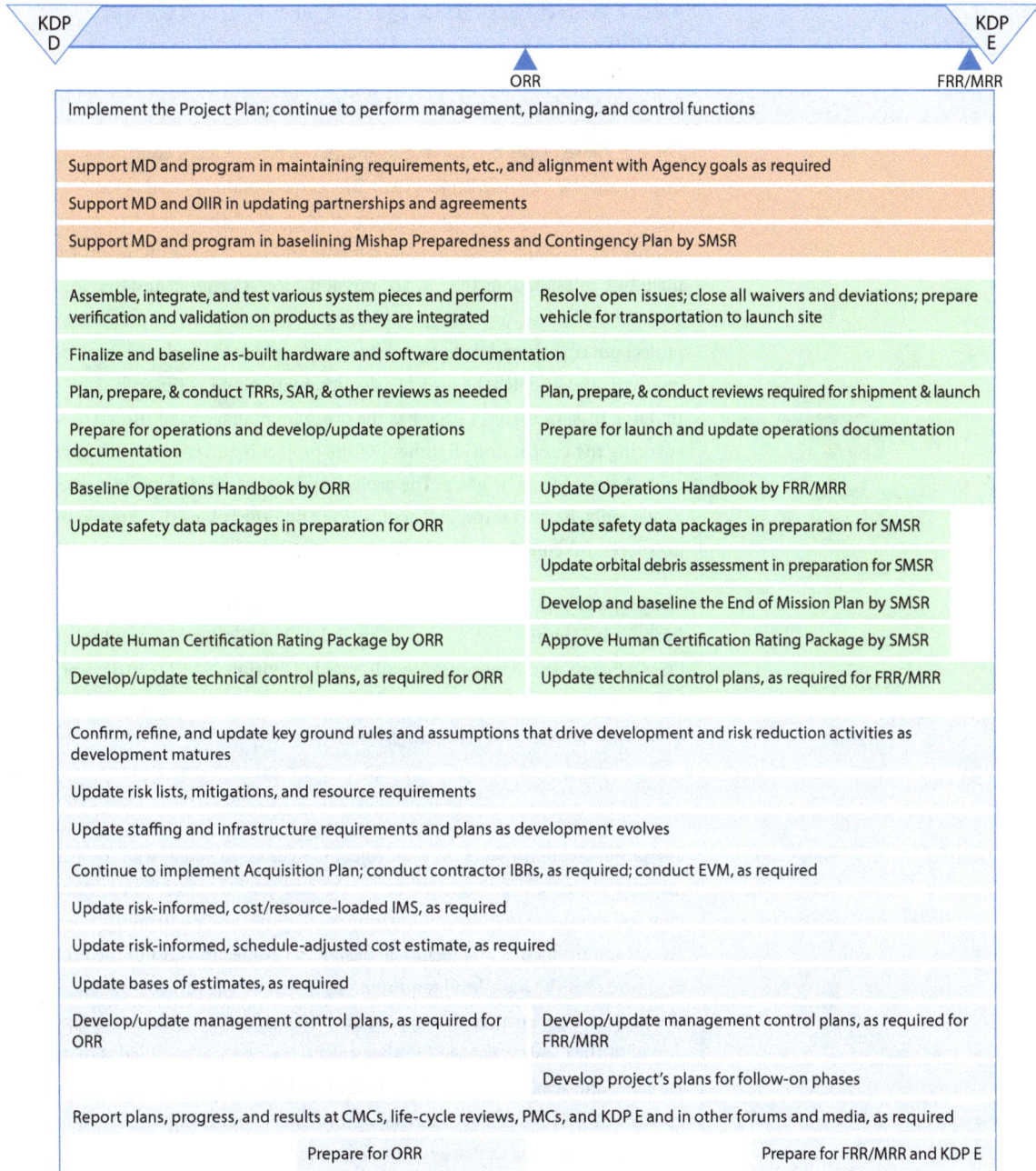

Legend: ☐ Project management, planning, and control tasks
☐ Work for which Headquarters is responsible but the project helps accomplish (e.g., international partnerships are a Headquarters responsibility, but the projects help develop and finalize those partnerships)
☐ Technical work the project is doing

Acronyms: MD = Mission Directorate; OIIR = Office of International and Interagency Relations.

Note: These are typical high-level activities that occur during this project phase. Placement of reviews is notional.

Figure 4-13 Project Phase D Flow of Activities

4.4.3.2 Project Phase D Management, Planning, and Control Activities

4.4.3.2.1 Supporting Headquarters Planning

During Phase D, the project manager and the project team continue to implement the baseline Project Plan. The project manager and project team continue to support the program manager and the MDAA in maintaining the baseline program requirements and constraints on the project, including mission objectives/goals; mission success criteria; and **driving mission, technical, and programmatic ground rules and assumptions**. The project obtains an update to these, if needed and particularly if a descope is required, and updates the project's documentation and plans accordingly.[25] In doing this, the project supports the program manager and the MDAA in ensuring the continuing alignment of the project requirements with applicable Agency strategic goals. The project updates, as needed, project external agreements, partnerships, and acquisition and other plans that are required for successful completion of this and remaining life-cycle phases.

The project supports the Mission Directorate in baselining the **Mishap Preparedness and Contingency Plan** and delivering the document to OSMA/Safety and Assurance Requirements Division (SARD) 30 days prior to the SMSR per *NPR 8621.1, NASA Procedural Requirements for Mishap and Close Call Reporting, Investigating, and Recordkeeping.*

4.4.3.2.2 Management Control Processes and Products

The project team implements the Project Plan as approved at KDP D. This includes utilizing the Technical, Schedule, and Cost Control Plan and management tools to guide monitoring, managing, and controlling the project requirements, and technical design, schedule, and cost of the project to ensure that the high-level requirements levied on the project are met. The project team ensures that appropriate infrastructure and, in coordination with the Centers engaged in the project, trained and certified staff are available and ready when needed to support the activities of this phase. It updates, as needed, project external agreements, partnerships, and acquisition and other plans that are required for successful completion of this and remaining life-cycle phases.

The project team implements its plans for acquisition in accordance with the approved Acquisition Plan. The project also updates, identifies, assesses, and mitigates (if feasible) supply chain risks, including critical or single-source suppliers needed to design, develop, produce, and support required

---

[25] Program requirements on the project are contained in the Program Plan.

capabilities at planned cost and schedule. The project reports risks to the program. For contracts requiring EVM (refer to the NFS), the project conducts any required IBRs.

For projects using EVM, the project reports EVM metrics to the program and the Mission Directorate as defined in the Project Plan.

As system integration begins, the project continues to identify, assess, and update the technical, cost, schedule, and safety risks that threaten the system development and risks that are likely to drive the project's cost and schedule estimates. The project maintains a record of accepted risks and the associated rationale for their acceptance, actively assesses open risks, and develops and implements mitigation plans. It updates resources being applied to manage and mitigate risks.

Project managers manage the project within the approved baselines identified in their Management Agreement. This includes the technical baseline, project's risk posture, Integrated Master Schedule (IMS), and baseline Life-Cycle Cost Estimate (LCCE), all consistent with the program requirements and constraints on the project, the key assumptions, workforce estimates, and infrastructure requirements.

The project maintains and updates, if required, the project baselines and Management Agreement under configuration management with traceability to the ABC approved at KDP C.

The project updates the CADRe (parts A, B, and C) consistent with the *NASA Cost Estimating Handbook* with a final version 30 to 45 days after SIR to reflect any changes from the SIR.

4.4.3.3 Project Phase D Technical Activities and Products

The project continues to perform the technical activities required in NPR 7123.1 for this phase. It plans, prepares for, and performs other reviews, as necessary and applicable. Examples of other reviews include Test Readiness Reviews (TRR) and System Acceptance Reviews (SAR). The project team conducts TRRs to ensure that the test articles (hardware/software), test facility, support personnel, and test procedures are ready for testing and data acquisition, reduction, and control. (See Table G-10 in Appendix G of the NPR 7123.1 for entrance and success criteria of the TRR.) The SAR is conducted to evaluate whether a specific end item is sufficiently mature to be shipped from the supplier to its designated operational facility or launch site. (See Table G-11 in Appendix G of the NPR 7123.1 for entrance and success criteria of the SAR.) If applicable, the project and the LSP finalize plans for the integration and test of the spacecraft at the launch site, preparations for launch (including readiness reviews, see Section 4.3.1.4), and launch and post-launch support.

As the various components and subassemblies arrive at the integration facility, the project:

- Begins to assemble, integrate, and test the various system pieces and complete V&V on the products as they are integrated.
- Prepares the preliminary V&V Report before ORR and then baselines the report by FRR/MRR.
- Transitions or delivers the final products and baselines the as-built hardware and software documentation.
- Prepares for operations, updates the Operations Concept (if needed), and baselines the Mission Operations Plan and the Operations Handbook at ORR.

The project updates safety data packages in accordance with *NPR 8715.7, Expendable Launch Vehicle Payload Safety Program* and local range requirements and NPR 7120.5 (for HSF). If required, the project updates the Human Rating Certification Package prior to ORR and submits the package for certification prior to the SMSR. The project also updates the SMA Plan prior to the SMSR.

The project baselines the Science Data Management Plan that was initially developed during Phase B at ORR and updates the plan at FRR/MRR, if required.

The project updates the Project Protection Plan, Education Plan, and Communications Plan by ORR. The project updates the project protection plan by MRR/FRR.

The project team documents and implements all technical, management, and operational security controls as required by NPR 2810.1 for IT systems during Phase D of the project. It ensures that IT security certification and accreditation requirements are met as specified in NPR 2810.1 for IT systems during Phase D of the project.

The project baselines the orbital debris assessment in accordance with NPR 8715.6 using the format and requirements contained in *NASA-STD-8719.14, Process for Limiting Orbital Debris* prior to the SMSR. If applicable, the project completes the initial Collision on Launch Analysis, per NPR 8715.5 and presents the analysis at the FRR. The project baselines the End of Mission Plan (EOMP) per NPR 8715.6 and NASA-STD-8719.14 Appendix B prior to the SMSR. The EOMP is a living document that grows with the project as it operates up to its inclusion in the Decommissioning/

Disposal Plan at KDP F. The format for EOMPs is described in *NASA STD 8719.14, Process for Limiting Orbital Debris*. The EOMP describes the project management approach and the mission overview; spacecraft description; assessment of spacecraft debris released during and after passivation; assessment of spacecraft potential for on-orbit collisions; assessment of spacecraft post mission disposal plans and procedures; assessment of spacecraft reentry hazards (all data added during flight); and assessment of hazardous materials contained on the spacecraft.

Once the hardware is shipped to the launch site, the project, with launch site support, begins the process of receiving and inspecting the hardware, reassembling the spacecraft as required, completing final spacecraft testing, and resolving any open issues that remain. The project supports launch rehearsals, participates in press conferences, and supports the launch approval process described below.

When the project is ready for launch, the project team obtains the approved documents required for launch. If applicable, the project manager ensures that the nuclear launch approval process has been properly completed and provides the OSMA Nuclear Flight Safety Assurance Manager with a listing of all radioactive materials planned for launch with the associated risk assessments 30 days prior to the SMSR in accordance with NPR 8715.3, Chapter 6.

Finally, the project documents lessons learned in accordance with NPD 7120.6 and the project's Knowledge Management Plan.

4.4.4 Launch Approval Process and Transition to Operations

This section applies to tightly coupled programs, single-project programs, and projects.

The process for completing KDP III (tightly coupled programs)/KDP E (projects and single-project programs) and obtaining approval for launch and early operations is complex and unique. The KDP III/KDP E decision to launch and conduct early operations includes approval for the transition to the operations phase of the life cycle, however, unlike other life-cycle phase transitions, the transition to operations does not occur immediately after the KDP III/KDP E. For robotic space flight programs and projects, this transition to operations occurs following a successful launch and on-orbit checkout. For human space flight programs and projects, this transition to

> The decision to launch and conduct early operations is a critical decision for the Agency. The KDP III (KDP E for projects and single-project programs) decision occurs before launch to provide Decision Authority approval for this decision. The KDP III/KDP E decision includes approval for the transition to the operations phase of the life cycle, however, unlike other life-cycle phase transitions, the transition to operations does not occur immediately after the KDP III/KDP E. This transition occurs after launch and checkout. The timing for this transition stems from the historical practice of funding missions through on-orbit checkout, transitioning from the development team to the operations team following on-orbit checkout, and funding mission operations separately.

operations occurs after initial operations[26] have been successfully completed, and all flight test objectives (including human rating) have been met. For the program or project to gain approval to launch and conduct early operations, the governing Program Management Council (PMC) meets to conduct a review of readiness for flight, at which the program or project is expected to demonstrate that it is ready for a safe, successful launch and early operations with acceptable risk within Agency commitments. For human space flight programs and projects, this review is the Agency Flight Readiness Review (FRR).[27] For robotic space flight programs and projects this review is the Mission Readiness Briefing (MRB). The KDP III/KDP E decision is made at the end of the Agency FRR for human space flight programs and projects, and at the end of the MRB for robotic space flight programs and projects. The details of the process for human and robotic space flight programs and projects to gain approval to launch and conduct early operations are described below.

4.4.4.1 Human Space Flight Programs and Projects

For human space flight programs and projects, preparation for KDP III (tightly coupled programs)/KDP E (projects and single-project programs) and approval for launch and early operations includes a series of reviews to establish and assess the program or project's readiness. These reviews include the Operations Readiness Review (ORR), programmatic pre-FRR(s) which may be conducted by the project, program, and Mission Directorate (MD), the Center pre-FRR,[28] and Safety and Mission Success Review (SMSR), and culminate with the Agency FRR. The KDP III/KDP E decision is made at the end of the Agency FRR. In the short timeframe between the Agency FRR and launch, the Launch Readiness Review (LRR) (also known as the L-1 day Mission Management Team (MMT) Review) is conducted for final review before launch.

A Post-Launch Assessment Review (PLAR) is conducted after launch to determine the program or project's readiness to begin the operations phase

When more than one launch and flight operations sequence, or more than one launch, flight and landing/recovery operations sequence is needed to successfully complete initial operations, this series of reviews, or a subset of the series, is repeated for each sequence. The Agency FRR is conducted for each sequence. However, the KDP III/KDP E decision is made only once, at the initial Agency FRR.

---

[26] Human space flight programs and projects develop flight systems that return to Earth, such as the Shuttle Program or MPCV Program, or develop flight systems that remain in orbit, such as the ISS Program. Initial operations for the former programs and projects may require one or more launch, flight and landing and recovery operations sequences to meet all flight test objectives. Initial operations for the latter programs and projects may include one or more launch and flight operations sequences, such as launch and assembly flights for the ISS Program. The initial operations timeline for both types of human space flight programs and projects may span multiple years.

[27] The human space flight Agency FRR is chaired by the MD AA, and attended by the DA.

[28] The Center pre-FRR may be conducted in conjunction with the program or project pre-FRR.

of the life cycle. For human space flight programs and projects that develop flight systems that return to Earth, the PLAR may be combined with the Post-Flight Assessment Review (PFAR), which is conducted after landing and recovery. Figure 4-14 depicts the series of reviews leading to KDP III/ KDP E and launch for human space flight programs and projects, and the PLAR and PFAR reviews.

The objectives of the ORR are to evaluate the readiness of the program, project, ground systems, personnel, procedures, and user documentation to assemble, integrate and test flight systems (using associated ground systems) during the development phase, as well as to operate the flight system and associated ground systems in compliance with defined program or project requirements and constraints during the operations phase.

Programmatic pre-FRRs may be conducted by the project, program and Mission Directorate, per agreements established by the Mission Directorate with the program and project. The objectives of these pre-FRRs are to determine the program or project's readiness for vehicle rollout to the launch pad, launch, and flight. The objectives of the Mission Directorate's

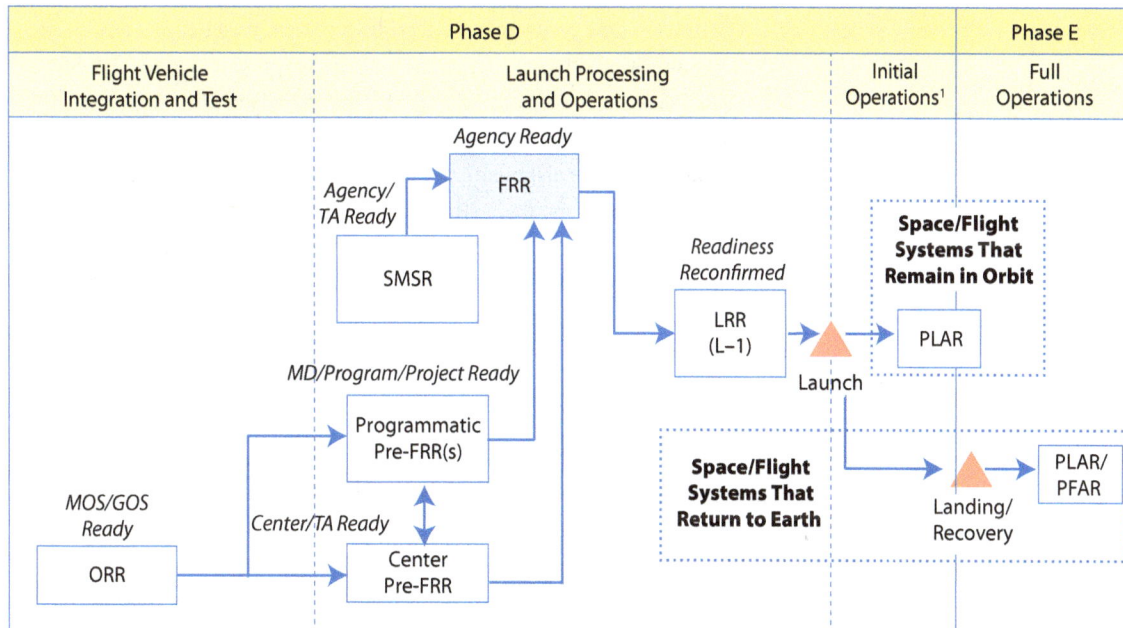

Legend: ☐ Review ☐ KDP III/KDP E

[1] Initial operations may include multiple launch, flight, and landing/recovery operations sequences.

Figure 4-14 KDP III/KDP E Flow Chart for Human Space Flight Programs and Projects

pre-FRR, if conducted, may also include determining the readiness of external supporting entities (e.g., SCaN, Eastern and Western Range). The program or project certifies the completion of all tasks and identifies any planned work required to prepare the flight/ground hardware/software, support facilities, and operations personnel to safely support launch and flight. This includes review of necessary data to ensure satisfactory closeout of all Mission Directorate flight readiness certification requirements, exceptions, and launch constraints, in sufficient detail to enable determination of flight readiness.

The objectives of the Center pre-FRR are for Center management and Technical Authorities to determine the readiness of the program or project and the Center institutional resources that support the program or project for vehicle rollout to the launch pad, launch, and flight.

The objectives of the SMSR are to prepare Agency safety and engineering management to participate in program final readiness reviews preceding flights or launches, including experimental/test launch vehicles or other reviews as determined by the Chief, Safety and Mission Assurance. The SMSR provides the knowledge, visibility, and understanding necessary for senior safety and engineering management to either concur or nonconcur in program decisions to proceed with a launch or significant flight activity.

The results of the programmatic pre-FRRs, Center pre-FRR, and SMSR, and the readiness of external supporting entities, are presented to the Agency FRR. The objectives of the Agency FRR are to evaluate the program or project and all supporting systems, including ground, hardware, and software systems, personnel, and procedures, for readiness for a safe and successful launch and flight. The KDP III/KDP E decision is made at the end of the initial Agency FRR. At KDP III/KDP E the program or project is expected to demonstrate that it is ready for a safe, successful launch and early operations with acceptable risk within Agency commitments. The Certification of Flight Readiness (COFR) is signed at the conclusion of the Agency FRR.

The LRR/L-1 Review is held no later than 1 day before launch.[29] The objectives of the LRR/L-1 Review are to update the vehicle, payload processing and mission status, close out actions from preceding reviews, including the Agency FRR, programmatic pre-FRRs, and Center pre-FRR, resolve any remaining issues, address any issues associated with weather, and authorize approval to proceed into launch countdown.

---

[29] The LRR/L-1 Review is typically chaired by the associate administrator for the Human Exploration and Operations Mission Directorate.

The PLAR is a non-KDP-affiliated review that is conducted after launch. The objectives of the PLAR, accomplished through the MMT meetings, are to evaluate the in-flight performance of the flight systems. More than one test flight (i.e., launch and flight operations sequence) may be required to successfully accomplish all flight test objectives, satisfy human rating requirements, and complete initial operations, and multiple PLARs may be conducted throughout the initial operations period, as determined by the MMT. A PLAR is conducted by the Mission Directorate following completion of initial operations to determine the program or project's readiness to begin the operations phase of the life cycle and to transfer responsibility to the operations organization. At this PLAR, the program or project is expected to demonstrate that it is ready to conduct mission operations with acceptable risk within Agency commitments. For human space flight programs and projects that develop flight systems that return to Earth, this PLAR may be combined and conducted in conjunction with the PFAR.

The PFAR is a non-KDP-affiliated review associated with human space flight programs and projects that develop flight systems that return to Earth. It is conducted after a launch, flight, and landing and recovery operations sequence is completed. The objectives of the PFAR are to evaluate accomplishment of flight test objectives, including satisfaction of human rating requirements. Accomplishments and any vehicle and mission support facility performance issues and anomalies are documented, and lessons learned are captured. More than one test flight (i.e., launch, flight, and landing and recovery operations sequence) may be required to successfully accomplish all flight test objectives, satisfy human rating requirements, and complete initial operations, and multiple PFARs may be conducted throughout the initial operations period.

4.4.4.2 Robotic Space Flight Programs and Projects

For robotic space flight programs and projects, preparation for KDP III (tightly coupled programs)/KDP E (projects and single-project programs) and approval for launch and early operations includes a series of reviews to establish and assess the readiness of the program or project's spacecraft and the launch vehicle. These reviews include the Operations Readiness Review (ORR), Mission Readiness Review (MRR), Launch Vehicle Readiness Review (LVRR), and Safety and Mission Success Review (SMSR), and culminate with the Mission Readiness Briefing (MRB). The KDP III/KDP E decision is made at the end of the MRB. The DPMC where the MRB is conducted constitutes the governing PMC for Category 2 and 3 projects and most

The Mission Readiness Briefing (MRB) is the Directorate Program Management Council meeting immediately preceding KDP E, where the MDAA for robotic programs (typically the Science Mission Directorate (SMD)) is presented with the results of the project's Operational Readiness Review (ORR), Mission Readiness Review (MRR), and Safety and Mission Success Review (SMSR) and, based on acceptable results, approves the project to proceed through the launch event into mission operations. The MRB constitutes the governing PMC for Category 2 and 3 projects and nonnuclear Category 1 projects (when delegated). The KDP E decision is made at the end of this meeting. Category 1 projects with nuclear power sources have a subsequent Agency Program Management Council (APMC), where the KDP E decision is made.

Category 1 projects.[30] The final launch decision is made at the Launch Readiness Review (LRR) where all involved parties provide their final readiness to launch. Figure 4-15 depicts the series of reviews leading to KDP III/KDP E and launch for robotic space flight programs and projects, and the PLAR, at which the program or project's readiness to begin the operations phase of the life cycle is determined.

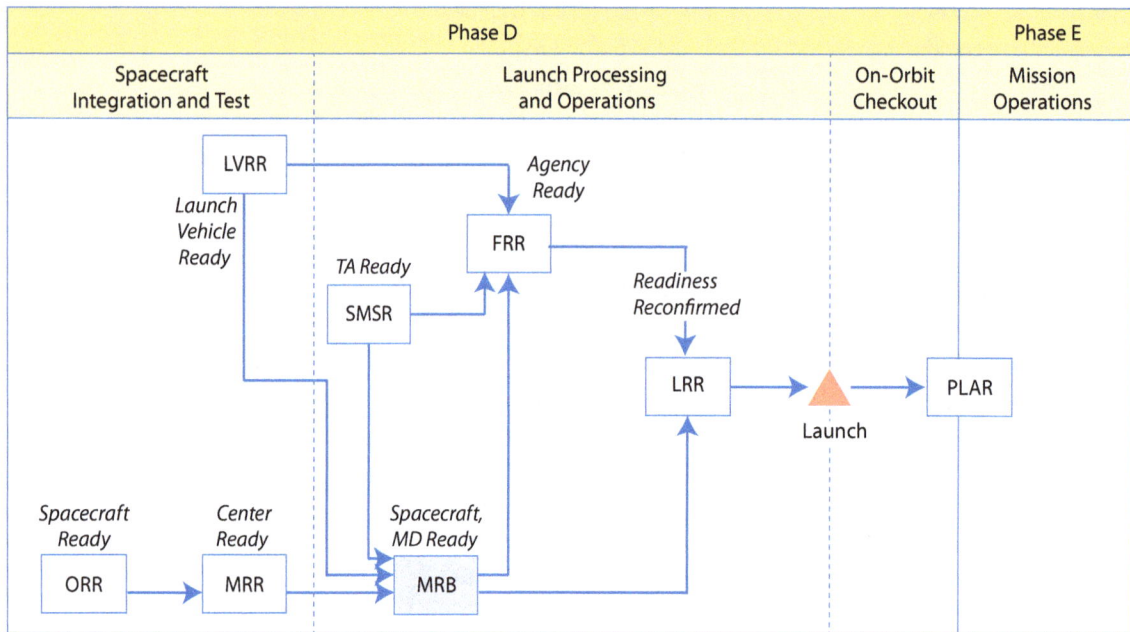

Legend: ☐ Review   ☐ KDP III/KDP E

Figure 4-15 KDP III/KDP E Flow Chart for Robotic Space Flight Programs and Projects

The objectives of the ORR are to evaluate the readiness of the spacecraft program, project, ground systems, personnel, procedures, and user documentation to operate the flight systems and associated ground systems in compliance with defined program or project requirements and constraints during the operations/sustainment phase.

The objective of the LVRR is to certify the readiness of the launch vehicle to proceed with spacecraft/launch vehicle integration activities.[31] Any launch

---

[30] Decision Authority for KDP E is usually delegated to the MDAA by the NASA Associate Administrator, except for projects with nuclear power sources. Category 1 projects with nuclear power sources have an APMC following the MRB.

[31] The LVRR is chaired by the Launch Services Program Manager.

vehicle anomalies/issues associated with the mission are reviewed. The LVRR is typically held prior to the MRR.

The objectives of the MRR are to evaluate the readiness of the program or project's spacecraft, ground systems, personnel, and procedures for a safe and successful launch and flight/mission.

The objectives of the SMSR are to prepare Agency safety and engineering management to participate in program or project final readiness reviews preceding flights or launches, including experimental/test launch vehicles or other reviews as determined by the Chief, Safety and Mission Assurance. The SMSR provides the knowledge, visibility, and understanding necessary for senior safety and engineering management to either concur or nonconcur in program or project decisions to proceed with a launch or significant flight activity.

At the MRB meeting (and APMC if applicable), the results of the ORR, MRR, and SMSR are presented. The objectives of the MRB (or APMC) are to evaluate the program or project and all supporting systems, including ground, hardware, and software systems, personnel, and procedures, for readiness for a safe and successful launch and flight/mission. The KDP III/ KDP E decision is made at the end of the MRB (or APMC). At KDP III/ KDP E the program or project is expected to demonstrate readiness for launch and early operations with acceptable risk within Agency commitments. Based on acceptable results, the MDAA (or NASA Associate Administrator) approves the program or project to proceed to launch.

The objective of the FRR is to status the readiness of the launch vehicle and spacecraft to enter into the final launch preparation.[32] The FRR is held about 5 days before launch to review the mission status and close out any actions from the LVRR, MRR, SMSR, and MRB that constrain launch.

The objective of the LRR is to provide final launch readiness status from all the mission elements, close out any actions from the FRR which constrain launch, authorize approval to initiate the launch countdown and sign the Certification of Flight Readiness (COFR). The LRR is held at the launch site no later than 1 day before launch.[33]

The PLAR is a non-KDP-affiliated review that is conducted after the mission has launched and on-orbit checkout has been completed. The objectives of the PLAR are to evaluate the in-flight performance of the program or project flight systems early in the mission and to determine the program or

---

[32] The FRR is chaired by the NASA Launch Manager.

[33] The LRR is chaired by the Human Exploration and Operations MD Director of Launch Services or may be delegated to the Launch Services Program manager.

project's readiness to begin the operations phase of the life cycle and transfer responsibility to the operations organization. At the PLAR, the program or project is expected to demonstrate that it is ready to conduct mission operations with acceptable risk within Agency commitments.

4.4.5 Completing System Assembly, Integration and Test, Launch and Checkout (Phase D) Activities and Preparing for Operations and Sustainment (Phase E)

4.4.5.1 Finalizing Plans for Phase E

The project develops and updates its plans for work to be performed during Phase E and F, including updates, if needed and particularly if a descope is required, to life-cycle review plans, the project IMS, details on technical work to be accomplished, key acquisition activities planned, and plans for monitoring performance against plan. The project incorporates the impact of performance against the plan established at KDP D.

The project prepares and finalizes Phase E work agreements. The work scope and price for Phase E contracts may be negotiated prior to approval to proceed into operations but not executed. (Once the project has been approved to proceed at KDP E and funding is available, the negotiated contracts may be executed, assuming no material changes.)

The project documents the results of Phase D activities and generates the appropriate documentation as described in NPR 7123.1 and NPR 7120.5E Tables I-4 and I-5 and Tables 4-6 and 4-7 at the end of this chapter.

4.4.5.2 Project Phase D Reporting Activities and Preparing for Major Milestones

4.4.5.2.1 Project Reporting

The project manager reports to the Center Director or designee, and supports the program executive in reporting the status of project Implementation at many other forums, including Mission Directorate monthly status meetings and the Agency's monthly Baseline Performance Review (BPR). Section 5.12 provides further information regarding potential project reporting.

4.4.5.2.2 Project Internal Reviews

Prior to life-cycle reviews, projects conduct internal reviews in accordance with NPR 7123.1, Center practices, and NPR 7120.5. These internal reviews are described in Section 4.3.2.2.2.

4.4.5.2.3 Preparing for Major Milestone Reviews

Projects support the life-cycle reviews described in Section 4.4.4 in accordance with NPR 7123.1, Center practices, and NPR 7120.5, ensuring that the life-cycle review objectives and expected maturity states defined in NPR 7120.5 have been satisfactorily met. Life-cycle review entrance and success criteria in Appendix G of NPR 7123.1 and the life-cycle phase and KDP information in Appendix E of this handbook provide specifics for addressing the six assessment criteria required to demonstrate the project has met the expected maturity state. The *NASA Standing Review Board Handbook* provides additional detail on this process for the ORR, which requires an independent SRB.

Projects plan, prepare for, and support the governing PMC review prior to KDP E and provide or obtain the KDP readiness products listed in Section 4.2.3.

Once the KDP has been completed and the Decision Memorandum signed, the project updates its documents and plans as required to reflect the decisions made and actions assigned at the KDP.

In tightly coupled and single-project programs, project(s) transition to KDP E in accordance with the plan for reviews documented in the Program or Project Plan.

4.4.6 Project Phase E, Operations and Sustainment Activities

4.4.6.1 Project Phase E Life-Cycle Activities

During Phase E, the project implements the Project Plan/Missions Operations Plan developed in previous phases. Mission operations may be periodically punctuated with Critical Event Readiness Reviews (CERR), e.g., a trajectory correction maneuver or orbit insertion maneuver. Human space flight missions may conduct PFARs specific to the project needs. (See "Sustainment and Sustainment Engineering" box for an explanation of sustainment activities.)

The mission operation phase ends with the Decommissioning Review (DR) and KDP F, at which time mission termination is approved. The DR may be combined with the Disposal Readiness Review (DRR) if the disposal of the spacecraft will be done immediately after the DR:

- The objectives of the CERR are to evaluate the readiness of the project and the flight system for execution of a critical event during the flight operations phase of the life cycle.

Sustainment and Sustainment Engineering

Sustainment generally refers to supply, maintenance, transportation, sustaining engineering, data management, configuration management, manpower, personnel, training, habitability, survivability, environment, safety, supportability, and interoperability functions.

The term "sustaining engineering" refers to technical activities that can include, for example, updating designs (e.g., geometric configuration), introducing new materials, and revising product, process, and test specifications. These activities typically involve first reengineering items to solve known problems, and then qualifying the items and sources of supply. The problems that most often require sustaining engineering are lack of a source (e.g., vendor going out of business), component that keeps failing at a high rate, and long production lead time for replacing items.

As parts age, the need and opportunity for sustaining engineering increase. The practice of sustaining engineering includes not only the technical activity of updating designs, but also the business judgment of determining how often and on what basis the designs need to be reviewed.

- The objectives of the PFAR when conducted during this phase are to evaluate how well mission objectives were met during a space flight mission and to evaluate the status of the returned vehicle.
- The objectives of the DR are to evaluate the readiness of the project to conduct closeout activities, including final delivery of all remaining project deliverables and safe decommissioning of space flight systems and other project assets and to determine if the project is appropriately prepared to begin Phase F.
- The objectives of the DRR are to evaluate the readiness of the project and the flight system for execution of the spacecraft disposal event.

At KDP F, the project team is expected to demonstrate that the project decommissioning is consistent with program objectives and the project is ready for safe decommissioning of its assets and closeout of activities, including final delivery of all remaining project deliverables and disposal of its assets.

A general flow of Phase E activities is shown in Figure 4-16.

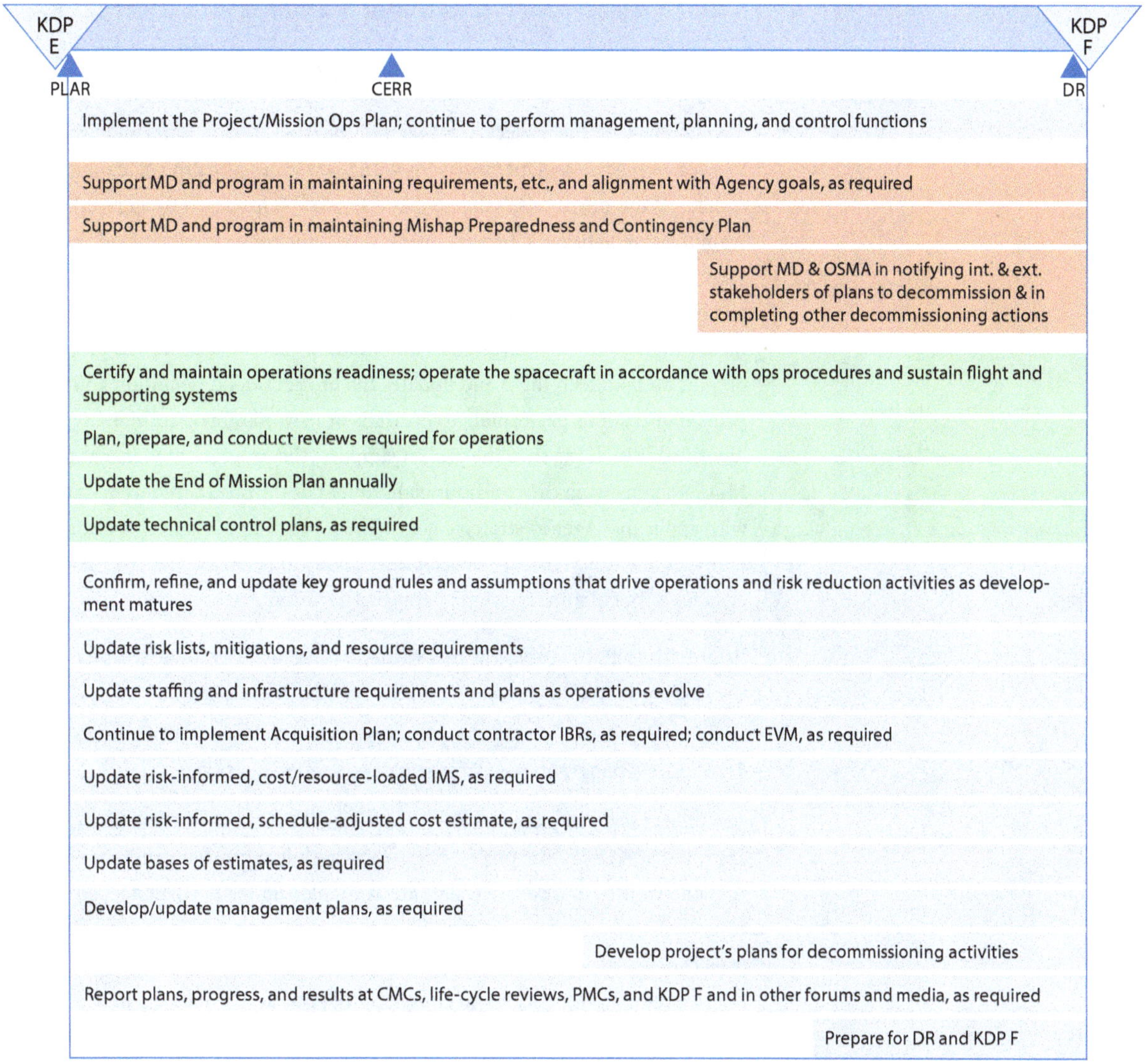

KDP E — PLAR — CERR — DR — KDP F

Implement the Project/Mission Ops Plan; continue to perform management, planning, and control functions

Support MD and program in maintaining requirements, etc., and alignment with Agency goals, as required

Support MD and program in maintaining Mishap Preparedness and Contingency Plan

Support MD & OSMA in notifying int. & ext. stakeholders of plans to decommission & in completing other decommissioning actions

Certify and maintain operations readiness; operate the spacecraft in accordance with ops procedures and sustain flight and supporting systems

Plan, prepare, and conduct reviews required for operations

Update the End of Mission Plan annually

Update technical control plans, as required

Confirm, refine, and update key ground rules and assumptions that drive operations and risk reduction activities as development matures

Update risk lists, mitigations, and resource requirements

Update staffing and infrastructure requirements and plans as operations evolve

Continue to implement Acquisition Plan; conduct contractor IBRs, as required; conduct EVM, as required

Update risk-informed, cost/resource-loaded IMS, as required

Update risk-informed, schedule-adjusted cost estimate, as required

Update bases of estimates, as required

Develop/update management plans, as required

Develop project's plans for decommissioning activities

Report plans, progress, and results at CMCs, life-cycle reviews, PMCs, and KDP F and in other forums and media, as required

Prepare for DR and KDP F

Legend: ☐ Project management, planning, and control tasks
☐ Work for which Headquarters is responsible but the project helps accomplish (e.g., international partnerships are a Headquarters responsibility, but the projects help develop and finalize those partnerships)
☐ Technical work the project is doing

Acronyms: MD = Mission Directorate; OIIR = Office of International and Interagency Relations.

Note: These are typical high-level activities that occur during this project phase. Placement of reviews is notional.

Figure 4-16 Project Phase E Flow of Activities

4.4.6.2 Project Phase E Management, Planning and Control Activities

4.4.6.2.1 Supporting Headquarters Planning

During Phase E, the project manager and the project team implement the Project Plan/Mission Operations Plan. In some cases, the project team that developed the mission is disbanded and Phase E is managed by a project team that specializes in mission operations. The project manager and project team continue to support the program manager and the MDAA in maintaining the baseline program requirements and constraints on the project, including mission objectives/goals and mission success criteria. The project obtains an update to these and updates the project's documentation and plans if operations performance shortfalls or new mission requirements are identified.[34] In doing this, the project supports the program manager and the MDAA in ensuring the continuing alignment of the project requirements with applicable Agency strategic goals. The project supports the program manager and the MDAA in developing options to resolve operations deficiencies or to enhance mission operations performance.

Prior to the DR, the project works with the Mission Directorate to update the Mishap Preparedness and Contingency Plan if necessary.

4.4.6.2.2 Management Control Processes and Products

The project team implements the Project Plan/Mission Operations Plan as approved at KDP E. The project team ensures that appropriate infrastructure and trained and certified staff are available and ready when needed to support the activities of this phase. The project team updates, as needed, project external agreements, partnerships, and acquisition and other plans that are required for successful completion of this and remaining life-cycle phases. As directed by the program manager, the project supports the development of Project Plan revisions to continue the mission into extended operations beyond the primary mission phase or beyond any extension previously included in the plan.

The project team implements acquisition activities in accordance with the approved Acquisition Plan. The project updates, identifies, assesses, and mitigates (if feasible) supply chain risks, including critical or single-source suppliers needed to design, develop, produce, and support required capabilities at planned cost and schedule. The project reports risks to the program. The project implements contract closeouts, as appropriate.

---

[34] Program requirements on the project are contained in the Program Plan.

As mission operations begin, the project continues to identify, assess, and update the **technical, cost, schedule and safety risks** that threaten the system operations and drive cost and schedule estimates. The project maintains a record of accepted risks and the associated rationale for their acceptance, actively assesses open risks, and develops and implements mitigation plans. It updates resources being applied to manage and mitigate risks.

Project managers manage the project within the approved baselines identified in the **Management Agreement**. The project maintains and updates, if required, the project baselines and Management Agreement under configuration management. As a minimum, the project does the following:

- Manages programmatic margins and resources to ensure successful completion of this and remaining life-cycle phases within budget, schedule, and risk constraints.
- Updates the IMS when changes warrant.
- Updates cost estimates and their basis when changes warrant.
- Assesses the adequacy of anticipated budget availability against phased life-cycle cost requirements and commitments, incorporating the impact of performance to date.

Projects provide an **updated CADRe (parts A, B, and C)** consistent with the *NASA Cost Estimating Handbook* within 90 days after the completion of spacecraft post-launch checkout. This CADRe is based on the "as built" launched baseline.

4.4.6.3 Project Phase E Technical Activities and Products

The project performs its operations technical activities as required in NPR 7123.1 for this phase. It certifies and maintains mission operations readiness, as required; operates the spacecraft in accordance with the operations procedures; sustains the spacecraft and supporting systems as the need arises; captures and archives mission technical results; and evaluates when it is ready for end of mission. It updates the **EOMP** as described in NPR 8715.6 and NASA-STD-8719.14, Appendix B, annually, as well as updating the **Security and Project Protection Plans** annually.

The project team conducts an annual IT security assessment of IT systems in conformance to the requirements of NPR 2810.1 during Phase E of the project.

Finally, the project team documents lessons learned in accordance with NPD 7120.4 and per NPD 7120.6 and the project's **Knowledge Management Plan**.

4.4.7 Project Decommissioning and Disposal

All projects will eventually cease as a natural evolution of completing their mission objectives. When this occurs, the Mission Directorate, program, and project need to be sure that all the products produced by the project (e.g., spacecraft, ground systems, test beds, spares, science data, operational data, returned samples) are properly dispositioned and that all project activities (e.g., contracts, financial obligations) are properly closed out. The project develops a Decommissioning/Disposal Plan to cover all activities necessary to close the project out and conducts a DR in preparation for final approval to decommission by the Decision Authority (or designee) at KDP F.

The decommissioning of a project with operating spacecraft requires that the project team ensure the safe and adequate disposal of the spacecraft. Figure 4-17 provides an overview of the disposal of a spacecraft, the various documents that are produced as part of this, and the order and timing of major activities and document deliveries.

The actual disposal of the spacecraft (reorbit, deorbit, and passivation) needs to meet Agency orbital debris requirements and is a critical event. As a result, this event requires a DRR. This review evaluates the readiness of the project and the flight system for execution of the spacecraft disposal event (see NPR 7120.5E and Table E-3 in this handbook). In many cases, such as small spacecraft, the decommissioning and disposal occur relatively close together. In these instances, the DR and DRR may be conducted together.

Decommissioning/disposal and Phase F end when the project funding is finally terminated.[35]

The Decommissioning/Disposal Plan is prepared by the project manager and approved by the program manager; Center Director; Chief, SMA (via Orbital Debris Program Manager); the MDAA; and the Decision Authority, if not the MDAA. This plan is approved at KDP F.

The Decommissioning/Disposal Plan contains the following:

- Updated EOMP, including method and location of disposal; planned status of spacecraft after disposal; and schedule, safety and environmental considerations;
- Updated Mishap Preparedness and Contingency Plan and predefined contingency/mishap scenarios;

---

[35] Funding for SMD projects covers the archival of the science data produced by the spacecraft (and the ancillary data for its interpretation) prior to project termination. This ensures that the science community will have access to this data for follow-on science research and data analysis.

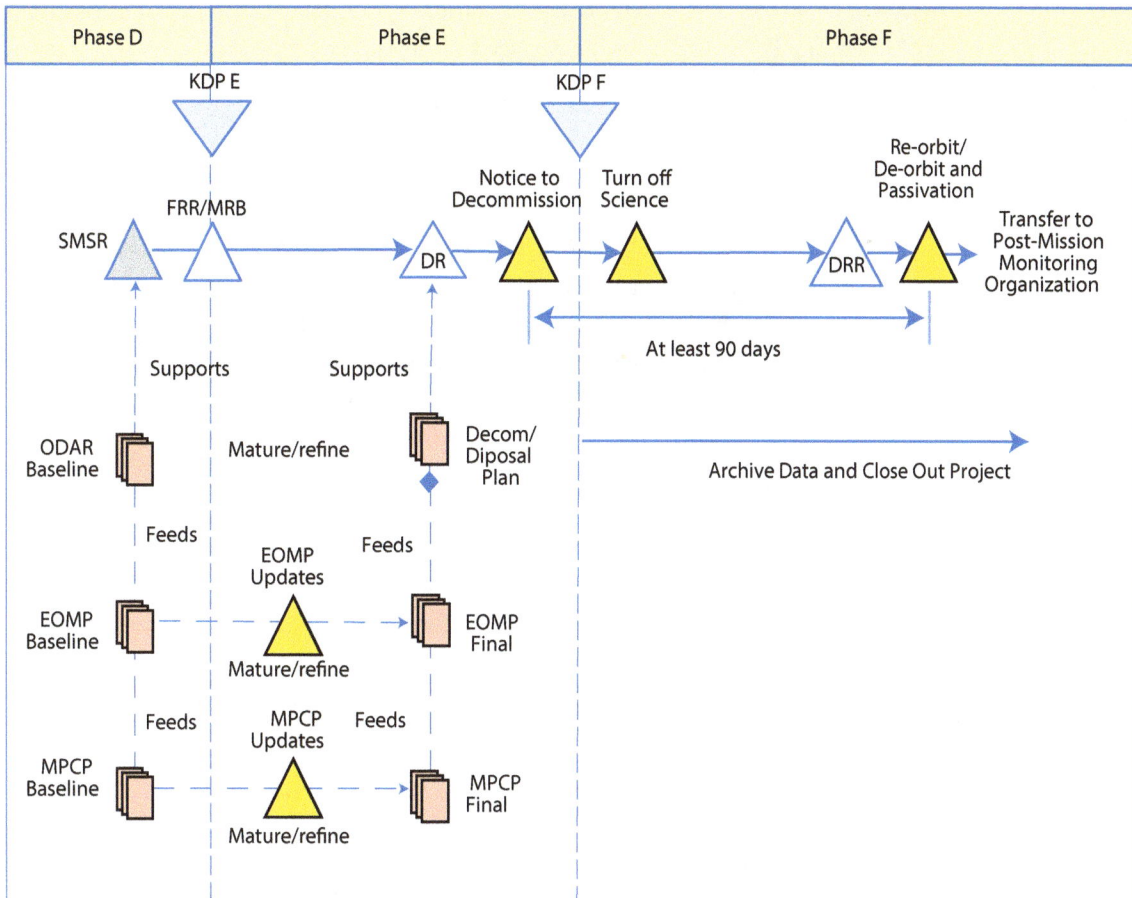

Legend: ▲ Events △ Life-Cycle Reviews △ Other Review ▯ Documents

Acronyms: DR = Decommissioning Review; DRR = Disposal Readiness Review; EOMP = End of Mission Plan; MPCP = Mishap Preparedness and Contingency Plan; ODAR = Orbital Debris Assessment Report; SMSR = Safety and Mission Success Review.

Figure 4-17 Spacecraft Disposal Process Flow

- Approach and plans for notifying stakeholders and customers of the intent to decommission the project and spacecraft as described in *NPD 8010.3, Notification of Intent to Decommission or Terminate Operating Space Missions and Terminate Missions*;

- Approach and plans for:
  - Archiving science, operations, and engineering data (e.g., methods, media, locations);
  - Maintaining communications security;
  - Dispositioning all hardware, software, and facilities remaining on the ground;

- Closing out contracts, financial obligations, and project infrastructure and transferring project personnel; and

- Long-term monitoring of spacecraft remaining on orbit.

4.4.7.1 Completing Operations and Sustainment (Phase E) and Preparing for Decommissioning and Closeout (Phase F)

The project develops and updates its plans for work to be performed during Phase F, including updates, if needed, to life-cycle review plans, project IMS, details on technical work to be accomplished, key acquisition activities planned, and plans for monitoring performance against plan. The project incorporates the impact of performance against the plan established at KDP E.

The project prepares and finalizes Phase F work agreements.

The project documents the results of Phase E activities and generates the appropriate documentation as described in NPR 7123.1 and NPR 7120.5E Tables I-4 and I-5 and Tables 4-6 and 4-7 at the end of this chapter.

4.4.7.2 Project Phase E Reporting Activities and Preparing for Major Milestones

4.4.7.2.1 Project Reporting

The project manager reports to the Center Director or designee, and supports the program executive in reporting the status of project Implementation at many other forums, including Mission Directorate monthly status meetings and the Agency's monthly Baseline Performance Review (BPR). Section 5.12 provides further information regarding potential project reporting.

4.4.7.2.2 Project Internal Reviews

Prior to the life-cycle reviews, projects conduct internal reviews in accordance with NPR 7123.1, Center practices, and NPR 7120.5. These internal reviews are described in Section 4.3.2.2.2.

4.4.7.2.3 Preparing for Major Milestone Reviews

Projects plan, prepare for, and support the CERR, PFAR, and DR (and DRR if combined with the DR) life-cycle reviews in accordance with NPR 7123.1, Center practices, and NPR 7120.5, ensuring that the life-cycle review objectives and expected maturity states defined in NPR 7120.5 have been satisfactorily met. Life-cycle review entrance and success criteria in Appendix G of NPR 7123.1 and the life-cycle phase and KDP information in Appendix E of this handbook provide specifics for addressing the six assessment criteria required to demonstrate that the project has met the expected maturity state.

Projects plan, prepare for, and support the governing PMC review prior to KDP F and provide or obtain the KDP readiness products listed in Section 4.2.3.

Once the KDP has been completed and the Decision Memorandum signed, the project updates its documents and plans as required to reflect the decisions made and actions assigned at the KDP.

In tightly coupled and single-project programs, project(s) transition to KDP F in accordance with the plan for reviews documented in the Program or Project Plan.

4.4.8 Project Phase F, Decommissioning/Disposal and Closeout Activities

4.4.8.1 Project Phase F Life-Cycle Activities

During Phase F, the project implements the Decommissioning/Disposal Plan developed and approved in Phase E. The project dispositions all spacecraft ground systems, data, and returned samples, including safe and adequate disposal of the spacecraft. The project team dispositions other in-space assets. The project team closes out all project activities in accordance with the Decommissioning/Disposal Plan. The project performs a Disposal Readiness Review (DRR) if it was not performed as part of the Decommissioning Review (DR).

The objectives of the DRR are to evaluate the readiness of the project and the flight system for execution of the spacecraft disposal event.

A general flow of Phase F activities is shown in Figure 4-18.

4.4.8.2 Project Phase F Planning, Control, and Technical Activities and Products

During Phase F, the project manager and the project team perform the technical activities required in NPR 7123.1 for this phase. They perform spacecraft and other in-space asset disposal and closeout and disposition ground systems, test beds, and spares. They monitor decommissioning and disposal risks, actively assess open risks, and develop and implement mitigation plans.

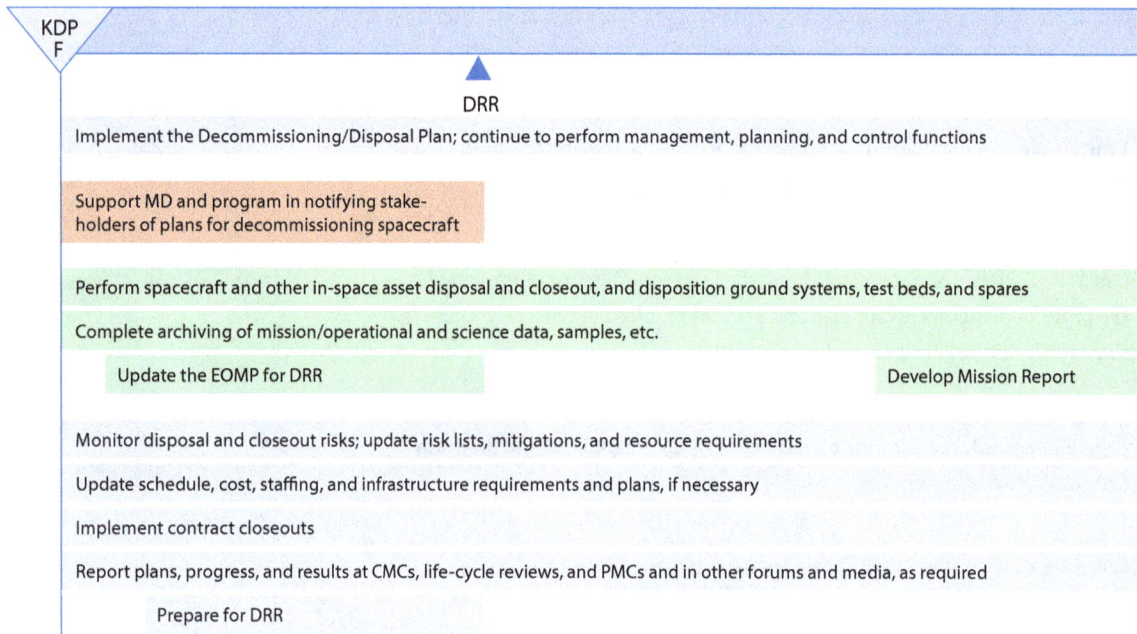

Legend: ☐ Project management, planning, and control tasks

☐ Work for which Headquarters is responsible but the project helps accomplish (e.g., international partnerships are a Headquarters responsibility, but the projects help develop and finalize those partnerships)

☐ Technical work the project is doing

Acronyms: MD = Mission Directorate; OIIR = Office of International and Interagency Relations.

Note: These are typical high-level activities that occur during this project phase. Placement of reviews is notional.

Figure 4-18 Project Phase F Flow of Activities

If the project's DRR was not performed as part of the DR, the project updates its Mishap Preparedness and Contingency Plan, EOMP, disposal portions of the Decommissioning/Disposal Plan, and SMA Plan prior to the DRR. The project also updates its technical, cost, schedule and safety risks and cost estimates prior to the DRR. In addition, the project continues updating the Security and Project Protection Plans annually.

The project team completes archiving mission/operational and science data and documents the results of Phase F activities. It completes storage and cataloging of returned samples and archives project engineering and technical management data. It implements contract closeouts, as appropriate. It develops the Mission Report and documents lessons learned in accordance with NPD 7120.4 and per NPD 7120.6 and the project's Knowledge Management Plan.

The project team provides the final update to the CADRe (part C) consistent with the *NASA Cost Estimating Handbook* within 60 days after the end of decommissioning and disposal. The purpose is to capture the content and cost of the decommissioning and disposal.

Decommissioning/disposal and Phase F end when project money is finally terminated.[36]

4.4.8.3 Project Phase F Reporting Activities and Preparing for Closeout

The project continues to report to the Center Director (or designee) and the Mission Directorate as required to report the status of decommissioning and disposal. The project manager will probably be required to report the status at many other forums, including Mission Directorate monthly status meetings and the Agency's monthly BPR.

The project plans, prepares for, and supports the project DRR life-cycle review (if needed) in accordance with NPR 7123.1, Center practices, and the requirements of this handbook, including the DRR objectives and expected maturity state defined in NPR 7120.5E and Table D-3 in this handbook. Life-cycle review entrance and success criteria in Appendix G of NPR 7123.1 and the life-cycle phase and KDP information in Appendix E of this handbook provide specifics for addressing the six assessment criteria required to demonstrate that the project has met the expected maturity state.

Prior to the DRR, the project conducts internal reviews in accordance with NPR 7123.1, Center practices, and NPR 7120.5 and performs an assessment of the project's readiness to proceed to the DRR.

4.5 Project Products by Phase

4.5.1 Non-Configuration-Controlled Documents

For non-configuration-controlled documents, the following terms and definitions are used in Tables 4-6 and 4-7:

- "Initial" is applied to products that are continuously developed and updated as the project matures.

The Mission Report is a summary of what the mission accomplished and is prepared at the end of a mission. It has also been called an End of Mission report, but this is not to be confused with the End of Mission Plan (EOMP). (See Section 4.4.3.3.) The Mission Report generally includes a summary of the mission accomplishments, science data/samples collected, and a summary of the results achieved. This report is prepared in conjunction with documenting the mission's lessons learned per NPD 7120.6, and the project's Knowledge Management Plan. Projects need to ensure that resources are allocated to develop the Mission Report and lessons learned. These provide a valuable historical record of NASA's accomplishments and the issues that were encountered and overcome as part of the mission.

---

[36] Funding for SMD projects covers the archival of the science data produced by the spacecraft (and the ancillary data for its interpretation) prior to project termination. This ensures that the science community will have access to this data for follow-on science research and data analysis.

- "Final" is applied to products that are expected to exist in this final form, e.g., minutes and final reports.
- "Summary" is applied to products that synthesize the results of work accomplished.
- "Plan" is applied to products that capture work that is planned to be performed in the following phases.
- "Update" is applied to products that are expected to evolve as the formulation and implementation processes evolve. Only expected updates are indicated. However, any document may be updated, as needed.

4.5.2 Configuration-Controlled Documents

For configuration-controlled documents, the following terms and definitions are used in Tables 4-6 and 4-7:

- "Preliminary" is the documentation of information as it stabilizes but before it goes under configuration control. It is the initial development leading to a baseline. Some products will remain in a preliminary state for multiple life-cycle reviews. The initial preliminary version is likely to be updated at subsequent life-cycle reviews but remains preliminary until baselined.

- "Baseline" indicates putting the product under configuration control so that changes can be tracked, approved, and communicated to the team and any relevant stakeholders. The expectation on products labeled "baseline" is that they will be at least final drafts going into the designated life-cycle review and baselined coming out of the life-cycle review. Baselining of products that will eventually become part of the Project Plan indicates that the product has the concurrence of stakeholders and is under configuration control. Updates to baselined documents require the same formal approval process as the original baseline.

- "Approve" is used for a product, such as Concept Documentation, that is not expected to be put under classic configuration control but still requires that changes from the "Approved" version are documented at each subsequent "Update."

- "Update" is applied to products that are expected to evolve as the formulation and implementation processes evolve. Only expected updates are indicated. However, any document may be updated, as needed. Updates to baselined documents require the same formal approval process as the original baseline.

Table 4-6 Project Milestone Products Maturity Matrix

Products	Pre–Phase A KDP A	Phase A KDP B		Phase B KDP C	Phase C KDP D		Phase D KDP E		Phase E KDP F	Phase F
	MCR	SRR	SDR/MDR	PDR	CDR	SIR	ORR	MRR/FRR	DR	DRR
Headquarters and Program Products[1]										
1. FAD	Baseline									
2. Program Plan	Baseline									
2.a. Applicable Agency strategic goals	Baseline	Update	Update							
2.b. Documentation of program-level requirements and constraints on the project (from the Program Plan) and stakeholder expectations, including mission objectives/goals and mission success criteria	Preliminary	Baseline	Update	Update						
2.c. Documentation of driving mission, technical, and programmatic ground rules and assumptions	Preliminary	Preliminary	Baseline	Update	Update	Update				
3. Partnerships and interagency and international agreements	Preliminary	Update	Baseline U.S. partnerships and agreements	Baseline international agreements						
4. ASM minutes		Final								
5. NEPA compliance documentation per NPR 8580.1				Final documentation per NPR 8580.1						
6. Mishap Preparedness and Contingency Plan				Preliminary		Update		Baseline (SMSR)	Update	Update

Table 4-6 Project Milestone Products Maturity Matrix

(continued)

Project Technical Products[2]

Products	Pre–Phase A KDP A — MCR	Phase A KDP B — SRR	Phase A KDP B — SDR/MDR	Phase B KDP C — PDR	Phase C KDP D — CDR	Phase C KDP D — SIR	Phase D KDP E — ORR	Phase D KDP E — MRR/FRR	Phase E KDP F — DR	Phase F — DRR
1. Concept Documentation	Approve	Update	Update	Update						
2. Mission, Spacecraft, Ground, and Payload Architectures	Preliminary mission and spacecraft architecture(s) with key drivers	Baseline mission and spacecraft architecture, preliminary ground and payload architectures. Classify payload(s) by risk per NPR 8705.4.	Update mission and spacecraft architecture, baseline ground and payload architectures	Update mission, spacecraft, ground and payload architectures						
3. Project-Level, System, and Subsystem Requirements	Preliminary project-level requirements	Baseline project-level and system-level requirements	Update Project-level and system-level requirements, Preliminary subsystem requirements	Update project-level and system-level requirements. Baseline subsystem requirements						
4. Design Documentation			Preliminary	Baseline Preliminary Design	Baseline Detailed Design	Update		Baseline As-built hardware & software		
5. Operations Concept	Preliminary	Preliminary	Preliminary	Baseline						
6. Technology Readiness Assessment Documentation	Initial	Update	Update	Update	Update					
7. Engineering Development Assessment Documentation	Initial	Update	Update	Update						
8. Heritage Assessment Documentation	Initial	Update	Update	Update						
9. Safety Data Packages				Preliminary	Baseline	Update	Update	Update		
10. ELV Payload Safety Process Deliverables				Preliminary	Preliminary	Baseline				
11. V&V Report							Preliminary	Baseline		

Table 4-6 Project Milestone Products Maturity Matrix

Products	Pre–Phase A KDP A	Phase A KDP B		Phase B KDP C	Phase C KDP D			Phase D KDP E	Phase E KDP F	Phase F
	MCR	SRR	SDR/MDR	PDR	CDR	SIR	ORR	MRR/FRR	DR	DRR
12. Operations Handbook						Preliminary	Baseline	Update		
13. Orbital Debris Assessment per NPR 8715.6	Preliminary Assessment			Preliminary design ODAR	Detailed design ODAR			Final ODAR (SMSR)		
14. End of Mission Plans per NPR 8715.6/NASA-STD 8719.14, App B								Baseline EOMP (SMSR)	Update EOMP annually	Update EOMP
15. Mission Report										Final
Project Management, Planning, and Control Products										
1. Formulation Agreement	Baseline for Phase A; Preliminary for Phase B		Baseline for Phase B							
2. Project Plan			Preliminary	Baseline						
3. Plans for work to be accomplished during next Implementation life-cycle phase				Baseline for Phase C		Baseline for Phase D		Baseline for Phase E	Baseline for Phase F	
4. Documentation of performance against Formulation Agreement (see #1 above) or against plans for work to be accomplished during Implementation life-cycle phase (see #3 above), including performance against baselines and status/closure of formal actions from previous KDP		Summary	Summary	Summary	Summary	Summary	Summary	Summary	Summary	
5. Project Baselines			Preliminary	Baseline	Update	Update	Update	Update		
5.a. Top technical, cost, schedule and safety risks, risk mitigation plans, and associated resources	Initial	Update	Update	Update	Update	Update	Update	Update	Update	Update
5.b. Staffing requirements and plans	Initial	Update	Update	Update	Update		Update			

Table 4-6 Project Milestone Products Maturity Matrix

(continued)

Products	Pre–Phase A KDP A — MCR	Phase A KDP B — SRR	Phase A KDP B — SDR/MDR	Phase B KDP C — PDR	Phase C KDP D — CDR	Phase C KDP D — SIR	Phase D KDP E — ORR	Phase D KDP E — MRR/FRR	Phase E KDP F — DR	Phase F — DRR
5.c. Infrastructure requirements and plans, business case analysis for infrastructure Alternative Future Use Questionnaire (NASA Form 1739), per NPR 9250.1	Initial	Update	Update Baseline for NF 1739 Section A	Update Baseline for NF 1739 Section B	Update					
5.d. Schedule	Risk informed at project level with preliminary Phase D completion ranges	Risk informed at system level with preliminary Phase D completion ranges	Risk informed at subsystem level with preliminary Phase D completion ranges. Preliminary Integrated Master Schedule	Risk informed and cost- or resource-loaded. Baseline Integrated Master Schedule	Update IMS	Update IMS	Update IMS			
5.e. Cost Estimate (Risk-Informed or Schedule-Adjusted Depending on Phase)	Preliminary Range estimate	Update	Risk-informed schedule-adjusted range estimate	Risk-informed and schedule-adjusted Baseline	Update	Update	Update	Update	Update	Update
5.f. Basis of Estimate (cost and schedule)	Initial (for range)	Update (for range)	Update (for range)	Update for cost and schedule estimate	Update	Update	Update	Update	Update	Update
5.g. Confidence Level(s) and supporting documentation			Preliminary cost confidence level and preliminary schedule confidence level	Baseline joint cost and schedule confidence level						
5.h. External Cost and Schedule Commitments			Preliminary for ranges	Baseline						
5.i. CADRe			Baseline	Update	Update	Update		Update[3]	Update	
6. Decommissioning/ Disposal Plan									Baseline	Update disposal portions

[1] These products are developed by the Mission Directorate.

[2] These document the work of the key technical activities performed in the associated phases.

[3] The CADRe for MRR/FRR is considered the "Launch CADRe" to be completed after the launch.

Table 4-7 Project Plan Control Plans Maturity Matrix

NPR 7120.5 Program and Project Plan— Control Plans[1]	Pre–Phase A MCR	Phase A KDP B SRR	Phase A KDP B SDR/MDR	Phase B KDP C PDR	Phase C KDP D CDR	Phase C KDP D SIR	Phase D KDP E ORR	Phase D KDP E MRR/FRR	Phase E KDP F DR
1. Technical, Schedule, and Cost Control Plan	Approach for managing schedule and cost during Phase A[2]	Preliminary	Baseline	Update					
2. Safety and Mission Assurance Plan		Baseline	Update	Update	Update			Update (SMSR)	Update
3. Risk Management Plan	Approach for managing risks during Phase A[2]	Baseline	Update	Update					
4. Acquisition Plan	Preliminary Strategy	Baseline	Update	Update					
5. Technology Development Plan (may be part of Formulation Agreement)	Baseline	Update	Update	Update					
6. Systems Engineering Management Plan	Preliminary	Baseline	Update	Update					
7. Information Technology Plan		Preliminary	Baseline	Update					
8. Software Management Plan(s)		Preliminary	Baseline	Update					
9. Verification and Validation Plan	Preliminary Approach[3]		Preliminary	Baseline	Update	Update			
10. Review Plan	Preliminary	Baseline	Update	Update					
11. Mission Operations Plan						Preliminary	Baseline	Update	
12. Environmental Management Plan			Baseline						
13. Integrated Logistics Support Plan	Approach for managing logistics[2]	Preliminary	Preliminary	Baseline	Update				
14. Science Data Management Plan				Preliminary			Baseline	Update	
15. Integration Plan	Preliminary approach[2]		Preliminary	Baseline	Update				
16. Configuration Management Plan		Baseline	Update	Update					
17. Security Plan			Preliminary	Baseline					Update annually
18. Project Protection Plan			Preliminary	Baseline	Update	Update	Update	Update	Update annually
19. Technology Transfer Control Plan			Preliminary	Baseline	Update				

Table 4-7 Project Plan Control Plans Maturity Matrix (continued)

NPR 7120.5 Program and Project Plan—Control Plans[1]	Pre–Phase A	Phase A KDP B		Phase B KDP C	Phase C KDP D		Phase D KDP E		Phase E KDP F
	MCR	SRR	SDR/MDR	PDR	CDR	SIR	ORR	MRR/FRR	DR
20. Knowledge Management Plan	Approach for managing during Phase A[2]		Preliminary	Baseline	Update				
21. Human Rating Certification Package	Preliminary approach[3]	Initial	Update	Update	Update		Update	Approve Certification	
22. Planetary Protection Plan			Planetary Protection Certification (if required)	Baseline					
23. Nuclear Safety Launch Approval Plan			Baseline (mission has nuclear materials)						
24. Range Safety Risk Management Process Documentation				Preliminary	Preliminary	Baseline			
25. Education Plan			Preliminary	Baseline	Update		Update		
26. Communications Plan			Preliminary	Baseline	Update		Update		

[1] See template in NPR 7120.5E, Appendices G and H, for control plan details.

[2] Not the Plan, but documentation of high-level process. May be documented in MCR briefing package.

[3] Not the Plan, but documentation of considerations that might impact the cost and schedule baselines. May be documented in MCR briefing package.

5 Special Topics

This chapter is devoted to more detailed explanation and exploration of particular policy topic areas. This handbook contains core information on these topics, but it is not the only resource. Many of these topics have additional expanding and changing information in various communities of practice. These communities are referenced, but are also likely places to find additional material that will be developed after this handbook is published.

5.1 NASA Governance

This section highlights key aspects of Governance that are particularly important to the management and execution of space flight programs and projects.

NASA's management structure focuses on safety and mission success across a challenging portfolio of high-risk, complex endeavors, many of which are executed over long periods of time. *NPD 1000.0, NASA Governance and Strategic Management Handbook* sets forth NASA's Governance framework—the principles and structures through which the Agency manages its missions and executes its responsibilities. Familiarity with NASA Governance provides an understanding of the fundamental principles for all individuals with a significant role in NASA programs and projects or their support. Appendix D provides a summary of the roles and responsibilities for key program and project management officials.

NASA Governance provides an organizational structure that emphasizes safety and mission success by taking advantage of different perspectives that different organizational elements bring to issues. The organizational separation of Programmatic and Institutional Authorities in NASA Governance is a cornerstone of NASA's system of checks and balances that supports mission success. (See Figure 2-3.)

- The Programmatic Authority resides with the Mission Directorates and their respective programs and projects.

- The Institutional Authority encompasses all those Headquarters and associated Center organizations and authorities not in the Programmatic Authority. The Institutional Authority includes the offices within the Mission Support Directorate and its associated organizations at the Centers, the Center Directors, and the Technical Authorities (TAs), who are individuals with specifically delegated authority in Engineering (ETA), Safety and Mission Assurance (SMA TA), and Health and Medical (HMTA).

NPR 7120.5 differentiates between "programmatic requirements" and "institutional requirements." Both categories of requirements ultimately need to be satisfied in program and project Formulation and Implementation.

Programmatic requirements focus on the products to be developed and delivered and specifically relate to the goals and objectives of a particular NASA program or project. These programmatic requirements flow down from the Agency's strategic planning process and are the responsibility of the Programmatic Authorities. Table 5-1 shows this flow down from Agency strategic planning through Agency, directorate, program, and project requirements levels to the systems that will be implemented to achieve the Agency goals.

Institutional requirements focus on how NASA does business and are independent of any particular program or project. These requirements are issued by NASA Headquarters (including the Office of the Administrator and Mission Support Offices (MSO)) and by Center organizations and are the responsibility of the Institutional Authorities. Institutional requirements may respond to Federal statute, regulation, treaty, or Executive Order. They are normally documented in the following:

- NASA Policy Directives (NPDs)—Agency policy documents that describe what is required by NASA management to achieve NASA's vision, mission, and external mandates and who is responsible for carrying out those requirements.

- NASA Procedural Requirements (NPRs)—Documents that provide the Agency's mandatory requirements for implementing NASA policy as delineated in an associated NPD.

- NASA Standards—Formal documents that establish a norm, requirement, or basis for comparison, a reference point to measure or evaluate against. A technical standard, for example, establishes uniform engineering or technical criteria, methods, processes, and practices. NASA standards include Agency-level standards as well as Center-level standards.

Table 5-1 Programmatic Requirements Hierarchy

Requirements Level	Content	Governing Document	Approver	Originator
Strategic Goals	Agency strategic direction	*NPD 1000.0, NASA Governance and Strategic Management Handbook*; *NPD 1001.0, NASA Strategic Plan*; and Strategic Programming Guidance	NASA Administrator	Support Organizations
Agency Requirements	Structure, relationships, principles governing design and evolution of cross-Agency Mission Directorate systems linked in accomplishing Agency strategic goals and outcomes	Architectural Control Document (ACD)	NASA Administrator	Host MDAA with inputs from Other Affected MDAAs
Mission Directorate Requirements	High-level requirements levied on a program to carry out strategic and architectural direction, including programmatic direction for initiating specific projects	Program Commitment Agreement (PCA)	NASA AA	MDAA
Program Requirements	Detailed requirements levied on a program to implement the PCA and high-level programmatic requirements allocated from the program to its projects	Program Plan	MDAA	Program Manager
Project Requirements	Detailed requirements levied on a project to implement the Program Plan and flow down programmatic requirements allocated from the program to the project	Project Plan	Program Manager	Project Manager
System Requirements	Detailed requirements allocated from the project to the next lower level of the project	System Requirements Documentation	Project Manager	Responsible System Lead

MDAA = Mission Directorate Associate Administrator; NASA AA = NASA Associate Administrator

- **Center Policy Directives (CPDs)**—Center-specific policy documents that describe requirements and responsibilities that apply only to the issuing Center and operations performed by NASA personnel at that Center. CPDs extend requirements delineated in associated NPDs and NPRs.

- **Center Procedural Requirements (CPRs)**—Center-specific procedural requirements and responsibilities for implementing the policies and procedural requirements defined in related NPDs, NPRs, or CPDs. CPRs apply only to the issuing Center and operations performed by NASA personnel at that Center.

- **Mission Directorate Requirements**—Requirements contained in Mission Directorate documentation that apply to activities, products, or services supporting program and project office needs, which could extend across multiple Centers.

Figure 5-1 shows the flow down from *NPD 1000.0, NASA Governance and Strategic Management Handbook through Program and Project Plans*. The figure identifies the five types of institutional requirements that flow down to these plans: engineering, program or project management, safety and mission assurance, health and medical, and MSO functional requirements. These terms are defined in Appendix A.

5.1.1 Programmatic Authority

Programmatic Authority flows from the Administrator through the Associate Administrator to the Mission Directorate Associate Administrator (MDAA), to the program manager, and finally to the project manager per NPD 1000.0. Because there are different types of programs that require different management approaches, the MDAA may delegate some of his/her Programmatic Authority to deputy associate administrators, division directors, or their equivalent such as program directors, depending on the Mission Directorate organizational structure, consistent with the following principles:

- As a general rule, the MDAA does not delegate responsibility beyond his/her immediate organization for strategic planning; policy formulation and approval; definition and approval of programs, projects, and missions; assignment of programs, projects, and selected managers; Mission Directorate budget development and allocation; and assessment and reporting of performance. Delegations are documented to ensure roles and responsibilities are understood and accountability is clear. The responsibilities and authority of the MDAA and those individuals with delegated Programmatic Authority are documented in the Program Plan and/or the Program Commitment Agreement (PCA) such that they are unambiguous with minimal overlap.

- The program manager is responsible for the formulation and implementation of the program as described in NPR 7120.5 and *NPR 7123.1, NASA Systems Engineering Processes and Requirements*. This includes responsibility and accountability for ensuring program safety; technical integrity; technical, cost, and schedule performance; and mission success; developing and presenting time-phased cost estimates, budget, and funding requirements; developing and implementing the Program Plan, including managing program resources; implementing a risk management process that incorporates Risk-Informed Decision Making (RIDM) and Continuous Risk Management (CRM); overseeing project implementation, including resolution of project risks by such means as allocation of margins to mitigate risks; periodically reporting progress to the Mission Directorate; and supporting Mission Directorate activities.

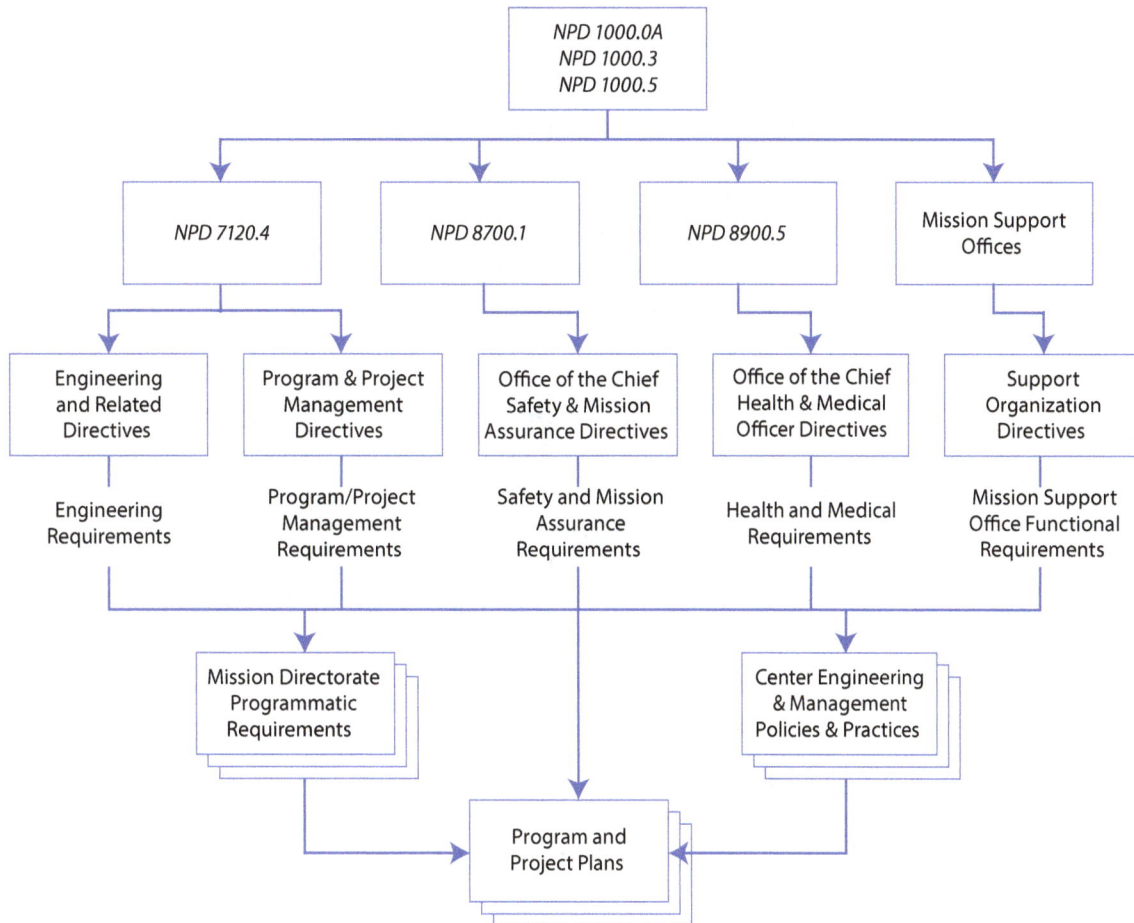

Figure 5-1 Institutional Requirements Flow Down

- The project manager reports to the program manager and both are supported by one or more NASA Centers (with facilities and experts from line or functional organizations). The project manager, however, is responsible for the formulation and implementation of the project as described in NPR 7120.5 and NPR 7123.1. This includes responsibility and accountability for the project safety, technical integrity, and mission success of the project, while also meeting programmatic (technical, cost, and schedule performance) commitments. To accomplish this, the project manager requires a breadth of skills, needing to be knowledgeable about governing laws; acquisition regulations; policies affecting program and project safety; training of direct-report personnel; risk management; environmental management; resource management; program- and project-unique test facilities; the health of the industrial base and supply chain supporting the program and project, including critical and single-source suppliers; software management; responding to external requests for audits (e.g., OMB); protecting intellectual property and technology; and other aspects of program and project management.

- The program and project manager coordinate early and often throughout the program or project life cycle with mission support organizations at NASA Headquarters through the sponsoring Mission Directorate and the implementing Centers. These mission support organizations include legal, procurement, security, finance, export control, human resources, public affairs, international affairs, property, facilities, environmental, aircraft operations, IT, planetary protection, and others. They provide essential expertise and ensure compliance with relevant laws, treaties, Executive Orders, and regulations. It is also important to ensure that organizations having a substantive interest (these might include supporting activities such as facilities and logistics) are integrated effectively into the program's or project's activities as early as appropriate and throughout the duration of the organizations' interest to include their needs, benefit from their experience, and encourage communication.

5.1.2 Institutional Authority

The Institutional Authority consists of those organizations not in the Programmatic Authority. As shown in Figure 2-3, this includes Engineering, Safety and Mission Assurance, and Health and Medical organizations, Mission Support Organizations (MSOs), and Center Directors. Technical Authorities for Engineering, Safety and Mission Assurance, and Health and Medical are a unique segment of the Institutional Authority. They support programs and projects in two ways:

1. They provide technical personnel and support and oversee the technical work of personnel who provide the technical expertise to accomplish the program or project mission.
2. They provide Technical Authorities, who independently oversee programs and projects. These individuals have formally delegated Technical Authority traceable to the Administrator and are funded independently of programs and projects. The Technical Authorities are described in Section 5.2.

Key roles and responsibilities within the Institutional Authority reside with the Mission Support Directorate (MSD) and the Center Director.

The MSD Associate Administrator establishes directorate policies and procedures for institutional oversight for mission support functional areas (e.g., procurement). As part of MSD, the MSOs are the "official voices" of their institutional areas and the associated requirements established by

NASA policy, law, or other external mandate. Their authorities are asserted horizontally (across Headquarters) and vertically (Headquarters to Centers and within Centers). The delegated responsibilities of MSOs vary depending on their functional areas: finance, procurement, information technology, legal, and facilities engineering. Common responsibilities of MSOs are to:

- Represent the institutional function and convey respective institutional requirements established by law, Agency policy, or other external or internal authority to program and project managers.

- Collaborate with program or project managers on how best to implement prescribed institutional requirements and achieve program or project goals in accordance with all statutory, regulatory, and fiduciary responsibilities.

- Ensure conformance to institutional requirements, either directly or by agreement with other NASA organizations.

- Disposition all requests for modification of prescribed institutional requirements[1] in their respective area of responsibility.

Because programs and projects are executed in NASA Centers, a Center has both execution and Institutional Authority responsibilities, and the Center Director needs to ensure that both of these functions operate within the Governance and management structure dictated by *NPD 1000.0, NASA Governance and Strategic Management Handbook.*

As part of the execution responsibility, the Center Director is responsible for ensuring that the Center is capable of accomplishing the programs, projects, and other activities assigned to it in accordance with Agency policy and the Center's best practices and institutional policies. In accomplishing this role, a Center Director:

- Establishes, develops, and maintains the institutional capabilities (processes and procedures, human capital—including trained and certified program and project personnel, facilities, and infrastructure) required for the execution of programs and projects. This includes sound technical and management practices, internal controls, and an effective system of checks and balances to ensure the technical and programmatic integrity of program or project activities being executed at the Center.

- Works with the Mission Directorate and the program and project managers, once assigned, to assemble the program and project team(s) that will accomplish the program or project.

In accordance with *NPR 7120.5*: "Center Directors are responsible and accountable for all activities assigned to their Center. They are responsible for the institutional activities and for ensuring the proper planning for and assuring the proper execution of programs and projects assigned to the Center." This means that the Center Director is responsible for ensuring that programs and projects develop plans that are executable within the guidelines from the Mission Directorate and for assuring that these programs and projects are executed within the approved plans. In cases where the Center Director believes a program or project cannot be executed within approved guidelines and plans, the Center Director works with the project manager, program manager and Mission Directorate to resolve the problem.

[1] A prescribed requirement is one levied on a lower organizational level by a higher organizational level.

- Supports programs and projects by providing needed Center resources; providing support and guidance to programs and projects in resolving technical and programmatic issues and risks; monitoring the technical and programmatic progress of programs and projects to help identify issues as they emerge; and proactively working with the Mission Directorates, programs, projects, and other Institutional Authorities to find constructive solutions to problems.

- Proactively works on cross-Center activities to benefit both the programs and projects and the overall Agency long-term health.

As part of the Institutional Authority responsibility, a Center Director assures that program and project teams at the Center accomplish their goals in accordance with the prescribed requirements and the Agency's and Center's procedures and processes. Institutional Authority responsibility also means that the Center Director has the responsibility to ensure that the programs and projects are accomplishing their work in accordance with the institutional (including technical) requirements. When the program or project violates institutional requirements, the Center can direct the program or project to correct the deficiency. As an example, if the program or project is not performing requirements flow down properly, the Center may direct the program or project to correct how requirements are established, documented, and traced. However, this authority does not mean that the Institutional Authority can direct a program or project to exceed the programmatic requirements and constraints when correcting deficiencies. When this situation occurs, the program or project, Center Director, and MDAA need to work together to resolve the issue(s). In accomplishing this, the Center Director:

- Is delegated Technical Authority in accordance with NPR 7120.5E Section 3.3 and concurs with the Center's Technical Authority implementation plan, and ensures that delegated institutional and technical authority is properly executed by programs and projects at the Center.

- Ensures that programs and projects properly follow institutional and technical authority requirements.

- Establishes and maintains ongoing processes and forums, including the Center Management Council (CMC), to monitor the status and progress of programs and projects at their Center and to provide a summary status at Baseline Performance Reviews (BPRs) and other suitable venues.

- Periodically reviews programs and projects to assure they are performing in accordance with the Center's and the Agency's requirements, procedures, and processes.

- Keeps the Decision Authority advised of the executability of all aspects of the programs and projects (programmatic, technical, and all others) along with major risks, mitigation strategies, and significant concerns.

- Concurs in the adequacy of cost and schedule estimates and technology assessments and the consistency of these estimates with planned Agency requirements, workforce, and other resources.

- Certifies that programs and/or projects have been accomplished properly as part of the launch approval process.

- Ensures that Center training and certification programs for program and project managers are in place and ensures that program and project managers have met the training requirements.

5.2 Technical Authority

This special topic discusses key aspects of NASA's policy for Technical Authority and provides additional information to clarify the policy, explain the rationale behind it, and provide a historical perspective on its origin. The flow of this section is as follows:

- The origin of the technical authority process—why we have a technical authority process.

- Technical Authority and NASA Governance—how Technical Authority flows through the NASA organization as part of NASA's checks and balances.

- Common general technical authority roles—the roles that are common to all TAs.

- Engineering Technical Authority (ETA)—ETA delegations and various roles from the NASA chief engineer down to the project. Provides examples of how ETA is implemented.

- Safety and Mission Assurance Technical Authority (SMA TA)—the SMA TA process. Points to the SMA documents that govern the process.

- Health and Medical Technical Authority (HMTA)—the HMTA process. Points to the HMTA documents that govern the process.

5.2.1 Overview

The Technical Authority process is one of the important checks and balances built into NASA Governance. It provides assistance and independent oversight of programs and projects in support of safety and mission success. In NPR 7120.5 and this document, the term "technical authority" is used to describe individuals with delegated levels of authority and to refer to elements of the TA process.

Technical Authorities (TAs) have formally delegated responsibility that is traceable to the Administrator and provide independent oversight of programs' and projects' technical activities. Technical Authorities are provided by the Engineering, Safety and Mission Assurance, and Health and Medical organizations.

Delegation of TA is not a process of abdicating authority. TAs who further delegate their technical authority do not give up the responsibility and authority with which they are entrusted. They still remain accountable and participate in the TA chain as described herein.

5.2.2 The Origin of the Technical Authority Process

After the loss of the space shuttle *Columbia*, NASA recognized that its system of checks and balances needed strengthening. The Columbia Accident Investigation Board (CAIB) recommended the "establishment of an independent Technical Engineering Authority that is responsible for technical requirements and all waivers to them and will build a disciplined, systematic approach to identifying, analyzing, and controlling hazards throughout the life cycle of the Shuttle System."[2]

NASA chose to take a comprehensive approach to strengthening its systems and processes supporting the safety and mission success of all programs and projects while also addressing the CAIB's shuttle system recommendations. The resulting changes included improvements in NASA Governance, a revised statement of Agency core values, formalization of improved principles and processes for providing relief from prescribed requirements (Tailoring Principles, Section 5.4), establishment of a formally recognized process for resolving serious dissent by any individual (see Section 5.3 for information on the Dissenting Opinion process), and establishment of the Technical Authority process to provide independent oversight of programs and projects in support of safety and mission success. Refer to the CAIB report for good insights that need to be familiar to those involved in space flight programs and projects.

---

[2] Refer to R7.5-1 of the CAIB report accessible in the NASA On-Line Directives Information System (NODIS) at http://nodis3.gsfc.nasa.gov/ under "Useful Links," "Initiatives, Reports, Plans, etc.," and "Columbia Accident Investigation Board (CAIB) Report."

5.2.3 Technical Authority and NASA Governance

All NASA programs and projects are required to follow the technical authority process established in Section 3.3 of NPR 7120.5. This policy stems from NASA's Governance policy. NASA Governance is documented in *NPD 1000.0, NASA Governance and Strategic Management Handbook* and defines the structure by which the Office of the Administrator and senior staff provide leadership across the Agency and the core values and the principles by which NASA manages. Key principles in this framework include having clearly defined roles and responsibilities and having an effective system of checks and balances to provide a firm foundation for the balance of power between organizational elements.

The Technical Authority process is one of the important checks and balances built into NASA program and project management in support of safety and mission success. The process is built on the organizational and financial separation of the Programmatic and Institutional Authorities.[3] (See Section 5.1.) The separation enables the roles of the Programmatic and Technical Authorities to be wired into the basic organizational structure in a way that emphasizes their shared goal of mission success while taking advantage of the different perspectives each brings to issues.

Technical Authority originates with the NASA Administrator and is then delegated to the NASA Associate Administrator and then to the NASA Chief Engineer for Engineering Technical Authority (ETA); the Chief, Safety and Mission Assurance for SMA Technical Authority (SMA TA); and then to the Center Directors for ETA and SMA TA. The Administrator delegates Health and Medical Technical Authority (HMTA) to the NASA Chief Health and Medical Officer (CHMO). HMTA may then be delegated to the Center Chief Medical Officer (CMO) with the concurrence of the Center Director. The Center Director (or designee) is responsible for establishing and maintaining Center Technical Authority policies and practices, consistent with Agency policies and standards.

Subsequent delegations down from the Center Director are made to selected individuals at specific organizational levels. Such delegations are formal and traceable to the Administrator and documented in the Center plan for Technical Authority implementation. The individuals with Technical Authority are funded independent of a program or project. Technical Authorities located at Centers remain part of their Center organization.

The process supports clearly defined Technical Authorities and ensures the independence of the Technical Authorities.

Two important principles related to Technical Authority that need to be kept in mind are:

- The responsibilities of the program or project manager have not been diminished by the implementation of Technical Authority. The program or project manager is still ultimately responsible for the success of the program or project.

- Nothing in the Technical Authority process is intended or may be construed to abridge or diminish the SMA power to "suspend work" granted in *NPD 1000.3, The NASA Organization.* The Chief, Safety and Mission Assurance is authorized to suspend any operation or project activity that presents an unacceptable risk to personnel, property, or mission success and provide corrective action.

---

[3] Programmatic Authority resides with the Mission Directorates and their respective programs and projects. The Institutional Authority includes the remaining Headquarters and Center organizations.

5.2.4 Common Technical Authority Roles

5.2.4.1 General TA Roles for Program- and Project-Level TAs

Individuals with delegated Technical Authority at the program or project level have common responsibilities as delineated below. These responsibilities are formalized in policy so that the Technical Authority's day-to-day involvement in program or project activities ensures that significant views from the Technical Authorities are available to the program and project in a timely manner and are handled during the normal program and project processes. TAs are expected to keep their discipline chain of authority informed of issues as they arise, including direct communication between the Center's engineering director, SMA director (or equivalent), and chief medical officer with their counterparts at NASA Headquarters. Common responsibilities include:

a. Serving as members of program or project control boards, change boards, and internal review boards.

b. Working with the Center management and other Technical Authority personnel, as necessary, to ensure that the quality and integrity of program or project processes, products, and standards of performance related to engineering, SMA, and health and medical reflect the level of excellence expected by the Center or, where appropriate, by the NASA Technical Authority community.

c. Ensuring that requests for waivers or deviations from Technical Authority requirements are submitted by the program or project to and acted on by the appropriate level of Technical Authority. (Refer to Section 5.4.)

d. Assisting the program or project in making risk-informed decisions that properly balance technical merit, cost, schedule, and safety across the system.

e. Providing the program or project with their view of matters based on their knowledge and experience, assisting the program or project in obtaining the Technical Authority community view of requirements or issues when needed, and raising a Dissenting Opinion (see Section 5.3) on a decision or action when appropriate significant, substantive disagreement exists.

f. Serving as an effective part of NASA's overall system of checks and balances.

5.2.4.2 Unique ETA and SMA TA Roles in Support of HTMA

Due to Center infrastructure differences, the flow down of HMTA processes and responsibilities from the CHMO varies between Centers. To assist the Chief Health and Medical Officer with HMTA implementation, the NASA Chief Engineer and the NASA Chief, Safety and Mission Assurance agreed to support Agency-wide HMTA implementation through the utilization of Engineering and SMA personnel as HMTA awareness and communication links at each Center (not including JSC due to its existing HMTA infrastructure).

The program/project level Engineering and SMA TAs already involved in the day-to-day program/project work will serve as the HMTA "awareness eyes and ears" at the Centers. The primary role of the Engineering TA and SMA TA in the HMTA process is to identify potential HMTA issues and ensure that they are flowed to the appropriate levels of HMTA in a timely manner. The Engineering TA and SMA TA will establish interfaces and communication flow paths with the appropriate level of HMTA (CMO, JSC POC or OCHMO POC) for identification and resolution of potential HMTA issues. (See Section 5.2.7.3 for more details).

5.2.5 Special Risk Acceptance Roles

In recognition of the importance of systems that are associated with human flight, the top-level documents developed by a program detailing Agency-level requirements for human-rated systems are signed by the Administrator or his/her formally delegated designee.

To ensure proper oversight, decisions related to technical and operational matters involving safety and mission success residual risk[4] require formal concurrence by the responsible Technical Authority(ies) (ETA, SMA TA, and/or HMTA). This concurrence is based on the technical (engineering and safety) merits of the case.

Residual risks to personnel or high-value hardware require not only TA concurrence, but also the concurrence of the cognizant safety organization.

For matters involving human safety risk (see *NPR 8000.4, Agency Risk Management Procedural Requirements*), the actual risk taker(s) (or official spokesperson(s) and their supervisory chain) must formally consent to taking the risk, and the responsible program, project, or operations manager formally accept the risk. (For requirements in policy see both NPD 1000.0

---

[4] "Residual risk" is the risk that remains after all mitigation actions have been implemented or exhausted in accordance with the risk management process. (See *NPD 8700.1, NASA Policy for Safety and Mission Success.*)

and NPR 7120.5 as well as *NPR 8705.2, Human Rating Requirement for Space Systems*.)

5.2.5.1 Derived Technical Authority Roles

The TAs have additional roles that are specified in NPR 7120.5 but are not specifically discussed in the Technical Authority Roles and Responsibilities of the NPR. These are:

- Dispositioning requests for a Non-Applicable designation for a prescribed requirement that a program or project has evaluated as being "not relevant" and/or "not capable of being applied" to the applicable to the program, project, system, or component when the requirement is specified for implementation at the level of the Technical Authority. (See Section 5.4.6 of this handbook.)

- Assisting the program or project manager in determining when the program or project is ready for a life-cycle review as part of readiness assessment. (See Section 5.10 of this handbook.)

5.2.6 Technical Authority and Dissent

Infrequent circumstances may arise when a Technical Authority disagrees with a proposed programmatic or technical action and judges that the issue rises to a level of significance that needs to be brought to the attention of the next higher level of management (i.e., a dissenting opinion exists). (See Section 5.3.) In such circumstances, resolution occurs prior to implementation of the action whenever possible. However, if the program or project manager considers it to be in the best interest of the program or project, he/she has the authority to proceed at risk in parallel with the pursuit of a resolution. The program or project manager informs the second-higher level of management of the decision to proceed at risk. Since in this case the disagreement is between the program or project manager and the TA, the notification would be to the second-higher level of both the Programmatic and Technical Authority.

Resolution is jointly attempted at successively higher levels of Programmatic Authority and Technical Authority until the dissent is resolved.

Final appeals are made to the NASA Administrator. The adjudication path (see Figure 5-2) for the resolution is essentially the opposite of the authority flow-down path from the Administrator. (See Section 5.3 for more details on the dissenting opinion process.)

Notification of the second-higher level of management is provided because of the importance of a Dissenting Opinion and its resolution. This is particularly important in this instance because the Programmatic Authority has decided to proceed at risk in the presence of a Dissenting Opinion. The second-higher level of management is notified to provide personnel at that level with the option of becoming involved. This is not intended to skip a management level in the resolution process so much as to position the second-higher level to be knowledgeable of the issue and to support expeditious resolution at that level if it becomes necessary.

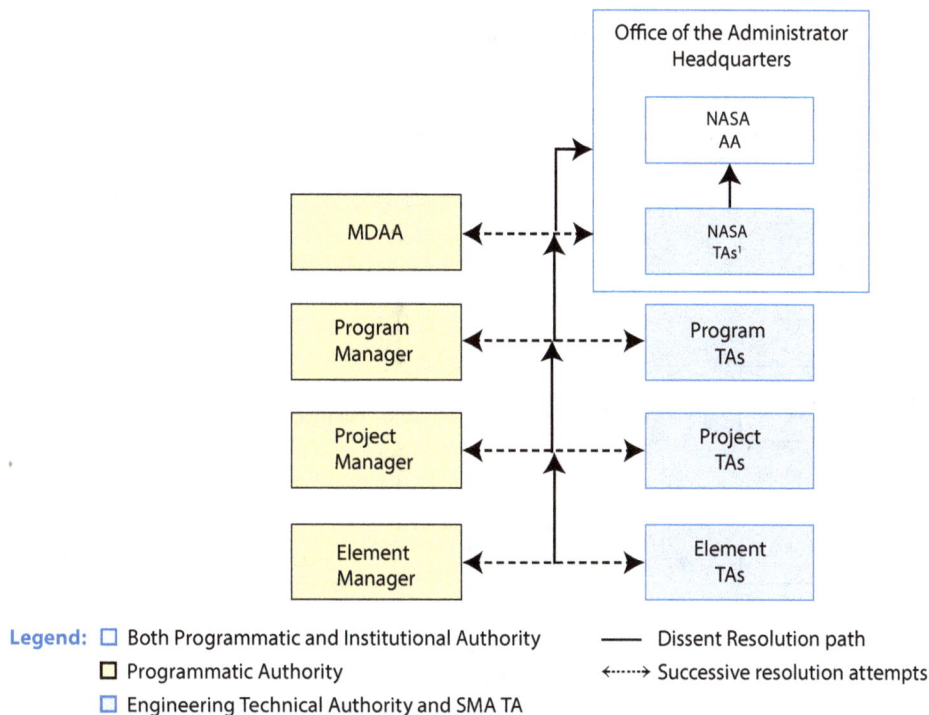

Legend: ☐ Both Programmatic and Institutional Authority — Dissent Resolution path
☐ Programmatic Authority ◄······► Successive resolution attempts
☐ Engineering Technical Authority and SMA TA

Note: This figure is a simplified representation of levels of dissent and does not necessarily depict all involved parties. Resolution is attempted at each level. If not resolved, the issue rises to the next level. The dissenting opinion process can start at any level.

[1] "NASA TAs" represents TAs above program level, including the NASA Chief Engineer and Center Directors, some of whom are at Headquarters.

Figure 5-2 Dissenting Opinion Resolution for Issues Between Programmatic Authority and Technical Authority

5.2.7 Specific Roles of the Different Technical Authorities

All Technical Authorities are part of the Institutional Authority and, as delineated in NPR 7120.5, provide technical oversight of and guidance to programs or projects.

5.2.7.1 Engineering Technical Authority

The ETA establishes and is responsible for the engineering design processes, specifications, rules, best practices, and other activities necessary to fulfill programmatic mission performance requirements.

Figure 5-3 provides a high-level illustration of the structure of ETA and its interface with the Programmatic Authority. Note that a Center may have more than one engineering organization and ETA is delegated to different areas as needed.

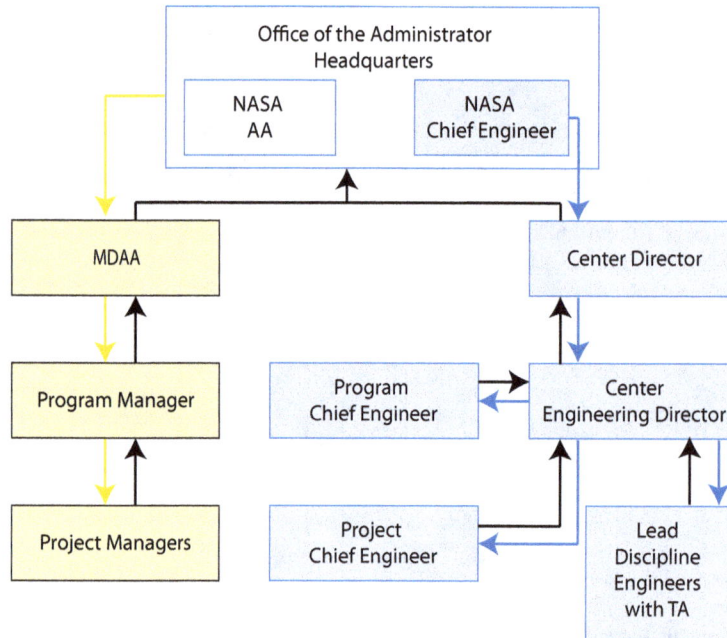

Legend:
☐ Both Programmatic and Institutional Authority   —— Direct report
☐ Programmatic Authority   —— Program Authority flow
☐ Engineering Technical Authority   —— Engineering Technical Authority flow

Note: This figure is a simplified representation and does not necessarily depict all involved parties.

Figure 5-3 Simplified Illustration of a Representative Engineering Technical Authority Structure

5.2.7.1.1 Roles of High-Level Engineering Technical Authorities

NASA Chief Engineer. The NASA chief engineer approves the appointment of the Center engineering directors (or equivalent), the appointment of ETAs on programs and Category 1 projects, and is notified of the appointment of other ETAs established by the Center Director.

Office of the Chief Engineer's (OCE's) Mission Directorate Chief Engineers. Mission Directorate chief engineers report to the NASA Chief Engineer and oversee the performance of all programs and projects in their assigned Mission Directorate. Mission Directorate chief engineers are not in the line of authority, but have an advisory responsibility to be aware, involved, and informed. They serve as advisors to the NASA Chief Engineer and the cognizant Mission Directorate. This includes assisting in the resolution of Dissenting Opinions that are elevated to Headquarters.

Center Director. The Center Director (or designee) develops the Center's ETA policies and practices, consistent with Agency policies and standards.

The Center Director delegates Center ETA implementation responsibility to an individual in the Center's engineering leadership.

The Center Director appoints, with the approval of the NASA chief engineer, individuals for the position of Center engineering director (or equivalent) and for the ETA positions down to and including program chief engineers and Category 1 project chief engineers (or equivalents).[5] The Center Director appoints Category 2 and 3 project chief engineers and lead discipline engineers.

Center Engineering Director (or Equivalent). The Center engineering director is responsible for supporting the Center Director in the establishment, implementation, and management of the ETA for the Center. In addition, the Center engineering director supports the program- and project-level Technical Authorities in processing changes to, waivers of, or deviations from requirements that are the responsibility of the ETA. This includes all applicable Agency and Center engineering directives, requirements, procedures, and standards. Centers further delegate ETA depending on the Center's organizational structure and management approach.

5.2.7.1.2 Roles of Program- and Project-Level Engineering Technical Authorities

Program/Project Chief Engineer (PCE). The PCE is the position to which the program- and project-level Engineering TA (ETA) has been delegated. The ETA at the program and project level manages the engineering activities—including systems engineering, design, development, sustaining engineering, and operations—and remains part of the Institutional Authority. The ETAs have access to the depth and breadth of expertise within the Center's engineering organization when needed.

Additionally, the program- and project-level ETAs serve as the HMTA "awareness eyes and ears" at the Centers. The primary role of the ETA in the HMTA process is to identify potential HMTA issues and ensure that they are flowed to the appropriate levels of HMTA in a timely manner. (See Section 5.2.7.3 for more details).

Lead Discipline Engineer (LDE). The LDE is a senior technical engineer in a specific discipline at the Center. Different Centers use different titles for this position. The LDE assists the program or project through direct involvement with working-level engineers to identify engineering requirements and develop solutions that comply with the requirements. The LDE works through and with the project-level ETA to ensure the proper application and

---

[5] Centers may use an equivalent term for these positions, such as Program/Project systems engineer.

management of discipline-specific engineering requirements and Agency standards. LDEs who are ETAs have formally delegated Technical Authority traceable to the Administrator and are funded independent of programs and projects.

To support the program or project while maintaining ETA independence and provide an effective check and balance:

- The program manager concurs in the appointment of the program-level ETA and the project manager concurs in the appointment of the project-level ETA.

- An ETA cannot approve a request for relief from a nontechnical derived requirement established by a Programmatic Authority. ETAs are expected to provide their recommendation(s).

- An ETA may approve a request for relief from a technical derived require-ment if the ETA ensures that the independent Institutional Authority subject matter expert (SME) who is the steward for the involved technical requirement concurs in the decision to approve the requirement relief. "Technical derived requirements" in this paragraph are those owned by the Technical Authority (policies, requirements, procedures, prac-tices, and technical standards of the Agency or Center). (Any party with a dispute regarding authority for granting relief of a technical derived requirement may raise a Dissenting Opinion. See Section 5.3 for details on the Dissenting Opinion process.) The rationale behind the second and third provisions for ETA is as follows:

  a. Without the second provision, the ETA (an Institutional Authority) could be put in the position of granting relief from a nontechnical requirement established by a Programmatic Authority. This would be noncompliant with Governance.

  b. Further, if the program or project ETA is, or acts as, the Decision Authority on matters related to granting requirement relief to a derived technical requirement, the Technical Authority check and balance system would be compromised for requirements derived at the program or project level. This is because the board is empowered to grant relief to requirements that it has established. Therefore, the TA (in this case the board chair) could not provide the independent over-sight that is fundamental to Technical Authority.

  c. In the case of granting relief from Technical Authority requirements, the third provision enables effective checks and balances to be main-tained. This is accomplished by ensuring that a second ETA agrees with the action to accept the tailoring of a requirement that is the responsibility of the Technical Authority.

5.2.7.2 Safety and Mission Assurance Technical Authority

The Safety and Mission Assurance Technical Authority (SMA TA) establishes and oversees implementation of the SMA processes, specifications, rules, and best practices necessary to fulfill safety and programmatic mission performance requirements.

SMA TA originates with the NASA Administrator and is formally delegated to the NASA AA and then to the Chief, Safety and Mission Assurance. SMA TA then flows from the Chief, Safety and Mission Assurance through the Center Director to the Center SMA director (see blue lines in Figure 5-4). The Center SMA director is responsible for establishing and maintaining institutional SMA policies and practices, consistent with Agency policies and standards. The Center SMA director is also responsible for assuring that programs and projects comply with the Center SMA and Agency SMA requirements and adhere to their SMA Plan. The program or project SMA plan serves as an agreement between the program or project and SMA TA, describing how the SMA requirements will be implemented and providing the basis for evaluation of SMA performance. The SMA plan can be either standalone or part of the Program or Project Plan with Center SMA concurrence. The Center SMA director also monitors, collects, and assesses institutional, program, and project SMA performance results.

SMA TA is assigned when new programs or projects are started. The Center SMA director, in consultation with the NASA Chief, SMA, appoints program- and project-level chief safety and mission assurance officers (CSOs) to exercise the TA role within programs and projects. SMA TA provides input to program or project planning; oversees any proposed technical or process changes or decisions that might increase risk to safety, quality, or reliability; and guides and advises program, project, or Agency management on handling this risk. The SMA TA also reviews and authorizes the closure of safety issues prior to flight and operations and for decommissioning and disposal of spacecraft in whole or in part. Depending on the level of risk and entity at risk (e.g., public and high-value assets), the Chief, SMA is consulted or his or her concurrence is obtained on the acceptance of increased risks. For example, NPR 8715.3 spells out good principles for risk acceptance decisions concerning radioactive material that may require a decision by the Chief, SMA. The chief safety officer also consults with the Chief, SMA when program risk decisions based on the program risk matrices are elevated to the NASA Administrator or AA. Center-specific, and some program- and project-level, SMA TA plans document the agreed upon responsibilities, reporting, processes, and deliverables of the Center SMA Technical Authorities to the Chief, SMA. The Center SMA directors obtain NASA Chief, SMA concurrence on these Center SMA TA plans.

NASA SMA TA includes safety (institutional and programmatic), reliability,maintainability, quality, and software assurance, as well as micro-meteoroid and orbital debris, launch and range safety, nuclear flight safety, nondestructive evaluation, workmanship, explosives, pressure vessels, metrology and calibration, and electrical, electronic, and electromechanical (EEE) parts assurance.

SMA requirements are both NASA specific and flowed down from a variety of sources, which include Federal laws and regulations and Presidential Directives. To ensure that NASA's compliance with these requirements, program and project requirements that impact safety and mission success, and other external SMA requirements and direction, the Office of the Chief, SMA (OSMA) has defined the delegation of authority for granting relief from requirements for which OSMA is responsible. This delegation authority is defined in *NPR 8715.3, NASA General Safety Program Requirements* and the process is defined in *NASA-STD-8709.20, Management of Safety and Mission Assurance Technical Authority (SMA TA) Requirements*. The Chief, SMA hears appeals of SMA decisions when issues cannot be resolved below the Agency level.

SMA TA works with the Engineering Technical Authority (ETA) and with the Health and Medical Technical Authority (HMTA). An assessment of the safety risks that may result from engineering changes is the minimum interaction that needs to take place. Early involvement of the SMA TA in the program and project and various boards, beginning with program and project solicitations and planning, and in evaluating tailoring (waiver and deviation requests) helps ensure mission success without unnecessary risk to NASA systems, personnel, and the public.

Additionally the program- and project-level SMA TAs serve as the HMTA "awareness eyes and ears" at the Centers. The primary role of the SMA TA in the HMTA process is to identify potential HMTA issues and ensure that they are flowed to the appropriate levels of HMTA in a timely manner. (See Section 5.2.7.3 for more details.)

5.2.7.3 Health and Medical Technical Authority

HMTA implements the responsibilities of the Office of the Chief Health and Medical Officer (OCHMO) to ensure that Agency health and medical standards are addressed in program and project management when applicable and appropriate. HMTA provides independent oversight of all human health, medical, and human performance matters that either arise in association with the execution of NASA programs or projects, or are embedded in NASA programs or projects. HMTA is not related to OCHMO's institutional responsibilities of Occupational Health (with some very specific exceptions) and Research Subject Protection, which are governed by laws, regulations, and requirements external to NASA. Programs and projects must recognize and understand HMTA owned requirements to ensure that the proper level of HMTA is involved with invoking, tailoring and waiving these requirements.

Consistent with Engineering and SMA Technical Authority, HMTA originates with the Administrator and is formally delegated to the NASA AA and then to the Chief Health and Medical Officer (CHMO). The Chief Health and Medical Officer delegates HMTA directly (not through the Center Director) to a Chief Medical Officer (CMO) at five Centers:

- Johnson Space Center (only Center with HMTA infrastructure and personnel)
- Kennedy Space Center
- Goddard Space Flight Center
- Dryden Flight Research Center
- Ames Research Center

The other five Centers do not have a CMO and therefore have no HMTA presence except as provided by the ETAs and SMA TAs.

Due to Center infrastructure differences, the flow down of HMTA processes and responsibilities from the CHMO varies between Centers. To assist the Chief Health and Medical Officer with HMTA implementation, the NASA Chief Engineer and the NASA Chief, Safety and Mission Assurance agreed to support Agency-wide HMTA implementation through the utilization of Engineering and SMA personnel as HMTA awareness and communication links at each Center (not including JSC due to its existing HMTA infrastructure).

The program-or-project level Engineering TAs and SMA TAs already involved in the day-to-day program/project work serves as the HMTA "awareness eyes and ears" at all Centers. The primary role of the Engineering TA and SMA TA in the HMTA process is to identify potential HMTA issues and ensure that they are flowed to the appropriate levels of HMTA in a timely manner. The Engineering TA and SMA TA establishes interfaces and communication flow paths with the appropriate level of HMTA (CMO, JSC POC, or OCHMO POC) for identification and resolution of potential HMTA issues.

To illustrate the implementation of this interface, the SLS Program (located at MSFC) has assigned the program SMA TA as its interface POC with JSC. JSC in turn has established an HMTA interface for HMTA issues that may be identified by the SLS Program. Figures 5-4 and 5-5 illustrate the HMTA flow paths for potential issues identified by the program-or-project level Engineering TAs and SMA TAs at NASA Centers.

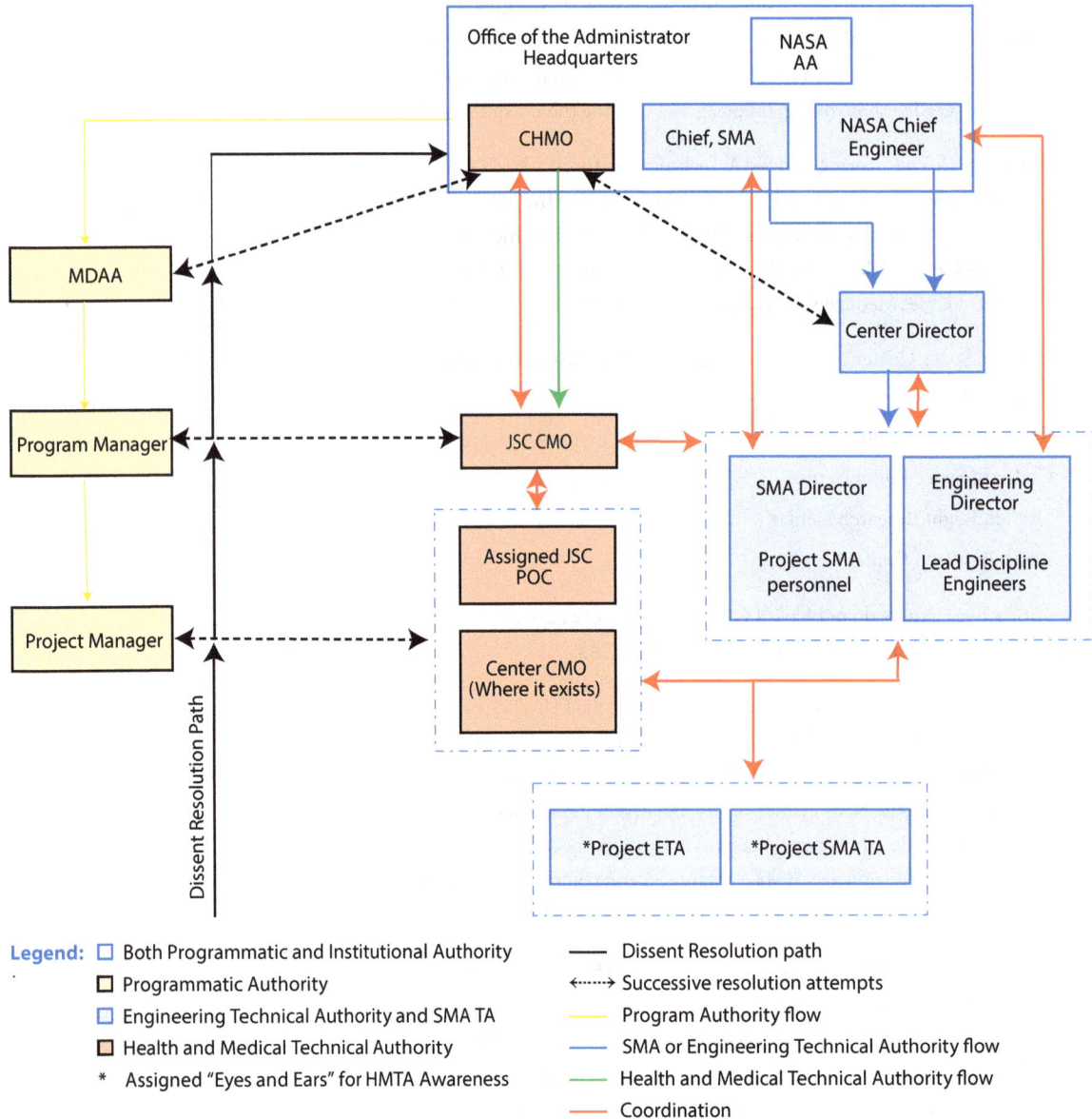

Legend:
- ☐ Both Programmatic and Institutional Authority
- ☐ Programmatic Authority
- ☐ Engineering Technical Authority and SMA TA
- ☐ Health and Medical Technical Authority
- * Assigned "Eyes and Ears" for HMTA Awareness

- —— Dissent Resolution path
- ←┈┈┈→ Successive resolution attempts
- —— Program Authority flow
- —— SMA or Engineering Technical Authority flow
- —— Health and Medical Technical Authority flow
- —— Coordination

Note: This figure is a simplified representation and does not necessarily depict all involved parties. Resolution is attempted at each level. If not resolved, the issue rises to the next level. The dissenting opinion process can start at any level.

Figure 5-4 Human Space Flight Health and Medical, Engineering, and SMA Flow of Technical Authority and Dissenting Opinion Resolution

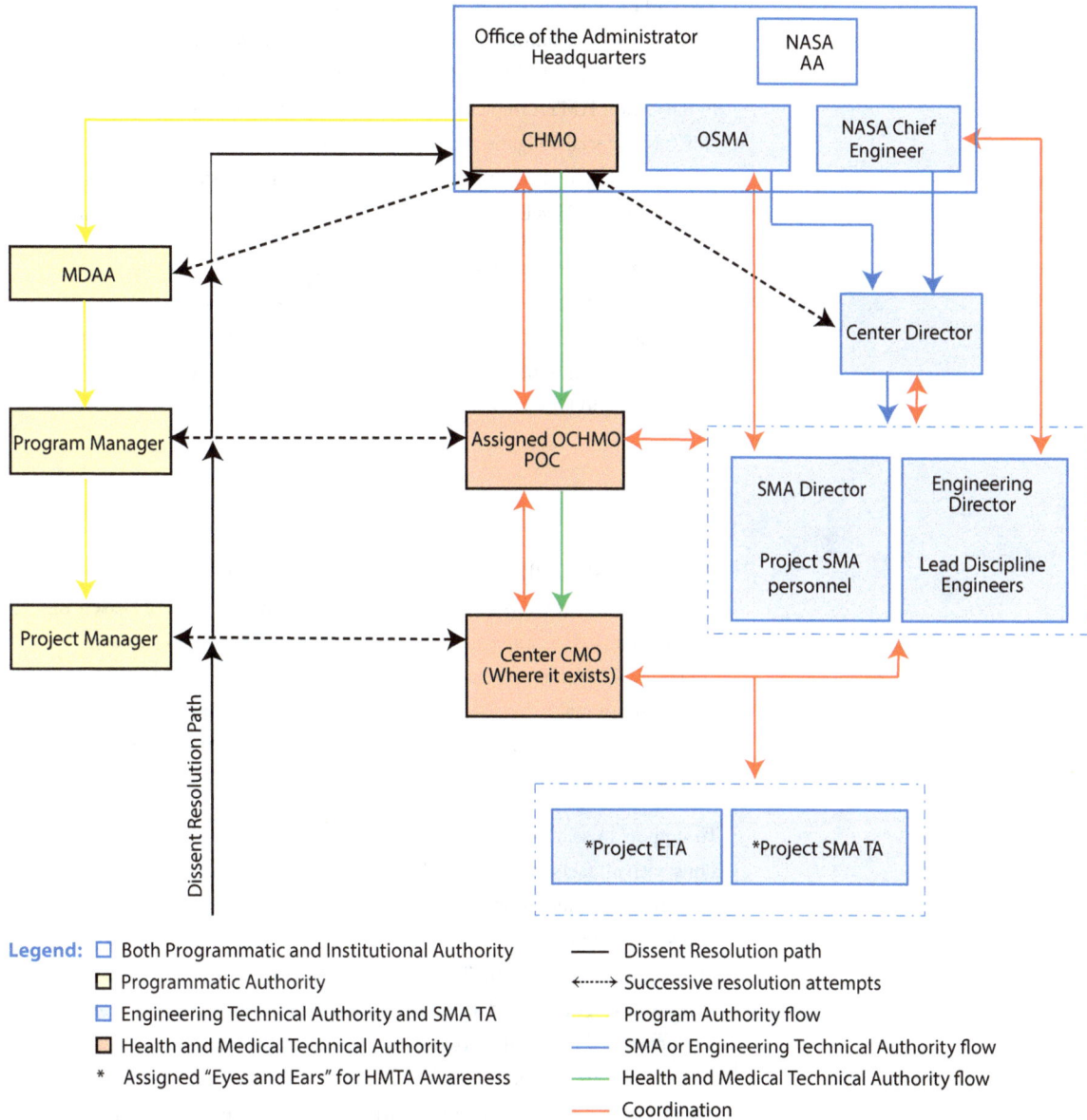

Figure 5-5 Robotic Health and Medical, Engineering, and SMA Flow of Technical Authority and Dissenting Opinion Resolution

Legend: □ Both Programmatic and Institutional Authority
□ Programmatic Authority
□ Engineering Technical Authority and SMA TA
□ Health and Medical Technical Authority
* Assigned "Eyes and Ears" for HMTA Awareness

— Dissent Resolution path
◄······► Successive resolution attempts
— Program Authority flow
— SMA or Engineering Technical Authority flow
— Health and Medical Technical Authority flow
— Coordination

Note: This figure is a simplified representation and does not necessarily depict all involved parties. Resolution is attempted at each level. If not resolved, the issue rises to the next level. The dissenting opinion process can start at any level.

The HMTA flowdown, delegation, and communications processes, including roles and responsibilities, are specified in *NPR 7120.11, NASA Health and Medical Technical Authority (HMTA) Implementation* and further described in the Center HMTA implementation plan. This NPR recognizes that medical staff has a special obligation to protect the handling and dissemination of an individual's medical information. These legal and ethical restrictions are managed by the HMTA and must be complied with by all Agency personnel.

The Program or Project Plan describes how the program or project will comply with HMTA requirements and processes as described in NPR 7120.11. The CHMO hears appeals of HMTA Dissenting Opinions when issues cannot be resolved below the Agency level as described in Section 5.3 and Figures 5-4 and 5-5.

5.3 Dissenting Opinion Process

NASA has historically supported the full airing of issues, including alternative and divergent views. There are numerous examples where a Dissenting Opinion has led to changes that enhanced safety and mission success. However, NASA has also had some notable examples where dissenting views did not make their way to decision makers at the appropriate level in a timely manner. Two examples can be found in the Shuttle accidents. (See "*Challenger* and *Columbia* Case Studies" box.)

To support mission success, NASA teams need to have full and open discussions with all facts made available to support understanding and objective assessment of issues to make the best possible decisions. Diverse views are to be fostered and respected in an environment of integrity and trust with no suppression or retribution. To support these goals, NASA has established a uniform, recognized, and accepted process for resolving serious dissent and has formalized it in policy. This is the Dissenting Opinion process which further empowers team members to provide their best input to decision makers on important issues and clearly defines the roles and responsibilities of both sides when there is a dissent. A Dissenting Opinion expresses a view that a decision or action, in the dissenter's judgment, needs to be changed for the good of NASA and requests a review by higher level management. In this context, "for the good of NASA" is to be read broadly to cover NASA, mission success, safety, the project, and the program.

The Dissenting Opinion process is based on a belief that each team member brings unique experience and equally important expertise to every issue and that the recognition of and openness to that unique experience, expertise, and insight improves the probability of identifying and resolving challenges

Challenger and *Columbia* Case Studies

Challenger (STS51-L). The night before the launch of the *Challenger* there were discussions between NASA and its contractor for the Solid Rocket Motor (SRM) and within the contractor's organization about the effect of the temperature predicted for launch. The predicted temperatures were lower than any previous launch, and the concern was that it would adversely affect the performance of the O-rings designed to seal the joints between the SRM segments that prevented hot gas leakage in the vicinity of the external tank. The initial recommendation by Morton Thiokol was not to launch.

The report of the Presidential Commission on the *Challenger* Accident (Rogers report) concluded:

> The decision to launch the *Challenger* was flawed. Those who made that decision were unaware of the recent history of problems concerning the O-rings and the joint and were unaware of the initial written recommendation of the contractor advising against the launch at temperatures below 53 degrees Fahrenheit and the continuing opposition of the engineers at Morton Thiokol after the management reversed its position. They did not have a clear understanding of Rockwell's concern that it was not safe to launch because of ice on the pad. If the decision makers had known all of the facts, it is highly unlikely that they would have decided to launch 51-L on January 28, 1986. (Vol. 1 Chapter 5)

> The unrelenting pressure to meet the demands of an accelerating flight schedule might have been adequately handled by NASA if it had insisted upon the exactingly thorough procedures that were its hallmark during the Apollo program. (Vol. 1 Chapter 7)

The process of arriving at the launch decision illuminates how serious safety concerns can be overridden by concerns for schedule...particularly when there is no effective check and balance by an authority that can speak with an equal voice to the Programmatic Authority.

Columbia STS107. Post-launch photographic analysis of *Columbia* showed a large piece and two smaller pieces of foam struck Columbia's underside and left wing. Analysis the day after launch indicated the large piece of foam (20–27 inches long and 12–18 inches wide) impacted the shuttle at a relative speed of 416–573 mph. As a result, the photo analysis team requested high-resolution photos be obtained by the Dept. of Defense to assist in the assessment and subsequent analysis. This was the first of three distinct requests for on-orbit imagery. Schedule pressure contributed to management declining to pursue the requests for imagery. (See CAIB report Vol. 1 Section 6.2 for more background information.)

to safety and mission success. NASA's core value of teamwork[6] captures this philosophy.

---

[6] NASA's most powerful tool for achieving mission success is a multidisciplinary team of diverse competent people across all NASA Centers. NASA's approach to teamwork is based on a philosophy that each team member brings unique experience and important expertise to project issues. Recognition of and openness to that insight improves the likelihood of identifying and resolving challenges to safety and mission success. NASA is committed to creating an environment that fosters teamwork and processes that support equal opportunity, collaboration, continuous learning, and openness to innovation and new ideas.

STS-121 Return to Flight Launch Decision

In June 2006 with the FRR for the mission approaching, the top engineering and SMA authorities at NASA determined that the residual risk was "probable catastrophic"—or unacceptable for mission execution. This was reported up through the SMA and engineering channels respectively. The Shuttle program manager reported this also to NASA Headquarters, but disagreed with the hazard categorization. He asked that a higher authority address the matter at the FRR scheduled for June 17, 2006.

After discussing the issue with technical authorities and the other review board members, NASA Chief Engineer Chris Scolese and Chief SMA Officer Bryan O'Connor decided to "nonconcur" in proceeding to launch, recommending that the mission be delayed until the faulty design could be improved. First and foremost came flight safety, and Scolese and O'Connor were concerned that the as of yet unresolved ice/frost-ramp problem could jeopardize the safe return of the orbiter and its crew.

In the FRR, Scolese and O'Connor were the last two to be polled. Everyone had been go for launch until Scolese said "no." "It wasn't easy being the only one to say no. I wasn't completely sure what Bryan would say after me. It wasn't fun, but I think it was the right thing to do, and it was the right way to do it."

The flight readiness endorsement document does not have an option for "nonconcurrence." On the forms, Scolese and O'Connor had to cross off "concurrence" and write in their own nonconcurrence and rationale.

Following two days of discussion, the FRR Chair and Associate Administrator for the Space Operations Mission Directorate, W.H. Gerstenmaier, believed that the risks were acceptable and decided to proceed with launch. Because of the Chief Engineer's and the SMA Chief's "nonconcurrence," the chair elevated the final decision to the Administrator, Michael Griffin. The following is Michael Griffin's assessment of the situation with regard to the "dissent" or nonconcurrence:

> Some of the senior NASA individuals responsible for particular technical areas, particular disciplines, expressed that they would rather stand down until we had fixed the ice/frost ramps with something better, whereas many others said, "No, we should go ahead."
>
> So, we did not have unanimity. Therefore, a decision had to be made. Now, one possible way of making decisions is that unless everybody feels that we should go, then we will stand down. In which case, I don't think for Shuttle flights or any other flights, we don't need an Administrator. We don't actually make decisions. We just make sure that no one is unhappy. That's not the method that we're using.[1]

Having carefully considered both sides of the story, Griffin agreed with the FRR chair that the risk was acceptable. He made the decision to proceed with the flight.

In the end, neither Scolese nor O'Connor asked him to reconsider. They believed that the mandatory requirement for safe haven and a crew rescue launch-on-need capability adequately mitigated the flight-crew safety risk.

---

[1] John Kelly, "NASA Chief Michael Griffin's STS-121 Flight Rationale Explained." *Florida Today*, (June 21, 2006). Reproduced in Space.com. Available at http://www.space.com/2525-nasa-chief-michael-griffin-sts-121-flight-rationale-explained.html.

[T]he two Agency officials said the foam loss will not threaten the crew because NASA has a plan for the astronauts to move into the International Space Station, if in-orbit inspections find serious damage to the spacecraft. The crew would await rescue 81 days later by a second space Shuttle.[2]

After the FRR meeting, Scolese and O'Connor issued a statement about their nonconcurrence in the decision process.

Crew safety is our first and most important concern. We believe that our crew can safely return from this Mission.

We both feel that there remain issues with the orbiter—there is the potential that foam may come off at the time of launch. That's why we feel we should redesign the ice/frost ramp before we fly this Mission. We do not feel, however, that these issues are a threat to the safe return of the crew. We have openly discussed our position in the Flight Readiness Review—open communication is how we work at NASA. The Flight Readiness Review Board and the Administrator have heard all the different engineering positions, including ours, and have made an informed decision and the agency is accepting this risk with its eyes wide open.[3]

Reflections on the Launch Decision

The Shuttle *Discovery* (*STS-121*) launched on July 4, 2006, and successfully concluded 13 days in space. The crew had spent the mission transferring cargo to the International Space Station and performing a variety of other tasks, including testing crack-repair methods in the reinforced carbon–carbon panels on the leading edge of the orbiter's wing. In the aftermath of *Columbia* and prior to *STS-121*, it had been noted that in theory:

Astronauts will be able to repair cracks as small as a fraction of an inch, or plug holes in the wings as big as 4 inches (10 centimeters). Anything bigger—the gash in Columbia's left wing was between 6 and 10 inches (10 to 25 centimeters)— and the Shuttle crew will have to move into the Station until [another Shuttle] can be launched to rescue them.[4]

After the fact, Scolese noted that he believed that the *STS-121* launch decision was an example that the review process works. He and O'Connor believed that in the process, Griffin had been made fully aware of the residual risks to both the orbiter and the flight crew, and that the decision process had been appropriately thorough, professional, and consistent with NASA's core values and Governance. He felt *STS-121* was a great success, showing that the NASA culture had changed in the wake of *Columbia*.

A year later, NASA Administrator, Michael Griffin, wrote the following in *ASK* magazine:

Generally speaking, decisions are the responsibility of line organizations, either programmatic or institutional. In some cases, where there is a substantial disagreement, decisions will be appealed by one side or the other. A good recent example is the launch decision for STS-121. In that case, programmatic authorities made the decision to launch, and institutional authorities appealed that decision in light of concerns about ice/frost ramp foam losses from the Shuttle's external tank. In that case, the appeal came to the level of the Administrator, because agreement could not be found at lower levels. And my belief is that decisions of that magnitude deserve the attention of NASA's top management, so our governance process worked well in that case.[5]

---

[2] Mike Schneider, "Shuttle Launch a Go Despite Damaged Foam." (July 4, 2006) *Washington Post*. Available at http://www.washingtonpost.com/wp-dyn/content/article/2006/07/03/AR2006070300996.html.

[3] NASA, NASA Statement on Decision to Launch Shuttle Discovery. (June 19, 2006) Available at http://www.nasa.gov/mission_pages/shuttle/news/121frr_oconnor_scolese.html. See Appendix 5 for a list of references.

[4] MSNBC, "NASA Says It's Fixed Shuttle Foam Problem." (August 31, 2004) Available at http://www.msnbc.msn.com/id/5831547/.

[5] Michael D. Griffin, "The Role of Governance." *ASK*, Issue 26 (Spring 2007).

In the team environment in which NASA operates, team members often have to determine where they stand on a decision. In assessing a decision or action, team members have three choices: agree, disagree but be willing to fully support the decision, or disagree and raise a Dissenting Opinion.

There are three parts to a Dissenting Opinion:

- A disagreement by an individual with a decision or action that is based on a sound rationale (not on unyielding opposition),
- An individual's judgment that the issue is of sufficient importance that it warrants a specific review and decision by higher level management, and
- The individual specifically requests that the dissent be recorded and resolved by the Dissenting Opinion process.

The decision on whether the issue in question is of the significance that warrants the use of the Dissenting Opinion process is the responsibility and personal decision of the dissenting individual.

5.3.1 Responsibilities of the Individual Raising a Dissenting Opinion

Individuals who raise a Dissenting Opinion have the following responsibilities:

- Be knowledgeable of the Dissenting Opinion process.
- Be competent in the matter involved in the dispute.
- Raise the concern and the basis and rationale for the concern in a professional and timely manner. (This normally is done during the team deliberations leading up to a decision to ensure that the decision maker has an understanding of all views before making the decision.)
- Support the joint resolution process.

5.3.2 Responsibilities of a Decision Maker

A decision maker has a responsibility to fully support NASA's "teamwork" core value. This includes conducting discussions, meetings, and boards in a professional manner that:

- Promotes full and open discussion of issues with all their associated facts and considerations,
- Fosters and respects diverse views,
- Invites thoughtful presentations of alternative ideas and approaches, and
- Ensures the team understands the basis for the decisions made.

Dissenting Opinion Process and SMA "Suspend Work":

- *NPD 1000.B* states "that in an extreme case that presents an unacceptable risk to personnel, property, or mission success, the Chief, Safety and Mission Assurance (SMA) (or his delegated representative) is authorized to suspend any operation or project activity and provide guidance for corrective action."

- Nothing in the Dissenting Opinion process is intended or may be construed to abridge or diminish this SMA responsibility.

Such an approach helps ensure that the decision maker has the best possible basis for the decision. It also minimizes the need for Dissenting Opinions. Note that the decision maker's responsibilities start before the Dissenting Opinion exists. When a Dissenting Opinion is raised, the decision maker receiving a Dissenting Opinion has an obligation to work to support the resolution process and to maintain an environment of integrity and trust with no suppression or retribution.

Unresolved issues of any nature (e.g., programmatic, safety, engineering, health and medical, acquisition, accounting) within a team need to be quickly elevated to achieve resolution at the appropriate level. The decision on whether the issue in question is of significance to warrant the use of the Dissenting Opinion process is the responsibility and personal decision of the dissenting individual. Supporting the resolution of the dissent is the responsibility of both parties and is a joint process involving representatives on both sides of the issue.

When time permits, the disagreeing parties jointly document the issue. This involves clearly defining the issue, identifying the agreed-to facts, discussing the differing positions with rationale and impacts, and documenting each party's recommendations. The joint documentation is approved by the representative of each view, concurred with by affected parties, and provided to the next-higher level of the involved authorities with notification to the second-higher level of management. This may involve a single authority (e.g., the Programmatic Authority) or multiple authorities (e.g., Programmatic and Technical Authorities). In cases of urgency, the disagreeing parties may jointly present the information stated above orally with all affected organizations represented, advance notification to the second-higher level of management, and documentation follow up.

Management's decision on the memorandum (or oral presentation) is documented and provided to the dissenter and to the notified managers and becomes part of the program or project's retrievable records. If the dissenter is not satisfied with the process or outcome, the dissenter may appeal to the next higher level of management. The dissenter has the right to take the issue upward in the organization, even to the NASA Administrator if necessary.

5.3.3 Appeal Path for Dissenting Opinions

Figure 5-6 illustrates potential appeal paths for Dissenting Opinions among various authorities in a single-Center environment. The three parts of the figure show different ways a Dissenting Opinion may be generated. The path on the left shows a Dissenting Opinion flow where the dissent is strictly

The emphasis on the joint process involving both parties to a Dissenting Opinion at all phases of the resolution process is intended to ensure that the authorities involved in resolving the dissent fully understand the position of both parties.

The preparation of a joint document is encouraged because of the clarifying effect that comes from writing things down. Experience has shown that the process of committing the issue to writing tends to depersonalize the issue and in many cases leads to a clearer understanding of the issue and the differing views. At times this has led to a resolution prior to elevating the issue up the management chain. Even if writing the document does not result in resolution, there is a secondary benefit; specifically, the document leads to an efficient presentation and decision process.

within the programmatic path. As the figure shows, the dissent flows up the programmatic chain until resolution is achieved. A simple example may be a project manager requiring an element manager to have a Preliminary Design Review (PDR) by a specific date, which the element manager determines is unreasonable due to a nontechnical issue. They would try to work this schedule conflict among themselves, but if they cannot resolve it, then it rises to the program manager. If the program manager cannot resolve the issue, then it rises to the MDAA and next to the Associate Administrator. Since the requirement owner is the OCE, the NASA chief engineer would be consulted along with the AA.

In Figure 5-6, the figure on the right is similar to the one on the left except that the Dissenting Opinion is now strictly within the TA and engineering chain of command. For example, if the project chief engineer and the LDE disagree on a waiver to a TA requirement and they cannot resolve it among themselves, then the dissent rises to the next higher level of management, in this case the program chief engineer and the Center engineering director. If they cannot resolve it, then it goes to the Center Director and the NASA CE.

Finally, the center figure shows the flow for a dissent between the Programmatic Authority and the TA. An example is an element manager who wants to waive a TA requirement for a lower factor of safety on a pressure vessel design to save cost. If the element manager and the element chief engineer cannot agree, then the dissent rises to the project manager and project chief engineer, then to the program manager and program chief engineer, and finally to the MDAA and NASA chief engineer if necessary.

Figure 5-7 illustrates a dissenting opinion resolution path in a multi-Center environment.

A Dissenting Opinion between authorities at the element level would rise to the project level with notification at the program level. Since two authorities are involved in the Dissenting Opinion process, both authorities would be involved at each step in the resolution path.

Before leaving Center B, resolution typically is attempted within Center B and a Center position typically is established. This does not mean that an individual raising the dissent could be overridden by the appropriate Center TA. If the appropriate Center TA did not agree with the position taken by the dissenter, this would become part of the information carried to the next level of the resolution process. After the project level, the next step in the resolution path would be at the program level with notifications to the MDAA.

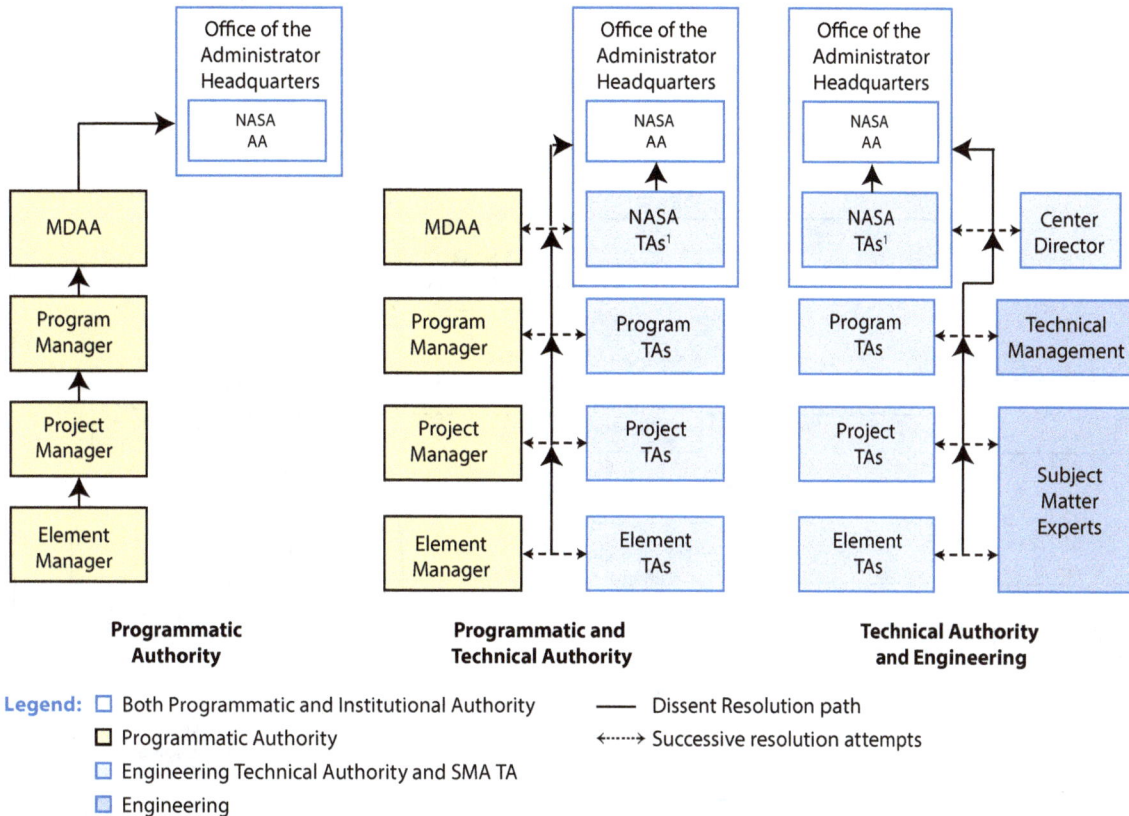

Figure 5-6 Simplified Potential Appeal Paths for Dissenting Opinion Resolution in a Single-Center Environment

Note that the process flow described above complies with policy and, for graphic simplicity, the web of communications among entities is not shown. The essential nature of these communications is recognized and helps in a timely resolution of the issue at hand.

Figure 5-8 illustrates the multi-Center communications framework for Orion/LAS ETA. The communications framework was an effective construct for day-to-day operations and execution. It also served as a method to ensure the ETA at various levels in the projects and Centers were informed and engaged in issue resolution at the right venue/time. In this figure, assume the LAS Deputy Chief Engineer has a Dissenting Opinion about a technical matter. The DCE first works with the LAS CE to resolve the issue and informs the next level up of the issue. If they cannot resolve it, they

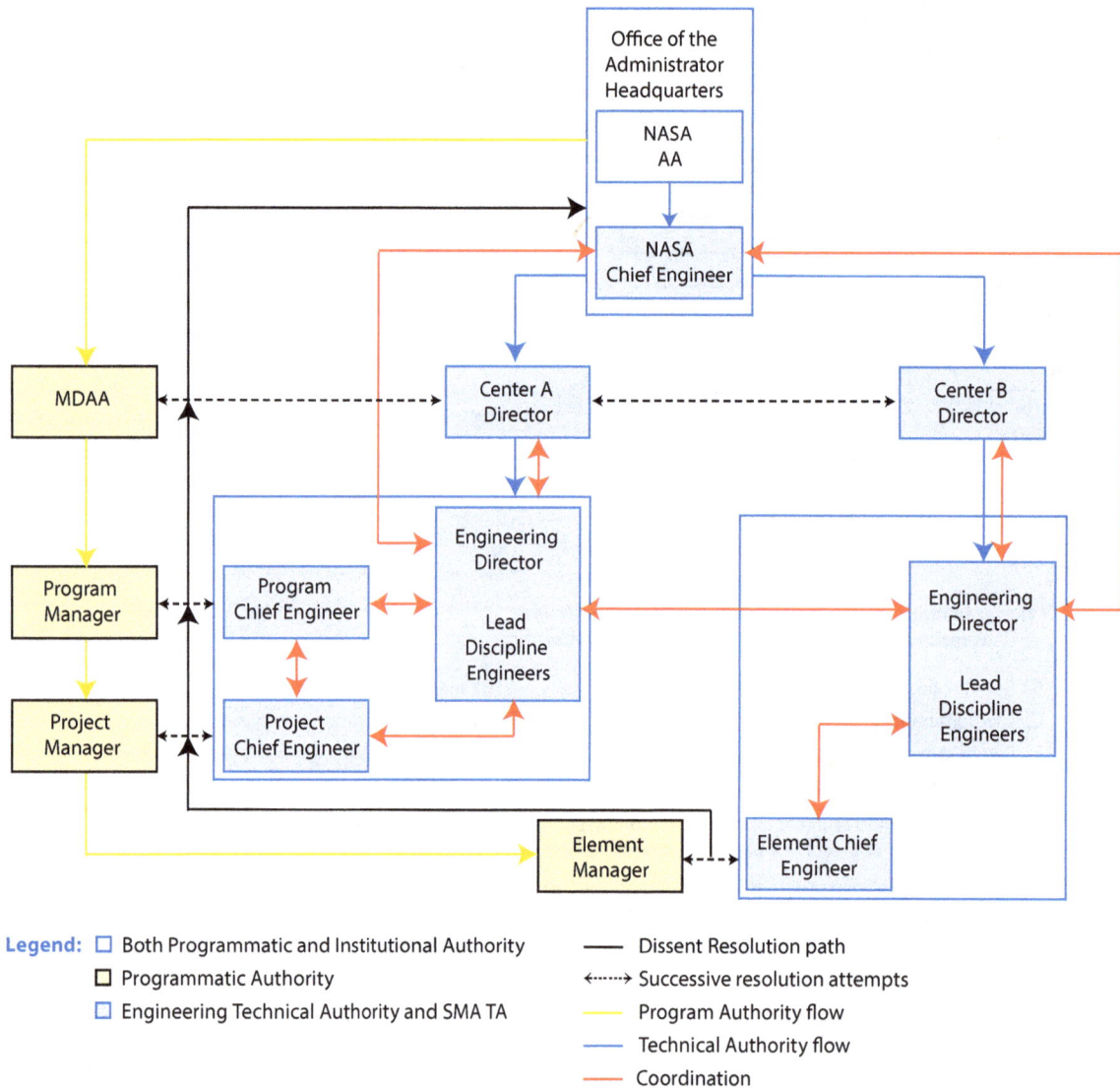

Figure 5-7 Dissenting Opinion Resolution Path in Multi-Center Environment

then meet with the Orion CE to discuss the issue and seek solutions. If the Joint LAS and Orion Chief Engineers can resolve the dissent at their level to the satisfaction of the originator of the dissenting opinion (the LAS DCE in this example), this is communicated to one level above them for information and actions are then executed to resolve the issue. If they cannot come to a joint agreement, the dissenting opinion is next presented to a Joint Engineering board constructed of senior engineering personnel from

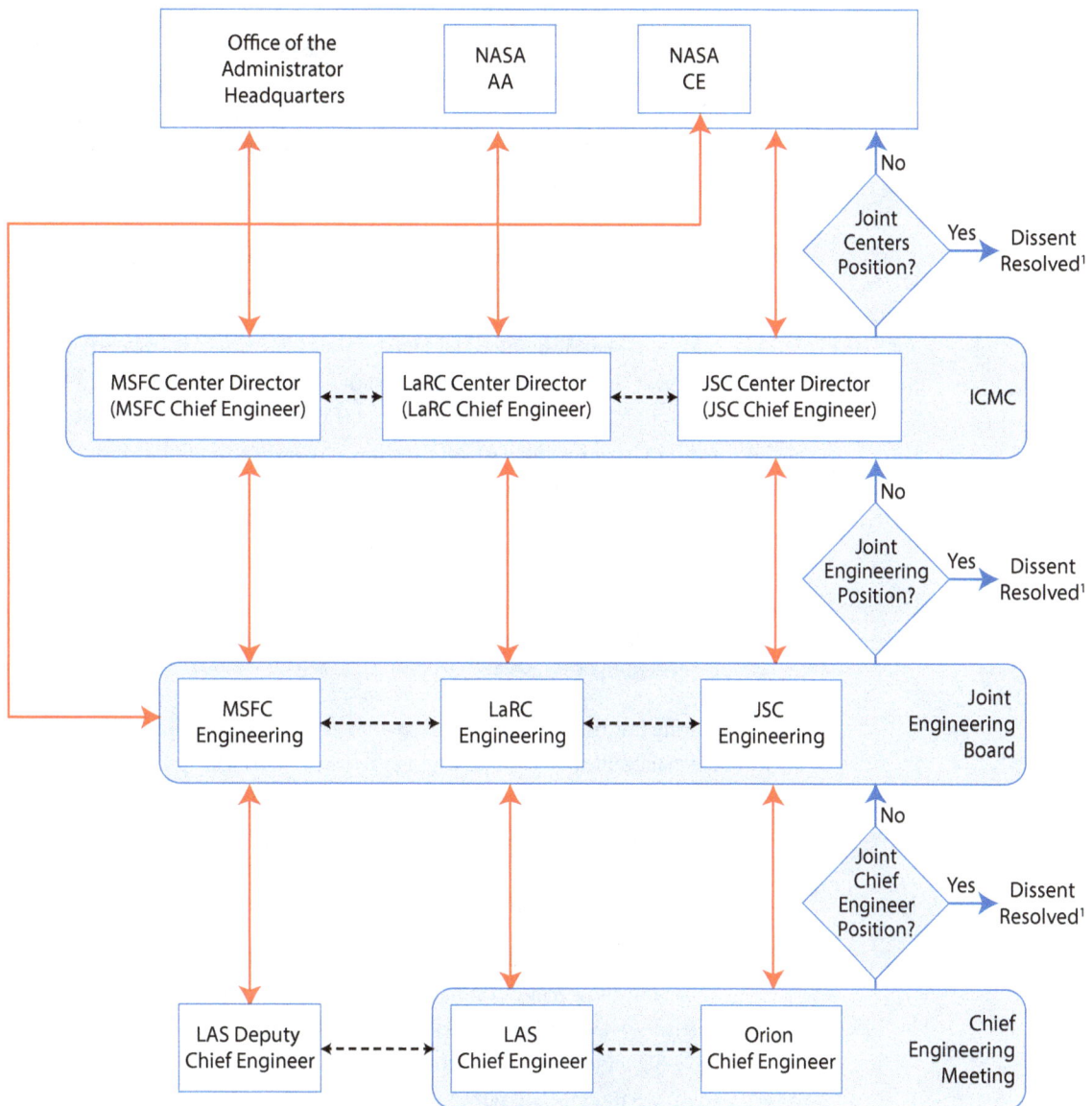

Legend: ◂┄┄▸ Dissent Resolution path   ── Coordination

Acronym: LAS = Launch Abort System, which is a subsystem of Orion.

[1] Inform all parties, including one level up from the deciding board and the original dissenter.

Figure 5-8 Dissenting Technical Opinion in a Multi-Center Environment (Orion/LAS ETA Communication Framework)

all the centers involved and the next level up is made aware of the issue. If the Joint Engineering Board can resolve the issue to the originators satisfaction, they informed the next level up and then execute the resolution. If they cannot resolve it at their level it is taken to an ICMC consisting of the Center Chief Engineers and Center Directors (as needed). If they can resolve it to the originator's satisfaction, they inform the next level up and execute the resolution. If they cannot resolve the issue it is taken to the Agency CE for discussion and resolution in a similar manner. If necessary, the issue may be brought to the NASA Administrator as a final arbitration.

HMTA dissenting opinions for human space flight programs follow the flow path shown on Figure 5-4. This path reflects the role of the JSC Chief Medical Officer (CMO) as the delegated HMTA for human space flight. For the Centers without a CMO, the JSC CMO will also fulfill that role in the process. For all other issues (including research and technology) the same path is used except the CHMO replaces the JSC CMO as shown in Figure 5-5.

5.3.4 Notifications During the Dissenting Opinion Resolution Process

During the dissenting opinion resolution process, NPR 7120.5 requires that the management of both dissenting parties be informed. Specifically, the level of management above the dissenting parties, are provided with the joint documentation developed by the dissenting parties (preferably written but may be oral in cases of urgency) of the disagreement, and the second-higher level of the involved authorities are notified of the specifics of the disagreement.

When the disagreement cannot be resolved at the level above the disagreeing parties, the second-higher level is provided the joint documentation, and the third-higher level notified of the specifics of the disagreement. This process continues until this disagreement is resolved.

When the disagreement reaches a Center's Director of Engineering or Safety and Mission Assurance or the Chief Medical Officer, if one is assigned, notification of the disagreement to the second-higher level includes both the Center Director and the NASA Headquarters TA counterpart (i.e., NASA Chief Engineer, Director, SMA or Chief HMO, depending on the discipline of the disagreement).

Once the disagreement has been resolved, management's decision on the dissent memorandum (or oral presentation) is documented and provided to the dissenter and to the managers involved in assessing and adjudicating the disagreement, including the level above the authority where the decision was ultimately resolved.

In the event an authority chooses to overrule a lower level authority's decision or non-concur with any dissenting opinion, transparency in decision making requires they explain it to the person raising the issue and those above them in the authoritative chain.

5.4 Tailoring Requirements

The tailoring process supports NASA's goal of Technical Excellence by providing and maintaining a sound basis for the requirements imposed on NASA's space flight programs and projects. The principles can be viewed as another piece of providing proper balance between organizational elements by having a check and balance system as described in *NPD 1000.0, NASA's Governance and Strategic Management Handbook*.

It is NASA policy to comply with all prescribed requirements, directives, procedures, and processes unless relief is formally granted by the designated party. However, NASA policy also recognizes the need to accommodate the unique aspects of each program or project to achieve mission success in an efficient and economical manner. Tailoring is the process used to adjust or seek relief from a prescribed requirement to accommodate the specific needs of a task or activity (e.g., program or project). Relief from a requirement may be granted in the form of a ruling that a requirement is non-applicable or in the form of a waiver or a deviation. Tailoring is both an expected and accepted part of establishing the proper requirements for a program or project. A secondary benefit of a formalized and disciplined approach to granting relief from prescribed requirements is that in time and with proper feedback, it will result in improved prescribed requirements.

5.4.1 Delegation of Tailoring Approval Authority

Delegation of tailoring approval authority is done formally. The individual with tailoring approval authority for a particular requirement has the responsibility to consult with the other organizations that were involved in the establishment of the specific requirement and to obtain the concurrence of those organizations having a substantive interest.

Three examples of how delegation of approval authority has been formally implemented for HQ-originated requirements are:

- The Office of the Chief Engineer (OCE) periodically issues a letter documenting the delegation of authority for granting relief from requirements

The two terms "waiver" and "deviation" provide a temporal indicator that allows separating requirement tailoring proposed before the requirement was put under configuration control ("seeking permission") from those made after ("seeking forgiveness"). Definitions of these two terms are:

- Waiver—A documented authorization releasing a program or project from meeting a requirement after the requirement is put under configuration control at the level the requirement will be implemented.

- Deviation—A documented authorization releasing a program or project from meeting a requirement before the requirement is put under configuration control at the level the requirement will be implemented.

for which OCE is responsible. This includes requirements contained within NASA Policy Directives (NPDs), NASA Procedural Requirements (NPRs), and Technical Standards. The delegation letter can be located on the OCE tab under the "Other Policy Documents" menu in the NASA On-Line Directives Information System (NODIS). Program and project managers can also work with the Center representative of the responsible organization (e.g., the NASA Headquarters Office of Safety and Mission Assurance (OSMA)) to determine if tailoring authority has been delegated to a Center person and, if so, who the delegated authority is.

- NASA's Safety and Mission Assurance (SMA) requirements come from a variety of sources which include Federal laws and regulations, inter-agency agreements, and Presidential Directives. To ensure NASA's compliance with these requirements and other external SMA require-ments and direction, the office of the NASA Chief, SMA has defined the process for determining the delegation of authority for granting relief from requirements for which the NASA Headquarters Office of SMA (OSMA) is responsible. This delegation authority is in *NPR 8715.3, NASA General Safety Program Requirements*, and the process is defined in *NASA-STD-8709.20, Management of Safety and Mission Assurance Technical Authority (SMA TA) Requirements*. This includes requirements contained within NPDs, NPRs, and Technical Standards.

- The Office of the Chief Health and Medical Officer (OCHMO) promul-gates mandatory human system technical standards for human space flight programs and projects. Approval authority for tailoring these stan-dards is delegated to the JSC Chief Medical Officer, who is the individual designated as HMTA for all human space flight programs and projects. Tailoring of HMTA requirements for other NPR 7120.5 programs must be discussed with OCHMO.

When a Center Director (or designee) formally delegates tailoring approval authority, the delegation is documented in accordance with Center processes. See "Types of Requirements" box for more information.

5.4.2 Tailoring NPR 7120.5 Requirements

NPR 7120.5 requires that all space flight programs and projects follow the tailoring process delineated in the NPR. The foundations for this process are the tailoring principles that flow down from *NPD 1000.0, NASA Governance and Strategic Management Handbook*.

The organization at the level that established the requirement approves the request for tailoring that requirement unless this authority has been formally delegated elsewhere. The organization approving the tailoring

Types of Requirements

Programmatic Requirements—focus on space flight products to be developed and delivered that specifically relate to the goals and objectives of a particular program or project. They are the responsibility of the Programmatic Authority.

Institutional Requirements—focus on how NASA does business independent of the particular program or project. They are the responsibility of the applicable Institutional Authority.

Allocated Requirements—established by dividing or otherwise allocating a high-level requirement into lower level requirements.

Derived Requirements—arise from:

• Constraints or consideration of issues implied but not explicitly stated in the higher level direction originating in Headquarters and Center institutional requirements or

• Factors introduced by the architecture and/or the design.

These requirements are finalized through requirements analysis as part of the overall systems engineering process and become part of the program/project requirements baseline.

Technical Authority Requirements—a subset of institutional requirements invoked by Office of the Chief Engineer, Office of Safety and Mission Assurance, and Office of the Chief Health and Medical Officer documents (e.g., NASA Procedural Requirements (NPRs) or technical standards cited as program or project requirements or contained in Center documents). These requirements are the responsibility of the office or organization that established the requirement unless delegated elsewhere.

Additional types of requirements are defined in Appendix A.

---

disposition consults with the other organizations that were involved in the establishment of the specific requirement and obtains the concurrence of those organizations having a substantive interest. (See the "Considering Other Stakeholders in Tailoring Requirements" box for more information.)

The involved management at the next higher level is informed in a timely manner of the request to tailor a prescribed requirement.

Each program and project is required by the NPR 7120.5 to complete and maintain a Compliance Matrix. The Compliance Matrix provides a streamlined process for documenting the program's or project's compliance with the NPR's requirements or how the program or project is tailoring the requirements in accordance with Paragraph 3.5 of NPR 7120.5E. The

The next higher level may be counting on the original requirement in a manner that is not known to the lower level (e.g., the requirement may have been used in a higher level analysis of which the lower level is not aware.)

Timely interaction among management levels supports a philosophy that contributes to mission success…specifically, the goal of "no surprises."

Considering Other Stakeholders in Tailoring Requirements

The organization that establishes a requirement (or formally delegated designee) is in the best position to know why the requirement was established and to assess a request for relief and its associated justification. In addition, this interaction of user and the party responsible for establishing a requirement provides important feedback to the organization responsible for the requirement that can be used to determine whether the requirement needs reassessment.

In many instances, several organizations may have played a significant role in establishing a requirement or may be affected by tailoring the requirement. Consultation with these organizations is essential to avoid adverse unintended consequences as these organizations may have background and/or insights that may not be readily apparent. The organization responsible for the document that contains the requirement being considered for tailoring is the organization from which tailoring approval is sought unless this authority has been formally delegated elsewhere. The organization with the tailoring authority is responsible for consulting with the other organizations involved in establishing the requirement and for obtaining the concurrence of those organizations having a substantive interest.

Compliance Matrix tailoring includes signatures from the organizations responsible for requirements that are not already required signatories to the Formulation Agreement or Program or Project Plan, including the Office of the Chief Engineer, which is responsible for the major part of the NPR 7120.5 requirements.

The Compliance Matrix is attached to the Formulation Agreement for projects in Formulation and/or the Program or Project Plan. Once the Formulation Agreement or Program or Project Plan is signed by the required signatories, the tailoring in the matrix is approved, and a copy is forwarded to the OCE. No other waiver or deviation documentation is required.

Tailoring of NPR 7120.5 requirements is dispositioned by the designated officials shown in Table 5-2, unless formally delegated elsewhere. Requests for NPR 7120.5 requirement tailoring may be submitted in the form of the Compliance Matrix or submitted as an individual waiver or as part of a group of waivers. Regardless of whether the waiver is approved as a stand-alone document or as part of the Compliance Matrix, the required signatures from the responsible organizations or their designee are obtained. If the Compliance Matrix changes or if compliance is phased for existing programs or projects, updated versions of the Compliance Matrix are incorporated into an approved updated Formulation Agreement or Program or Project Plan revision. (See NPR 7120.5 for phasing requirements.)

Table 5-2 Waiver or Deviation Approval for NPR 7120.5 Requirements

	Project Manager	Program Manager	Center Director	MDAA	Chief Engineer	NASA AA
Programs		Recommends	Concurs[2]	Recommends	Approves	Informed
Category 1, 2, and 3 Projects	Recommends	Recommends	Concurs[2]	Recommends	Approves	Informed
Reimbursable Space Flight Projects	Recommends		Concurs[2]	Recommends[1]	Approves	Informed
Waivers or deviations with dissent						Approves

[1] As applicable.

[2] Unless otherwise delegated.

5.4.3 Tailoring Process Documentation

If programs or projects find a need to submit a waiver or deviation later in the life cycle, the attributes and data needed for tracking, which follow in the list and Table 5-3 below, are included in requests to expedite processing and support requirement compliance tracking. If the Compliance Matrix is used to request tailoring, the process is streamlined upfront when requirements are flowed down; however, inclusion of attributes in the documentation of tailoring is still helpful. The specific format or form in which the attributes and tracking data are submitted is the responsibility of the requesting activity, but must be usable by the receiving organization. All requirement relief requests (deviations or waivers) are also copied to the SMA TA at the program or project level for risk review. Additional process, requirements, and required data elements for requesting tailoring of Agency-level SMA TA requirements can be found in NPR 8715.3 and NASA-STD-8709.20. For applicable requirements, the tailoring process results in an entry in the

Table 5-3 Tracking Data

Requirement originates from: NPR, NPD, NID, CPR, CPD, CPC, Center Work Instructions (CWI) Mandatory Technical Standard Non-mandatory Technical Standard Other/don't know (specify)	Rating (to be defined by the program/project/activity and properly documented): Critical Major Minor Additional information is attached
Type: Non-applicable (not relevant or not capable of being applied) Technically equal or better Requires acceptance of additional risk Involves nonconforming product Involves noncompliant requirement	Other: Permanent requirement relief Temporary requirement relief Recurring request for relief There is a need for corrective action to prevent recurrence

Note: All characteristics that apply are to be listed. A Center, program, or project may break the specified categories into additional logical subcategories. A Center, program, or project may recommend additional characteristics to the NASA Chief Engineer at any time.

Compliance Matrix for NPR 7120.5 requirements and/or the generation of deviations or waivers.

Minimum attributes for requests for requirement relief include:

- Descriptive title and date for requirement relief request.
- Unique identifier for the source of requirement relief request.
- Name of Center, program, project, and contractor involved in request, as applicable.
- Activity responsible for request (include contact information).
- Complete identification of requirement for which relief is requested.
- Description of the requirement(s), specification(s), drawing(s), and other baselined configuration, documentation, or product(s) affected due to this request.
- Description of the scope, nature, and duration of this request (e.g., identification of the system, parts, lot, or serial numbers).
- Identification of other organizations, systems, or components that may be affected.
- Justification for acceptance and reference to all material used to support acceptance.
- If appropriate, description of, or reference to, the corrective action taken or planned to prevent future recurrence.
- Risk evaluation. If acceptance increases risk, identify the names with signatures of the Technical Authority(ies) who has(have) agreed that the risk has been properly characterized and is acceptable and the names with signatures of the Programmatic Authority(ies) who has(have) agreed to accept the additional risk.

5.4.4 Tailoring a Derived Requirement

"Derived requirements" are requirements established by a Programmatic Authority arising from:

- Constraints or consideration of issues implied but not explicitly stated in the higher level direction originating in Headquarters and Center institutional requirements or
- Factors introduced by the architecture and/or the design.

The tailoring principles apply to derived requirements, so a programmatic authority at the level that established the derived requirement approves a request for tailoring the derived requirement unless this authority has been formally delegated elsewhere.

An organizational entity seeking relief from a derived requirement submits a request for a waiver or deviation to the organization at the level that established the derived requirement or to its designee. If the source organization established the derived requirement, it has the authority to disposition the request for the derived requirement relief. However, if the source organization was flowing the derived requirement down from a higher authority and was not delegated the authority to grant relief from the derived requirement, the source organization forwards the request to the higher authority for dispositioning. This process illustrates the need for programs and projects to be able to trace the origin of their requirements.

An example illustrating the tailoring of a derived requirement is in the "Tailoring of a Derived Requirement: Example" box.

Tailoring of a Derived Requirement: Example

A project determines that it needs to specify a pressure vessel. In the design implementation, the project decides to use a composite overwrapped pressure vessel (COPV) that meets the Agency-level requirement for a safety factor of N. Because of a perceived technology risk, the project decides to impose a higher safety factor of $N + m$.

The extra increment on the safety factor is a derived requirement and is the responsibility of the Programmatic Authority at the level that established it.

If the project decides to change the design from a COPV to a metallic pressure vessel, the associated changes in specified requirements can be approved at the project level with notification to the next higher level and to others who would be impacted by the change.

Similarly, if the project decides to eliminate the extra added safety factor ($+ m$), the requirement can be changed at the project level as this is the organization and level that established the requirement.

However, if the project proposes that the new metallic tank need only meet a safety factor of $N - x$ (less than the Agency requirement), the tailoring principles would require the approval of the appropriate Technical Authority.

5.4.5 Tailoring a Technical Authority Requirement

Technical Authority requirements are invoked by OCE, OSMA, or Office of the Chief Health and Medical Officer (OCHMO) documents (e.g., in NPDs, NPRs, and/or NASA standards), usually flowed down in Center institutional documents. Tailoring of these requirements is the responsibility of the office or organization that established the requirement unless delegated elsewhere.

Technical Authorities at the program or project level ensure that the approval for tailoring Technical Authority requirements is obtained from the appropriate Technical Authority that established the requirement (or designee). It follows from basic principles that a program- or project-level Technical Authority cannot approve relief from a Technical Authority requirement unless they have been formally delegated this authority.

5.4.6 Non-Applicable Prescribed Requirements

A prescribed requirement that is not relevant and/or not capable of being applied to a specific program, project, system, or component can be characterized as non-applicable and can be approved by the individual who has been delegated oversight authority by the organization that established the requirement. This approval can be granted at the level where the requirement was specified for implementation; e.g., the project-level Engineering Technical Authority (ETA) could approve a non-applicable designation for an engineering requirement applicable to the project level. The request and approval documentation becomes part of the retrievable program or project records. No other formal deviation or waiver process is required.

5.4.7 Request for a Permanent Change

A request for a permanent change to a prescribed requirement in an Agency or Center document that is applicable to all programs and projects is submitted as a "change request" to the office responsible for the requirement's policy document unless formally delegated elsewhere. No special form or format for a change request is specified in NPR 7120.5. No special form or format is required to enable existing Center forms and processes to be used.

5.5 Maturing, Approving, and Maintaining Program and Project Plans, Baselines, and Commitments

This special topic discusses key aspects of NASA's policy for developing and managing a well-defined baseline state for space flight programs and projects. It describes the Agency Baseline Commitment and outlines the maturation of the program or project cost and schedule estimates during Formulation and the establishment of the program or project baseline at Key Decision Point (KDP) I (tightly coupled programs) and KDP C (projects and single-project programs). Detailed discussions of enabling and supporting topics are also provided, including the Decision Authority, the Decision

The non-applicable prescribed requirement provision was included to provide an efficient means to grant and document relief from a specific class of requirements for which the need for relief is obvious and the judgment is likely to be the same regardless of who makes the determination. The criteria of being "nonrelevant" or "not being capable of being applied" were selected to identify non-applicable requirements. This criterion allows approval to be handled by the designated oversight authority at the level the requirement was specified for implementation. Required documentation was also simplified for non-applicable prescribed requirements. The documentation of the decision (including identification by parties involved) is recorded for completeness.

Memorandum, and the Management Agreement. The flow of this section is as follows:

- The Agency Baseline Commitment (ABC).

- Maturing the program or project life-cycle cost (LCC) and schedule estimates during Formulation, and establishing the program or project baseline at KDP I/KDP C.

- Relationships between the LCC, ABC, unallocated future expenses (UFE), and Management Agreement.

- Changing the program or project cost plan and ABC.

- The Decision Authority, the Agency's responsible individual who makes the KDP determination on whether and how a program or project proceeds through the life cycle and authorizes the key program cost, schedule, and content parameters that govern the remaining life-cycle activities.

- The Decision Memorandum, which documents important Agency-level decisions related to programs and projects at and between KDPs.

- The Management Agreement, documented in the Decision Memorandum, which defines the parameters, including cost and schedule, and authorities for which the program or project manager has management control and accountability.

This special topic references several specific cost terms used by the Agency including formulation costs, development costs, JCL, life-cycle cost, and Agency Baseline Commitment. Table 5-4 depicts the scope in terms of life-cycle phases for each of these terms.

Unallocated Future Expenses (UFE) are the portion of estimated cost required to meet the specified confidence level that cannot yet be allocated to the specific WBS sub-elements because the estimate includes probabilistic risks and specific needs that are not known until these risks are realized. (For programs and projects that are not required to perform probabilistic analysis, the UFE should be informed by the program or project's unique risk posture in accordance with Mission Directorate and Center guidance and requirements. The rationale for the UFE, if not conducted using a probabilistic analysis, should be appropriately documented and be traceable, repeatable, and defensible.) UFE may be held at the project level, program level, and the Mission Directorate level.

Table 5-4 Phases Included in Defined Cost Terms

Definition	Formulation			Implementation				
	Project Phases							
	Pre–Phase A	A	B	C	D	E	Extended Operations	F
Formulation Cost		■	■					
Development Cost				■	■			
JCL Scope				■	■			
Life-Cycle Cost		■	■	■	■	■		■
Agency Baseline Commitment		■	■	■	■	■		■

Note: The ABC is not established until KDP C and will include the actual Phase A and B costs.

The Agency Baseline Commitment (ABC) is an integrated set of program or project requirements, cost, schedule, technical content, and when applicable, the joint cost and schedule confidence level. The ABC cost is equal to the program or project LCC approved by the Agency at approval for Implementation. The ABC is the baseline against which the Agency's performance is measured during the Implementation Phase of a program or project. Only one official baseline exists for a program or project, and it is the ABC. The ABC for projects with an LCC of $250 million or more and the ABC for tightly coupled and single-project programs form the basis for the Agency's external commitment to the Office of Management and Budget (OMB) and Congress and serve as the basis by which external stakeholders measure NASA's performance for these programs and projects. Changes to the ABC are controlled through a formal approval process. (An ABC is not required for loosely coupled and uncoupled programs.)

5.5.1 The Agency Baseline Commitment

Managing and overseeing a program or project requires establishing a known reference or baseline state by which future performance and future states can be measured and compared. Program and project baselines consist of an agreed-to set of requirements, technical content, Work Breakdown Structures (WBSs), life-cycle cost (LCC), including all unallocated future expenses (UFE) (held within and outside the program or project), joint cost and schedule confidence level (JCL), when applicable, schedules, and other resources such as workforce and infrastructure.

LCC is the total of the direct, indirect, recurring, nonrecurring, and other related expenses both incurred and estimated to be incurred in the design, development, verification, production, launch/deployment, prime mission operation, maintenance, support, and disposal of a program or project, including closeout, but not extended operations.[7] The LCC of a program or project or system can also be viewed as the total cost of ownership over the program or project or system's planned life cycle from Formulation (excluding Pre–Phase A) through Implementation (excluding extended operations). The LCC includes the cost of the launch vehicle.

A program or project baseline, called the Agency Baseline Commitment (ABC), is established at approval for Implementation (KDP I for tightly coupled programs, KDP C for projects and single-project programs). (An ABC is not required for uncoupled and loosely coupled programs.) The baseline forms the foundation for program or project execution and reporting done as part of NASA's performance assessment and Governance process.

The program or project develops or updates the LCC in preparation for each life-cycle review that immediately precedes a KDP. Prior to the KDP for approval for Implementation (KDP I for tightly coupled programs, KDP C for projects and single-project programs), the program or project develops the ABC. The ABC and/or LCC are assessed, along with other key parameters, during the life-cycle review process, and are authorized as part of the KDP. The authorized ABC and/or LCC are documented in the KDP Decision Memorandum.

The NASA Associate Administrator (AA) approves all ABCs for programs requiring an ABC, and projects with an LCC greater than $250 million. The

---

[7] For long-duration (decades) programs such as human space flight programs, it is difficult to establish the duration of the life-cycle scope for the purposes of determining the LCC. Under these circumstances, programs define the life-cycle scope in the Formulation Authorization Document (FAD) or Program Commitment Agreement (PCA). Projects that are part of these programs document their LCC in accordance with the life-cycle scope defined in their program's Program Plan, PCA or FAD, or the project's FAD.

NASA Administrator's agreement is required for all program and project ABCs with an LCC greater than $1 billion, and for all Category 1 projects.

NASA uses the term "baseline" in many different contexts. "Baseline" as used in the context of the ABC is different from "baseline" used in a different context such as configuration management. As defined in *NASA-STD-0005, NASA Configuration Management (CM) Standard*, "A configuration baseline identifies an approved description of the attributes of a product at a point-in-time and provides a known configuration to which changes are addressed." While the configuration management context often allows for approval of baseline changes at a project or program level configuration control board, baseline changes in the context of the ABC require approval from the Decision Authority.

See Section 5.5.3 for a more detailed discussion of the relationship between the LCC and ABC. See Section 5.5.4 for a more detailed discussion of processes and procedures for changing the cost plan (replanning), and changing the ABC (rebaselining).

5.5.2 Maturing the Program or Project LCC and Schedule Estimates during Formulation, and Establishing the Program or Project ABC

At the beginning of Formulation, there is a relative lack of maturity and broad uncertainties regarding the program or project's scope, technical approach, safety objectives, acquisition strategy, implementation schedule, and associated costs. During Formulation, these program or project parameters are developed and matured. A major objective of the Formulation phase for tightly coupled programs, single-project programs, and projects is to develop high fidelity cost and schedule estimates that enable the program or project to establish a sound, achievable baseline for Implementation at KDP I (tightly coupled programs) or KDP C (projects and single-project programs). The expected states of the program or project LCC and schedule at KDPs 0 and I (tightly coupled programs) or KDPs A, B, and C (projects and single-project programs) reflect this maturation process.

Major objectives of the Formulation phase for loosely coupled and uncoupled programs are to develop credible cost and schedule estimates, supported by a documented Basis of Estimate (BoE), that are consistent with the available funding and schedule profile, and to demonstrate that proposed projects are feasible within available resources. The expected states of the program cost and schedule estimates at KDPs 0 and I reflect this maturation process.

The joint cost and schedule confidence level (JCL) is the product of a probabilistic analysis of the coupled cost and schedule to measure the likelihood of completing all remaining work at or below the budgeted levels and on or before the planned completion of the development phase. The JCL is required for all tightly coupled and single-project programs, and for all projects with an LCC greater than $250 million. The JCL calculation includes consideration of the risk associated with all elements, regardless of whether or not they are funded from appropriations or managed outside of the program or project. JCL calculations include the period from approval for Implementation (KDP I for tightly coupled programs, KDP C for projects and single-project programs) through the handover to operations. Per NPR 7120.5, Mission Directorates plan and budget tightly coupled and single-project programs (regardless of life-cycle cost) and projects with an estimated life-cycle cost greater than $250 million based on a 70 percent JCL or as approved by the Decision Authority. Mission Directorates ensure funding for these projects is consistent with the Management Agreement and in no case less than the equivalent of a 50 percent JCL.

5.5.2.1 Project and Single-Project Program Formulation

The Formulation Agreement is developed during Pre-Phase A. At KDP A, the Formulation Agreement is finalized and approved for Phase A, and preliminary for Phase B. It identifies the activities necessary to characterize the complexity and scope of the project or program, increase understanding of requirements, and identify and mitigate significant risks. It identifies and prioritizes the work required to determine and mitigate high-risk drivers. This work enables the development of high-fidelity LCC and schedule estimates, or high-fidelity LCC and schedule range estimates (projects with an LCC greater than $250 million and single-project programs), at KDP B and high-fidelity LCC and schedule commitments at KDP C.

The LCC is provided as a preliminary estimate or range estimate. The schedule is risk-informed at the project level and includes a planned date for KDP B and a preliminary date or range for Phase D completion. Internal planned dates for other project milestones may also be included. Once authorized by the Decision Authority, the preliminary LCC estimate or range estimate, the preliminary schedule estimate or schedule range estimate, and the Management Agreement are documented in the KDP A Decision Memorandum. The cost in the Management Agreement is the authorized formulation cost. (See Section 5.5.6 for a more detailed description of the Decision Memorandum and Management Agreement.)

At KDP B, the Formulation Agreement is finalized and approved for Phase B, and a preliminary version of the Program or Project Plan is provided. High-fidelity LCC and schedule estimates or range estimates are provided. The LCC estimate or range estimate is risk-informed and schedule-adjusted. The schedule is risk-informed at the subsystem level and includes a preliminary date or range for Phase D completion. A preliminary Integrated Master Schedule (IMS), and preliminary cost confidence and schedule confidence levels (projects with an LCC greater than $250 million and single-project programs) are also provided. Once authorized by the Decision Authority, the preliminary LCC and preliminary schedule estimates or range estimates, cost and schedule confidence levels (if required), and the Management Agreement are documented in the KDP B Decision Memorandum. When applicable, the LCC range estimate serves as the basis for coordination with the Agency's stakeholders. The cost in the Management Agreement is the authorized formulation cost.

At KDP C, the Program or Project Plan is finalized and approved. The work identified in the Formulation Agreement has been completed, enabling the program or project to define high-fidelity LCC and schedule estimates. The LCC is a risk-informed, schedule-adjusted single number. The schedule is risk-informed and cost- or resource-loaded and is no longer provided as a range. An IMS and JCL (if required) are also provided. The fidelity of the LCC and schedule estimates and the maturity of the program or project planning enable the establishment of the program or project ABC baseline. Once authorized by the Decision Authority, the ABC, including the LCC and schedule, the JCL (if required), and the Management Agreement are documented in the KDP C Decision Memorandum.

Figure 5-9 illustrates the development, approval, and documentation of Decision Memoranda, Management Agreements, Formulation Agreements, Program or Project Plans, LCC ranges, the LCC, and the ABC throughout the life cycle for projects and single-project programs.

5.5.2.2 Tightly Coupled Program Formulation

Tightly coupled programs contain projects that have a high degree of organizational, programmatic, and technical interdependence and commonality. This requires the program to ensure that the projects are synchronized and well integrated throughout their respective life cycles, both with each other and with the program. Since the program is intimately tied to its projects, the program's Formulation Phase activities mirror those of the project life

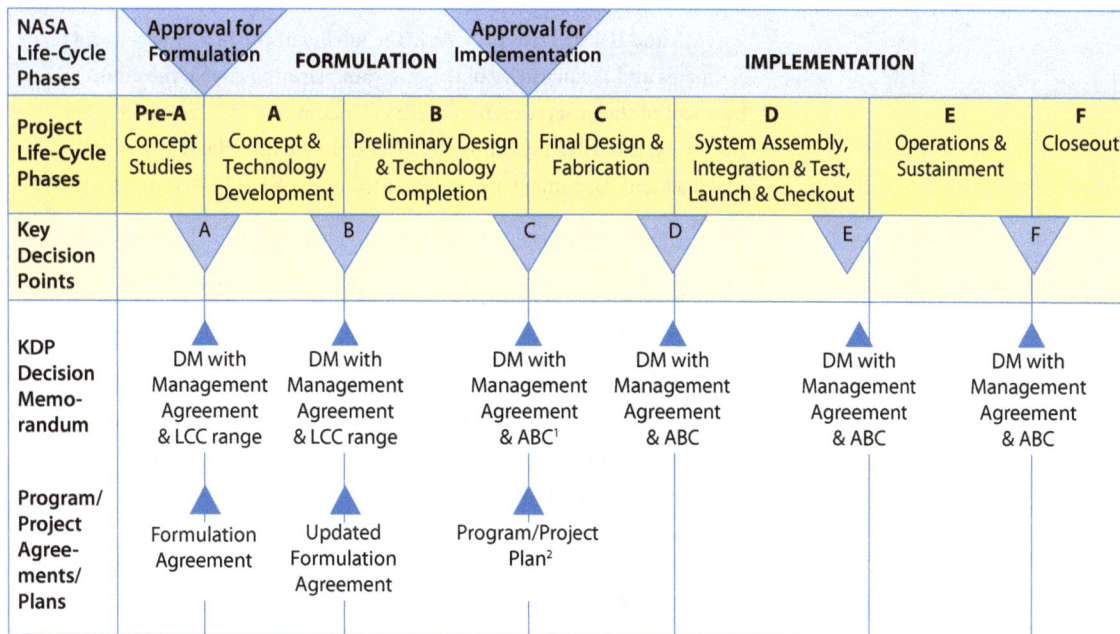

[1] Changes in the ABC after this point may require a rebaseline review.

[2] Program/Project Plans are updated as needed during implementation.

Figure 5-9 Approval of Plans and Baselines

cycle, and program formulation products such as the LCC and schedule estimates and the ABC are informed by the analogous project formulation products; e.g., the LCC and schedule estimates and ABCs.

During program Formulation, the Program Plan is finalized and approved and a credible LCC and schedule ranges are provided. The LCC and schedule ranges are supported by a documented Basis of Estimate (BoE). The LCC range estimate is risk-informed and schedule-adjusted. The schedule is risk-informed. A preliminary IMS and preliminary cost confidence and schedule confidence levels are also provided. If KDP 0 is required, once authorized by the Decision Authority, the preliminary LCC range estimate, preliminary schedule range, preliminary cost and schedule confidence levels, and Management Agreement are documented in the KDP 0 Decision Memorandum. The LCC range estimate serves as the basis for coordination with the Agency's stakeholders. The cost in the Management Agreement is the authorized formulation cost.

Program approval (KDP I) for Implementation occurs after the program-level Preliminary Design Review (PDR), which allows for a more developed definition of the preliminary design before committing to the complete scope of the program. At KDP I, the Program Plan is updated. The LCC is a risk-informed, schedule-adjusted single number. The schedule is risk-informed and cost- or resource-loaded and is no longer provided as a range. An IMS and JCL are also provided. The fidelity of the LCC and schedule estimates and the maturity of the program planning enable the establishment of the program ABC baseline. Once authorized by the Decision Authority, the ABC, including the LCC and schedule, the JCL, and the Management Agreement are documented in the KDP I Decision Memorandum.

5.5.2.3 Projects and Tightly Coupled and Single-Project Program Implementation

At KDP II (tightly coupled programs), and KDP D (project and single-project programs), and subsequent KDPs, the LCC estimate and the IMS are updated. During the Implementation phase, tightly coupled programs continue to have program life-cycle reviews tied to the projects' life-cycle reviews to ensure that program implementation products such as LCC estimates and schedules are informed by the analogous project implementation products. The Agency expects a program or project to meet the commitments it made at KDP I/KDP C, and for the LCC and ABC authorized at KDP I/KDP C to remain the same throughout Implementation. For tightly coupled programs, single-project programs, and projects with a LCC greater than $250 million, development cost or schedule growth that exceeds development cost or schedule in the ABC may trigger external reporting requirements and may require the ABC to be rebaselined. (See Section 5.5.4.)

5.5.2.4 Loosely Coupled and Uncoupled Program Formulation and Implementation

During program Formulation, the Program Plan is finalized and approved, initial cost and schedule estimates are developed, and the program develops credible risk-informed program implementation options that fit within the desired schedule and available funding profile. Instead of an LCC range, the cost estimate may be represented merely as an annual funding limit consistent with the budget. The program is not required to develop program cost and schedule confidence levels. If KDP 0 is required, once authorized by the Decision Authority, the initial cost and schedule estimates and the Management Agreement are documented in the KDP 0 Decision Memorandum.

At KDP I, credible cost and schedule estimates, supported by a documented Basis of Estimate (BoE), are established. These estimates are consistent with driving assumptions, risks, system requirements, conceptual designs, and the available funding and schedule profile. The program demonstrates that proposed projects are feasible within available resources. The program is not required to develop a JCL or an ABC. Once authorized by the Decision Authority, the cost and schedule estimates and the Management Agreement are documented in the KDP I Decision Memorandum.

During program Implementation at KDP II (and subsequent KDPs), the program provides updated, credible cost and schedule estimates that are supported by a documented BoE and are consistent with driving assumptions, risks, project implementation, and the available funding and schedule profile.

During Formulation and Implementation, the program provides analysis that provides a status of the program's risk posture. This status is presented to the governing PMC as each new project reaches KDP B and C or when a project's ABC is rebaselined.

5.5.3 Relationships Between the LCC, ABC, UFE, and Management Agreement

Figure 5-10 illustrates the constituent cost elements of a project's LCC estimate developed for Formulation and Implementation, and the relationship between the LCC estimate, the ABC, UFE, and the Management Agreement. The constituent elements are analogous for programs.

For projects with an LCC greater than $250 million, both a low and a high estimate are developed in recognition of the relative lack of maturity and broad uncertainties regarding the technical approach and associated costs at this early stage of a project. This range is refined as Formulation proceeds, making trades and improving estimates, and helps support the establishment of a sound achievable cost estimate for Implementation at KDP C.

The left side of Figure 5-10 shows the constituents of the project's LCC range estimate during Formulation. The bottom of the left hand side of the figure shows the authorized Formulation cost,[8] which is the total authorized cost for Formulation activities required to get to KDP C. When the Formulation Agreement is approved at KDP A, this is the authorized cost for Phase A and Phase B. At KDP B, the Formulation cost includes the actual cost for Phase A and the updated cost for Phase B. Since not all costs can be explicitly identified in Formulation, an allowance may be included for UFE, generally at the project level during Formulation. The Formulation cost and the UFE constitute the project's Management Agreement during Formulation. The final constituent is the LCC range estimate. During Formulation, the project develops both a low and a high estimate for the project's LCC. The expectation is that the final LCC will fall within this LCC estimate range.

The right side of Figure 5-10 shows the constituents of the project's LCC estimate during Implementation. At KDP C for Implementation, the Formulation cost is actual cost and is shown at the bottom of the right-hand side. The remaining LCC is divided between the estimated cost that can be allocated to a specific WBS sub-element, and the unallocated future expenses (UFE), which are those costs that are expected to be incurred but cannot yet be allocated to a specific WBS sub-element. The UFE is divided into UFE included in the Management Agreement under the project manager's control and UFE managed above the project (e.g., the program and MDAA). The estimated LCC is equal to the project's Management Agreement plus the UFE managed above the project, and this estimated LCC becomes the cost part of the ABC at KDP C.

5.5.4 Replanning and Rebaselining

NASA has established policies and made a series of management improvements to strengthen its baseline performance. For example, it has introduced the JCL and UFE and has established links between NPR 7120.5 requirements and future budgeting decisions. The Agency expects a program or project to meet the commitments it makes at KDP I (tightly coupled programs) or KDP C (projects and single-project programs), and that the LCC and ABC authorized at these KDPs will remain the same throughout Implementation. Failure to meet these commitments may impact the Agency's portfolio.

---

[8] Formulation cost is defined as the total of all costs incurred while the program or project is in Formulation, even if some of the individual project elements have initiated development activities. Pre-Formulation costs (i.e., Pre–Phase A costs) are not included in Formulation costs.

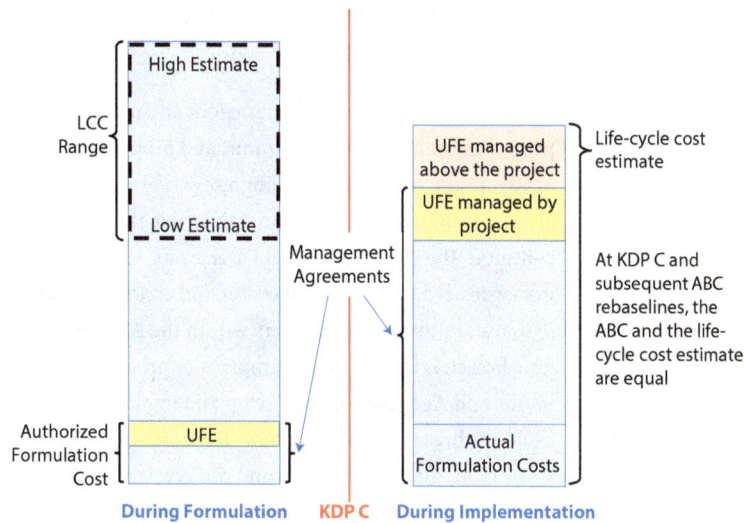

Note: Figure is notional and not drawn to scale.

Figure 5-10 Constituent Parts of a Project's Life-Cycle Cost Estimate for Formulation and Implementation

Replanning[9] and rebaselining, in the context of this section, are driven by changes in program or project cost parameters. Replanning and rebaselining are differentiated by the magnitude of the changes in cost parameters, in particular in the program or project's development cost, and by the program or project's life-cycle phase at the time the cost growth is identified. Replanning may occur during any life-cycle phase, including Formulation. Rebaselining occurs only in Implementation after the program or project has baselined the ABC (KDP I (tightly coupled programs) or KDP C (projects and single-project programs)).

- Certain changes in program or project cost parameters that do not require changes to the program or project ABC, LCC, or development cost are not considered cost growth. Replanning is the process by which a program or project implements and documents this type of change. An example of this type of change is reallocation or distribution of UFE to a WBS account, whether that UFE is within the Management Agreement or outside of the Management Agreement.

- Cost growth that results in exceeding the ABC after KDP I/KDP C may necessitate a replan if the development cost growth is between 15 and

---

[9] The program or project manager may also replan for many other reasons unrelated to cost that could involve workforce, schedule or other resources or organization. These other types of replanning are not addressed in this section.

30 percent or a rebaseline of the ABC if the development cost growth exceeds 30 percent.

The need to rebaseline is an anomalous situation, and for tightly coupled programs, single-project programs, and projects with a LCC greater than $250 million, is reported to Congress as a breach. In such cases, congressional reauthorization is required to enable the program or project to continue. The Agency, Mission Directorate, Center, and program or project manager need to vigilantly monitor and control the scope and performance to maintain the cost parameters within the ABC. As soon as the potential for a breach is identified, the program or project, Center, Mission Directorate, and Agency need to develop and implement corrective actions to avoid the breach. Periodic reviews (e.g., monthly reviews, BPR, etc.) have a role in monitoring program and project performance and identifying corrective actions to mitigate the risk of breaching.

Growth in LCC or development cost[10] may trigger external reporting requirements. For projects with an LCC greater than $75 million, a 10 percent growth in LCC triggers external reporting. Growth of 15 percent of the development cost in the ABC or an extension in schedule of 6 months or more (based on the schedule in the ABC) may also trigger additional external reporting. (See Section 5.12 for more detail on external reporting.)

Figure 5-11 illustrates different scenarios involving changes in project cost parameters that require either replanning or rebaselining. (These scenarios are also applicable to programs.)

- The left-most portion of Figure 5-11 illustrates the original KDP C Decision Memorandum, with project UFE within the Management Agreement and UFE held above the project level.

- Going from left to right, the second portion of Figure 5-11 illustrates distribution of UFE within the Management Agreement to WBS accounts. This replan does not require a change to the project's Management Agreement.

- The third portion of Figure 5-11 illustrates distribution of UFE held above the project to WBS accounts. This replan requires a change to the project's Management Agreement since responsibility for additional UFE in the ABC has been transferred to the project's control. The change to the project's Management Agreement requires an amendment to the

---

[10] Development cost is defined as the total of all costs from the period beginning with approval to proceed to Implementation (KDP I for tightly coupled programs, KDP C for projects and single-project programs) through operational readiness at KDP III (tightly coupled programs) or the end of Phase D (projects and single-project programs).

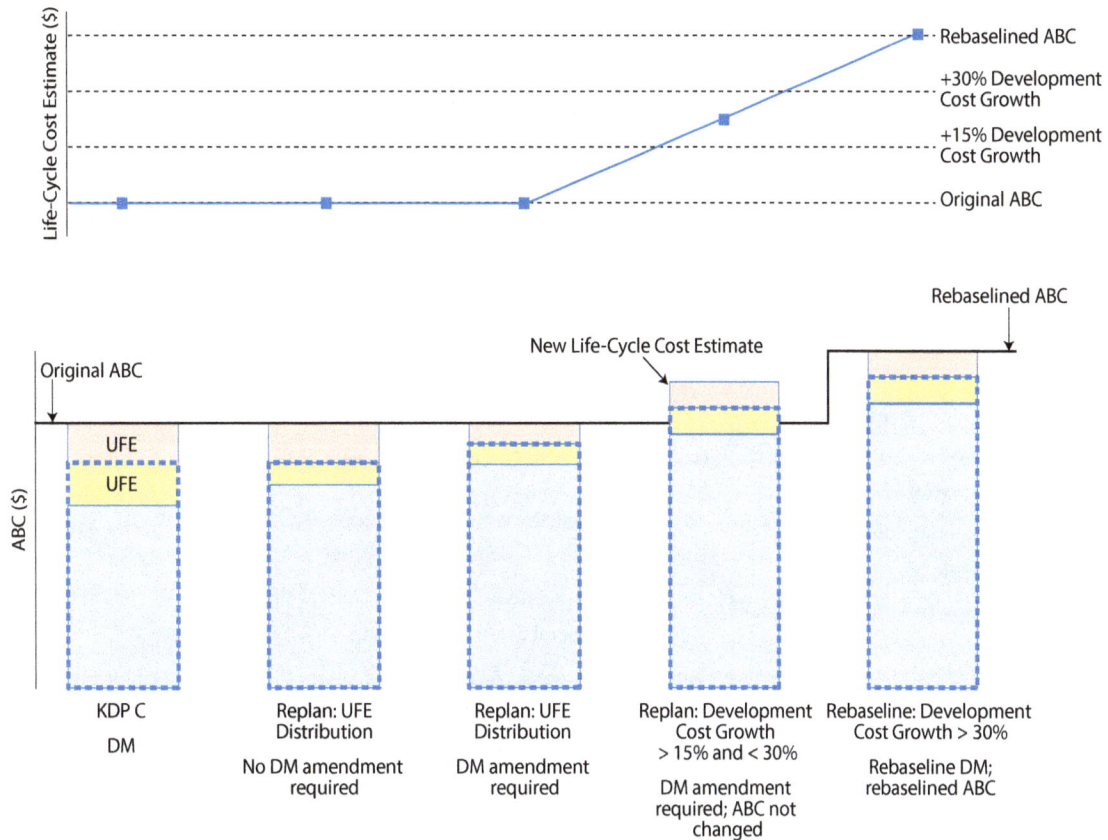

Legend: ⬚ Managed by project per Management Agreement

Note: Figure is not drawn to scale.

Figure 5-11 Distribution of UFE Versus Cost Growth Scenarios

Decision Memorandum. (The Decision Memorandum is amended by the signing parties (including the Decision Authority) between KDPs, if necessary, to reflect changes to the Management Agreement.) The replan Decision Memorandum records any changes to scope, schedule, cost, or cost profile.

• The fourth portion of Figure 5-11 illustrates a scenario in which development cost exceeds the development cost in the ABC by less than 30 percent but more than 15 percent (see upper part of figure). This increase in development cost is tracked as cost growth, necessitates a replan, and requires a change to the project's Management Agreement, since additional funding has been added to the project's control. The change to the project's Management Agreement requires an amendment to the Decision Memorandum. The replan Decision Memorandum records a new,

The Decision Authority is the individual authorized by the Agency to make important decisions on programs and projects under their purview. The Decision Authority makes the KDP decision by considering a number of factors, including technical maturity; continued relevance to Agency strategic goals; adequacy of cost and schedule estimates; associated probabilities of meeting those estimates (confidence levels); continued affordability with respect to the Agency's resources; maturity and the readiness to proceed to the next phase; and remaining project risk (safety, cost, schedule, technical, management, and programmatic). The NASA AA signs the Decision Memorandum as the Decision Authority for programs and Category 1 projects at the KDP. The MDAA signs the Decision Memorandum as the Decision Authority for Category 2 and 3 projects at the KDP. This signature signifies that, as the approving official, the Decision Authority has been made aware of the technical and programmatic issues within the program or project, approves the mitigation strategies as presented or with noted changes requested, and accepts technical and programmatic risk on behalf of the Agency.

increased project LCC, but the project ABC is not increased. The Decision Memorandum also records any changes to scope, schedule, cost, or cost profile.

- The right-most portion of Figure 5-11 illustrates a scenario in which development cost exceeds the development cost in the ABC by more than 30 percent (see upper part of figure). Cost growth of this magnitude necessitates a rebaseline of the project's ABC. If the project's LCC is greater than $250 million, congressional reauthorization is also required. The criteria and process for rebaselining an ABC and the associated documentation requirements are described in the next section.

5.5.4.1 Rebaseline Review

Rebaselining the ABC is required under the following circumstances:

a. The estimated development cost exceeds the development cost portion of the ABC LCC by 30 percent or more;

b. The NASA Associate Administrator judges that events external to the Agency make a rebaseline appropriate; or

c. The NASA Associate Administrator judges that the program or project scope defined in the ABC has been changed or a tightly coupled program or project has been interrupted.

ABCs are not rebaselined to reflect cost or schedule growth that does not meet one or more of these criteria.

Rebaseline reviews are conducted when the ABC needs to be rebaselined. In order to establish a new baseline, the Decision Authority institutes a review to examine the previously baselined gate products. The Standing Review Board (SRB), at the discretion of the Decision Authority, participates in the review per NASA SRB procedures. The objective of the review is to determine if the program or project can proceed to a new baseline. The Decision Authority determines the scope and depth of the Rebaseline Review for the extant phase to be reexamined. As part of this process, an independent cost and schedule assessment is performed. The results of the Rebaseline Review are documented and presented to the Decision Authority. If the rebaseline is approved by the Decision Authority, a new Decision Memorandum records the new ABC and any changes to project scope, schedule, LCC, cost profile, and Management Agreement.

5.5.5 Decision Authority

The Decision Authority is the Agency individual who is responsible for making the KDP determination on whether and how a program or project proceeds through the life cycle and for authorizing the key program cost, schedule, and content parameters that govern the remaining life-cycle activities, including, for tightly coupled and single-project programs, and for projects, the ABC baseline at KDP I or KDP C.

For programs and Category 1 projects, the Decision Authority is the NASA Associate Administrator. The NASA Associate Administrator may delegate this authority to the Mission Directorate Associate Administrator (MDAA) for Category 1 projects. For Category 2 and 3 projects, the Decision Authority is the MDAA. (See Chapter 4 for more information on categorization.) The MDAA may delegate to a Center Director Decision Authority to determine whether Category 2 and 3 projects may proceed through KDPs into the next phase of the life cycle. However, the MDAA retains authority for all program-level requirements, funding limits, launch dates, and any external commitments.

All delegations are documented and approved in the Program Commitment Agreement (PCA) or Program Plan, depending on which Decision Authority is delegating.

The Decision Authority's role during the life cycle of a program and project is covered in more detail in NPR 7120.5 Section 2.3 Program and Project Oversight and Approval, and in Chapters 3 and 4 of this handbook.

> The limitation on delegation by the MDAA to a Center Director is necessary to preserve the separation of the roles of the Programmatic and Institutional Authorities as required by NASA Governance.

5.5.6 Decision Memorandum

The Decision Memorandum and associated documentation provide a summary of key decisions made by the Decision Authority at a KDP, or, as necessary, in between KDPs. Its purpose is to ensure that major program or project decisions and their basis are clearly documented and become part of the retrievable records. The Decision Memorandum also provides the basis for NASA to meet various internal and external cost and schedule tracking and reporting requirements.

When the Decision Authority approves a program or project's entry into the next phase of its life cycle at a KDP, the Decision Memorandum documents this approval, the key program or project cost, schedule, and content parameters authorized by the Decision Authority that govern the remaining life-cycle activities, and any actions resulting from the KDP. These parameters include the LCC and schedule estimates, or when applicable, cost and schedule range estimates, cost and schedule confidence levels, and the JCL. The UFE and schedule margin held by the project or program, and the UFE and schedule margin held above the project or program level are also included. The Decision Memorandum also describes the constraints and parameters within which the Agency and the program or project manager will operate in the next phase of the life cycle and the extent to which changes in plans may be made without additional approval. If the Decision Authority determines that the program or project is not ready to proceed to the next life-cycle phase, the Decision Memorandum documents the Decision Authority's direction relative to the way forward.

The Decision Memorandum documents two key agreements: the Management Agreement and, when applicable, the Agency Baseline Commitment (ABC).

The Management Agreement is documented in the Decision Memorandum at every KDP. It defines the parameters and authorities over which the program or project manager has management control. The Management Agreement includes the schedule and cost (by year) at which the Agency agrees that funding[11] will be made available to the program or project and at which the program or project manager and the Center agree to deliver the content defined in the Program or Project Plan. UFE and schedule margin available within the Management Agreement are also documented. The Management Agreement should be viewed as a contract between the Agency and the program or project manager. Both the Agency and the program or project manager are accountable for compliance with the terms of the agreement. The Management Agreement may be changed between KDPs as the program or project matures and in response to internal and external events. This requires an amendment to the Decision Memorandum.

The ABC is documented in the Decision Memorandum at approval for Implementation (KDP I for tightly coupled programs, KDP C for projects and single-project programs) and subsequent KDPs. The UFE and schedule margin held above the project by the program and/or the Mission Directorate and, for programs, the UFE and schedule margin held above the program by the Mission Directorate are documented in the Decision Memorandum and constitute the difference between the Management Agreement and the ABC. (An example of schedule margin held above the project level would be a Launch Readiness Date (LRD) in the ABC that is later than the LRD in the Management Agreement. This provides the Agency with flexibility to adjust launch manifests, to adapt to changing priorities, or to mitigate unanticipated technical issues.) During planning and execution of the program or project as risks are realized, the UFE or schedule margin may be released to the program or project through a change to the Management Agreement, which requires amending the Decision Memorandum.

---

[11] Agency policy does not permit Mission Directorates to hold back portions of these amounts.

The Decision Memorandum may be amended by the signing parties, including the Decision Authority, between KDPs to reflect changes to the Management Agreement, LCCE, or ABC. This includes changes in the estimated cost or schedule associated with the approved scope, changes in the budget or funding profile that may drive a change in schedule or cost, or a change to the program or project scope.[12] Amendments to the Decision Memorandum also identify any significant changes in program or project risk. The NASA Associate Administrator is notified of Decision Memorandum amendments that reflect a growth in the program or project LCC, development cost, or schedule estimate beyond the ABC for any programs and projects that are subject to external reporting. Section 5.5.6.1 describes the content required in the Decision Memorandum and Section 5.5.6.2 describes the process for preparing and completing the Decision Memorandum.

5.5.6.1 Decision Memorandum Contents

The content prescribed by Decision Memorandum templates supports compliance with Decision Memorandum requirements in NPR 7120.5 Section 2.4.1:

- Summary—Which program or project, which KDP, which governing PMC, date of meeting, and which governing NPR.

- Decision—Whether the program or project is approved, conditionally approved, or disapproved to proceed to the next phase and any specific direction to the program or project.

- Technical content—Content as described in the Formulation Authorization Document (FAD), the Program or Project Plan, and/or the KDP briefings, as modified by actions issued at the KDP.

- Cost and Schedule Tables—Approved cost and schedule estimates or range estimates, ABC (if applicable), cost and schedule within the Management Agreement, cost phased by year, and, if applicable, any associated confidence levels (cost and schedule) or JCL.

- Key Assumptions—Supporting data and information to support the basis of estimates, including but not limited to applicable definitions, methodology, tools, scope, allowances, exclusions, and any tailoring deviations.

- Actions—Any actions resulting from the KDP.

- Signatures—Concurrence signatures of NASA officials responsible for relevant policies and requirements; approval signature of the Decision Authority.

---

[12] "Project scope" encompasses the approved programmatic content and deliverables.

Supporting Datasheet. This document provides a supporting breakout of the cost and schedule information in the Decision Memorandum as well as key contracts. It ensures that everyone is on the same page at the start of each phase and provides a basis for tracking during the phase. It also provides a means of providing Congress and OMB with correct, up-to-date cost information as required:

- **Cost**—The cost plan by year, by phase, and by WBS breakout is provided, as well as project- and Mission Directorate-held UFE by year. These tables also break out any Construction of Facilities (CoF) costs, which are part of the project cost estimate but are reported in a separate programmatic CoF budget to Congress.

- **Schedule**—Key NPR 7120.5E schedule milestones along with key procurement, delivery, integration, and/or testing milestones.

- **Contract**—Provides the current value of key contracts and contract options.

Note 1: Decision Memorandum datasheets record program or project costs associated with Pre–Phase A and extended operations to maintain traceability to the financial records. These costs are not included in the program or project LCC estimate.

Note 2: Construction of Facilities (CoF) cost is usually included by projects in the most relevant WBS element. The Decision Memorandum datasheet, however, provides for breakout of CoF costs because, while CoF is included in the project's LCC estimate, it *is not* included in the project's budget as presented to Congress.

Baseline Report (for projects with LCC greater than $250 million). This is a narrative that provides a high-level description of the approved project plan. It is simply an update of the project pages in the most recent NASA budget to Congress (a link to which is found at the bottom of every NASA web page). If the project has not been featured in the budget (projects are typically not featured until they reach KDP B), the format for the Baseline Report is the same as the budget pages. In these cases, the Baseline Report serves as the basis for the project pages in future budgets.

5.5.6.2 Preparing and Completing the Decision Memorandum

The Decision Memorandum process and supporting templates are managed by the Office of the Chief Financial Officer (OCFO) Strategic Investments Division (SID). The SID point of contact assists the Mission Directorate's Program Executive (PE) or equivalent in navigating this process. The Decision Memorandum information required varies for each KDP reflecting the changing requirements for each KDP. Current templates may be found at NASA's OCFO community of practice site (https://max.omb.gov/community/pages/viewpage.action?pageId=646907686), which can be reached via the NASA Engineering Network (NEN) Project Planning and Control (PP&C) community of practice site.

The Decision Memorandum templates are designed to support an array of NASA policy requirements and management strategies with respect to program or project life cycles, planning and replanning, and baselining; Work Breakdown Structure (WBS); cost estimation; cost and schedule confidence levels (if applicable), and UFE. In addition, the datasheet facilitates the comparison of cost estimates with project budget/financial systems.

Decision Memorandum content includes very high-level summaries of the detailed Program or Project Plan (including the schedule and cost plan) assessed during the life-cycle review preceding each KDP. The PE or equivalent is responsible for preparing and updating the Decision Memorandum and obtaining signatures. For programs and projects that do not have a PE, the Mission Directorate identifies the person responsible for developing and coordinating the Decision Memorandum. Even though it is not signed until reviewed by the governing Program Management Council (PMC), preparation of the Decision Memorandum and supporting materials is initiated at the beginning of the life-cycle review and KDP process:

- While preparing for the life-cycle review, the PE meets with SID to determine whether any desired tailoring of the templates can be approved.

- The PE completes the Decision Memorandum template summarizing information contained in the Program or Project Plan. The information used is consistent with what is provided to the SRB (or other reviewing body) prior to the life-cycle review leading up to the KDP. The PE provides the draft Decision Memorandum to the project, program, SRB, and SID.

- The PE updates the Decision Memorandum draft, if necessary, to reflect any changes to the Program or Project Plan, schedule, cost estimate, and if applicable, confidence levels as a result of the life-cycle review or ensuing management briefings.

- The PE shares the updated Decision Memorandum draft with the signatories or their points of contact at least two weeks prior to the governing PMC meeting and addresses any questions individual signatories may have. This advanced discussion facilitates signatory agreement on the Decision Memorandum at the governing PMC meeting.

The PE provides the completed Decision Memorandum materials to the governing PMC Executive Secretary along with other materials for the KDP meeting. The Decision Memorandum is nominally signed at the end of the governing PMC meeting. Some changes to the Decision Memorandum may be required during the meeting based on the discussions that take place at the meeting. If required changes are extensive or additional discussion/information is needed before the governing PMC members sign, the PE makes the necessary changes, pre-coordinates the changes with the signatory points of contact, and acquires the Decision Memorandum signatures after the meeting. If the Decision Authority determines that the program or project is not ready to proceed to the next phase, the Decision Memorandum documents the Decision Authority's direction with respect to the program or project's next steps. The signed document is provided to SID to be archived.

5.5.6.3 Decision Memorandum Signatories and Their Commitments

The NASA AA signs the Decision Memorandum in the case of programs and Category 1 projects as the Decision Authority approving the Program or Project Plan at the specified KDP. This signature signifies that the approving official has been made aware of the technical and programmatic issues within the program or project, approves the mitigation strategies as presented or with noted changes requested, and accepts technical and programmatic risk on behalf of the Agency.

The Mission Directorate AA (MDAA) signs the Decision Memorandum in the case of Category 2 and 3 projects as the Decision Authority approving the Project Plan at the specified KDP. In the case of a Category 2 project, this signature signifies that the approving official has been made aware of the technical and programmatic issues within the program or project, accepts the mitigation strategies as presented or with noted changes requested, and accepts technical and programmatic risk on behalf of the Mission Directorate and Agency.

In all cases, the MDAA signs the Decision Memorandum to certify that the proposed program or project satisfies the requirements of the underlying mission and can execute the mission within the resources provided, and to commit funding for the mission at the proposed levels in all future budgeting exercises.

The Chief Engineer signs the Decision Memorandum to certify that the programmatic and engineering policies and standards of the Agency have been followed in bringing the program or project to the governing PMC, and that the technical and programmatic risk are acceptable.

The Chief, Office of Safety and Mission Assurance (OSMA) signs the Decision Memorandum to certify that all Agency policies and standards related to safety and mission assurance have been followed by the program or project, and that the residual safety and mission success risks are acceptable.

If the program or project involves areas and issues under the auspices of the Health and Medical Technical Authority (HMTA), the Chief Health and Medical Officer (CHMO) signs the Decision Memorandum to certify that all Agency policies and standards related to human health and medical care have been followed by the program or project, and that the residual health risk is acceptable.

The Director of the Office of Evaluation signs the Decision Memorandum to certify that independent analysis of programmatic risk has been conducted and used in a fashion consistent with Agency policies, and further, that this analysis was presented and used in a way that informed the Agency decision process.

The Chief Financial Officer signs the Decision Memorandum to certify that any description of past funding, present obligations, commitments on budgets, schedules, LCC estimates, and JCL estimates provided to entities outside of NASA (e.g., OMB, Congress) are accurate and consistent with previous commitments, that the decision is clear and unambiguous with respect to the financial commitment being made, and that it complies with all authorization and appropriation law and other external reporting requirements.

The host Center Director signature reflects a commitment to provide the necessary institutional staffing and resources to make the program or project successful. This signature certifies that the appropriate Agency and Center policies, requirements, procedures, practices, and technical standards are in place and are being met. Further, this signature reflects concurrence with all aspects of the plan approved at the governing PMC. The Center Director's signature also represents the consent to accept residual institutional safety risk in accordance with established Center procedures and policies. In the event the host Center is not the sole implementing Center, the implementing Center Director(s) signature(s) conveys consent to accept residual institutional safety risk in accordance with all participating Centers' procedures and policies.

If the mission is one led by a Principal Investigator (PI), the PI signs the Decision Memorandum certifying that the proposed mission concept and mission systems will meet the Level 1 Requirements. This signature also represents a commitment to execute within the approved cost and schedule given the identified risks.

The program and project manager's signatures represent a commitment to execute the plan approved at the governing PMC.

5.6 Cost and Schedule Analysis Work to Support Decisions

5.6.1 Cost and Schedule Estimates

Cost and schedule estimates have an essential role in program and project management and must have a sound documented basis. All programs and projects develop cost estimates and planned schedules for the work to be performed in the current and following life-cycle phases. As part of developing these estimates, the program or project documents the Basis of Estimate (BoE) in retrievable program or project records. The BoE documents the ground rules, assumptions, and drivers used in cost and schedule estimate development and includes applicable model inputs and outputs, rationale/justification for analogies, and details supporting bottom-up cost and schedule estimates. The BoE is contained in material available to the Standing Review Board (SRB) and management as part of the life-cycle review and Key Decision Point (KDP) process. Good BoEs are well-documented, comprehensive, accurate, credible, traceable, and executable. Sufficient information on how the estimate was developed needs to be included to allow review team members, including independent cost analysts, to reproduce the estimate if required. Types of information can include estimating techniques[13] (e.g., bottom-up, vendor quotes, analogies, parametric cost models), data sources, inflation, labor rates, new facilities costs, operations costs, sunk costs, etc.

Program and project planning must be consistent with:

- Coverage of all costs associated with obtaining a specific product or service, including:
  - Costs such as institutional funding requirements, technology investments, and multi-Center operations;
  - Costs associated with Agency constraints (e.g., workforce allocations at Centers); and
  - Costs associated with the efficient use of Agency capital investments, facilities, and workforce.
- Resources projected to be available in future years based on the Agency's strategic resource planning. This includes the periodic portfolio reviews and resulting direction and the NASA budget process (i.e., Planning, Programming, Budgeting, and Execution (PPBE)).

---

[13] See *NASA-HDBK-2203* (http://swehb.nasa.gov) for information on software cost estimating and cost analysis.

- A risk-informed schedule at KDP 0/KDP A/KDP B and a risk-informed cost- or resource-loaded schedule at all other KDPs. (See Section 5.7 for discussion on resource-loaded versus cost-loaded schedules.)

- Decisions and direction documented in the program or project's approved Decision Memorandum.

- Unallocated future expenses (UFE) as approved in the program or project's Management Agreement, and funded schedule margin.

- Evaluation of suppliers' qualifications and past performance and the realism embodied in the suppliers' cost and schedule proposals.

- Reconciled estimates (differences are understood and their rationale documented) when independent estimates are required by the Decision Authority.

5.6.1.1 Cost by Year

Federal agencies have a unique three-step process for spending money, with funds having to be appropriated and obligated before being spent. Cost estimates are made on the basis of the content to be completed—and therefore paid for—in each fiscal year. Cost estimates are captured in the Cost Analysis Data Requirement (CADRe) and other project life-cycle review documents based on the expected year of expenditure. The Decision Memorandum and datasheet, however, are designed to ensure that the cost estimate is phased based on when NASA needs to request New Obligation Authority (NOA) so that there is time to get the funds obligated before they are spent. This will typically require a slight shift to the left of the cost profile (see chart of curves in Figure 5-12) in the Decision Memorandum compared to the CADRe and other cost estimation profiles.

5.6.1.2 CADRe (Cost Analysis Data Requirement)

The CADRe is a formal project document that describes the programmatic, technical, and life-cycle cost and cost and schedule risk information of a project. A 2005 initiative, the CADRe is NASA's unique response to the need to improve cost and schedule estimates during the formulation process, providing a common description of a project at a given point in time. The CADRe is prepared by NASA Headquarters' Cost Analysis Division (CAD) using existing project data prepared during the life-cycle review process. By capturing key information, the CADRe tracks and explains changes that occur from one milestone to the next, which helps the project manager record in an Agency document all the events, both internal and external, that occurred during the project.

Margins are the allowances carried in budget, projected schedules, and technical performance parameters (e.g., weight, power, or memory) to account for uncertainties and risks. Margins are allocated in the formulation process, based on assessments of risks, and are typically consumed as the program or project proceeds through the life cycle.

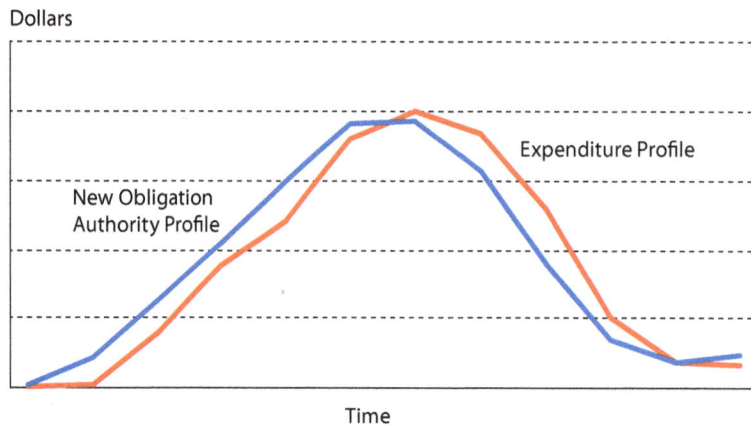

Figure 5-12 Example New Obligation Authority Profile in a Decision Memorandum Compared to the CADRe and Other Expenditure Profiles

Completed CADRes are available on the NASA Safety Center Knowledge Now (NSCKN) site[14] and are incorporated into the One NASA Cost Engineering (ONCE) database, a secure, web-based application which allows for easy retrieval and fast analysis of CADRe data across multiple projects and milestone events. Utilization of CADRe data helps project managers analyze important attributes of projects, and enables project managers to develop improved cost and schedule estimates and deliver projects within cost and schedule, and technical margins.

Composed of three parts, the CADRe captures detailed programmatic, technical, and cost data in a standardized format. The document is prepared six times during the life cycle of a project at major milestones (SRR, PDR, CDR, SIR, launch, End of Mission (EOM)). See Figure 5-13.

The three parts of a CADRe are as follows:

- PART A describes a NASA project at each milestone (SRR, PDR, CDR, SIR, launch, and End of Mission), and describes significant changes that have occurred. This section includes essential subsystem descriptions, block diagrams, and heritage assumptions needed for cost analysis purposes. The templates for all three parts for robotic or human space flight missions can be found at: http://www.nasa.gov/offices/ooe/CAD.html.

---

[14] See https://nsckn.nasa.gov to request access to the NSCKN site. To access the ONCE database go to the ONCE website www.oncedata.com and click on the "request access" link on that page. The key requirement for access is to have a NASA identity in NASA's IDMAX system.

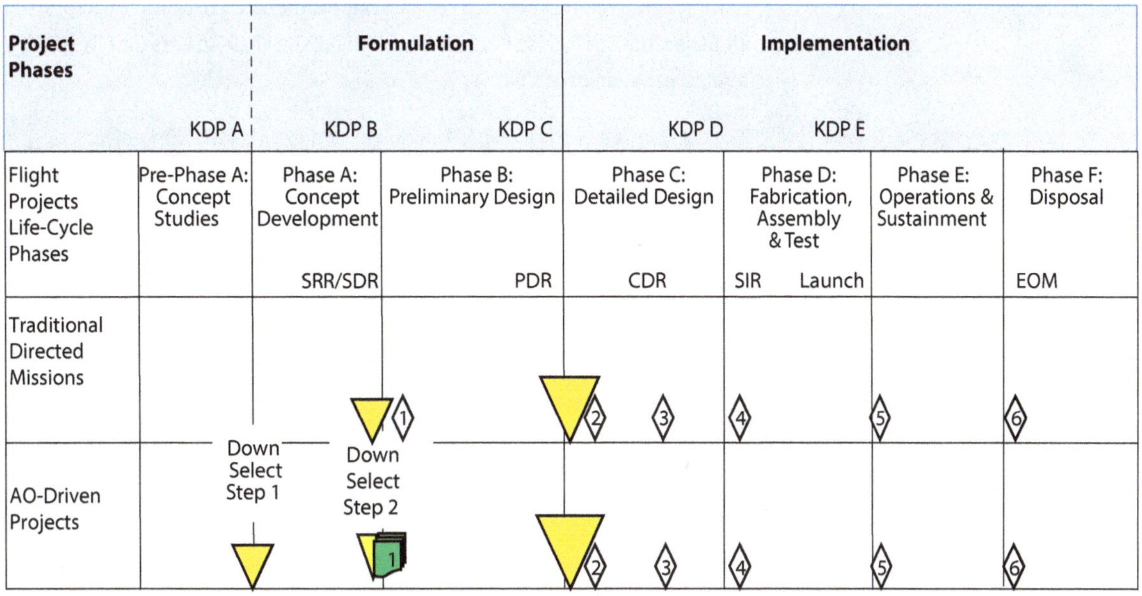

Project Phases		Formulation			Implementation			
	KDP A	KDP B		KDP C	KDP D	KDP E		
Flight Projects Life-Cycle Phases	Pre-Phase A: Concept Studies	Phase A: Concept Development	Phase B: Preliminary Design	Phase C: Detailed Design	Phase D: Fabrication, Assembly & Test	Phase E: Operations & Sustainment	Phase F: Disposal	
		SRR/SDR	PDR	CDR	SIR Launch		EOM	

Note: ▽ Key Decision Point (KDP)   ▣ CADRe delivered; based on Concept Study Report and winning proposal.

[1] All parts of CADRe due 30–45 days after KDP B.

[2] All parts of CADRe due 30–45 days after KDP C using PDR material.

[3] Update as necessary 30–45 days after CDR using CDR material.

[4] Update as necessary 30–45 days after KDP D using SIR material.

[5] All parts due 90 days launch, based on as built or as deployed configuration.

[6] Update Part C only after the end of decommissioning and disposal.

Figure 5-13 Frequency of CADRe Submissions

- PART B contains standardized templates to capture in an Excel Workbook key technical parameters that are considered to drive cost such as mass, power, data rates, and software metrics. The formats of this template follow standard NASA terminology such as Current Best Estimates (CBE) and CBE Plus Contingency.

- PART C captures in an Excel Workbook the NASA project's cost estimate and actual life-cycle costs within the project's Work Breakdown Structure (WBS) and a NASA Cost Estimating WBS. This section also captures the project schedule, risks, and ground rules and assumptions.

The CADRe program satisfies a foundational cost-estimating need, which is to provide historical cost data that are vital to performing estimates for future missions. The CADRe provides information to support an Independent Cost Estimate (ICE) as well as actual cost and technical information so that estimators can do a better job of projecting the cost and schedule of future analogous projects. This way, important data are captured across all major flight projects at NASA, including major instruments that fly on foreign partner spacecraft.

The CADRe is a project-owned document and is signed by the project manager; therefore, it does not include any independent assessments or evaluations or opinions about the project. It simply records the known configuration at specific milestones. HQ/CAD provides the necessary funding and support to prepare the document on behalf of the project using existing project documentation prepared during the life-cycle review process. In the few cases where a CADRe is prepared for a previously launched mission, CAD will make the determination whether there are enough data. If so, CAD will prepare a single launch or EOM CADRe. These CADRes also are very useful for historical benchmarking and understanding cost, schedule, and technical trends over time.

The process of preparing a CADRe is as follows: after a short kickoff with the project manager, CAD will collect all relevant existing documentation approximately 60–90 days before the life-cycle review milestone. Concurrently with the life-cycle review process, CAD prepares the CADRe using the most recently available data and existing project documentation that provides descriptive information, mass statements, power statements, schedules, risk list, and life-cycle cost estimates, and any other technical parameters that tend to drive costs. CAD delivers the document for the project manager's review and signature shortly after the capstone KDP briefing, such as the APMC or DPMC, when the cost and schedule positions are finalized.

Since CADRes represent snapshots of a project at successive key milestones, the ONCE database captures all the changes that occurred to previous projects and their associated cost and schedule impacts. The result provides enhanced insight and management of historical cost and technical data, which are helping to advance costing practices and analysis across the Agency. With a large historical archive of project data, it is possible to determine trends that can be very useful to project managers. Here are some examples:

- Cost engineers use CADRe to estimate the cost of future systems based on known technical parameters such as mass and power. The CADRe data are also used to help evaluate proposals from contractors on new missions.
- System engineers use CADRe information to perform mass architecture trades earlier in concept design by using time-tagged mass data on all major NASA projects.

- CADRe data can be used to conduct research to help understand cost and schedule trends and patterns over time and across projects. The results of this research are already helping project managers better understand how to plan for cost and schedule risks in their projects. For example, recent analysis of CADRe data has shown that schedule growth on instruments is a significant factor that increases the total cost. This recent analysis of instruments developed between August 1990 and November 2009 shows the average instrument development schedule growth was 33 percent or about 10 months. (See Figure 5-14.)

- In another example, research of the mass data in CADRe is showing that actual instrument mass often exceeds the planned mass contingencies that are routinely used. The analysis of over 30 NASA instruments showed that the mass contingency established at SRR was not enough to protect against mass growth. With this information, future project managers will be able to program more appropriate levels of mass contingency tailored for the type of instrument that is being developed.

These are just a few examples of how CADRe data can be used to help program and project managers. The use of CADRe has captured data of key historical missions looking back approximately 10 years, where the data were available, and has supported several NASA studies. As the number

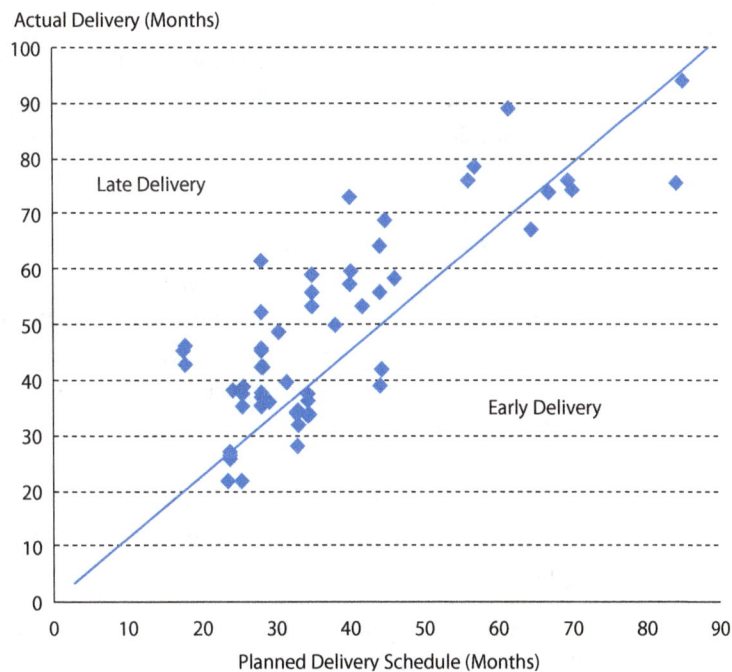

Figure 5-14 Planned versus Actual Instrument Delivery Schedules

of CADRes continues to grow, CAD can perform more robust analyses resulting in more advanced costing practices and tools.

5.6.2 Probabilistic Analysis of Cost and Schedule

Probabilistic analysis of cost and/or schedule estimates is required for tightly coupled programs and single-project programs (regardless of life-cycle cost), and projects with an LCCE greater than $250 million. When the probabilistic analysis is developed for only one parameter (i.e., cost or schedule) or when generally referring to a probabilistic assessment of the level of confidence of achieving a specific goal, the analysis is referred to merely as a "confidence level." When the probabilistic analysis is developed to measure the likelihood of meeting both cost and schedule, the analysis is referred to as a joint cost and schedule confidence level (JCL). A JCL is defined as the probability that actual cost and schedule will be equal to or less than the targeted cost and schedule. For example, a 70 percent JCL is the point on the joint cost and schedule probability distribution curve where there is a 70 percent probability that the project or program will be completed at or lower than the estimated cost and on or before the estimated schedule. (See *NASA Cost Estimating Handbook* at www.ceh.nasa.gov.)

5.7 Realistic Cost and Schedule Estimating and the JCL

A joint cost and schedule confidence level (JCL) is a quantitative probability statement about the ability of a program or project to meet its cost and schedule targets. (See "Risk Acceptance" box for more information.) Put simply, the JCL is the probability that a project or program's actual cost will be equal to or less than the targeted cost and its schedule will be equal to or less than the targeted schedule date. The process of developing a JCL requires that the program or project combine its cost, schedule, and risk into a complete, integrated quantitative picture that helps the decision makers understand the program or project's prospects for success in achieving its cost and schedule goals. A JCL is actually much more than just an output confidence level; it is a systematic framework process for integrating a program or project's cost, schedule, and risk artifacts. The technique identifies specific risks and allows decision makers to better understand those risks and the context for the program or project's phased funding requirements.

More than just a policy requirement, the JCL is also a valuable management tool that helps enforce some best practices of program and project

Risk Acceptance

This is a probabilistic world. Even a trip to the grocery store presents a risk that there will not be any milk. However, mitigation measures are not taken to compensate for this risk because the probabilities that the milk may not be available in the store are very low; so the risk is accepted. Also, when taking a road trip, departure time is scheduled based on the risks that are likely to happen. The more complicated the travel plan (this risk can be measured using factors such as length of road trip and number of major cities to travel through), the more margin is included in the estimated time to get there. The importance of arriving at the destination also influences the decision. For example, more time may be added if the trip is for an interview than if it is just a joy ride. Granted, this is performed mentally without thinking about it, but the process is there. So, if this type of mental exercise is performed for a road trip, why would a similar exercise not be performed for a multi-hundred-million-dollar mission? This is the thought process behind the current NASA policy. NASA deals with complex one-of-a-kind spacecraft that have many associated risks and uncertainties. It only makes sense to quantify these risks and uncertainties to help NASA management determine the appropriate amount of acceptable risk.

management, planning, and control as well as potentially enhancing vital communication among various stakeholders.

5.7.1 History of JCL

To understand the current situation, it is necessary to go back to 2002. In 2002, the United States GAO issued a report that identified major causes of cost growth that included incomplete cost risk assessment and flawed initial program planning. The GAO completed a detailed examination of NASA's cost estimating processes and methodologies for various programs. The report recommended that NASA establish a standard framework for developing LCCEs that included a cost risk assessment identifying the level of uncertainty inherent in the estimate. Formal probabilistic estimating guidance was first mentioned in February 2006 in an email from the NASA Administrator directing NASA's largest program at the time, the Constellation Program, to budget to a 65 percent confidence level. A month later in March 2006 at a strategic management meeting, the NASA Administrator established budgeting projects at a 70 percent cost confidence level based on the independent cost estimate as a NASA standard practice.

Several issues arose from the initial guidance. First, the lack of formally documented policy guidance hindered effective implementation. Second, by omitting schedule risk in the confidence level, a vital programmatic variable was being inconsistently utilized. Last, the reconciliation process between projects and the Agency's non-advocate groups was tedious. In January

2009, NASA updated its cost estimating policy to address those issues. Policy was inserted in the NASA Governance structure and was expanded to specify a joint cost and schedule confidence level, now known as a JCL, for tightly coupled and single-project programs regardless of life-cycle cost and for projects with an estimated LCC greater than $250 million.

Although the tools, techniques, and methodologies were well understood and had been demonstrated in certain industries, much of the analysis had not been done traditionally on a highly uncertain complex development process. NASA continues to hone the associated best practices and understanding for JCL analysis.

5.7.2 Intention of Policy

Currently, NASA uses a variety of cost analysis methodologies to formulate, plan, and implement programs and projects.

In the Formulation Phase, specifically for KDP 0/KDP B, NASA calls for programs and projects subject to the JCL requirement to provide probabilistic analyses on both their cost and schedule estimates. This analysis is then used to determine a high and a low estimate for cost and for schedule. There are two good candidate methodologies for producing the risk estimates and associated results: (1) complete parametric estimates of cost and schedule, or (2) complete a JCL. Completing a JCL at KDP 0/KDP B is not recommended. This is primarily because programs and projects typically do not have detailed plans available to support an in-depth JCL model and, by design, the confidence level requirement is intended to "bound the problem" at this point in the program or project life cycle. Conducting a parametric estimate of schedule and cost utilizes the historical data and performance of the Agency and provides a valuable estimate of the range of possibilities. Attempting a JCL at KDP 0/KDP B for these reasons is therefore not recommended, although it would fulfill the requirements if it were completed.

The rationale for conducting a JCL in support of KDP I/KDP C is for two primary reasons:

- The program's or project's plan is well defined, at this stage; and
- This is the timeframe in which NASA is committing to external stakeholders.

NASA uses a single approach for completing a JCL: Probabilistic Cost-Loaded Schedule (PCLS).

NASA actively implements PCLS to link its commitment probabilistically to the program's or project's specific plan. The Agency uses this assessment when considering its external commitment (KDP I/KDP C) as one means of ensuring the program or project has a robust plan with costs linked to schedule where both are informed by risks.

Once a solid baseline is approved, NASA policy does not require a program or project to maintain the artifacts used to calculate the JCL but uses a variety of performance metrics to assess how well the program or project is performing against its plan. If these metrics show that a program's or project's performance varies significantly from its plan, the program or project may need to replan, but Agency policy only requires a repeat calculation of the JCL in the event of a rebaseline. The JCL can be used as a management tool; however, currently it is the Agency requirement at KDP I/KDP C and is being used as a planning tool.

5.7.3 Policy Clarifications

Three general areas of NASA's JCL policy need to be clarified.

5.7.3.1 Resource-Loaded Terminology

NASA policy clearly states that projects are required to generate a resource-loaded schedule. This terminology can be confusing and deserves some attention. NASA defines resource loading as the process of recording resource requirements for a schedule task/activity or a group of tasks/activities. The use of resource loading implies to many people that the tasks need to be loaded with specific work or material unit resources. This is not the intent of the policy. In general, the term resource-loaded schedule can be used interchangeably with cost-loaded schedule. The intent of the JCL policy is not to recreate the lower level management responsibilities of understanding and managing specific resources (labor, material, and facilities) but instead to model the macro tendencies and characteristics of the project. To do this, cost loading an effort is sufficient.

As a simple example, if two individuals were needed to perform a task (Person A and Person B), to resource load each person to that task, identify how many hours each person would put on that task and their associated labor rate. However, with regard to cost loading, the only interest is in the total effort measured in dollars for the entire team (Persons A and B).

5.7.3.2 Risk Informed

NASA policy states that a project needs to perform a risk-informed probabilistic analysis to produce a JCL. The term risk-informed can be ambiguous. NPR 7120.5 defines "risk" as "the potential for performance shortfalls, which may be realized in the future, with respect to achieving explicitly

The joint cost and schedule confidence level is the product of a probabilistic analysis of the coupled cost and schedule to measure the likelihood of completing all remaining work at or below the budgeted levels and on or before the planned completion of the development phase. The JCL is required for all tightly coupled and single-project programs and for all projects with an LCC greater than $250 million. Small Category 3, Class D projects with a life-cycle cost of under $150 million should refer to guidance on tailoring NPR 7120.5 requirements from the NASA AA, which can be found on the OCE tab in NODIS under "Other Policy Documents" at http/nodis3.gsfc.nasa.gov/OCE_docs/OCE_25.pdf. The JCL calculation includes consideration of the risk associated with all elements, regardless of whether or not they are funded from appropriations or managed outside of the program or project. JCL calculations include the period from approval for Implementation (KDP I for tightly coupled programs, KDP C for projects and single-project programs) through the handover to operations. Per NPR 7120.5, Mission Directorates plan and budget tightly coupled and single-project programs (regardless of life-cycle cost) and projects with an estimated life-cycle cost greater than $250 million based on a 70 percent JCL or as approved by the Decision Authority. Mission Directorates ensure funding for these projects is consistent with the Management Agreement and in no case less than the equivalent of a 50 percent JCL.

established and stated performance requirements." Typically, from a risk management perspective, discrete risks are identified and tracked, and mitigation plans are formulated. By **risk-informed**, the policy is stating that all appropriate discrete risks are modeled and that various uncertainties (that may not be discretely managed in the risk management system) are accounted for. Formal definitions within the context of a JCL on risk and uncertainty are discussed later and are summarized in Section 5.7.4.5, Implementation of Uncertainty Analysis.

5.7.3.3 Life-Cycle Costs and Schedule

The cost and schedule aspects of the JCL are required only through Phase D. By definition, this is not the total LCC or operational life of a project.

5.7.4 JCL Process Flow (Overview)

"JCL," as referred to in this special topic, refers to a probabilistic cost-loaded schedule (PCLS).[15] The reason the Agency focuses on the PCLS stems from the fact that the method forces the program or project and the review entity to focus on the program's or project's plan. This improves program or project planning by systematically integrating cost, schedule, and risk products and processes. It also facilitates transparency with stakeholders on expectations and the probabilities of meeting those expectations. Lastly, it provides a cohesive and holistic picture of the program or project's ability to achieve cost and schedule goals and enables the determination of UFE and funded schedule margins required by the program or project.

In summary, JCL helps set the foundation to answer fundamental questions such as:

- Does the program or project have enough funds?
- Can the program or project meet the schedule?
- What are the areas of risk affecting successful execution of the program or project?
- What risk mitigation strategies provide the best program or project benefit?

In general, there are five fundamental steps in building a JCL with one prerequisite step—identify goals for the JCL:

1. Build a JCL schedule/logic network.

---

[15] For more on the mechanics of convolving an independent cost and schedule distribution, refer to Garvey, P.R., *Probability Methods for Cost Uncertainty Analysis: A Systems Engineering Perspective*, New York, Marcel Dekker, 2000.

2. Cost load the schedule.

3. Implement the risk list.

4. Conduct an uncertainty analysis.

5. View the results and iterate.

Very simple illustrations depict the various steps, starting with Figure 5-15 below.

5.7.4.1 Identify Goals of JCL

As stated previously, a JCL is a policy requirement. But it can also be a valuable management tool. While certain quality standards must be met to satisfy policy, depending on goals and expectations of the JCL analysis, the JCL analysis may be set up to assist and be synergistic with other products and processes. When setting up the JCL process, especially the schedule, it is important to think about what questions the JCL should answer, who the primary users and beneficiaries will be, and what fundamental insight is desired. The program or project manager, as a primary user and beneficiary, must be engaged in the set up of the JCL process to understand and shape the underlying programmatic assumptions, including the BoE, to understand characteristics of the JCL analyses techniques, including the potential for double counting of risk (see Sections 5.7.4.2 and 5.7.4.5); and to identify the questions and insights to be addressed by the JCL.

5.7.4.2 Schedule Network

The backbone of the entire JCL analysis is the schedule. Having a quality schedule (with logic networking) is key to a successful JCL. Figure 5-15 shows a simple schedule with two parallel activity streams, one with three

The JCL is a valuable management tool. While the JCL is a methodology to quantify the amount of program or project budget and UFE that will be required to achieve a certain confidence level, the process of developing a JCL encourages communication between the programmatic planners, the technical community, and management as assumptions and risks are documented. It encourages communication between Agency leadership and the program or project management, affording leadership an opportunity to consider the underlying programmatic assumptions; to discuss the analysis techniques; and ultimately, to build consensus around the conclusions (budget levels, amount of UFE, risks involved, probability of meeting commitments, etc.). The JCL is a tool to help people understand the implications of the calculations and assumptions and make adjustments.

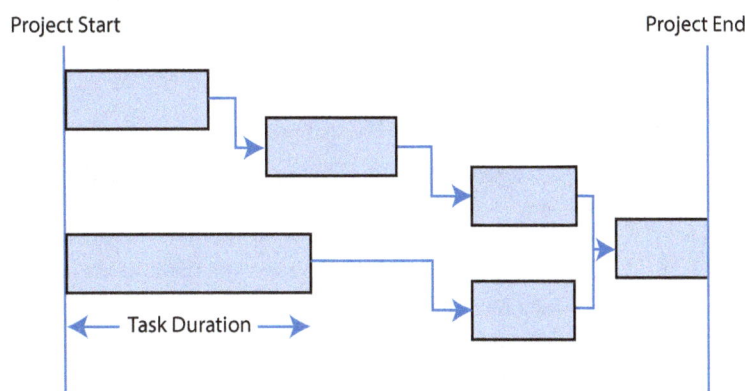

Figure 5-15 A Simple Schedule with Two Parallel Activity Streams

It is a recommended practice that schedule margin, based on risks, duration uncertainty, and historical norms, be clearly identifiable when included within the IMS. Schedule margin may also be referred to as "schedule contingency." The program or project manager owns and controls the schedule margin to the extent designated in his or her Management Agreement. Factors that may contribute to determining the amount of schedule margin are: a) expert judgment, b) rules of thumb, c) percentage of overall program or project (or activity) duration, and d) calculation by the expected value of risk impacts or through insight gained from a probabilistic schedule risk assessment. For clarification, it should be understood that schedule float (slack), which is a calculated value based on network logic, should not be considered as schedule margin. The analyst should consider schedule margin and slack separately to be vigilant against double counting risk.

activities and one with two activities, converging on a single integration activity. Once that integration activity is complete, the project is complete.

The schedule is logically linked, meaning that the predecessors and successors can be seen for every task. The project's milestone, in this case Project End, is linked into the schedule network, allowing an understanding of how the completion of that milestone is impacted when the duration of a predecessor changes.

5.7.4.3 Cost Loading

Once a robust schedule that accurately portrays the project work flow is available, the schedule can be cost-loaded. "Cost loading" refers to mapping cost to schedule. The cost effort for each activity needs to be loaded in groups of activities. To do this, cost is differentiated into two characteristics: Time Dependent (TD) and Time Independent (TI) costs.

TD costs are associated with program or project effort that is based on the duration of an activity. In cost estimating vernacular, TD costs are sometimes called "Fixed Costs" in that their periodic (i.e., daily, monthly, quarterly, and annual) values are fixed in nature and the resulting total cost is the total duration multiplied by the appropriate periodic value (burn rate). Many activities on a program or project display this behavior. Common examples are rent, utilities, facility maintenance, sustaining operations, program management, system engineering, quality assurance, other periodic fixed expenses, and other activities that display a level-of-effort (LOE) nature.

TI costs are associated with the total effort required for an activity, without regard for overall duration. This term refers to the behavior of the cost type and not to any impact that the costs have on time; in fact, for TI costs, the causal relationship is inverse to TD costs. The overall duration of TI costs is primarily a factor of three variables:

- Scope of work to be conducted;
- Productivity of the staff performing the work; and
- Achievable staffing level based on resource and fund availability.

Thus for TI elements, the overall duration of the task is determined by the effort required for its completion and the costs are not a function of time but rather scope; while for TD elements, cost is a direct function of duration. Many activities on a program or project display TI element behavior. Common examples are materials, completion-form tasks, design and development activities, tests, and one-time expenses.

TD costs can actually spread over separate tasks. An example is shown in Figure 5-16.

This example shows two sets of TD costs. One set expands across the entire project, which implies that there is a "standing army" of personnel that will follow the project regardless of where it is in the life cycle (i.e., project management). Another observation is that the two tasks that do not have TI costs still have TD costs, and it shows that these tasks are level-of-effort tasks that are executed by the TD resources or costs.

5.7.4.4 Inclusion of Risk in the Analysis

So far, the schedule represents the baseline plan for the project (cost and schedule). All durations and cost assumptions may have risk mitigation costs and schedule imbedded in the plan, but risk realization from the risk management system has not been incorporated. Traditionally, NASA programs and projects use their risk management system to help populate the risk activities; however, a JCL analysis does not have to be limited by what is currently being managed in the risk management system. For example, there may be a programmatic risk that does not "make it" into the risk management system but is still of concern to the project manager. The JCL analysis allows the project to model the programmatic consequences and expected value of these risks.

Figure 5-17 demonstrates how discrete risks are incorporated into the system. From a schedule perspective, a risk event is treated the same way as an activity; however, in the schedule, the risk event activity only occurs within a certain amount of time. Capturing risks and adding them into the schedule introduces the first probabilistic aspect of a JCL. From a static viewpoint, it looks like the risk is just an activity; but when simulations begin, the risk event will only occur x percent of the time. When the risk event does not occur, the activity and associated dollars will essentially default to zero; however, when the risk does occur, the activity takes on a duration and dollar impact. The duration impact when the risk occurs can be considered the duration consequence of that risk. There may be only TI associated costs with the risk. These costs would be the direct cost impact of the occurring risk. The duration impact of the risk affects the start date of the successor task. This impact could cause the timeframe of the TD costs on the bottom to expand. This potential expansion captures the indirect risk dollars associated with the discrete risk. When a project identifies risks for a JCL analysis, it is important that it identifies the activities that the risks affect, the probability of occurrence of the risk, and the consequence (in both direct cost and direct schedule) of the risk happening. Having a quality schedule with tasks that are linked logically is key to a successful JCL.

Risk is included in the JCL calculation by describing the uncertainty for each activity (for example, a triangle distribution showing optimistic, pessimistic, and most likely values for the cost and schedule inputs), and also by including discrete risks by making use of known liens and threats. Liens and threats come under the category of "known unknowns." They are currently causing an impact on the project or are anticipated, though the full cost may not yet be known. Some examples of liens are: workforce levels that are not adequate to meet the schedule, additional tasks added to the development process, rework of failed components, replacement of damaged hardware, and additional testing. Threats are events that have a potential negative impact on the project cost and schedule that may happen and can be considered based on the probability they will occur. The primary difference between liens and threats is that liens are happening or expected and threats have a lesser probability of happening. Examples of a threat include: cost impact associated with potential failed tests, failed technology development and design changes, or potential launch vehicle changes/impacts. Threats and liens are entered into and managed in a program or project's risk management and budget systems, usually with an associated probability.

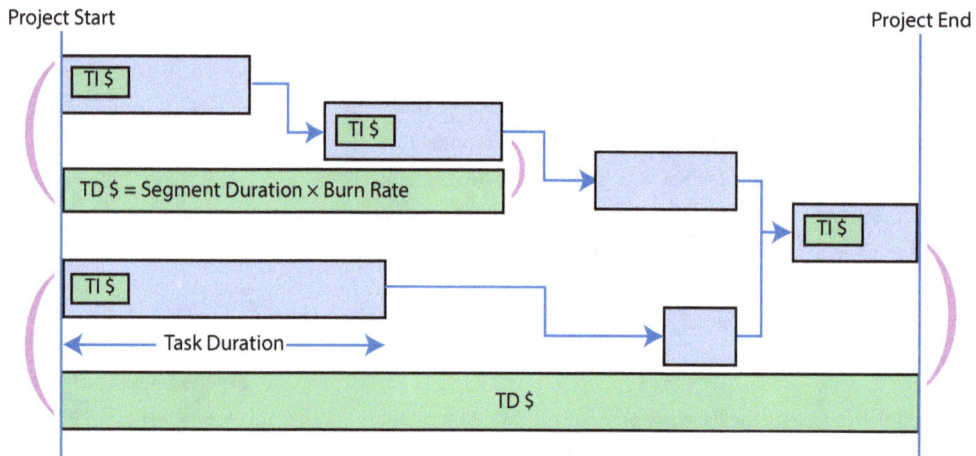

Project Start   Project End

TI $

TI $

TD $ = Segment Duration × Burn Rate

TI $

TI $

←———— Task Duration ————→

TD $

Legend: ▣ Tasks with both TI and TD costs   ☐ Task with only TD costs   ☐ Rolled-up TD costs

Notes: Time-Dependent Cost (TD $) is equivalent to task duration × burn rate; this increases if the schedule slips (e.g., level-of-effort tasks and "standing army" costs). Time-Independent Cost (TI $) does not change if the schedule slips (e.g., materials).

Figure 5-16 Example of Costs Spread Over Separate Tasks

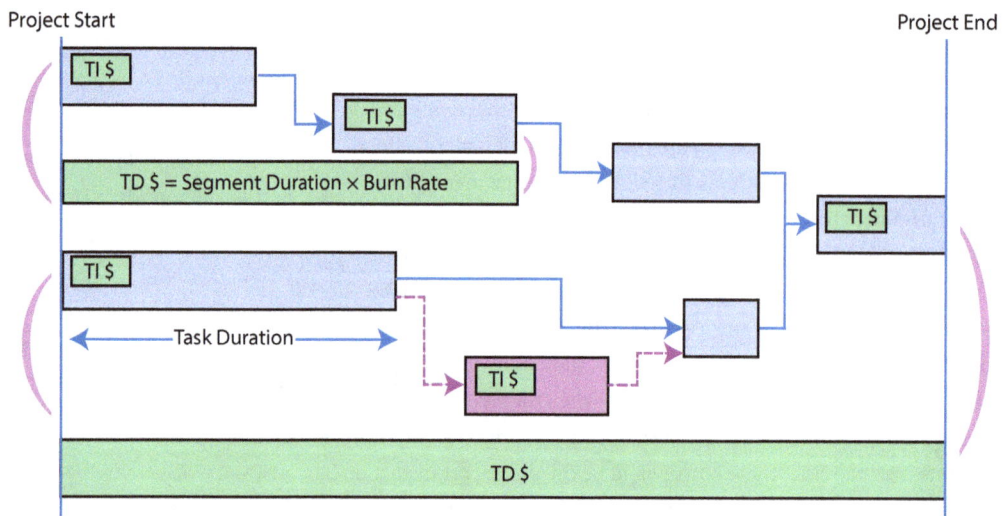

Project Start   Project End

TI $

TI $

TD $ = Segment Duration × Burn Rate

TI $

TI $

←———— Task Duration ————→

TI $

TD $

Legend:

▣ Tasks with both TI and TD costs   ☐ Task with only TD costs

☐ Rolled-up TD costs   ▣ Discrete risk modeled as an activity, with a defined probability of occurrence

Notes: Time-Dependent Cost (TD $) is equivalent to task duration × burn rate; this increases if the schedule slips (e.g., level-of-effort tasks and "standing army" costs). Time-Independent Cost (TI $) does not change if the schedule slips (e.g., materials).

Figure 5-17 Demonstration of How Discrete Risks are Incorporated into the System

5.7.4.5 Implementation of Uncertainty Analysis

The next step in performing a JCL is identifying and implementing the uncertainty.

Up to this point in the JCL process, the primary driver of the JCL results is the quantitative risk assessment and the effect it has on the risk-adjusted cost and schedule. While the risk assessment provides a snapshot in time of potential future events that may cause the project to overrun, it does not account for two key facets that have the ability to drive cost and schedule:

- **Unknown-unknowns**—Although NASA's Continuous Risk Management (CRM) process aims to create as comprehensive a risk register as possible, it is not feasible to predict all events that could possibly increase cost or schedule.

- **Uncertainty in the baseline estimate**—Disregarding risks altogether, it is impossible to precisely predict the time or budget required to complete various segments of space-vehicle research, development, and production.

Recognizing these two facets, JCL analysts need to account for uncertainty in their baseline cost and schedule plans.

Risk and uncertainty are distinct inputs to the JCL model. The two terms overlap. The indefiniteness about a project's baseline plan is partially caused by risks to the project. In traditional, input-based cost-risk analysis, discrete risks are not included as inputs since they would likely cause double counting when uncertainties in the technical inputs and cost outputs are accounted for. In JCL analysis, however, risks from the project's risk register are modeled alongside uncertainties applied to the baseline plan. This is done to increase the usefulness of the JCL analysis to a project manager: being able to discern the effect each risk has on a project's cost and schedule allows for the development of risk mitigation plans.

To avoid double counting, JCL analysts need to segregate uncertainty caused by risks already being modeled in the JCL simulation from the underlying uncertainty of the project's plan once these risks have been discounted. Although this segregation can never account for all aspects of double counting, the benefit to project managers of seeing their risk outweighs the potential for slight errors in the analysis. History and experience has shown that the variance in a typical JCL model is driven significantly more by the uncertainty inputs than the discrete risks. With this said, it is essential to consider uncertainty when conducting a JCL analysis.

There are various methods for selecting and applying cost and schedule uncertainty distributions to the JCL model. Typically, uncertainty is modeled using a three-point estimate. The low value represents the low

For JCL analysis, risk and uncertainty are defined as follows:

- Cost or Schedule Risk—A scenario that may (with some probability) come to pass in the future causing an increase in cost or schedule beyond a project's plan.

- Uncertainty—The indefiniteness about a project's baseline plan. It represents the fundamental inability to perfectly predict the outcome of a future event.

extreme of the cost or duration associated with the uncertainty, the middle value represents the "most likely" value, and the high value represents the high extreme. Although the baseline plan may not be any one of these numbers (low, middle, or high), it needs to be within the range of low and high. (Refer to Figure 5-18 for a visual representation.)

5.7.4.6 Visualization and Results

The process shown in Figure 5-18 is considered iterative. However, at any point in the JCL iteration process, the final and key step is interpreting the results of the analysis. Although an exhaustive list of possible output reports is not shown for brevity purposes, it is important to explain the most commonly used JCL chart, the scatter plot. A JCL calculation result, commonly referred to as a scatter plot, is often graphically depicted as shown in Figure 5-19.

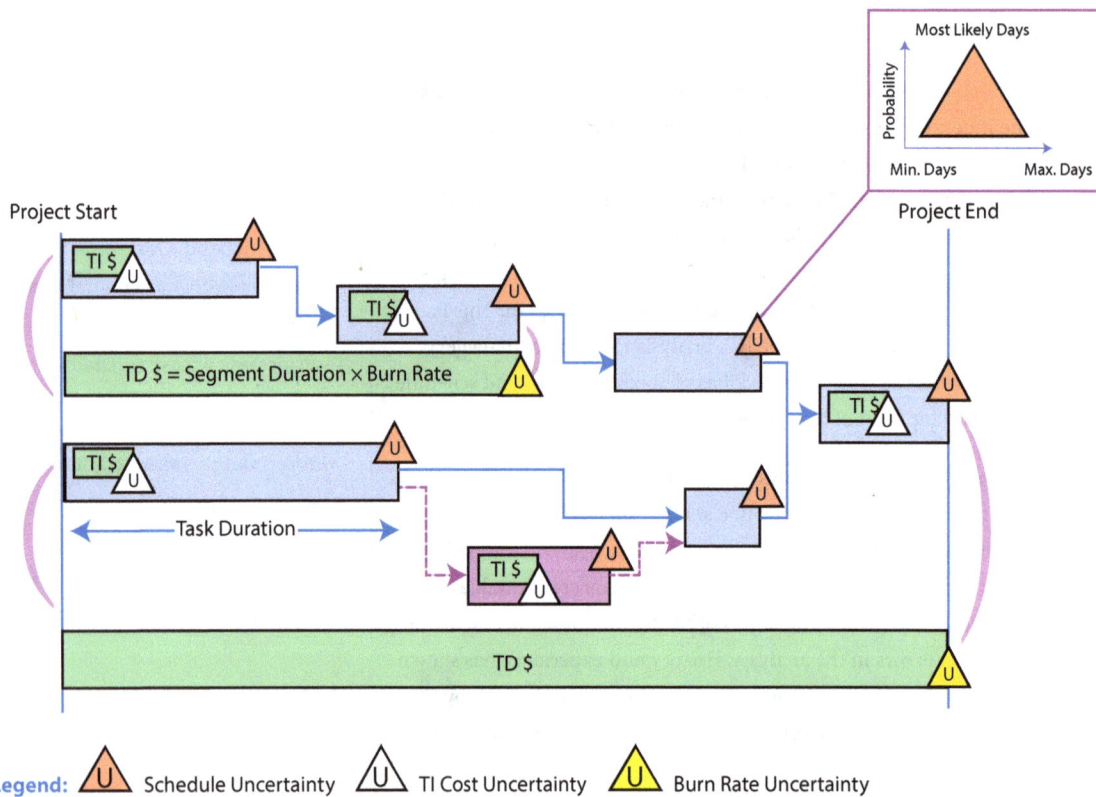

Legend: Schedule Uncertainty   TI Cost Uncertainty   Burn Rate Uncertainty

Notes: Time-Dependent Cost (TD $) is equivalent to task duration × burn rate; this increases if the schedule slips (e.g., level-of-effort tasks and "standing army" costs). Time-Independent Cost (TI $) does not change if the schedule slips (e.g., materials).

Figure 5-18 Visual Representation of an Uncertainty Model Using a Three-Point Estimate

🤔

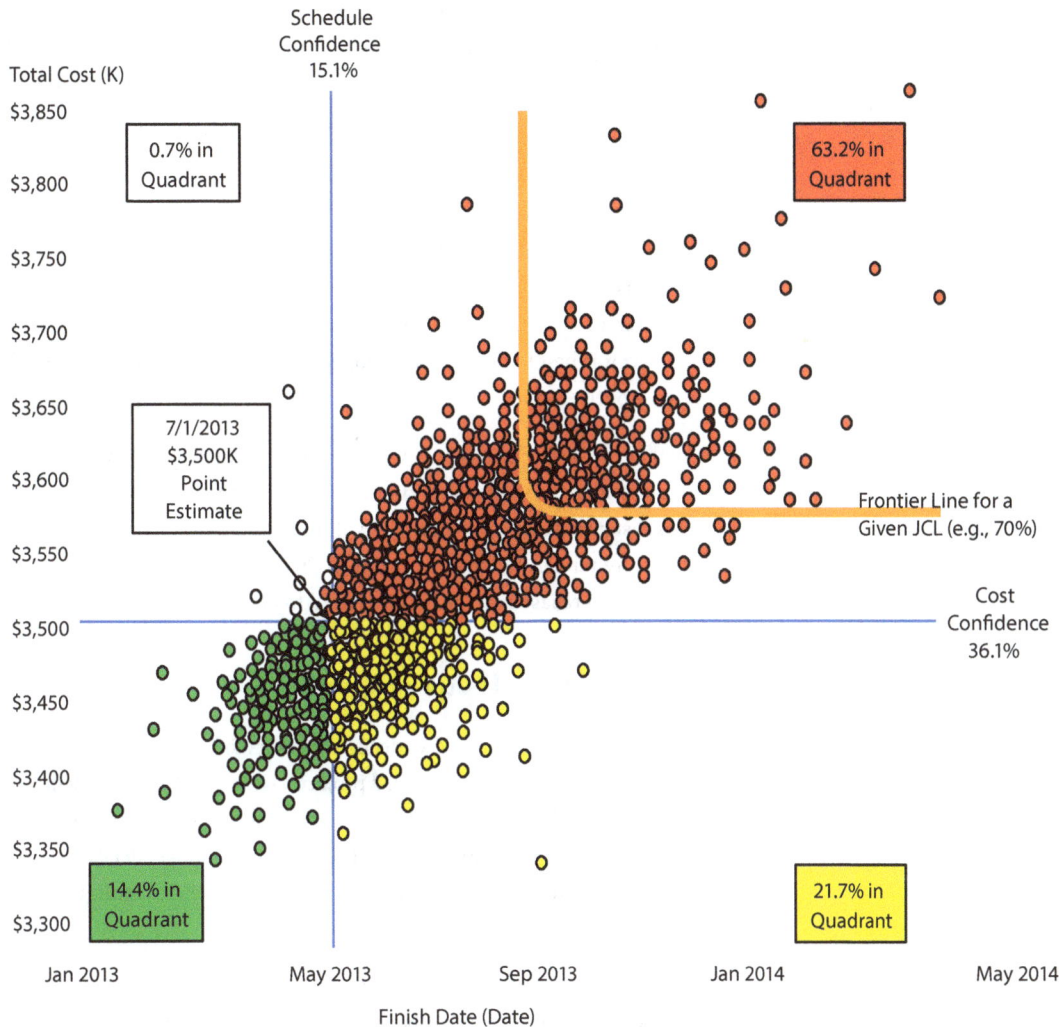

Figure 5-19 JCL Calculation Result, or Scatter Plot

The scatter plot shows iterations of cost and schedule risk analysis. Each dot in the scatter plot represents a specific result, or scenario, from the simulation calculation (cost and schedule). The x-axis represents the final completion date, and the y-axis represents the final cost through that completion date. In this example, the blue lines (the cross-hairs) intercept at the project's point estimate (baseline plan). To the bottom left, the red dots represent all the scenarios that are at or below the baseline cost and schedule. If the red dots are divided by the total number of dots, the result would be 19.6 percent of the dots are within cost and schedule—or put another way, a JCL of 19.6 percent. The cross-hairs can be moved to another date and cost to obtain the JCL for that combination. The horizontal bar of the cross-hairs indicates the cost confidence level, whereas the vertical bar of the cross-hairs indicates the schedule confidence level.

The yellow line in Figure 5-19 represents the "frontier line" or indifference curve, which specifies all the cost/schedule combinations that will meet a targeted JCL. In this example, the frontier curve represents a JCL of 50 percent. As a cautionary note, the asymptotic tails shown are purely academic; it is best to be as close as possible to the centrode of the cluster for a given frontier curve.

Note that the scatter plot is only valid for the current project baseline plan and is considered a snapshot in time. Changes to the project baseline plan due to cost growth or a schedule slip will fundamentally change the project's risk posture rendering the JCL invalid. If changes to the project's ABC result in the need to rebaseline, the JCL will need to be recalculated.

5.7.5 Unallocated Future Expenses

The development of a JCL allows decision makers to better understand the probability of success for a proposed project baseline and enables them to visualize the amount of risk that they are being asked to take with the proposed baseline cost and schedule. They can make budget decisions considering the individual risks and the context of the risk within the entire portfolio of programs.

Any reductions to the UFE will reduce the ability of the project to achieve its cost and schedule targets. When the UFE is a product of the probabilistic JCL analysis, any reduction in the UFE will reduce the probability of achieving the project cost and schedule targets in a manner that can be explicitly quantified. The UFE approach typically results in a more informed dialog between both external and internal decision makers and the project.

For programs and projects that are not required to perform probabilistic analysis, the UFE should be informed by the program or projects unique risk posture in accordance with Mission Directorate and Center guidance and requirements. The rationale for the UFE, if not conducted via a probabilistic analysis, should be appropriately documented and be traceable, repeatable, and defendable.

5.8 Federal Budgeting Process

NASA's program and project budget planning process is shaped by the Federal budgeting process. There is only one job that Congress must do every year, and that is appropriate the Federal budget, per the Constitution. (See Figure 5-20 for an example of the Federal budget cycle.)

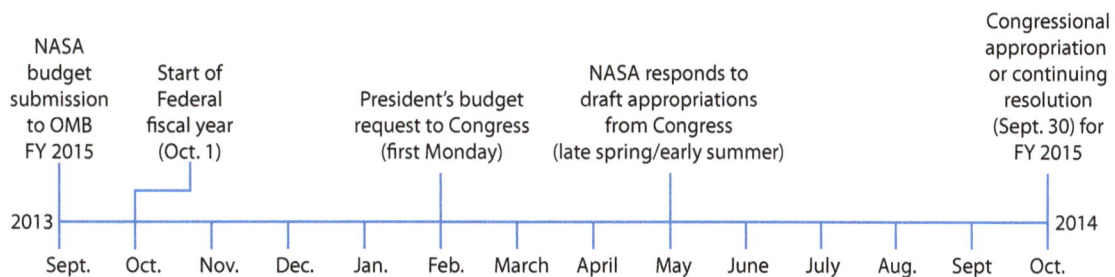

Figure 5-20   Example of the Federal Budget Cycle for FY 2015 Budget

The Federal budgeting process sets national priorities for the money the Government spends. Much national policy, and hence priority, is demonstrated by the President's and Congress's response to different items in NASA's budget request.

The Federal budgeting process can be seen as a one-year process. It starts with the delivery of the President's Budget Request (PBR) to the Congress and ends with the appropriation provided to the Federal agency. But the budget process at NASA starts well before that. To get to the President's Budget Request, there is a long process both at the Agency and at the Office of Management and Budget (OMB).

The Executive Branch (the President and Federal agencies) submits a budget request to Congress on or before the first Monday in February. The President's Budget Request includes funding requests for all Federal agencies and cabinet departments for the coming Federal Fiscal Year (FY), which begins on October 1 after Congress appropriates funds. NASA submits its portion of the PBR to OMB in September, preceding the President's February budget submittal. So NASA submits its budget request in September for the fiscal year that starts the following October 1. (For example, in September 2015 NASA submits its portion of the PBR that starts October 2016 for FY 2017.)

The big external drivers to NASA's budget process are:

- Producing a budget request to go into the PBR to Congress in February and
- The appropriation of funds every October 1st to begin a new fiscal year.

Without the annual appropriation of funds from Congress to do what it is authorized to do (whether by continuing resolution or appropriations bill), NASA would shut down. The Federal appropriations timeline calls for the final conference on appropriations in September after Congress' recess in August. All appropriations bills are to be signed before the October 1 beginning of the Federal fiscal year. In recent years, however, the final budget has not passed on time. If Congress reaches the start of the fiscal year without a budget in place, it usually passes a Continuing Resolution (CR) that temporarily funds the Federal Government at the level of the previous fiscal year. If Congress has not passed a final budget and does not pass a CR, the Federal government shuts down. The program or project manager needs to consider the possibility of a CR, which might mean working to the previous year's funding. This may result in decreased funding, or in funding provided later in the fiscal year than planned. A CR, especially one that lasts an entire year, may be particularly problematic for a program or project that was planning on increased funding, e.g., a program or project transitioning from Formulation to Implementation. Depending on the magnitude of the decrease in funding, or the length of the delay in funding, the Decision Memorandum and Management Agreement may need to be renegotiated and amended.

5.8.1 NASA's Interface with the Federal Budget Process

NASA's budget planning process takes into account that at any given point in time, NASA is involved in multiple budget years. Each winter (January/February) just preceding the release of the President's Budget Request from the White House to the Congress, all Federal agencies develop funding requirements for work that will be performed two fiscal years in the future. For example, in January 2016, Federal agencies focus on working on the budget request that will be submitted in September 2016 for funds for the fiscal year that starts October 2017 (FY 2018). That is two years ahead of the current "year of execution" or "performance year," which started October 2015. So, while work is being executed during FY 2015 (October 1, 2014, to September 30, 2015), NASA starts work in January 2015 to develop the budget request they will submit in September 2015 that goes into the PBR submitted in February 2016 for FY 2017. NASA's internal processes and products are aligned with this Federal cycle and justifying that request to Congress.

The full NASA budgeting process is the Planning, Programming, Budgeting, and Execution (PPBE) process. The PPBE process takes into account differing time spans, the complex interactions of external and internal requirements, external and internal assessments, and the specific needs of a multifaceted organization. The full PPBE process is explained in *NPR 9420.1, Budget Formulation*. (See also Section 5.8.3.)

5.8.2 OMB Passback

NASA's budget planning process has a 5-year horizon. The planning process starts with the OMB passback for the previous fiscal year and covers the budget year and four additional outyears.

Each year, OMB provides guidelines on the content of NASA's proposed budget through the passback in late November. The passback gives Federal agencies guidance on what the White House will and will not accept for inclusion in the President's Budget Request. NASA manages projects across their multiyear life cycle, but for the budget cycle, it submits a one-year request plus four years out. OMB comments on the single year but is also sensitive to the full program or project life cycle. In the passback, OMB provides control numbers for NASA budget accounts for a 5-year span. In the next budget cycle, these control numbers provide the starting point for the new budget development cycle.

As shown in Figure 5-21, when NASA receives the passback from OMB in November 2010, the Agency is engaged in three phases of the budget planning cycle in parallel:

[1] Includes planning for four out years (i.e., 2013–17).

[2] This request will reflect any change to the trajectory from previous year.

Figure 5-21 Simultaneous Multiyear Budget Process

- Spending the money appropriated for the execution year FY 2011 (October 2010 to September 2011),

- Negotiating the budget for NASA that will appear in the PBR to Congress in February 2011 (for FY 2012), and

- Planning the budget for FY 2013 and the four out years beyond that.

The process involved to submit NASA's budget request to OMB in September until receiving the actual funding appropriated by Congress in October of the following year is a year and a half, longer if the appropriation is delayed. The amount of the budget request may be altered by other considerations at any point in the process. When the appropriated funds are received, any difference in the amount appropriated from what was requested requires an immediate revision to Operations Plans,[16] which impacts the budget request being negotiated for the next year.

5.8.3 PPBE Process

The NASA PPBE process consists of four phases: Planning, Programming, Budgeting, and Execution. Figure 5-22 shows an overview of this process. The following sections provide a high-level view of each of these phases. For more information refer to *NPR 9420.1, Budget Formulation*.

5.8.3.1 Planning

NASA's PPBE process starts with planning. All of NASA's budget planning flows from NASA's strategic mission planning with the goal of acquiring/procuring the funding to either start or continue working on NASA's mission, programs, and projects and their supporting capabilities and infrastructure. Setting strategy is an iterative, interactive process where mission ideas are rolled up into goal statements and feed resource requests, and strategy is translated down into programs and projects to execute the mission. Mission planning precedes resource requests and detailed planning follows resource allocation at both the mission level and the program or project level. In addition to the programmatic planning, the institutional

---

[16] The Congressional Operating Plan (COP), and Agency Operating Plan (AOP) are used as the basis for ensuring that appropriated funds are used in compliance with Agency intent and Congressional mandates. The COP sets forth a high-level plan for how NASA intends to apply Agency financial resources during the fiscal year to fulfill its mission. Typically, the COP is at the program level. While not subject to statutory controls, the COP establishes a common understanding between NASA, OMB, and the Congress. The AOP is an internal plan based on the COP that provides greater detail and includes all programs and projects. When Agency programs and projects are changed or when new requirements become known, the AOP must be revised to reflect the new direction. If the change exceeds the limitations established in the current COP, NASA must submit a new plan to Congress.

Planning	Programming	Budgeting	Execution
Internal/External Studies and Analysis	Program and Resource Guidance	Programmatic and Institutional Guidance	Operating Plan and Reprogramming
NASA Strategic Plan	Program Analysis and Alignment	OMB Submit	Funds Distribution and Control
Annual Performance Goals	Agency Issues Book	President's Budget	Analysis of Performance
Implementation Planning			Reporting Requirements
Strategic Planning Guidance	Program Decision Memoranda	Appropriation	Performance and Accountability Report

Legend: ☐ Steps during Planning Phase   ☐ Steps during Programming Phase
☐ Steps during Budgeting Phase   ☐ Steps during Execution Phase

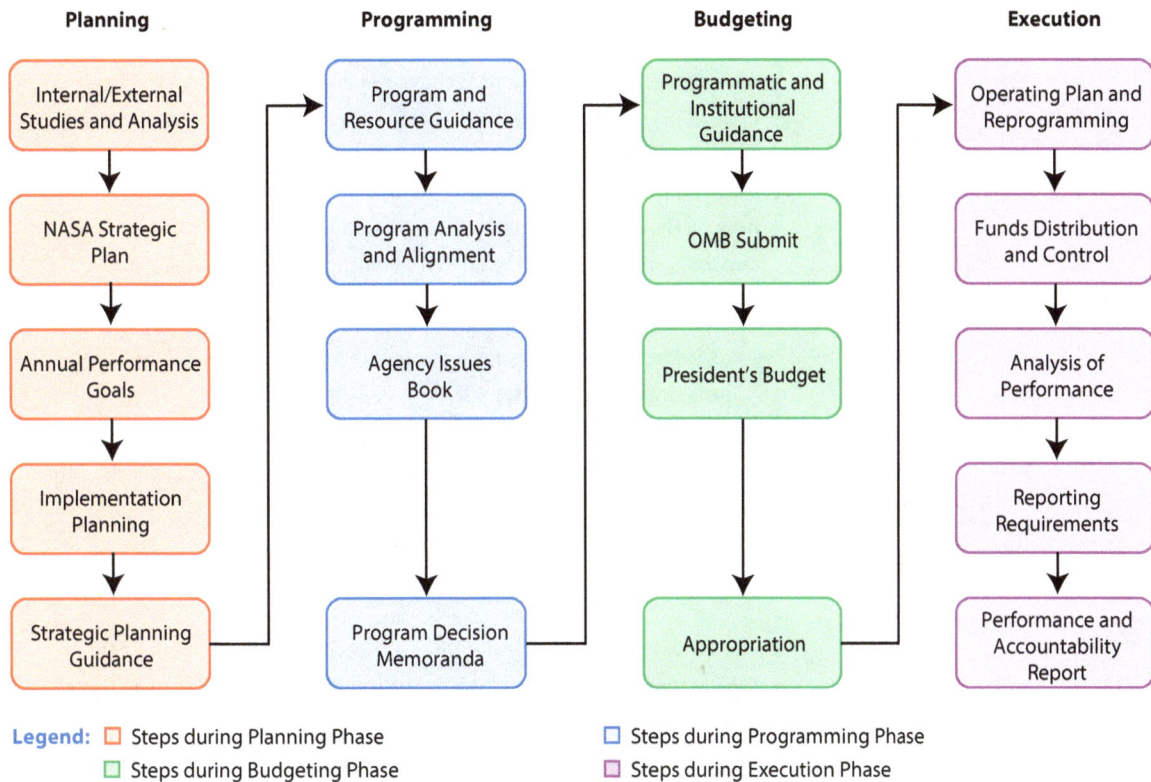

Figure 5-22 Annual PPBE Phases and Steps

side of the Agency plans what capabilities will be needed to support the mission.

At the highest level, strategic planning produces the NASA Strategic Plan (NPD 1001.0), which guides all other Agency planning. It is updated every three years and defines NASA's vision and the strategic goals that support, drive, and justify NASA's mission execution and research and development activities.

The Agency-level Strategic Implementation Planning (SIP) process guides specific budget and acquisition decisions. The SIP process allows the evaluation, short-term assessment, and long-term alignment of issues such as the appropriate application of White House priorities, Agency strategic planning, and new initiatives in a portfolio of programs and projects in the context of budget availability. The SIP process is implemented through select reviews conducted at the direction of the NASA Administrator, resulting in guidance to inform the strategic acquisition process. This guidance is incorporated into the Strategic Programming Guidance (SPG) and further applied to decisions made in the Agency Acquisition Strategy Meeting

(ASM). The SPG is produced annually and consolidates all the strategic information that will be used to develop the NASA budget and allocate resources across the Agency. Agency-level planning also includes the development of the Operations Plans that are generated after NASA receives its appropriation. These plans adjust resources in the current execution year based on the funding actually appropriated. The SPG is translated into planning on the Mission Directorate, mission support, and institutional level and into program- and project-level planning to execute the mission:

- Planning at the Mission Directorate level develops input for the Strategic Plan and supports resource allocation to the Mission Directorate's programs over their life cycle.

- Planning at the mission support and institutional level includes the infrastructure necessary to execute programs and projects over their life cycles.

- Program- and project-level planning encompasses all life-cycle planning done by programs and projects to support the execution of their mission.

The annual Strategic Programming Guidance (SPG) is the official, Strategic Management Council (SMC)-controlled high-level guidance for use in developing the Agency's portion of the PBR. The SPG includes both programmatic and institutional guidance, consolidating the information from the Strategic Plan, existing implementation plans, priorities, studies, assessments, and performance measures. Publication of the SPG officially kicks off the process whereby NASA builds the Agency's budget request to OMB and Congress and the subsequent management of resources allocated to programs and projects. The SPG consolidates all relevant strategic guidance for developing a programmatic and financial blueprint for the budget year plus four out years.

The SPG provides uniform strategic guidance for all involved in the budget process. This includes the Control Account Managers (CAMs) or managers within the program or project with responsibility to manage the inputs to the NASA budget process and directors of Mission Support Offices or Administrator staff offices with cross-cutting responsibilities that address the institutional infrastructure. (See Appendix A for a definition of Control Account Manager.)

OCFO manages the SPG development, which begins after the OMB passback for the prior budget year and is finalized after completion of the President's budget in early February. (See Figures 5-21 and 5-23.) The SPG is developed with the input and involvement of the Mission Directorates, Mission Support Offices, and Centers.

The SPG provides high-level funding and civil service (Full-Time Equivalent (FTE)) control totals by Center. The development of the SPG is roughly concurrent with the issuance of the Programming and Resource Guidance (PRG), which is another key piece of guidance needed for the programming phase of the PPBE process.

5.8.3.2 Programming

The programming phase of the PPBE process involves the analysis and strategic alignment of mission, constraints, and resources. This phase starts with the development of the following products:

- Programming and Resource Guidance (PRG), which translates the SPG guidance into programmatic guidance more relevant for the program or project managers and the Centers.
- Program Analyses and Alignment (PAA), which converts strategy into resourced programs/projects. The CAMs identify what their programs/projects intend to accomplish, identify any surplus or deficit capabilities and capacities, and identify the impact of funding reductions or any need for funding increases. The PAA is completed in mid-May.

Then Centers have an opportunity to analyze the SPG, PRG, and PAA information to determine possible institutional infrastructure issues. Any issues will be raised with the SMC through the Issues Book for decision before the budgeting phase begins. This step begins in mid-May and is completed in early June:

- Program Review/Issues Book reviews all previous guidance, inputs, analysis, and issues to identify critical issues that need to be brought to the SMC for a decision.
- The Decision Memorandum reflects the Executive Council (EC) decisions on the issues that were discussed at an SMC. The decisions document resource levels and FTE control totals for subsequent development of the budget.

5.8.3.3 Budgeting

In the budgeting phase, the OMB budget submission is developed under the guidance of the Office of the Chief Financial Officer (OCFO):

- **Programmatic and Institutional Guidance (PAIG)**—CAMs allocate resources at the project-level detail necessary for Centers to begin formulating the NASA full-cost budget.
- **OMB Budget**—CAMs develop the OMB budget submission under the guidance of the OCFO. This is the first step in the PPBE process in which

information is distributed outside of NASA. However, it is still predecisional data and is provided to OMB only.

- **President's Budget**—The OCFO coordinates with the Mission Directorates their responses to OMB questions on the budget submission, coordinates hearings with Mission Directorates, receives and responds to the OMB passback (OMB's formal response to the NASA budget submittal), and works appeals and settlement. Then the OCFO also manages the development of input to the PBR documents. The PBR, also known as the Congressional Budget Justification, is the annual NASA budget document that includes budget estimates at the program and project level, description and justification narratives, performance data, and technical descriptions.

- **Congressional Appropriation**—As discussed previously, this phase concludes with NASA receiving the resources and legislative guidance and adjusting its Operations Plans as necessary to respond to differences from the original budget request.

5.8.3.4 Execution

The execution phase in the budget process involves the implementation of the plans with associated monitoring, analysis, and control. In the context of programs and projects, execution is conducting the authorized work in accordance with the applicable 7120 series NPR.

5.8.4 Linkage Between Life Cycles

Figure 5-23 illustrates the points of connection between the program and project life-cycle planning and the budgeting cycle. (See Section 5.8.3.1 for more information on the relationship between program and project planning and the PPBE planning phase.) The program and project life cycle is not tied to a specific timeline but evolves with the development of the concept, mission, and technology. The program or project manager carries the new program or project idea forward along this cycle. The budget timeline is tied to specific annual events. Procurement personnel such as the CAMs and the Resource Management Officer (RMO) take the program or project forward along this timeline. (See Appendix A for a definition of RMO.) Points of synchronization include the initial conception: the program or project is vetted at the first Agency-level strategic planning meeting, which feeds into the SPG in January. There is also formal linkage when the program or project enters the Implementation Phase of the life cycle (KDPI/KDP C). At this point, the program or project is subject to external oversight. The Program Commitment Agreement (PCA) is

Program and Project Life Cycle

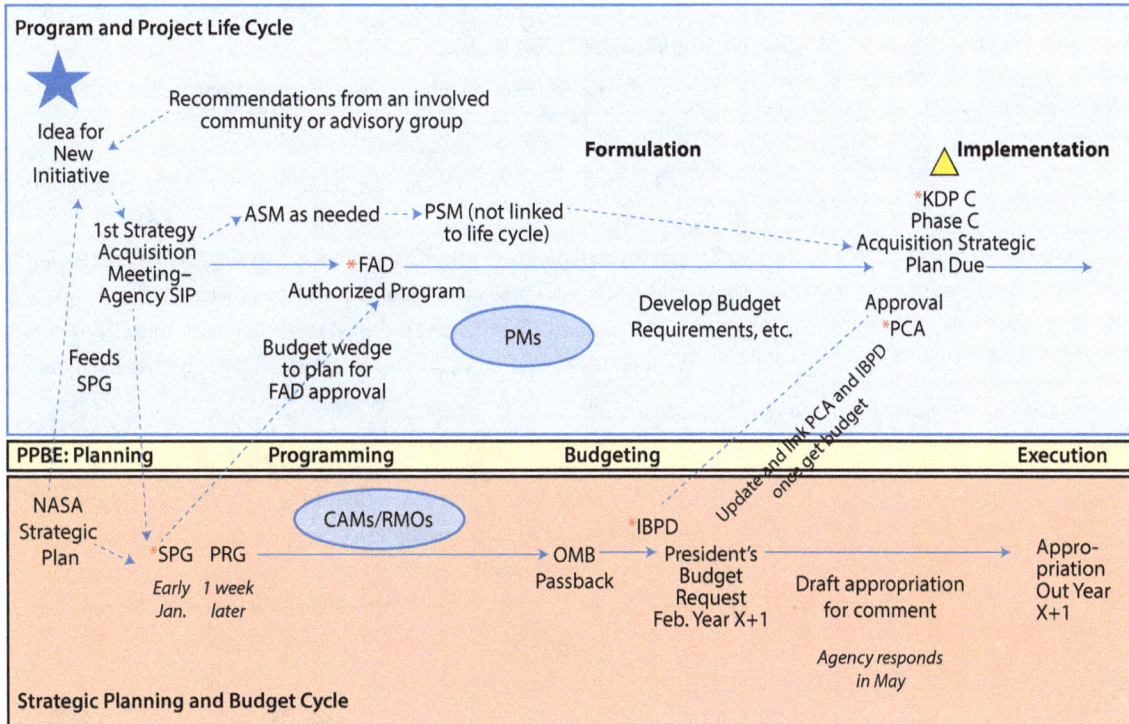

Figure 5-23 Linkages Between the Program and Project Life Cycles and the Budget Cycle

Legend: *   Points of synchronization across communities
→   Timelines
⋯→   Lines of communication/flow of information

Acronyms: ASM = Acquisition Strategy Meeting; CAM = Control Account Manager; FAD = Formulation Agreement Document; IBPD = Integrated Budget Performance Document; KDP = Key Decision Point; OMB = Office of Management; PBR = President's Budget Request; PCA = Program Commitment Agreement; PDM = Program Decision Memorandum; PM = Program and/or Project Manager; PRG = Programming and Resource Guidance; PSM = Procurement Strategy Meeting; RMO = Resource Management Officer; SIP = Strategic Implementation Planning; SPG = Strategic Planning Guide.

approved (or not); the Integrated Budget Performance Document (IBPD) is developed; and the Acquisition Plan is updated.

5.8.5 Program and Project Involvement in the Budgeting Process

Mission Directorates guide program and project involvement in the budget process based on the four general steps a Mission Directorate takes in developing its budget:

- Develop budget guidelines for the Centers (programs and projects);
- Conduct program or project reviews of the Center submittals;

- Develop budget recommendations for the Mission Directorate Associate Administrator (MDAA); and
- Develop Mission Directorate budget recommendations for the NASA Administrator.

The Mission Directorate develops budget guidelines for the Centers in accordance with Agency-level strategic programming guidance. The MDAA defines the overall program priorities and budget strategy for the upcoming process. The Mission Directorate then prepares narrative and numeric guidance to the Centers (programs and projects) consistent with this direction. The final budget guidance is entered into the Agency budget database known as N2. The narrative guidance is usually posted on an Agency-level site where it can be seen by the Centers.

Once the program and projects receive this guidance from their Center financial office, they begin to develop their program or project's submission based on this guidance while incorporating any changes needed as a result of the previous year's performance. Each Center will have a different process for developing budgets, and program and project teams need to work with the appropriate Center staff as directed by the Center management. Depending on unique Center policies, the budgets may be submitted to the Mission Directorate by the Center directly or by the program or project team. A Center may request project teams to submit their budgets through their program office.

Once received by the Mission Directorate, Mission Directorate personnel conduct reviews of Center program or project submittals. These assessments may include an on-site program or project review and may occasionally include visits to contractors and other facilities. Data from the formal Center budget submittals combined with the information garnered from the program or project reviews are used to identify and resolve issues. Issues may include variances in the budget relative to the guidelines, milestone changes, technical problems, contract or subcontract growth, and UFE status. These issues form a basis for further investigation and analysis. Programs, projects, and Centers may be asked to provide additional options to resolve the issues.

Once all issues are resolved, Mission Directorate personnel develop budget recommendations for the NASA Associate Administrator, who then submits them to the NASA Administrator.

5.9 The Work Breakdown Structure and Its Relationship to Agency Financial Processes

The Work Breakdown Structure (WBS) is a key element of program and project management. The purpose of a WBS is to divide the project into manageable pieces of work to facilitate planning and control of cost, schedule, and technical content. NPR 7120.5 requires that projects develop a product-based WBS in accordance with the Program and Project Plan templates. Figure 5-24 shows the standard template for those space flight projects conducted under the auspices of NPR 7120.5.

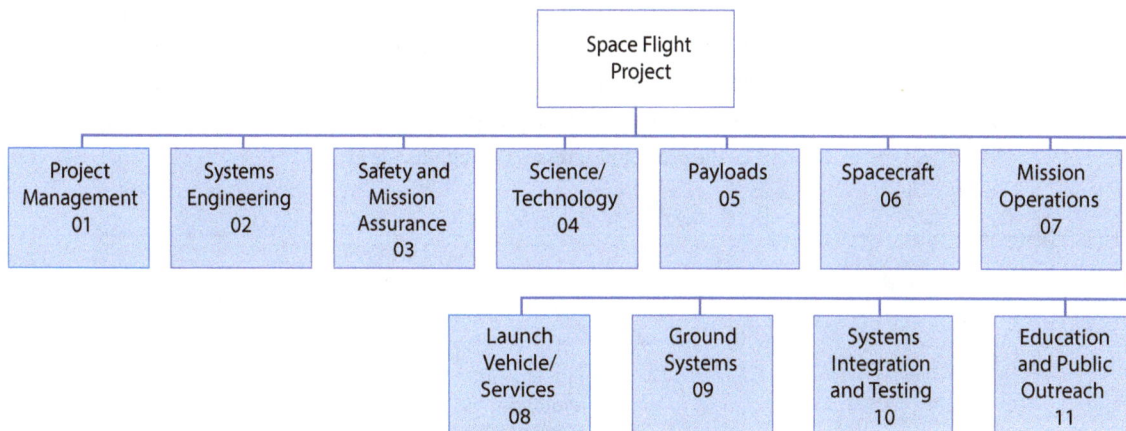

Figure 5-24 Standard Level 2 WBS Elements for Space Flight Projects

The WBS is developed as part of the Formulation activities to characterize the complexity and scope of the project after the FAD is issued at the end of Pre–Phase A. Developing the WBS is part of establishing the internal management control functions. Pre-Formulation activities are typically initiated by Mission Directorate Associate Administrators (MDAAs) or Center Directors or sometimes a program office and are not formally part of Formulation. Initial resources for pre-Formulation activities like Pre–Phase A concept studies are usually provided by the initiating organization and are not included in the Life-Cycle Cost (LCC), nor do they have their own unique project-level WBS element.

The WBS is a product-oriented family tree that identifies the hardware, software, services, and all other deliverables required to achieve an end project objective. The WBS then consists of the product tree plus the other enabling activities such as project or element management, systems engineering, safety and mission assurance, and others as necessary for completing the work. This generic structure can be depicted as shown in Figure 5-25. This structure subdivides the project's work content into increasing levels of detail down to the work package or product deliverable level. The enabling activities can be applied to each of the product layers as needed to fully characterize the major work elements. Developed early in the project life cycle, the WBS identifies the total project work to be performed, which includes not only all NASA in-house work, but also all work to be performed by contractors, international partners, universities, or any other performing entities. All work considered part of the project needs to be represented in the project WBS.

The elements of the project WBS are fundamental elements in many aspects of internal project management control. They form the basis for project funding and are the building blocks for cost estimating and analysis. Starting a project under a logical, accurate, and complete hierarchy that reflects the work of the project facilitates all aspects of project management as the project progresses through its life cycle.

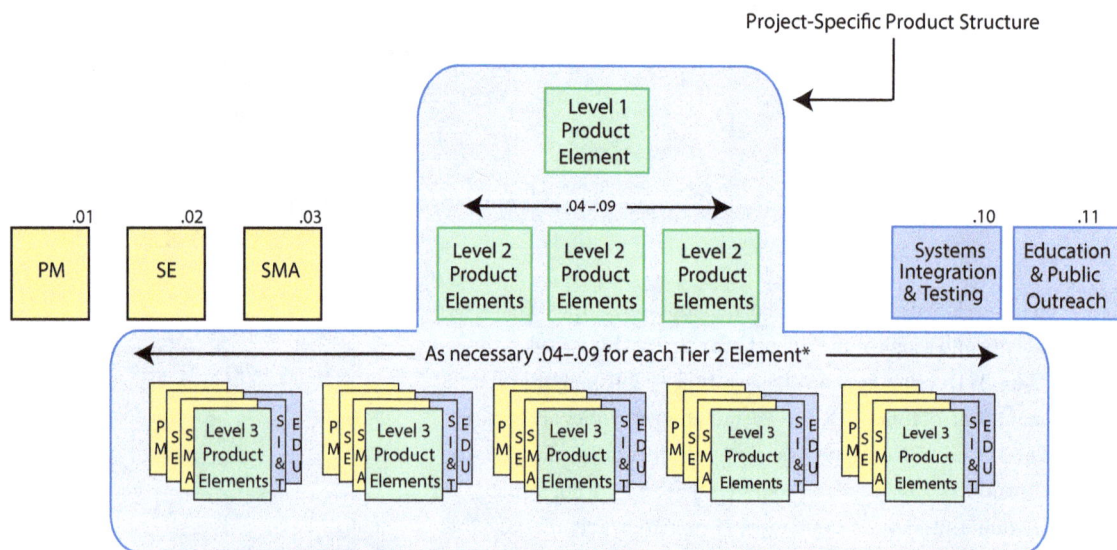

[1] PM, SE, SMA, SI&T, and EDU are as needed for this Product Tier.

Figure 5-25 WBS Structure from Products and Enabling Activities

5.9.1 Developing the Program WBS

A program WBS is a product-oriented hierarchical decomposition encompassing the total scope of the program, and includes deliverables to be produced by the constituent components including projects and activities. The program WBS includes, but is not limited to, program management artifacts such as plans, procedures, standards, and processes, the major milestones for the program, program management deliverables, and program office support deliverables.

The program WBS is a key to effective control and communication between the program manager and the managers of constituent projects: the program WBS provides an overview of the program and shows how each project fits in. The decomposition should stop at the level of control required by the program manager. Typically, this will correspond to the first one or two levels of the WBS of each constituent project. In this way, the program WBS serves as the controlling framework for developing the program schedule, and defines the program manager's management control points that will be used for earned value management, if applicable, as well as other purposes. The complete description of the program WBS components and any additional relevant information is documented in the program WBS dictionary, which is an integral part of the program WBS.

The program WBS does not replace the WBS required for each project within the program. Instead, it is used to clarify the scope of the program, help identify logical groupings of work for components including projects and activities, and identify the interface with operations and products. It is also a place to capture all non-project work within the program office, external deliverables such as public communications, and end-solution deliverables overarching the projects, such as facilities and infrastructure upgrades.

5.9.2 Developing the Project WBS

The subdivisions of work in the project WBS need to reflect a logical, accurate, and compatible hierarchy of work. Level 1 of the project WBS is the name of the project. No Level 1 (the project) element can be put in place without a program above it. Project managers make the Level 2 and below elements correspond to the project products plus other enabling activities necessary for completing the work. Depending on the type of project being conducted, these elements may be required to conform to a standard template. Figure 5-24 shows the standard template for those space flight projects conducted under the auspices of NPR 7120.5. Additional guidance is provided by the *NASA Work Breakdown Structure (WBS) Handbook and NASA/SP-2011-3422, NASA*

The standardization of Level 2 WBS elements for space flight projects is driven by the need for consistency, which enables more effective cost estimating and assessment of project work across the Agency. When the program and project management tools align, it facilitates strategic thinking, increases NASA's credibility in answering Congress, aids program and project management, and enables people to ask the right questions and get an answer. The standard WBS is intended to apply only to space flight projects, not programs.

Risk Management Handbook, which can be found on the Office of the Chief Engineer (OCE) tab under the "Other Policy Documents" menu in the NASA Online Directives Information System (NODIS).

Standard Level 2 elements that are not relevant to a particular project do not need to be used in the project WBS. If project content does not fit into the content of any existing standard Level 2 WBS element, new WBS elements may be requested through OCE and the Office of the Chief Financial Officer (OCFO) through the Metadata Manager (MdM) as part of submitting the WBS. Below WBS Level 2, the subordinate (children) WBS elements (Level 3 and lower) are determined by the project. The Level 3 and lower elements may differ from project to project but need to roll up to the standard WBS dictionary definition of the Level 2 element.

Regardless of structure, all project WBSs have the following characteristics:

- Apply to the entire life cycle of the project, including disposal and decommissioning.
- Support cost and schedule allocation down to a work package or product deliverable level.
- Integrate both Government and contracted work.[17]
- Allow for unambiguous cost reporting.
- Allow project managers to monitor and control work package/product deliverable costs and schedule, including Earned Value Management (EVM) and cost reporting.
- Capture both the technical and the business management and reporting.

An example of a Level 2 and 3 Space Flight WBS is provided in Figure 5-26 for the James Webb Space Telescope (JWST) project.

5.9.3 Space Flight Project Standard WBS Dictionary

When constructing the WBS, a dictionary or explanation of what is included in each of the elements is necessary to ensure all work is accounted for in a consistent and relevant manner. This dictionary should be widely available and understood by all reporting organizations. Examples of standard elements of the dictionary for a space flight project include:

Element 1: Project Management. The business and administrative planning, organizing, directing, coordinating, analyzing, controlling, and approval processes used to accomplish overall project objectives that are

---

[17] Project managers should work with industry/international partners to ensure consistent WBSs.

411672

JWST

.01 PM

.02 SE

.03 SMA

.04 Science/Tech

.05 Integrated Science Instrument Module

.06 JWST Space-craft

.07 Un-used

.08 Launch Vehicle/ Services

.09 Ground Systems

.10 Systems Integration & Testing

.11 Education & Public Outreach

.99 Agency Cost Assessment

.01 PM
.01 GSFC HQ Grants
.02 GSFC PM General
.03 GSFC Project Contractor Support
.04 GSFC Project IT Support
.05 GSFC General Business Support
.06 GSFC Project Office Housing
.07 GSFC Project Support General
.08 JPL JWST IRT Support
.09 JPL JWST Fee
.10 HQ PM
.11 GSFC JWST Historian
.12 ARC JwSt Project Support
.13 GSFC JWST PAO & EPO
.20 Agency Service Pool Ful Cost Assessment
.25 Agency FFS Full Cost Assessment
.50 GSFC GAO Reviews and Support
.88 GSFC Code 400 Taxes
.95 GSFC JWST IT Support to Code 500

.02 SE
01 GSFC Observatory Systems
.02 GSFC Modeling
.03 GSFC Contamination
.04 GSFC Observatory Thermal System
.05 GSFC Observ. Mech System
.06 GSFC Observ. SW
.07 GSFC Observ. OTE SE
.08 GSFC Observ. Electrical System
.09 GSFC Observ. Stray Light System
.10 GSFC Observ. GN&C System
.11 MSFC Plasma Environ. Analysis Support
.12 MSFC Online Materials Database
.13 GSFC JWST SEED Support
.14 MSFC JWST DTA Charging
.15 MSFC JWST Radiation Testing
.16 MSFC Sunshield Coupon Thermal Cycling
.17 GSFC JWST Cryo Shock Risk Mitigation
.18 GSFC Mech Sys West Coast Support
.19 LaRC Mechanical SW Support
.20 MSFC Kapton SC Cryo Arc Flare Testing
.66 GSFC JWST 1 FTE Supplement

.03 SMA
01 GSFC SMA Pool
.02GSFC Project ITA
.03GSFC ISIM ITA Assessment
.04 JWST SRB
.05 JWST Systems Review Board
.06 GSFC JWST Materials
.07 GSFC JWST Parts
.08 JWST SRB & IIRT (JPL)
.09 GRC NRC A3

.04 Science/Tech
01 Science Working Group
.02 GSFC Science Team
.03 Technology Development

.05 Integrated Science Instrument Module
01 GSFC ISIM Management
.02 ISIM SE
.03 ISIM Integration & Test
.04 GSFC ISM Grd Sys I&T Management
.05 NIRCAM Instrument
.06 NIR Spec
.07 MRI
.08 GSFC Fine Guidance Sensor
.09 ISIM Structure Subsystem
.10 ISIM Thermal
.11 ISIM IC&DH and IRSU
.12 GSFC ISIM Flight SW
.13 GSFC ISIM Electrical Harness Development
.14 GSFC ISIM NIRCam Notes
.15 GSFC ISIM OSIM
.16 ISIM Optics
.17 ISIM Electronics Compartment
.18 ISIM Harness Radiator System
.19 GSFC JWST Instrument Detectors
.20 GSFC NIRCAM Post Delivery Support

.06 JWST Space-craft
01 GSFC NGST Prime Contract
.02 GSFC Spacecraft Support
.03 Optical Telescope Element
.04 GSFC ISM Grd Sys I&T Management
.05 NIRCAM Instrument

.08 Launch Vehicle/ Services
01 GSFC Launch Vehicle
.02 KSC JWST Launch Services Support

.06 JWST Spacecraft
.04 XRCF Support (MSFC)
.05 Proton Energy Testing
.06 Cryocooler
.07 ARRA JWST G Contract

.09 Ground Systems
01 GSFC Ground Systems Development

.10 Systems Integration & Testing
01 GSFC I&T/General
.02 GSFC I&T SSDIF
.03 GSFC OPTICS System Eng Verif & Test
.04 GRC JWST Support
.05 JWST Support (JSC)
.06 GSFC JWST Transporter (Wallops)
.07 HQ APL/JWST Shipping Container Support
.08 GSFC JWST OTIS Support
.09 JWST OTIS Support (GSFC

.11 Education & Public Outreach
01 GSFC JWST Education & Public Outreach
.02 LaRC NIA Activity 2934

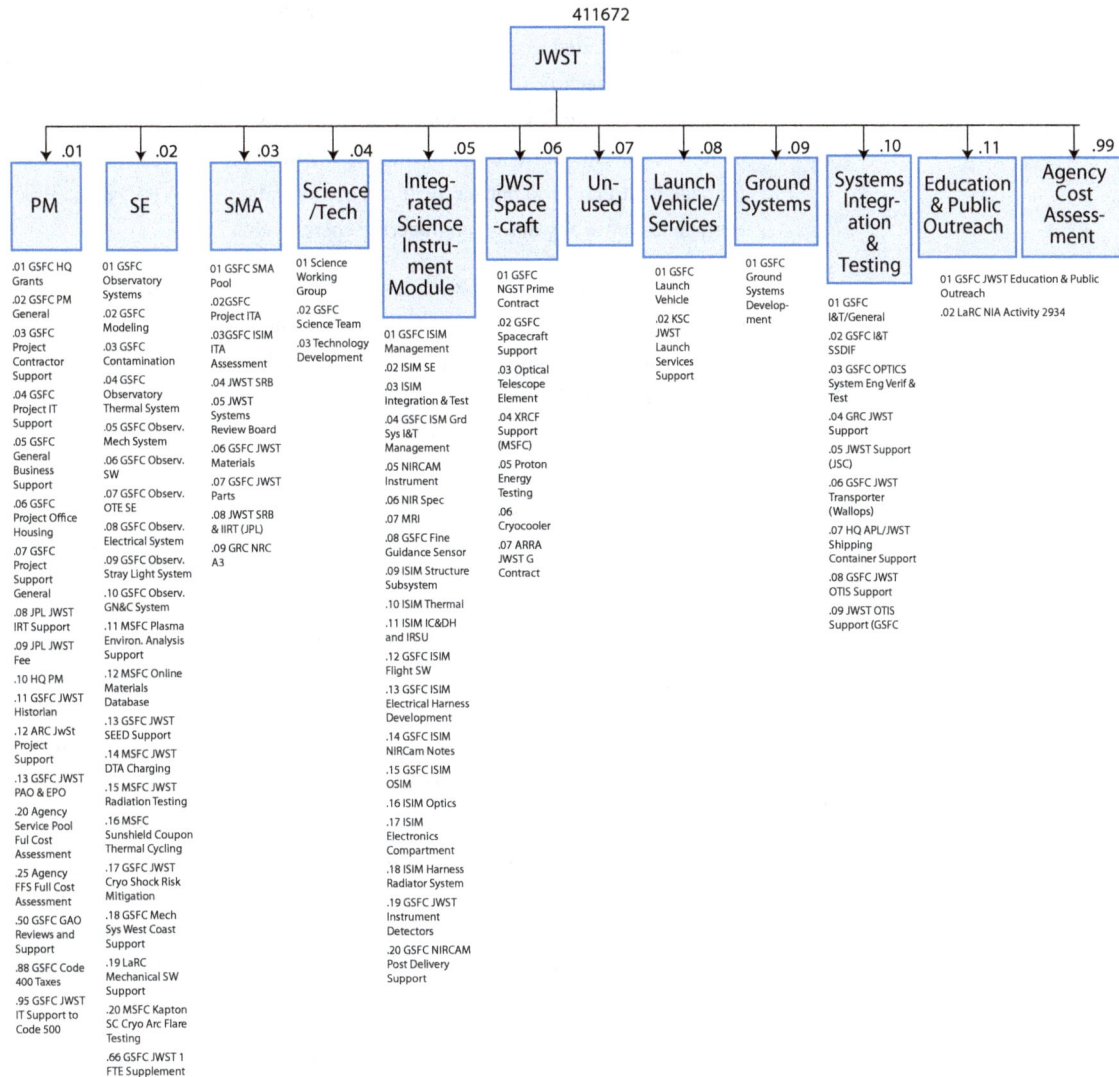

Figure 5-26 JWST 3 Level WBS Example

not associated with specific hardware or software elements. This element includes project reviews and documentation, non-project-owned facilities, and project UFE and funded schedule margins. It excludes costs associated with technical planning and management and costs associated with delivering specific engineering, hardware, and software products.

Element 2: Systems Engineering. The technical and management efforts of directing and controlling an integrated engineering effort for the project. This element includes the efforts to define the project space flight vehicle(s) and ground system and conduct trade studies. It includes the integrated planning and control of the technical program efforts of design engineering, software engineering, specialty engineering, system architecture development and integrated test planning, system requirements writing, configuration control, technical oversight, control and monitoring of the technical program, and risk management activities. Documentation products include requirements documents, Interface Control Documents (ICDs), Risk Management Plan, and master Verification and Validation (V&V) plan. This element excludes any design engineering costs.

Element 3: Safety and Mission Assurance. The technical and management efforts of directing and controlling the SMA elements of the project. This element includes design, development, review, and verification of practices and procedures and mission success criteria intended to assure that the delivered spacecraft, ground systems, mission operations, and payloads meet performance requirements and function for their intended lifetimes. This element excludes mission and product assurance efforts directed at partners and subcontractors other than a review/oversight function, and the direct costs of environmental testing.

Element 4: Science/Technology. This element includes the managing, directing, and controlling of the science investigation aspects, as well as leading, managing, and performing the technology demonstration elements of the project. It includes the costs incurred to cover the Principal Investigator, Project Scientist, science team members, and equivalent personnel for technology demonstrations. Specific responsibilities include defining the science or demonstration requirements; ensuring the integration of these requirements with the payloads, spacecraft, ground systems, and mission operations; providing the algorithms for data processing and analyses; and performing data analysis and archiving. This element excludes hardware and software for onboard science investigative instruments/payloads.

Element 5: Payload(s). This element includes the equipment provided for special purposes in addition to the normal Government standard equipment (GSE) integral to the spacecraft. This includes leading, managing, and implementing the hardware and software payloads that perform the scientific, experimental, and data gathering functions placed on board the spacecraft as well as the technology demonstration for the mission.

Element 6: Spacecraft. The spacecraft that serves as the platform for carrying payloads, instruments, humans, and other mission-oriented equipment in space to the mission destinations to achieve the mission objectives. The spacecraft may be a single spacecraft or multiple spacecraft/modules (i.e., cruise stage, orbiter, lander, or rover modules). Each spacecraft/module of the system includes the following subsystems, as appropriate: Crew, Power, Command and Data Handling, Telecommunications, Mechanical, Thermal, Propulsion, Guidance Navigation and Control, Wiring Harness, and Flight Software.

This element also includes all design, development, production, assembly, test efforts, and associated GSE to deliver the completed system for integration with the launch vehicle and payload. This element does not include integration and test with payloads and other project systems.

Element 7: Mission Operations. The management of the development and implementation of personnel, procedures, documentation, and training required to conduct mission operations. This element includes tracking, commanding, receiving/processing telemetry, analyses of system status, trajectory analysis, orbit determination, maneuver analysis, target body orbit/ephemeris updates, and disposal of remaining end-of-mission resources. This element does not include integration and test with the other project systems. (The same lower level WBS structure is often used for Mission Operation Systems during operations with inactive elements defined as "not applicable.")

Element 8: Launch Vehicle/Services. The management and implementation of activities required to place the spacecraft directly into its operational environment, or on a trajectory towards its intended target. This element includes launch vehicle, launch vehicle integration, launch operations, any other associated launch services (frequently includes an upper-stage propulsion system), and associated ground support equipment. This element does not include the integration and test with the other project systems.

Element 9: Ground System(s). The complex of equipment, hardware, software, networks, and mission-unique facilities required to conduct mission operations of the spacecraft systems and payloads. This complex includes the computers, communications, operating systems, and networking equipment needed to interconnect and host the mission operations software. This element includes the design, development, implementation, integration, test, and the associated support equipment of the ground system, including the hardware and software needed for processing, archiving, and distributing telemetry and radiometric data and for commanding the spacecraft. This element also includes the use and maintenance of the project test beds and project-owned facilities. This element does not include integration and test with the other project systems and conducting mission operations.

Element 10: Systems Integration and Testing. This element includes the hardware, software, procedures, and project-owned facilities required to perform the integration and testing of the project's systems, payloads, spacecraft, launch vehicle/services, and mission operations.

Element 11: Education and Public Outreach. This element provides for the education and public outreach (EPO) responsibilities of NASA's missions, projects, and programs in alignment with the Strategic Plan for Education. It includes management and coordinated activities, formal education, informal education, public outreach, media support, and website development.

For major launch or mission operations ground development projects, the WBS may be different than for projects centered on a spacecraft. For example, the spacecraft element may be changed to reflect the ground project major deliverable product (such as a facility). Elements that are not applicable such as payload, launch vehicle/services, ground system(s), and mission operations (system) might not be used. A technology development space flight project's WBS would also look different.

5.9.4 Developing Level 3 and Subsequent Elements for the Project WBS

The elements that make up the third and subsequent levels of the project WBS should be selected to classify all the work associated with the Level 2 element from which it derives. There is no standard template for these lower levels, so the project should develop a structure that will fully characterize the work. The project-specific product structure will be used to further breakdown the product elements of the standard Level 2 elements. The following paragraphs are provided to give an idea of how to further break down the work in the PM and SE elements.

Level 3 Elements to consider for the PM element (taken from the PMI's PM Body of Knowledge):

- Scope Management (includes project initiation, scope planning, scope definition, scope verification, and scope change control)
- Time Management (includes activity definition, activity sequencing, activity duration estimating, schedule development, and schedule control)
- Cost Management (includes resource planning, cost estimating, cost budgeting, and cost control)
- Integration Management (includes project plan development, project plan execution, and overall change control)
- Quality Management (includes quality planning, quality assurance, and quality control). Note that this element might instead be accounted for under the SMA element.
- Human Resource Management (includes organizational planning, staff acquisition, and team development)
- Communication Management (includes communications planning, information distribution, performance reporting, and administrative closure)

- Risk Management (included risk identification, risk qualification, risk response development, and risk response control)
- Procurement Management (includes procurement planning, solicitation planning, solicitation, source selection, contract administration, and contract closeout)

Level 3 elements to consider for SE can be taken from those described in NPR 7123.1:

- Stakeholder expectations and concept development
- Technical requirement definition
- Logical decomposition
- Technical solution definition
- Product implementation
- Product integration
- Product verification
- Product validation
- Product transition
- Technical planning
- Requirements management
- Interface management
- Technical risk management
- Technical configuration management
- Technical data management
- Technical assessment
- Technical decision analysis

5.9.5 Translating Work Breakdown into Funds

The following steps enable the project to translate work elements into the Agency's financial system:

- The Mission Directorate authorizes the project by issuing the FAD.
- The project team develops a high-level WBS, consistent with the NASA standard WBS, and documents the WBS in the Formulation Agreement.
- The project team inputs the project WBS into the Metadata Manager (MdM) system database, which initiates the WBS process for approval of the WBS and allocation of funds.

WBS approval enables resource management teams to allocate funds to specific WBS elements.

5.9.6 The Metadata Manager System

OCE is responsible for the official listing of all NASA programs and projects in MdM in accordance with *NPD 7120.4, NASA Engineering and Program/Project Management Policy.*

MdM is the Agency's official web-based tool for integrating master data across financial platforms. The codes representing all NASA programmatic and institutional WBS elements of programs and projects are established in MdM to be recognized as official NASA structures. The MdM system is a web-based Enterprise application that contains the Agency's official NASA Structure Management (NSM) data elements and associated attributes and codes. As the Agency's Enterprise repository for NSM data, MdM is used for identifying, creating, tracking, organizing, and archiving mission, theme, program, project, and WBS levels 2–7 NSM structural elements. As shown in Figure 5-27, MdM supplies NSM codes to the Agency's core financial system (Systems Application Products (SAP)), budget formulation system (N2), project management software system, and funds distribution systems (Work Instruction Management System) as they require coding structure data. WBS always refers to a structure starting with a 6-digit code, which occurs at Level 2 and below, within the Agency financial system.

When the project has developed its WBS, it inputs the WBS in the CFO's software interface, eBudget (budget.nasa.gov). eBudget can be accessed by contacting the MdM Support Line at (202) 358-1515 or by email at MdMHelpDesk@hq.nasa.gov. Approval for new WBS elements (including the project (Level 1)) is requested when the WBS is submitted. WBS requests to add new Level 1 and 2 elements are reviewed by several offices to ensure compliance with policy, guidance, and best practice. Programmatic WBS elements are governed by OCE (for NPR 7120.5 compliance) and the OCFO Budget Office. Institutional WBS elements are governed by the Mission Support Directorate and the OCFO Budget Office. Each new WBS Level 1 or 2 request is routed to representatives from the affected Mission Directorate, the OCE, and then OCFO. Requests for MdM changes to WBS levels 3-7 are approved by the affected Mission Directorate.

WBS approval enables resource management teams to allocate funds to specific WBS elements, making them available for obligation of funds. The NSM data starts from the several appropriations and flows to the Mission Directorates, then themes, programs, and down to projects. Once funding is available, the business management software for NASA's financial transactions, the Systems Application Products (SAP) software, and other Agency systems will recognize financial transactions to allocate funds to the WBS cost elements and enable civil servant labor and travel charges and acquisitions to proceed.

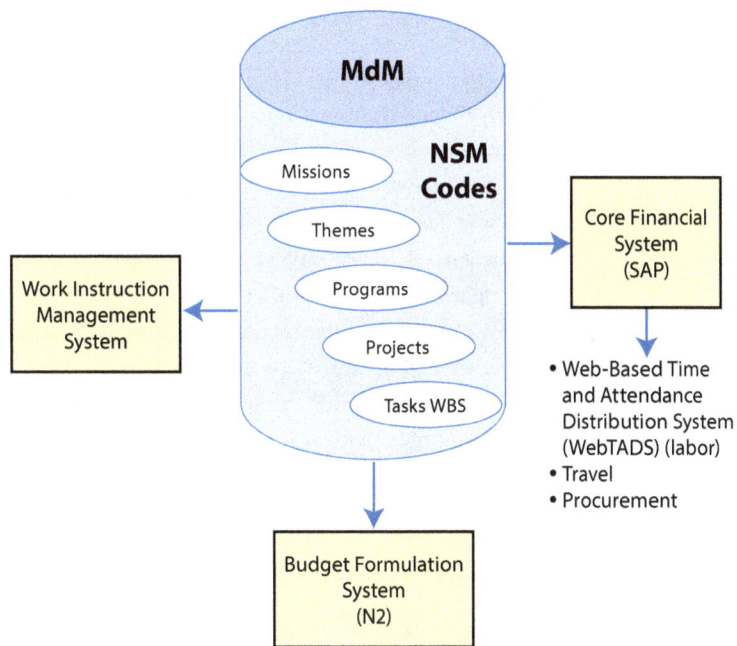

Figure 5-27 NASA's Central Repository for Enterprise Data, the Metadata Manager

Table 5-5 shows the hierarchy of NSM coding. All obligations and costs can only be allocated at or below the 6-digit code level. As a way of identifying the difference between a project and an activity, the Agency financial system has the project manager, or equivalent, designate whether it is a project or activity when setting up a 6-digit code.

Table 5-5 NSM Coding Hierarchy and Description

NSM Code	NSM Specifications	NSM Example
Mission Code	4 alpha's, smart Code*	ESMD
Theme Code	2–4 alpha's, smart Code	PROM
Program Code	4 alpha numeric (at least one alpha)	384A**
Project Code	6 digits, not smart coded	564815
WBS Level 2 Code	2 digits starting period delimiter Smart coded per NPR 7120.5	564815.11
WBS Level 3 Code	2 digits starting period delimiter	564815.11.01
WBS Level 4 Code	2 digits starting period delimiter	564815.11.01.13
WBS Level 5 Code	2 digits starting period delimiter	564815.11.01.13.21
WBS Level 6 Code	2 digits starting period delimiter	564815.11.01.13.21.09
WBS Level 7 Code	2 digits starting period delimiter	564815.11.01.13.21.09.02

* Smart codes refer to when the digits of the code have a meaning (e.g., are an acronym) rather than being random.

** Note there is no correlation between the program 4-digit codes and the first four digits of its project code.

5.9.7 Program WBS Work Elements

Programs are represented in the Agency financial system by 4-digit codes and do not have their own lower level structure. The best practice for funding ongoing program functions is for the program to establish a separate 6-digit activity. There are no standard WBS data elements for activities. Program offices are appropriately different across the Agency depending on the missions. However, WBS guiding principles should be applied when a program manager establishes a carefully anticipated and planned program office WBS to support its functions. The program office can use the Level 2 (.01) code for "Program Management" or "Program Office." Common subordinate elements include: Program Integration, Future Missions, Program Education/ Public Outreach, and Program Studies. Additional elements can always be added to a WBS. A WBS dictionary should be created to communicate the scope for each program WBS element.

5.10 Independent Standing Review Boards and Life-Cycle Reviews

The independent Standing Review Board (SRB) is a fundamental component of the Agency's checks and balances process. As former NASA Administrator Mike Griffin said, "You cannot grade your own homework." Independent experts review program and project "homework" with program and project members to find weaknesses that could turn into problems.

This special topic is intended to provide program and project teams with an overview of the independent Standing Review Board (SRB), the life-cycle review process, the SRB's roles and responsibilities in conducting certain life-cycle reviews performed at specific points in the program or project's life cycle, and the process for reporting the SRB's assessment of the program or project to the Decision Authority, typically in support of a Key Decision Point (KDP). NPR 7120.5E requires the program or project and an independent SRB to conduct most, but not all, of the life-cycle reviews, as indicated with a red triangle in the life-cycle figures for each type of program and project in NPR 7120.5, and in Figures 3-1, 3-2, 3-3, and 4-1 in this handbook.[18],[19] The program or project's Decision Authority, the MDAA, or the Center Director may also request an SRB to perform special reviews, such as rebaseline reviews or termination reviews. The Agency assigns responsibility for independent SRBs to two different organizations, the Independent Program Assessment Office (IPAO) and the Centers. For all programs the independent SRB is the responsibility of IPAO. For projects, the responsibility for the independent SRB is determined by the Decision Authority. The responsibility for the independent SRB for projects over

---

[18] Life-cycle reviews that do not require an independent SRB are conducted by the Center Director or designee in accordance with Center practices.

[19] The ORR is the last LCR the SRB routinely conducts. For supporting briefings after the ORR that lead to the KDP E, the SRB chair represents the SRB regarding the results of the ORR assessment.

$250 million LCC is typically assigned to IPAO. The responsibility for the independent SRB for projects with a life-cycle cost less than $250 million is typically assigned to a Center independent review team at the project's host Center. (See guidance from the NASA AA for small Category 3, Class D projects with a life-cycle cost of under $150 million on using an independent review team to perform independent assessments of the project in place of an SRB. Guidance can be found on the OCE tab in NODIS under "Other Policy Documents" at http/nodis3.gsfc.nasa.gov/OCE_docs/OCE_25.pdf.)

The program or project manager, with assistance from the Mission Directorate and Technical Authorities, determines when the program or project will hold the life-cycle review. The life-cycle review assessment is based on the six assessment criteria defined in NPR 7120.5 and Sections 3.1.2 and 4.1.2 in this handbook, life-cycle review entrance and success criteria defined in Appendix G in *NPR 7123.1, NASA Systems Engineering Processes and Requirements*, life-cycle products listed in Appendix I in NPR 7120.5E, and Tables 3-2, 3-3, 3-4, 3-5, 3-6, 4-6 and 4-7 in this handbook, and expected maturity states described in NPR 7120.5 and Appendix E of this handbook. Figure 5-28 provides an overview of the SRB formation, program or project internal activities leading up to the life-cycle review, and the life-cycle review reporting activities leading up to the Key Decision Point (KDP).

Additional information on the SRB and on life-cycle reviews conducted by the SRB is provided in the *NASA Standing Review Board Handbook*, which provides guidance to the NASA program and project communities and the SRBs regarding the expectations, timelines, and working interfaces with NASA Mission Directorates, Centers, review organizations, and the program or project. The *NASA Standing Review Board Handbook* provides the philosophy and guidelines for the setup, processes, and products of the SRB. The *NASA Standing Review Board Handbook* is available on the OCE website. It can be accessed by going to NODIS, Office of the Chief Engineer's section under the "Other Policy Documents" tab.

5.10.1 Standing Review Board

5.10.1.1 SRB Role and Responsibilities

The SRB is a fundamental component of the Agency's checks and balances Governance. The SRB is an independent advisory board in that it is chartered to assess programs and projects at specific points in their life cycle and to provide the program or project, the designated Decision Authority, and other senior management with a credible, objective assessment of how the program or project is doing relative to Agency criteria and expectations. The

Program/project initiates internal review process	Program/project conducts internal system/project reviews in accordance with approved review plan and Center practices; these internal reviews are typically the subsystem reviews for projects or integrated discipline and mission phase reviews for programs	Program/project prepares summary package(s) for presentation at the life-cycle review (SRR, SDR/MDR, PDR, CDR, etc.)

Convening Authorities

Decision Authority

NASA Chief Engineer[1]

Center Director

MDAA[2]

Director, Office of Evaluation[3]

Convening authorities:
- Jointly convene SRB
- Establish terms of reference
- Approve/concur SRB chair

SRB chair selects SRB members:
- Independent of program/project
- Some independent of host Center
- Must have representative experience in management, technical, & SMA
- Approved/concurred by convening authorities[4]

Life-cycle review

NPR 7123.1 review (e.g., PDR) | Independent integrated life-cycle review assessment (e.g., integrated PDR assessment)

May be one- or two-step[5]

Governing PMC makes recommendation to Decision Authority[7]

Convene governing PMC to consider:
- CMC/TA recommendations
- SRB Final Management Briefing Package
- Program/project disposition of SRB findings

Program/project dispositions SRB findings

CMC[6] assessment(s)

SRB reports out to project, program, Center, Mission Directorate, and Decision Authority

KDP

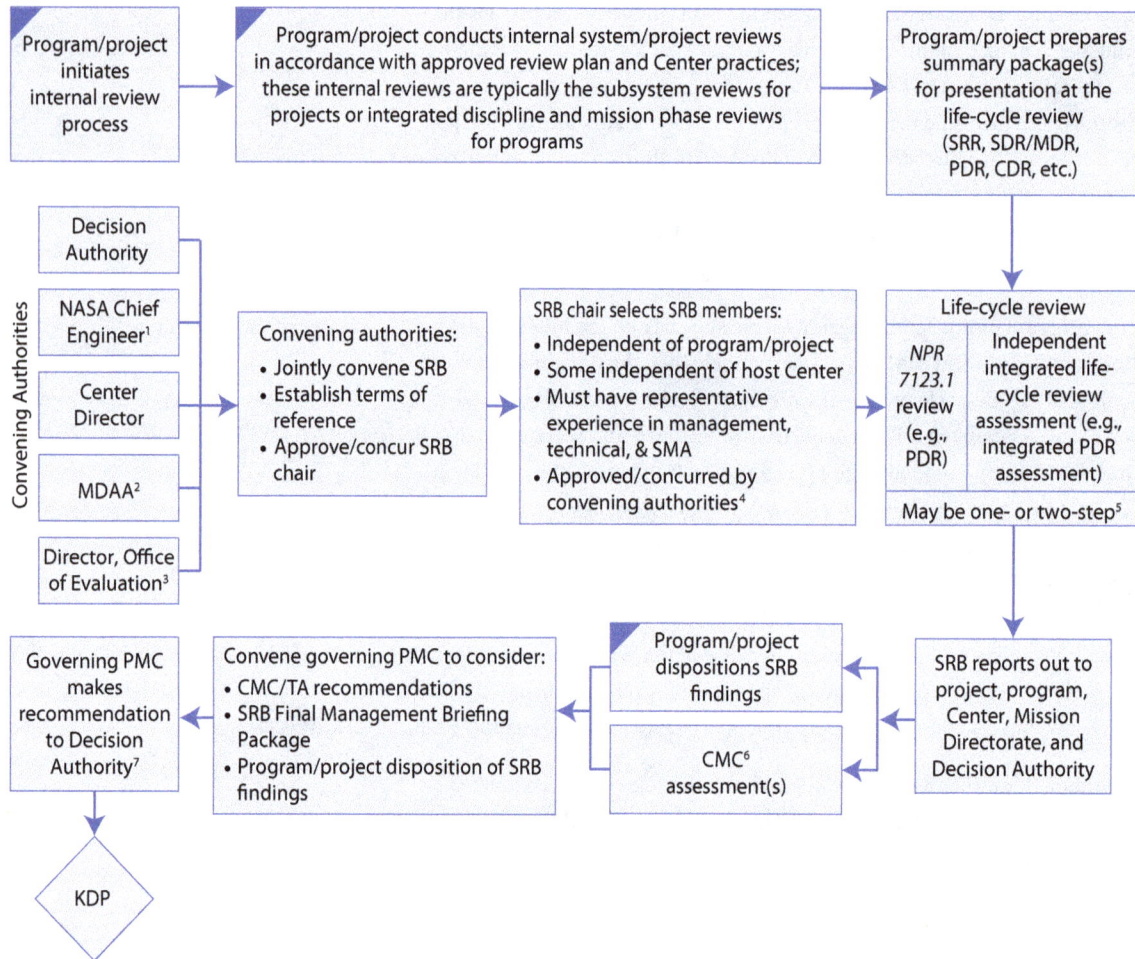

Legend: ▼ Program/project activity

[1] The NASA Chief Engineer is not a Convening Authority for Category 3 projects.

[2] The MDAA acts as a Convening Authority only when not already acting as the Decision Authority.

[3] For programs and Category 1 and 2 projects with a life-cycle cost exceeding $250 million.

[4] When applicable and at the request of the OCE, the OCHMO/HMTA determines the need for health and medical participation on the SRB.

[5] See Figures 5-29 and 5-30 for details.

[6] May be an Integrated Center Management Council when multiple Centers are involved.

[7] Life-cycle review is complete when the governing PMC and Decision Authority complete their assessment.

Figure 5-28 Overview of Life-Cycle Review Process

independent review also provides vital assurance to external stakeholders that NASA's basis for proceeding is sound.

The SRB is convened by the convening authorities specified in NPR 7120.5, Section 2.2.5.2. The SRB is responsible for conducting assessments of the program or project at life-cycle reviews based on criteria defined in NPR 7120.5 and NPR 7123.1 and any additional criteria imposed by the convening authorities. The SRB is responsible for meeting all of the evaluation objectives of the convening authorities at each life-cycle review. The SRB's role in life-cycle reviews is assessment; it does not have authority over any program or project. The SRB's involvement with the programs and projects is minimal between life-cycle reviews. The program or project provides the SRB with a list of future internal reviews planned before the next life-cycle review.

5.10.1.2 Forming the SRB

Each SRB has a chair, a review manager, board members, and in some instances, expert consultants-to-the-board. The chair and review manager are the primary interfaces with the program or project. The chair, review manager, board members, and expert consultants are carefully chosen and need to be competent, current, free from conflicts of interest, and acceptable to each senior manager that is convening the SRB, typically NASA's Associate Administrator, NASA's Chief Engineer, the Director of the Office of Evaluation (OoE), and the responsible Center Director.

The process of identifying the proposed SRB chair, board members, and expert consultants involves the Independent Program Assessment Office (IPAO) (when IPAO is responsible for the SRB), the Mission Directorate (usually its Program Executive (PE)), and the responsible Center. The program or project does not have an official voice in the selection of these members. However, the program and projects have a right to voice their opinions, particularly if the program or project believes that the board members are not competent and current.

5.10.1.3 Conflict of Interest for the SRB

To maintain the integrity of the independent review process and the SRB's Final Management Briefing Package, and to comply with Federal law, the conflict of interest procedures detailed in the *NASA Standing Review Board Handbook* are to be strictly adhered to in selecting the SRB chair, board members, and expert consultants to the board. Conflicts of interest may be personal, based on the personal interests of the individual (personal conflict of interest) or organizational, based upon the interests of the individual's employer (organizational conflict of interest). The SRB chair, review manager, board members, and expert consultants need to be free and remain free of conflicts of interest.

5.10.1.4 Terms of Reference

The scope, requirements, and assessment criteria for each life-cycle review are documented in the Terms of Reference (ToR) approved by the convening authorities. The program or project works with the SRB in developing the ToR. For each life-cycle review, the ToR describes program or project's products that the SRB will use or review as part of its assessment, and the timing of delivery of the products. The ToR also specifies the type of review, one-step or two-step (see next section). (See the *Standing Review Board Handbook* for a template of the ToR.)

5.10.2 Life-Cycle Reviews and Independent Assessment

Life-cycle reviews are designed to provide the program or project an opportunity to ensure that it has completed the work of that phase and provide an independent assessment of the program or project's technical and programmatic progress, status, and health against Agency criteria. The independent assessment serves as a basis for the program or project and management to determine if the work has been satisfactorily completed and if the plans for the following life-cycle phases are acceptable. If the program or project's work has not been satisfactorily completed or its plans are not acceptable, the program or project addresses the issues identified during the life-cycle review or puts in place the action plans necessary to resolve the issues. The program or project finalizes its work for the current phase during the life-cycle review. In some cases, the program or project uses the life-cycle review meeting(s) to make formal programmatic and technical decisions necessary to complete its work. In all cases, the program or project uses the results of the independent assessment and the resulting management decisions to finalize its work.

5.10.2.1 Determining the Type of Review—One- or Two-Step

All life-cycle reviews assess the program or project's technical maturity, programmatic posture, and alignment with the Agency's six assessment criteria. The full assessment can be completed in one step, called a one-step review, or divided into two separate steps, called a two-step review. The program or project manager has the authority to determine whether to hold a one-step review or a two-step review. This determination usually depends on the state of the program or project's cost and schedule maturity as described below. Any life-cycle review can be either a one-step review or a two-step review. The program or project manager documents the review approach in the program or project review plan.

Descriptions of the one-step and two-step life-cycle review processes are provided in Figures 5-29 and 5-30. (This section is written from the perspective of life-cycle reviews conducted by a program or project and an SRB. For life-cycle reviews that do not require an Agency-led SRB, the program or project manager will work with the Center Director or designee to prepare for and conduct the life-cycle review in accordance with Center practices and a Center-assigned independent review team. Small Category 3, Class D projects with a life-cycle cost of under $150 million should refer to guidance on using an independent review team to perform independent assessments of the project in place of an SRB. Guidance can be found on the OCE tab in NODIS under "Other Policy Documents" at http/nodis3.gsfc.nasa.gov/OCE_docs/OCE_25.pdf. When the life-cycle review is conducted by the program or project and a Center independent review team, rather than an Agency-led SRB, the remaining references to SRB need to be replaced with Center independent review team.)

- In a **one-step review**, the program or project's technical maturity and programmatic posture are assessed together against the six assessment criteria. In this case, the program or project has typically completed all of its required technical work as defined in NPR 7123.1 life-cycle review entrance and success criteria and has aligned the scope of this work with its cost estimate, schedule, and risk posture before the life-cycle review. The life-cycle review is then focused on presenting this work to the SRB. Except in special cases, a one-step review is chaired by the SRB chair. The SRB assesses the work against the six assessment criteria and then provides an independent assessment of whether or not the program or project has met these criteria. Figure 5-29 illustrates the one-step life-cycle review process. (Note: A one-step review for a program is analogous to a one-step review for a project.)

- In a **two-step review**, the program or project typically has not fully integrated the cost and schedule with the technical work. In this case, the first step of the life-cycle review is focused on finalizing and assessing the technical work described in NPR 7123.1. However as noted in Figure 5-30, which illustrates the two-step life-cycle review process, the first step does consider the preliminary cost, schedule, and risk as known at the time of the review. This first step is only one half of the life-cycle review. At the end of the first step, the SRB will have fully assessed the technical approach criteria but will only be able to determine preliminary findings on the remaining criteria since the program or project has not yet finalized its work. Thus, the second step is conducted after the program or project has taken the results of the first step and fully

There are special cases, particularly for human space flight programs and projects, where the program or project uses the life-cycle review to make formal decisions to complete their technical work and align it with the cost and schedule. In these cases, the program or project manager may co-chair the life-cycle review since the program or project manager is using this forum to make program or project decisions, and the SRB will conduct the independent assessment concurrently. The program or project manager will need to work with the SRB chair to develop the life-cycle review agenda and agree on how the life-cycle review will be conducted to ensure that it enables the SRB to fully accomplish the independent assessment. The program or project manager and the SRB chair work together to ensure that the life-cycle review Terms of Reference (ToR) reflect their agreement and the convening authorities approve the approach.

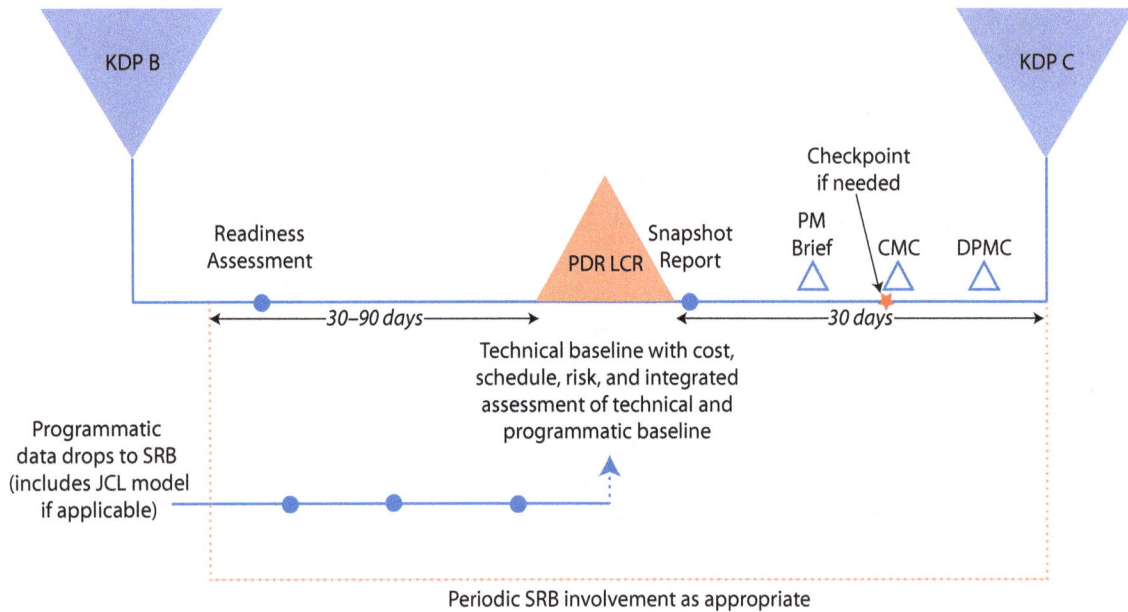

Notes: A one-or two-step review may be used for any life-cycle review. Section 5.10 and the *NASA Standing Review Board Handbook* provide information on the readiness assessment, snapshot reports, and checkpoints associated with life-cycle reviews. Time is not to scale.

Figure 5-29 One-Step PDR Life-Cycle Review Overview

integrated the technical scope with the cost, schedule, and risk, and resolved any issues that arose as a result of this integration. The period between steps may take up to six months depending on the complexity of the program or project. In the second step, which may be referred to as the Independent Integrated Life-Cycle Review Assessment, the program or project typically presents the integrated technical, cost, schedule, and risk, just as is done for a one-step review, but the technical presentations may simply update information provided during the first step. The SRB then completes the assessment of whether or not the program or project has met the six assessment criteria. In a two-step life-cycle review, both steps are necessary to fulfill the life-cycle review requirements. Except in special cases, the SRB chairs both steps of the life-cycle review. (Note: A two-step review for a program is analogous to a two-step review for a project.)

5.10.2.2 Conducting the Life-Cycle Review

As a prerequisite for scheduling a life-cycle review, the program or project manager, the SRB chair, and the Center Director or designated Engineering Technical Authority (ETA) representative mutually assess the program

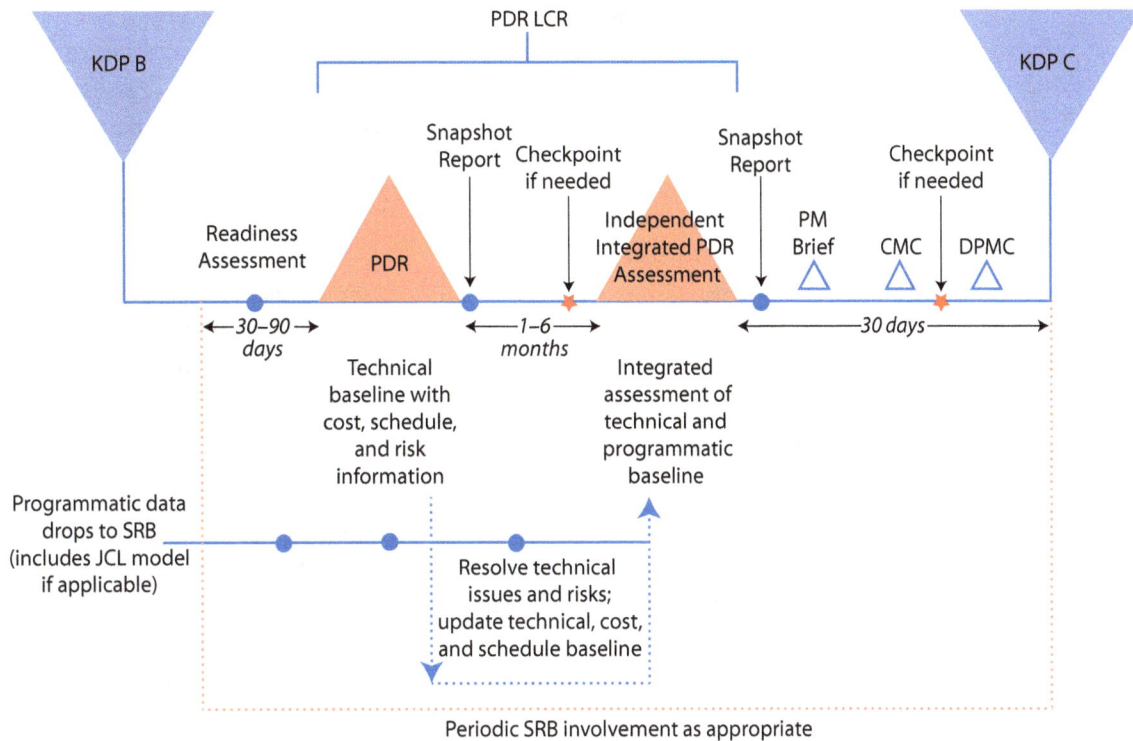

Notes: A one-or two-step review may be used for any life-cycle review. The *NASA Standing Review Board Handbook* provides information on the readiness assessment, snapshot reports, and checkpoints associated with life-cycle reviews. Time is not to scale.

Figure 5-30 Two-Step PDR Life-Cycle Review Overview

or project's expected readiness for the life-cycle review. This is a discussion, not a review. This assessment is conducted to ensure that the program or project is likely to reach the required state of maturity by the proposed date for the review. The program or project manager, the SRB chair, and the Center Director or designated ETA representative discuss the program or project's maturity with respect to entry criteria, gate products, and the expected states of maturity. The SRB chair's determination of readiness and any disagreements are reported to the Decision Authority for final decision. When the program or project manager judges that extenuating circumstances warrant proceeding with the life-cycle review, even though some maturity expectations will not be met by the time of the review, the program or project manager is responsible for providing adequate justification to the Decision Authority for holding the life-cycle review on the recommended date. The readiness assessment occurs approximately 30 to 90 calendar days prior to the proposed date for the life-cycle review.

In preparation for the life-cycle review, the program or project generates the appropriate documentation per NPR 7120.5 Appendix I, NPR 7123.1, and Center practices as necessary to demonstrate that the program or project's definition and associated plans are sufficiently mature to execute the follow-on life-cycle phase(s) with acceptable safety, technical, and programmatic risk.

During the life-cycle review, the program or project presents its status through sequential briefings for each agenda topic, typically given by the program or project lead. The life-cycle review is chaired in accordance with Section 5.10.2.1. The presenters answer questions from the SRB members in real time if possible. If further detail is required, the program or project may offer to provide the necessary information later in the review or arrange a splinter session in parallel with additional presentations.

The depth of a life-cycle review is the responsibility of the program or project manager and the SRB. The depth needs to be sufficient to permit the SRB to understand whether the design holds together adequately and whether the analyses, development work, systems engineering, and programmatic plans support the design and key decisions that were made. The SRB reviews the program or project's technical and programmatic approach, cost and schedule estimates, risk, performance, and progress against plans, and status with respect to success criteria and expected maturity states in NPR 7120.5E, NPR 7123.1, and this handbook.

5.10.2.3 Reporting the Results of the Life-Cycle Review

Rapid reporting to the convening authorities and the Decision Authority following the life-cycle review is essential to an efficient and effective review process. As a result, the SRB chair provides a summary of his/her preliminary findings to the Decision Authority no later than 48 hours after the life-cycle review is concluded. This summary is known as the snapshot report. The SRB chair provides a draft of the snapshot report to the program or project manager prior to the snapshot teleconference so they are informed and can be prepared to comment or respond. For a one-step review process, one snapshot report is required. For a two-step review process, a snapshot report is required after both the first step and the second step.

After the snapshot report, the SRB finalizes its findings and recommendations. The SRB's fundamental product is its assessment of whether the program or project meets the six assessment criteria or not. With this comes the recommendation to advance the program or project into the next life-cycle phase or to hold the program or project in the current phase. If the SRB recommends advancing the program or project with qualifications, the SRB will explain the qualifications and why these areas need not delay advancing the program or project to the next life-cycle phase. If the SRB does not recommend advancing the program or project, the SRB provides the rationale. The SRB provides its final findings and recommendations to the program or project manager, and the program or project manager prepares his or her final responses to the SRB's findings and recommendations. The program or project manager's response includes concurrence or nonconcurrence with the SRBs findings, associated rationale, and plans for addressing SRB findings.

Prior to presentation to the program or project's governing PMC in support of the KDP, the SRB and the program or project present the SRB findings and recommendations, and the program or project responses, to the responsible Center/CMC. For programs and for projects whose governing PMC is the APMC, the SRB and program or project also present to the Mission Directorate/DPMC.

The SRB findings and recommendations and the program or project response are provided to the convening authorities and Decision Authority prior to the KDP. If the KDP scheduled date is significantly more than 30 days after the life-cycle review concludes, a checkpoint may be required. At a checkpoint, the program or project manager describes to the Decision Authority the detailed program or project plans for significant decisions, activities, and commitments. The Decision Authority provides the program or project with interim authorization, guidance, and direction. For a one-step review, the Decision Authority may require a checkpoint when the KDP is estimated to be more than 30 days after the conclusion of the life-cycle review. For a two-step review, the Decision Authority may require a checkpoint when the KDP is estimated to be significantly more than 30 days after the second step, or when the second step is estimated to occur more than 6 months after the first step. During the period between the life-cycle review and the KDP, the program continues its planned activities unless otherwise directed by the Decision Authority.

The SRB findings and recommendations, and the program or project response, are presented to the program or project's governing PMC in support of the KDP. The Decision Authority reviews all the materials and briefings at hand, including briefings from the program or project team and the SRB, to make the KDP decision about the program or project's maturity and readiness to progress through the life cycle and authorizes the content, cost, and schedule parameters for the ensuing phase(s). (See Sections 3.2.3 and 4.2.3 for a more detailed description of a KDP.)

A life-cycle review is complete when the governing PMC and Decision Authority complete their assessment and sign the KDP Decision Memorandum.

5.11 Other Reviews

Special reviews may be convened by the Office of the Administrator, Mission Directorate Associate Administrator (MDAA), the Technical Authorities (TAs), or other convening authority. (See Section 5.2 for more information on Technical Authorities.) Special reviews may be warranted for projects not meeting expectations for achieving technical, cost, or schedule requirements; not being able to develop an enabling technology; or experiencing some unanticipated change to the project baseline. Special reviews include a Rebaseline Review and Termination Review. In these cases, the authorizing official(s) forms a special review team composed of relevant members of the Standing Review Board (SRB) and/or additional outside expert members, as needed. The chair for these reviews is determined by the convening authority. The convening authority provides either Terms of Reference (ToR) or a Memorandum of Understanding (MOU) to the chair of the review to govern the review. The process followed for these reviews is the same as for other reviews unless modified in the ToR or MOU. The special review team is dissolved following resolution of the issue(s) that triggered its formation. For more detail on Rebaseline Reviews, see Section 5.5.4.1.

Other reviews are part of the regular management process. For example, Safety and Mission Assurance (SMA) Compliance/Verification reviews are spot reviews that occur on a regular basis to ensure projects are complying with NASA safety principles and requirements (see Section 5.11.2). Program Implementation Reviews (PIRs) are intermittent SRB reviews requested by the Decision Authority to assess program progress and the program's continuing relevance to the Agency's Strategic Plan (see Section 5.11.3.)

Programs and projects may be subject to other reviews by organizations internal and external to NASA, for example, procurement, the Office of the Inspector General (OIG), and the Government Accountability Office (GAO).

5.11.1 Termination Review

There are a number of different ways a program or project can come to an end, but if a Decision Authority, MDAA, or Program Executive believes it may not be in the Government's best interest to continue funding a program or project, they can recommend a special Termination Review (presented to the governing PMC) to the Decision Authority. Circumstances such as the anticipated inability of the program or project to meet its commitments, an unanticipated change in Agency strategic planning, or an unanticipated change in the NASA budget may trigger a Termination Review. A Termination Review may be called a Confirmation or Continuation Review for a program or project in Formulation, a Cancellation Review for a program or project in development, or a Termination Review for a program or project in operations. Top-level requirements and criteria specific to the program or project that if not met might trigger a Termination Review need to have been defined in the Program and Project Plan.

Initiating a termination decision process generally includes an independent evaluation of the program or project by the SRB or a specially appointed independent team of experts. The Decision Authority will notify NASA's Associate Administrator and the Associate Administrator for Legislative and Intergovernmental Affairs prior to conducting the review. In addition to an internal independent assessment, the Decision Authority may also request an independent assessment by an outside organization (for example, an independent cost analysis by The Aerospace Corporation). The Termination Review is convened by the Decision Authority. At the Termination Review, the SRB or specially appointed independent team and the program and/or project team(s) present status, including any material requested by the Decision Authority. If a separate, external independent assessment is commissioned, the results of that assessment are also reported. In addition, a Center TA (see Section 3.3) presents an assessment. For tightly coupled programs with multiple Centers implementing the projects, an Office of the Chief Engineer (OCE) assessment is presented by the TA. Appropriate support organizations are represented (e.g., procurement, external affairs, legislative affairs, Office of the Chief Financial Officer (OCFO), and public affairs), as needed.

Termination Reviews are not undertaken lightly. The Decision Authority may give the program or project time to address deficiencies. He or she may allow a program or project to proceed to its Implementation Key Decision Point (KDP) (I or C) and allow the decision to be part of the KDP decision, which always includes termination or cancelation as an option. Termination after Implementation has greater implications than before Implementation.

A decision to terminate a program or project is recorded in a termination Decision Memorandum. (If projects are terminated, this would also be reflected in the Program Commitment Agreement (PCA).) Whether the termination decision occurs as part of a KDP or part of a special review, the memorandum documenting the decision to terminate needs to include a signature page indicating that all signatories acknowledge the decision, without necessarily agreeing to it. The decision and the basis for the decision are fully documented and generally reviewed with the NASA Administrator prior to final resolution.

Programs or projects might not go forward for different reasons. In the case of the Spectroscopy and Photometry of the Intergalactic Medium's

Diffuse Radiation (SPIDR) Small Explorer project, the principal investigator determined during Phase B that the project would not be able to meet the Level 1 requirement for resolution on their proposed data collection. For Gravity and Extreme Magnetism Small Explorer (GEMS), cost overruns and schedule slips plagued the project. Efforts to resolve technical issues were unsuccessful through Phase B, and the project was not approved to go to Implementation.

When a decision to terminate is made, several steps need to be followed. The decision is communicated to mission stakeholders. Generally, the NASA Administrator and Associate Administrator (who is the Decision Authority for programs) are already involved in the process. Where decision authority resides at the MDAA level, if the NASA Administrator or Associate Administrator has not yet been involved in the process, s/he needs to be informed. For all program and project missions in operations, across directorates, termination is handled in accordance with *NPD 8010.3, Notification of Intent to Decommission or Terminate Operating Space Missions and Terminate Missions*. For an operating mission, the NPD requires that the NASA Administrator be notified at least 90 days in advance of the termination of intent to terminate. (For additional details, see NPD 8010.3 and *SSSE MH2002, The Science Mission Directorate Enterprise Management Handbook*.)

The Chief Financial Officer and Associate Administrator for Legislative and Intergovernmental Affairs may also have participated in the process. If not, they need to be informed of the intent to terminate. The Office of Legislative and Intergovernmental Affairs is responsible for meeting the Agency's obligations to Congress in this situation. The reprogramming requirements laid out in Section 505 of the General Provisions of annual Commerce, Justice, Science, and Related Agencies' appropriations acts require that NASA notify the House and Senate Committees on Appropriations of a decision to terminate a program or project 15 days in advance of the termination of a program or project. The Office of Legislative and Intergovernmental Affairs is responsible for notifying the Committees on Appropriations pursuant to this reprogramming requirement. Protocol dictates, and it is in the Agency's interest, that such notification to the Committees on Appropriations, and expiration of the 15-day notification period, take place before there is public release of information regarding any termination of a program or project.

Once these official communications have been handled, it is important to ensure all other affected parties are informed, potentially including partners, members of international or interagency partnerships, parties to MOUs in effect, mission science team partners, and mission operations team partners. The program or project executive (or equivalent) needs to ensure that other program or project executives (or equivalent) are notified and can inform their projects and that the appropriate lessons learned are captured in an archive such as the on-line Lessons Learned Information System.

The program or project needs to have in place a Decommissioning Plan for disposal of program or project assets. This plan will need to be reviewed and finalized in accordance with the directions accompanying the termination decision and with approval of the MDAA, program and project managers, and/or program or project executive (or equivalent). For programs or projects in operations, on-orbit elements of the plan are reviewed and concurred with by the Office of Safety and Mission Assurance (OSMA) for orbital debris and other risk components.

5.11.2 SMA Compliance/Verification Reviews

NASA Headquarters SMA has a process that provides independent compliance verification for the applicable NASA SMA process and technical requirements within the program or project safety and mission assurance plan, the program or project baseline requirements set, and appropriate contract documentation. (See *NPR 8705.6, Safety and Mission Assurance (SMA) Audits, Reviews, and Assessments* for more detail.) This process includes the following SMA audits and assessments:

- Quality Audit, Assessment, and Review—This audit provides independent verification that each NASA Center, program, and project is in compliance with the applicable NASA SMA quality assurance requirements.

- Requirement Flow Down and SMA Engineering Design Audits and Assessments—This assessment provides independent verification of the flow down of SMA requirements to the NASA Centers, programs, and projects, including requirements flow down to NASA contracts, and provides independent evaluation of the NASA SMA requirements implemented on programs and projects for system safety, reliability and maintainability (R&M), risk analysis, and risk management.

- Safety and Mission Success Review (SMSR)—This review prepares Agency safety and engineering management to participate in program or project management preoperations or major milestone review forums. The SMSR provides the knowledge, visibility, and understanding necessary for senior Agency safety and engineering management to concur or nonconcur with program or project decisions to proceed.

5.11.3 Program Implementation Review (PIR) Guidance

As discussed in Chapters 2 and 3, programs follow a life cycle that requires various life-cycle reviews and key decision points (KDPs). Once a program is in Implementation, the Decision Authority may request that the program go through periodic Program Implementation Reviews (PIRs) followed by a KDP where the results of the review are considered and the program authorized to continue to the next phase in Implementation.

The PIR is an independent life-cycle review that is conducted by a Standing Review Board (SRB) following the standard independent review process protocols described in Section 5.10. The purpose of the PIR is to periodically evaluate the program's continuing relevance to the Agency's Strategic Plan, assess performance with respect to expectations, and determine the program's ability to execute the implementation plan with acceptable risk within cost and schedule constraints. The results of the review are reported to the APMC and the NASA AA to show whether or not the program still meets Agency needs and is continuing to meet Agency commitments as planned.

Programs within NASA vary significantly in scope, complexity, cost, and criticality and as a result, the scope of the PIR varies depending on the program type—uncoupled, loosely coupled, tightly coupled, and single-project programs. Each PIR will be tailored to best enhance the probability of success for the program undergoing review and to enable the SRB to gather the required information. The tailored review content results from a collaborative process that includes the program, the SRB, and the convening authorities.

The program tables in Appendix I of NPR 7120.5 and Tables in Section 3.5 of this handbook show the minimum products that are expected at PIRs for uncoupled and loosely coupled programs. For tightly coupled and single-project programs, the products in the MRR/FRR column, along with the PCA and interagency/international agreements, are generally used as the basis for the review. However, not all of the program products and control plans will be applicable to every NASA program. Thus the nature and extent of these documents varies with program type and total life-cycle cost. There may be additional, important program products that would be captured in the Program Plan. These products are used by the SRB as part of its review using the six assessment criteria. Additional information on PIRs can be found on the EEPMB Community of Practice.

5.12 External Reporting

This special topic describes some of the ongoing, high-level reporting to the White House and Congress of program and project decisions, technical performance, baselines, and cost and schedule estimates.

The quality and consistency of NASA's technical, cost, and schedule reporting is critical to the Agency's budget and its future. Federal agencies, including NASA, are part of the Executive Branch and report on their performance to the White House through the Office of Management and Budget (OMB). Federal agencies are also required to report on their performance directly to Congress in various ways, including through their budget submissions. The U.S. Government Accountability Office (GAO), as the audit, evaluation, and investigative arm of the Congress, assesses NASA technical, cost, and schedule performance along with that of other Federal agencies.

Because reporting requirements change over time and data can be requested by Congress, OMB, GAO, or the NASA Office of the Inspector General (OIG) at any time, the reporting described in this section is not a complete description of all the reporting that might be required of programs and projects.[20]

Section 5.12.1 provides background information on the conditions which led to many of NASA's external reporting requirements. Section 5.12.2 outlines NASA's integrated data collection and reporting process, and includes a description of the data that projects and programs provide in support of the external reporting requirements. Section 5.12.3 describes the major reports that NASA provides to Congress, GAO, and OMB. Section 5.12.4 discusses NASA's internal use of the data collected in support of external reporting requirements.

5.12.1 Conditions Leading to External Reporting Requirements

A 2004 GAO study[21] concluded that a lack of disciplined project cost estimating at NASA was resulting in project management problems, schedule slippage, and cost growth. In reaction, Congress created an external

---

[20] The Office of the Chief Financial Officer (OCFO) Strategic Investments Division (SID) maintains a Cost and Schedule community of practice page with updated information (including external reporting) at: https://max.omb.gov/community/x/TQePJg. Contact OCFO to request access.

[21] NASA: Lack of Disciplined Cost-Estimating Processes Hinders Effective Program Management [GAO-04-642].

reporting requirement in the NASA Authorization Act of 2005, i.e., the Major Program Annual Report (MPAR). MPAR requires NASA to report on projects in development with estimated life-cycle cost (LCC) exceeding $250 million. Projects of this size in Formulation are also subject to this report if they have awarded contracts of $50 million or more with development content. Congressional appropriations language also requires NASA and some other agencies to report if the LCC of projects with an LCC greater than $75 million grows by 10 percent or more.[22]

As a result of the congressional action, in part, the National Security Presidential Directive (NSPD) 49[23] establishes OMB responsibility for assessing technical, cost, and schedule performance for major space projects. In addition, all appropriations since FY 2008 have included direction for GAO to "identify and gauge the progress and potential risks associated with selected NASA acquisitions."[24] This has resulted in GAO's annual "Assessment of Large-Scale NASA Programs and Projects," the audit known internally as the Quick Look Book.

Some reporting requirements, such as the Annual Performance Plan (APP), are Government-wide to meet guidance in *OMB Circular A-11, Preparation, Submission and Execution of the Budget.*

5.12.2 Integrated Technical, Cost, and Schedule Data Collection and Reporting Process

NASA's Chief Financial Officer (CFO) is responsible for ensuring that the Agency meets its congressional and White House program and project performance reporting requirements. The Office of the Chief Financial Officer (OCFO) works with Congress, GAO, and the Office of Management and Budget (OMB) to align those organizations' technical, cost, and schedule reporting requirements with NASA's existing processes to facilitate streamlined reporting. For example, NASA has established a standard

---

[22] Section 530 of the appropriations language requires managers of projects with an LCC over $75 million that are in the Departments of Commerce or Justice, the National Aeronautics and Space Administration, or the National Science Foundation to report the increase. NASA must notify the House and Senate Committees on Appropriations within 30 days, including the date on which such determination was made; a statement of the reasons for such increases; the action taken and proposed to be taken to control future cost growth of the project; changes made in the performance or schedule milestones and the degree to which such changes have contributed to the increase in total program costs or procurement costs; and new estimates of the total project or procurement costs.

[23] National Security Presidential Directive 49: U.S. National Space Policy, 31 August 2006.

[24] Audit of NASA large-scale programs and projects (FY 2008 House Appropriations Report H.R. 2764 (P.L. 110-161)). Refer to the House report for the details.

basis for the congressional MPAR and OMB NSPD-49 reports. NASA also utilizes the KDP Decision Memorandum and a single quarterly data call to the Mission Directorates to collect the information needed to generate the various required reports. A number of reports are incorporated directly into NASA's budget submission to Congress to minimize workload.

Figure 5-31 depicts NASA's integrated process for collecting project technical, cost, and schedule data and developing reports:

- KDP Decision Memoranda and accompanying documents (datasheet and KDP report) are provided to OCFO's Strategic Investments Division (OCFO/SID) and serve as the starting point for reporting.

- OCFO/SID issues a quarterly data call to collect updates to the datasheet information as required for one or more of the required reports. This data call provides guidance to the Mission Directorates, which collect and verify project submissions and forward the submissions to OCFO.

- OCFO/SID extracts the specific rolled-up information required for each report. If a more detailed report is required for an individual project because it entered Implementation or exceeded a key threshold during the previous quarter, SID supports the Mission Directorate in preparing the more detailed report.

- OCFO transmits reports that go to OMB and GAO.

- For threshold reports and any other reports that will go to Congress, OCFO/SID provides the final report to the Office of Legislative and Intergovernmental Affairs (OLIA), which transmits the signed report to Congress. OLIA also transmits breach notifications to Congress.

OCFO/SID maintains a record of data and reports provided to Congress, GAO, or OMB on its Cost and Schedule community of practice page.

5.12.2.1 Quarterly Data Call

The quarterly data call utilizes a datasheet to collect core data common to many reports and to collect data necessary for explaining any differences between a project's cost estimate and its budget request. OCFO/SID modifies the datasheet if necessary when external reporting or Agency policy changes. The core data elements collected through the quarterly data call are as follows:

- Current Estimate—The project's life-cycle cost (LCC), which includes Phase A through Phase F costs. For projects with an LCC greater than $250 million, the LCC is initially reported as an estimated range at KDP B. At KDP C, the LCC is the Agency Baseline Commitment (ABC). Costs are broken out by year and by whether they are Formulation

```
┌───────────────────────────────────────────────────────────┐
│              ┌─────────────────────────────────────┐        │
│              │ SID collects KDP Decision Memoranda and │     │
│              │ accompanying datasheets to serve as the │     │
│              │         starting point for reporting.    │     │
│              └─────────────────────────────────────┘        │
│  Data              ┌──────────────────────────────────┐     │
│  Call              │ OCFO issues a quarterly data call to │   │
│              │ collect information required by one or more │  │
│              │   reports over the next quarter.          │    │
│              └──────────────────────────────────┘          │
│                    ┌──────────────────────────────────┐     │
│              │ Mission Directorates request that programs │   │
│              │ and projects complete requested data and │     │
│              │ submit quarterly data and datasheets to OCFO. │ │
│              └──────────────────────────────────┘          │
└───────────────────────────────────────────────────────────┘
```

Data Call
- SID collects KDP Decision Memoranda and accompanying datasheets to serve as the starting point for reporting.
- OCFO issues a quarterly data call to collect information required by one or more reports over the next quarter.
- Mission Directorates request that programs and projects complete requested data and submit quarterly data and datasheets to OCFO.

Data Integration
- The data from the sources above and other sources feed the following reports generated by OCFO and transmitted to customers.

Report Generation
- Quarterly Report, NSPD-49 Reports
- MPAR Reports, Sec. 509 Reports, APP and APR Reports, and Budget Reports
- "Quick Look" Data Collection instruments

External Customers
- OMB
- Congress
- GAO

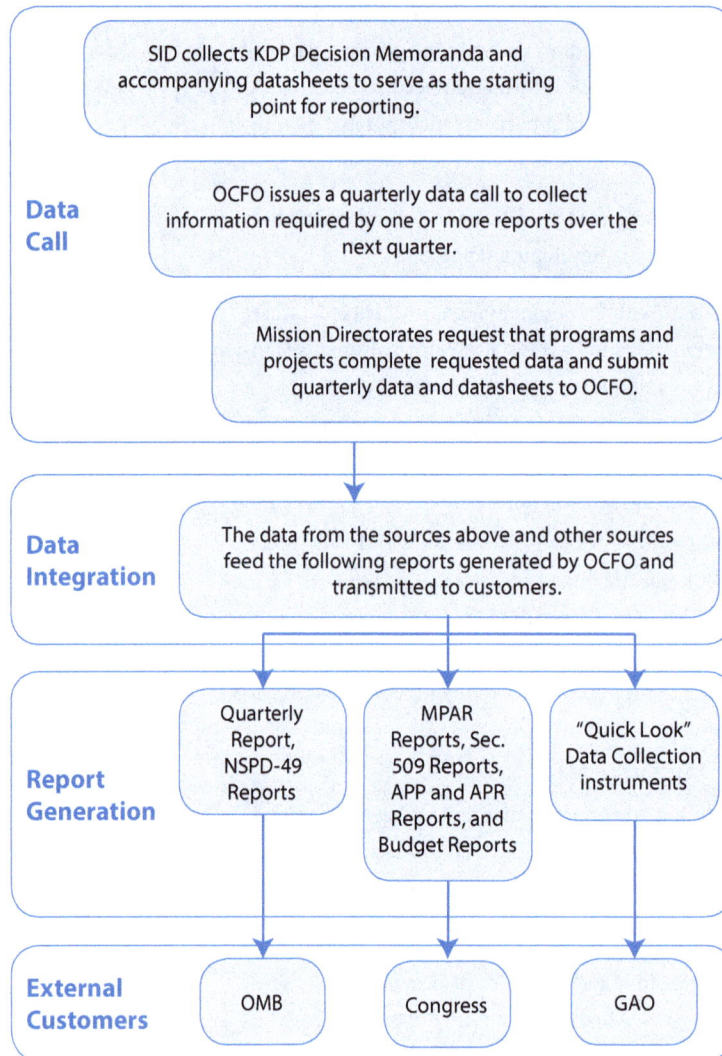

Figure 5-31 Integrated Technical, Cost, and Schedule Data Collection and Reporting Process

(Phases A and B), development (Phases C and D), or operations (Phases E and F) costs.

- Baseline—LCC/ABC at KDP C.
- Development Cost—The project's costs while the project is in Phase C or D. Costs are by Work Breakdown Structure (WBS) element as well as by year during development.
- Schedule—Key milestones, including KDPs and life-cycle reviews.
- Contract Value—Total award value and current value for awarded contracts with development content within exercised options. The value of contract options is included separately.

5.12.2.2 Additional Data Collected from Projects

Specific projects may be required to provide additional information for the GAO Quick Look Book and for other external reporting purposes such as baseline and threshold reports.

The GAO uses its Data Collection Instrument (DCI) to gather data for its Quick Look Book. There are five separate GAO DCIs for each project in the Quick Look audit; Cost, Schedule, Project, Contract, and Software. SID completes the Cost and Schedule DCIs. Projects complete the Project DCI and, in conjunction with OCE, the Software DCI. The Office of Procurement completes the Contract DCI. (See Table 5-8.) Agency coordination of audit activities is provided by the Mission Directorate Audit Liaison Representative (ALR) and the NASA audit lead in the NASA Office of Internal Controls and Management Systems (OICMS).

When additional information is required, the rules of engagement are negotiated with GAO at the beginning of each audit. OCFO/SID works with GAO to ensure that the Cost and Schedule DCI reporting and additional GAO reporting are closely coordinated. Requests for technical data are issued directly to the projects with notification to the Program Executive, the Mission Directorate ALR, and the NASA audit lead. Requests for baseline and threshold reports are issued to the NASA audit lead.

Cost information reported to Congress and OMB includes all UFE, whether it is held and managed at the project level or above. While UFE and schedule margin are not broken out in the DCIs, GAO does receive this information separately.

5.12.3 Major Cost and Schedule Reports Provided to Congress and OMB

Table 5-6 identifies major reports provided to Congress (MPAR, 10 Percent Cost Growth Report, Threshold Report, KDP B Cost Range Report, and OMB Circular A-11) and OMB (NSPD-49). Table 5-7, External Reporting Requirements for GAO, identifies major reports provided to GAO (Quick Look Book). The tables include details on report contents, when the reports are required, and applicable projects. The MPAR and NSPD-49 Reports include common components: Current Estimate and Baseline.

5.12.3.1 External Reports

Major Program Annual Report (MPAR). Report components include Current Estimate and Baseline. Congress requires these reports for projects in development (whether or not they are space flight projects) with an

Table 5-6 External Reporting Requirements for Congress and OMB

Report Name	Report Component	Cost and Schedule Content	Technical Content	Sources of Data	Congress	OMB
Major Program Annual Report (MPAR) (Applicable to projects in development with LCC > $250 million)	Current Estimate (Annually with budget submission)	Current estimated cost and schedule after KDP C, phased by WBS down to Level 2, with changes to baselines for LCC, development costs, key life cycle milestones and risks.	Project purpose, major systems, contributions from participating partners, Center project management roles, acquisition strategy, risk management, with changes to risks and technical parameters	Datasheet, Quarterly Data Call	Annually (included in NASA's annual budget to Congress)	Reviewed by OMB
	Baseline (KDP C)	(1) ABC at KDP C	Project purpose, major systems, contributions from participating partners, Center project management roles, acquisition strategy, risk management	(1) KDP C Decision Memorandum, (2) Datasheet, Quarterly Data Call for Contract Baseline	Annually (included in NASA's annual budget; if a project rebaselines, report may be required before next budget)	Reviewed by OMB as part of NSPD-49 submission (see below)
NSPD-49 Report (Applicable to (1) projects in development with LCC > $250 million; (2) projects in Formulation with LCC > $250 million and awarded contract of >$50 million with development content.)	Current Estimate	(1) Same as MPAR (2) Contract values	None	Quarterly data call		Quarterly
	Baseline (1) KDP C (2) Contract award date.	(1) Same as MPAR (2) Contract value.	(1) Same as MPAR (2) None	(1) KDP C DM & supporting documentation (2) Quarterly data call		(1) Quarter following KDP C (2) Quarter following award of contract.
Threshold Report (Applicable to projects in development with LCC >$250 million and satisfies NSPD-49 and MPAR requirements)	Notification (When development cost growth >15 percent of development cost in the ABC, or schedule slip >6 months based on the ABC schedule)	Changes in cost and schedule, detailed explanation or reasons for cost or schedule growth, mitigation actions planned and/or taken, expected outcomes of actions planned, and impacts on other programs	Detailed project overview and scope, including management and acquisition strategies, technical performance requirements, data products, mission success criteria, and description and analysis of alternatives	Mission Directorate works with project to develop report	When needed (Congressional notification followed by detailed report; and a detailed reporting timetable)	Reviewed by OMB as part of NSPD-49 submission (see above)
	Breach (When development cost growth >30 percent of development cost in the ABC)	Changes in cost and schedule, detailed explanation or reasons for cost or schedule growth, mitigation actions planned and/or taken, expected outcomes of actions planned, and impacts on other programs	Detailed project overview and scope, including management and acquisition strategies, technical performance requirements, data products, mission success criteria, and description and analysis of alternatives	Mission Directorate works with project to develop report	When needed	When needed
KDP B Cost Range (Applicable to projects in Phase B with LCC estimates >$250 million)	KDP B Cost Estimate	KDP B date and estimated LCC range, estimated launch readiness date or other key milestone	Same as for annual budget submission to Congress	KDP B DM and supporting documentation	Included in project pages in the annual budget to Congress	Reviews before submission.

Table 5-6 External Reporting Requirements for Congress and OMB (continued)

Report Name	Report Component	Cost and Schedule Content	Technical Content	Sources of Data	Congress	OMB
10 Percent Cost Growth Report (Applicable to projects with LCC ≥$75M)	Threshold (LCC growth >10 percent)	Cost growth	Explanation of cost growth	Mission Directorate works with project and OCFO/SID to develop report	When needed	When required
OMB Circular A-11 Performance Reporting	Management & Performance (M&P) section of the Congressional Justification, Annual Performance Plan, Annual Performance Reports	Varies by program area	Varies by program area	Developed by MDs as part of annual budget process	Provided in annual budget to Congress	

Table 5-7 External Reporting Requirements for GAO

Report Name	Report Component	Cost and Schedule Content	Technical Content	Sources of Data	GAO
Quick Look Book (Usually applicable to projects required to file MPAR and NSPD-49 reports)	Cost, schedule, project, contract and software DCIs (GAO's datasheets)	Current & baseline estimated cost and schedule, breakout by project phase, UFE.	Technical scope, progress, and risk, including critical and heritage technologies, drawing releases, parts quality issues, software TLOC, and technical leading indicators	Cost, schedule, & contract data from Integrated Quarterly Data Call; technical completed by project.	See Table 5-8
	Project documents	As required by NASA policy for project documents	As required by NASA policy for project documents	Project documents	See Table 5-8
	Site visits	May include specific GAO questions.	May include specific GAO questions.	Prepared responses to GAO questions	Annually

estimated Life-Cycle Cost Estimate (LCCE) exceeding $250 million.[25] The KDP C Decision Memorandum and supporting documentation serve as the basis for data included in the next annual MPAR report published in the congressional justification (annual budget request). The Department of Defense (DoD) and the National Oceanographic and Atmospheric Administration (NOAA) file similar reports.

NSPD-49 Report. Report components include Current Estimate and Baseline. NASA worked with OMB to make NSPD-49 apply to the same projects

[25] Pursuant to Section 103 of the NASA Authorization Act of 2005 (P.L. 109-155).

already included in the MPAR report. For projects in Formulation, reporting is limited to projects with an estimated LCC greater than $250 million and with awarded contracts of $50 million or more that include development content. NASA's OMB examiners receive quarterly updates on project technical, cost, and schedule performance during the course of the year for those projects covered by NSPD-49. All agencies involved in space flight file these reports. Cost and schedule reporting to OMB is common across the Federal Government.

Threshold Report. Report components include Notification and Breach. Commensurate with both NSPD-49 and MPAR requirements for projects with an LCC greater than $250 million, notifications are required for exceeding 15 percent of development costs in the ABC or 6 months schedule slippage based on the ABC schedule. If a breach occurs by exceeding development costs in the ABC by 30 percent, then congressional reauthorization and a new baseline (ABC) are required for continuation.

KDP B Cost Range. The 2012 Appropriations Act requires NASA to provide a cost range for projects with an LCC greater than $250 million in Phase B included in the Agency's annual budget to Congress. This is provided as a simple table within the budget pages for these projects.

10 Percent Cost Growth Report. NASA's annual congressional appropriations bills require NASA to report on projects with an LCC greater than $75 million that encounter a 10 percent LCC growth.[26]

OMB Circular A-11 Performance Reporting. The reporting components include the Management and Performance (M&P) section of the congressional justification, the Annual Performance Plan (APP), and the Annual Performance Report (APR). It includes performance goals and Annual Performance Indicators (APIs) that align to NASA's strategic framework as outlined in the Agency Strategic Plan and M&P section. Developing and reporting of these measures is coordinated between OCFO and the Mission Directorates.

Quick Look Book. The components include DCIs, project documents, and site visits. GAO has generally chosen to review projects already required to file MPAR or NSPD-49 reports and publishes its results annually.[27] GAO conducts site visits and receives project documents along with standardized cost, schedule, contract, and technical information. The Quick Look Book focuses on changes in project cost and schedule and provides GAO's explanations for these changes. Beyond the cost, schedule, and contract data

---

[26] Section 522 of Consolidated and Further Continuing Appropriations Act 2013 [P.L. 113-6].

[27] http://gao.gov/search?search_type=Solr&o=0&facets=&q=NASA+Assessments+of+Selected+large-scale+projects&adv=0.

provided in conjunction with the integrated quarterly call described above, GAO requests additional data to help them assess design stability, critical technologies, and technical maturity. Data elements provided to GAO in support of GAO's 2012 Quick Look Book are listed in Table 5-8. In addition to understanding project performance, GAO seeks to verify that NASA follows its acquisition, program or project management, and related policies. GAO also produces Quick Look Books on DoD projects.[28]

Table 5-8 provides a sample of data elements provided to GAO. As these data elements may change, contact the NASA lead for the GAO Quick Look audit for the latest list.[29]

Table 5-8 DCI 2012 Data Elements Provided to GAO

Data Category	Element	Frequency
Technical (collected in the project-level data collection instrument (DCI))	Design Stability	Annual + Updates
	Critical Technologies	Annual + Updates
	Heritage Technologies	Annual + Updates
	Software Complexity	Annual
	Quality Parts Issues	Annual
Cost (collected in the cost DCI)	MPAR Baseline	As occurs
	KDP B Estimated LCC Range	Semi-annual
	KDP C Baseline	As occurs
	JCLs completion date	As occurs (see KDP C docs)
	Project-held UFE	Monthly in Monthly Status Reviews (MSR)
Schedule (collected in the schedule DCI)	Key Milestones	Semi-annual
Contracts (collected in the contracts DCI)	Basic Information	Semi-annual
	Award Fee Structure	Semi-annual
Documentation	FAD/PCA	As occurs
	Project Plan	As occurs
	Control Plans	As occurs
	PDR/CDR Packages	As occurs
	SRB Final Briefing Package	As occurs
	KDP C, D, Replan, and Rebaseline Decision Memos	As occurs
	KDP C, D, Replan, and Rebaseline Datasheets and briefing charts	As occurs, includes all supporting documents
	MSR Presentations	Monthly

---

[28] Portfolio-level rollups of project-specific technical and cost and schedule performance are also provided to GAO as part of NASA's Corrective Action Plan responding to GAO's 'High Risk' audit. This reporting is not described here because it does not report on individual projects or require additional data from individual projects.

[29] NASA works with GAO to ensure that sensitive but unclassified (SBU) data, although shared with GAO, is not published.

5.12.4 NASA Management and Use of Data

All working files and final products for external reports, including submissions from Mission Directorates, technical performance, cost and schedule documents, and transmission emails are archived by SID by project, quarter, and report type. These files are available to NASA employees with approved access through the OCFO Cost and Schedule community of practice site (https://max.omb.gov/community/pages/viewpage.action?pageId=646907686). In addition, guidance materials and other resources are available on this site.

Program analysts use this information to better understand performance on an Agency-wide or portfolio basis, using tools such as cost and schedule trend analyses. These analyses help the Agency understand how changes in policies and practices affect performance. Beginning in 2007, NASA put a series of cost-management policy changes in place. NASA's record since 2007 indicates significant improvement in cost and schedule performance.

5.13 NASA Required and Recommended Leading Indicators

5.13.1 Background

The NASA Office of the Chief Engineer (OCE) undertook an effort to determine and implement a set of common metrics or leading indicators to assess project design stability and maturity at key points in a project's life cycle. These leading indicators enable NASA to objectively assess design stability and minimize costly changes late in development.

NASA's approach comprises three actions:

- NASA identified three leading indicators (common to almost every program or project) required to be reported by all programs and projects. These are mass margin, power margin, and Requests for Action (RFAs) (or other means used by the program or project to track review comments). These three leading indicators became policy in NASA Interim Directive (NID) to *NPR 7123.1A, NASA Systems Engineering Processes and Requirements* and then NPR 7123.1B and are required in NPR 7120.5E. Trending of these leading indicators shows the use of margin (estimated to actual) for mass and power and the timely closeout of RFAs. This trending helps the program or project manager understand how stable a design is as well as whether the design is maturing at the expected rate.

- NASA identified and piloted a set of augmented entrance criteria for both the Preliminary Design Review (PDR) and the Critical Design Review (CDR) to enhance the project's and NASA leadership's understanding of the project's maturity at those critical reviews. (See NPR 7123.1B.)

- NASA identified a common set of programmatic and technical leading indicators to support trending analysis throughout the life cycle. These leading indicators are highly recommended in the Program and Project Plan and Formulation Agreement templates of NPR 7120.5E.

Many NASA programs and projects already employ some or all of the leading indicators discussed here. The intent of codifying these leading indicators as a requirement is to ensure consistent application of this "best practice" across all projects. Through the process of considering, developing, measuring, assessing, and reporting these leading indicators, project teams gain additional insight or understanding into their programmatic and technical progress, and management is in a better position to make informed decisions.

Margins are the allowances carried in budget, projected schedules, and technical performance parameters (e.g., weight, power, or memory) to account for uncertainties and risks. Margins are allocated in the formulation process, based on assessments of risks, and are typically consumed as the program or project proceeds through the life cycle.

5.13.1.1 Applicability

All current and future space flight programs and projects, which fall under the authority of NPR 7120.5, that are in Phase A through Phase D are required to report on the three leading indicators. There may be cases where one or more of these leading indicators may not be applicable to a program or project. For example, purely software projects will not be required to report mass margin or power margins, but will be required to report the RFA/Review Item Discrepancy (RID)/Action Item burndown. Hardware that is not powered will not be required to report power margins. Programs or projects typically indicate their intention to follow or seek a waiver from these reporting requirements in their Formulation Agreement and/or Program or Project Plan. Agreement between the program and project manager and the NASA Chief Engineer typically is obtained prior to key decision point (KDP) B.

5.13.2 Introduction to Leading Indicators

This section discusses in general what leading indicators are and how they are used. It also provides general guidance on how the three required leading indicators are gathered and reported.

5.13.2.1 Leading Indicator Definition

A leading indicator is a measure for evaluating the effectiveness of how a specific activity is applied on a program or project. It provides information about impacts likely to affect the system performance objectives. A leading indicator may be an individual measure or collection of measures predictive of future system and project performance before the performance is realized. The goal of the indicators is to provide insight into potential future states to allow management to take action before problems are realized. (See the *Systems Engineering Handbook, NASA/SP-2007-6105 Rev 1*.)

Leading indicators are a subset of all the parameters that might be monitored by a program or project that have been shown to be predictive of performance, cost, or schedule in the later life-cycle phases. These leading indicators are used to determine both the maturity and stability of the design, development, and operational phases of a project. Three leading indicators—mass margins, power margins and RFA/RID/Action Item burndown—are considered critical and therefore are required to be tracked by programs and projects.

5.13.2.2 Application

Leading indicators are always viewed against time to determine the trend is progressing so that the product's stability and maturity can be assessed. The trend may be depicted in tabular form, or perhaps more usefully in graphical form. The leading indicators may also be plotted against planned values and/or upper or lower limits as defined by the Center, program, or project based on historical information.

By monitoring these trends, the program or project managers, systems engineers, other team members, and management can more accurately assess the health, stability, and maturity of their program or project and predict future problems that might require their attention and mitigation before the problems become too costly.

By combining the periodic trending of these leading indicators with the life-cycle review entrance and success criteria in NPR 7123.1, the program or project team will have better insight into whether the program and project products are reaching the right maturity levels at the right point in the life cycle, as well as an indication of the stability of those designs. The entrance criteria, in particular, address the maturity levels of both the end product as well as the project design documentation. However, just looking at maturity levels is not sufficient. If, for example, the product is appropriately designed to a CDR maturity level but there are still significant changes in the requirements, the project cannot be considered stable. The leading indicators aid in the understanding of both the maturity and stability of the program or project.

5.13.2.3 Gathering Data

These three leading indicators are to be gathered and reported throughout the program or project life cycle, starting after the System Requirements Review (SRR) and continuing through the System Acceptance Review (SAR) or the pre-ship review. They are normally gathered by the project at one or more levels within the product hierarchy. NASA is required to report this information externally at the project level; however, it is useful for the project team to gather the data one or more levels deeper within their product hierarchy so that the team may determine how best to allocate its power and mass resources. Teams for tightly coupled programs will need to gather the leading indicators from the various projects and provide a rolled-up set of parameters for reporting.

Note that the leading indicators may first be estimated and then, as more information is available, be refined or converted into actual measured values. For example, early in Phase A, the mass of a product may be estimated through modeling methods; whereas later in the life cycle, when the product is being built, measured masses can be used. Wherever the program or project is in its life cycle, the goal is to provide the most current and accurate information possible.

5.13.2.4 Reporting the Data

Leading indicator information is to be provided as part of the monthly Baseline Performance Review (BPR) submittal to Headquarters. Examples of spreadsheets and graphs are shown in Tables 5-9, 5-10, and 5-11, and in Figures 5-32, 5-33, and 5-34, but the exact format for reporting is left to the program or project or their management.

As a minimum, the following characteristics are provided as part of the report:

- Data are to be presented as a trend in a graph or table reported against time (month) for multiple time periods (not just the leading indicators for the current month) as appropriate for the program or project. Graphs should be accompanied by data tables.
- Milestone reviews should be provided as reference points on the graph or as part of the table.
- Graphs or tables should be annotated as needed to explain key features.

At the BPR, tables and graphs are included as backup unless an issue or anomaly associated with the indicators needs to be briefed explicitly, in which case they move into the main body of the BPR report.

5.13.3 Required Leading Indicators

The following descriptions of the required leading indicators discuss why trends in this area are important and describe the specific leading indicator measurements that need to be gathered, monitored, and reported.

5.13.3.1 RID/RFA/Action Item Burndown per Review Leading Indicator

During a life-cycle review, comments are usually solicited from the reviewing audience. Depending on the program or project, the comments may be gathered as RFAs, RIDs, and/or action items. How review comments are to be requested and gathered typically is determined by the program or project early in Phase A and documented in their Formulation Agreement and/or Program or Project Plan. Typically, if RFAs or RIDs are to be reported, action items are not. If the program or project team is only tracking action items, then they are expected to be reported.

The information that needs to be gathered for this leading indicator is:

- Total number of RFAs, RIDs, or action items (whichever the program or project uses).
- Number of open RFAs, RIDs, or action items (whichever the program or project uses).
- The planned rate for addressing and resolving or "burndown rate" of these items.

Table 5-9 shows an example spreadsheet tracking the number of RFAs for a given life-cycle review. Note that data are gathered monthly, not just at the next life-cycle review. Trending works best if the leading indicator measurements are taken regularly.

The planned burndown rate of the RFAs is included. While it can be difficult to predict how negotiations with submitters will progress, having a plan is important to communicate the desire of the program or project to resolve issues in a timely manner.

This information can perhaps be better seen graphically as shown in Figure 5-32. Note that a data table is produced with the graphical plot for additional information. The expectation is that the trend is reported over several months, whatever is appropriate for the program or project. The appropriate number of reporting months on any given graph needs to be agreed to by the program or project manager, Mission Directors, and OCE, and documented in the Formulation Agreement and/or Program or Project Plan.

5.13.3.2 Mass Margin Leading Indicator

For space flight programs or projects that contain hardware, mass margin is a required leading indicator to be gathered and reported. Leading indicators such as mass margin may also be considered Technical Performance Measures (TPMs) as referenced in NPR 7123.1. Leading indicators help the program or project team keep track of the most critical parameters. These are parameters that can drastically affect its ability to successfully provide the desired product. They are usually shown along with upper and/or lower tolerance bands, requirement levels, and perhaps stretch goal indications. When a leading indicator goes outside of the tolerance bands, more attention may be required by the program or project team to understand the underlying cause and to determine if corrective action is warranted.

For virtually every program or project that produces a product that is intended to fly into space, mass is a critical parameter. Programs or projects may track raw mass, but mass margin is considered to be a more informative indicator. As the concepts are fleshed out, a determination of how much a given launch vehicle can lift to the desired orbit/destinations will be

Table 5-9 Example Spreadsheet for Tracking Number of Open RFAs per Review

Date	MCR		SRR		PDR	
	Plan	Actual	Plan	Actual	Plan	Actual
Total	35		40		55	
10/07	30	30				
11/07	28	29				
12/07	26	27				
1/08	24	25				
2/08	22	24				
3/08	20	22				
4/08	18	20				
5/08	16	19				
6/08	14	15				
7/08	12	10	40	40		
8/08	10	8	38	37		
9/08	8	7	36	33		
10/08	6	5	34	30		
11/08	4	4	32	28		
12/08	2	3	30	23		
1/09	0	2	28	20		
2/09		2	26	18		
3/09		1	24	16		
4/09		0	22	15		
5/09		0	20	15		
6/09		0	18	13		
7/09		0	16	12	50	50
8/09			14	11	48	48
9/09			12	10	46	46
10/09			10	10	44	45
11/09			8	9	42	44
12/09			6	9	40	42
1/10			4	8	38	40
2/10			2	7	36	37
3/10			0	7	34	33
4/10				6	32	30
5/10				5	30	28
6/10				4	28	26
7/10				4	26	25
8/10				3	24	24
9/10				2	22	22
10/10				2	20	20
11/10				2	18	18
12/10				2	16	16
1/11				2	14	15
2/11				2	12	14
3/11				2	10	13
4/11				2	8	12
5/11				2	6	12
6/11				1	4	11
7/11				1	2	10

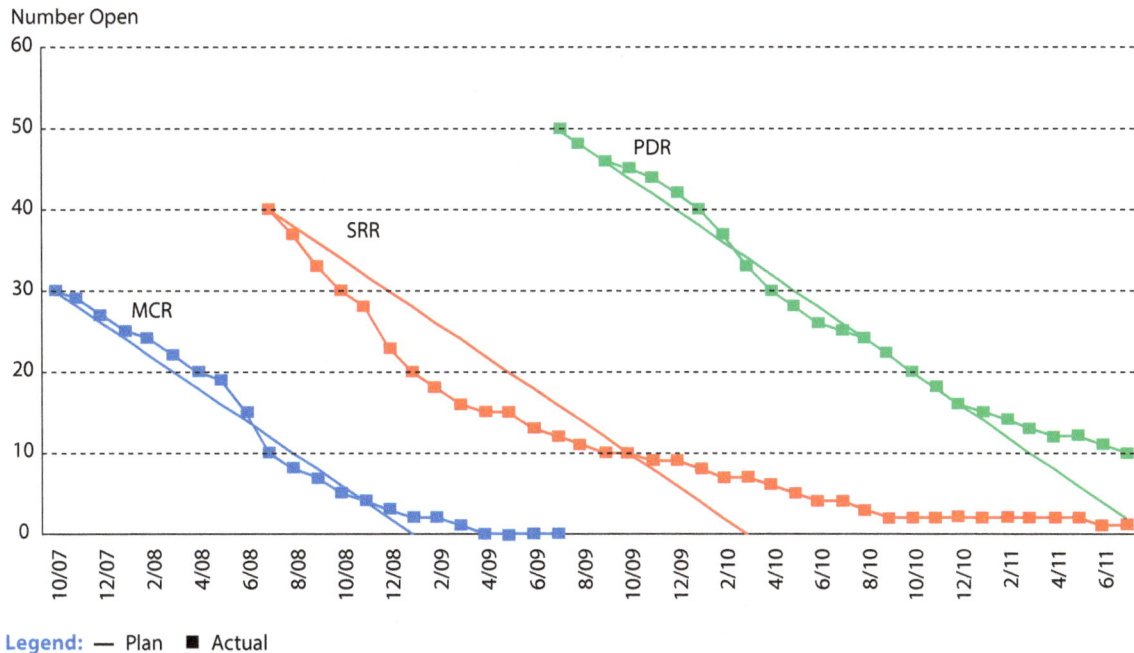

Legend: — Plan ■ Actual

Note: The data plotted in this figure are displayed in Table 5-9.

Figure 5-32 Example Plot for Number of Open RFAs per Review

determined. This in turn will place limits on the products that the program or project is providing. Mass parameters will be flowed down and allocated to the systems, subsystems, and components within the program or project's product. Tracking the ability of the overall product and its lower level subsystems and components to accomplish their mass goals becomes critical to ensure the success of the program or project. How far down into the product hierarchy the program or project decides to track these leading indicators is left up to the program or project. As a minimum, the rolled up mass margin indicator will be reported as part of the BPR status.

The information that needs to be gathered for this leading indicator is::

- Current estimated and/or measured mass.
- Not-to-exceed mass.
- Mass allocation.

Table 5-10 is an example of a spreadsheet that is tracking the mass margin of a project. The margins are estimated prior to the actual manufacturing of the product, and measured after it is produced. When reporting at the overall program or project level, the leading indicator may be a combination of estimates for subsystems/components that have not yet been produced and measured mass of those that have been produced.

Table 5-10 Example Spreadsheet for Tracking Mass Margin (kg)

Date	Not to Exceed (NTE)	Allocation	Current Mass
6/03	72	71	62
7/03	72	71	63
8/03	72	71	64
9/03	78	73	64.5
10/03	78	73	64
11/03	78	73	67
12/03	78	73	68
1/04	78	73	70
2/04	78	73	69
3/04	86	77	71.5
4/04	86	77	71
5/04	86	77	71
6/04	81	77	73
7/04	81	77	74
8/04	81	77	76
9/04	81	77	76
10/04	82	77	75
11/04	82	77	75
12/04	82	77	74
1/05	82	77	74
2/05	82	77	74
3/05	82	77	74
4/05	82	77	75
5/05	82	77	76
6/05	82	77	76
7/05	82	77	77.5
8/05	82	77	77.5
9/05	84	78	77.5
10/05	84	78	77.5
11/05	82	78	77
12/05	82	78	77
1/06	82	78	77
2/06	82	78	77
3/06	82	78.5	77
4/06	82	78.5	76
5/06	82	78.5	76
6/06	82	78.5	76
7/06	82	78.5	76
8/06	82	78.5	77.5
9/06	82	78.5	77.5
10/06	82	78.5	77.5
11/06	82	78.5	77.5
12/06	82	78.5	77.5

Perhaps a more effective way of displaying the mass margin indicator is graphically. Figure 5-33 is an example of displaying the mass margins for a particular program or project. Note that symbols are included on the graph to show where the life-cycle reviews were held. This helps put the information into the proper context. Also, enough information needs to be on the graph or in the legend to properly identify what mass is being tracked (e.g., dry mass, wet mass, mass of just the instrument, or mass of the entire vehicle).

5.13.3.3 Power Margin Leading Indicator

As with the mass margin leading indicator, tracking the power margin of a program or project that uses power is usually also a critical factor and is also considered a TPM. Availability of solar, nuclear, battery, or other power sources will also place a limit on how much power the various systems may consume. The program or project team will determine how deep within the product hierarchy these leading indicators will need to be allocated and tracked. As a minimum, the rolled-up power margin leading indicator will be reported as part of the BPR status.

The information that needs to be gathered for this leading indicator is:

- Current estimated and/or measured power consumption.
- Not-to-exceed power consumption (source limit).
- Power allocation.

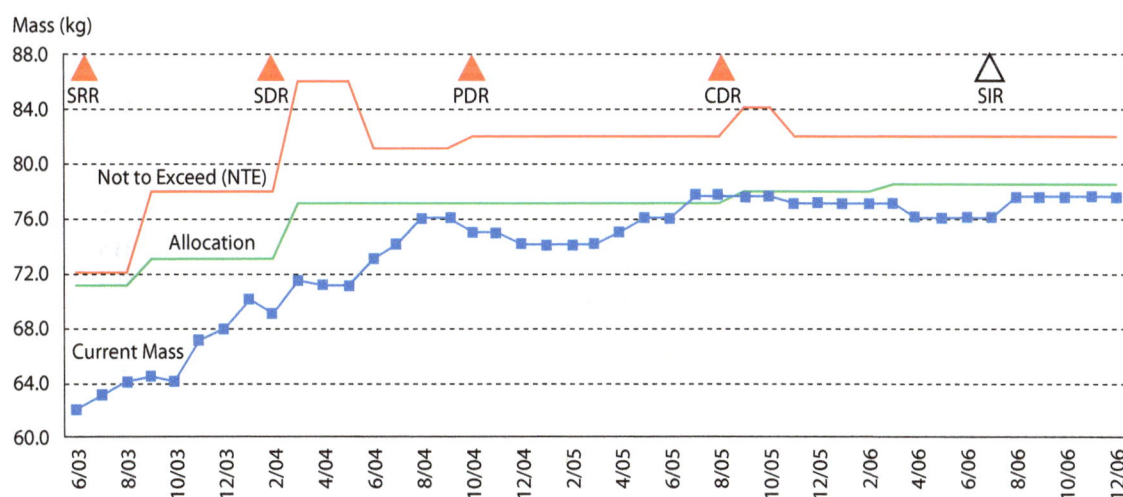

Note: The data plotted in this figure are displayed in Table 5-10.

Figure 5-33 Example Plot for Mass Margin Indicator

Table 5-11 Example Spreadsheet for Tracking Power Margin (W)

Date	Not to Exceed (NTE)	Allocation	Consumption
6/03	43	32	28
7/03	43	32	28
8/03	43	32	28
9/03	43	33	27
10/03	43	32	28
11/03	43	32	27.5
12/03	43	32	30
1/04	39	32	29
2/04	39	32	29
3/04	39	32	29
4/04	39	32	29
5/04	39	32	31
6/04	39	32	31
7/04	39	32	31
8/04	39	37	34
9/04	39	37	34.5
10/04	39	37	35.5
11/04	39	37	35.5
12/04	39	37	35.5
1/05	39	37	35.5
2/05	39	37	35.5
3/05	39	37	35.5
4/05	39	37	35.5
5/05	39	37	35.5
6/05	39	37	35.5
7/05	39	37	35.5
8/05	39	37	35.5
9/05	39	37	35.5
10/05	39	37	35.5
11/05	39	37	35.5
12/05	39	37	35.5
1/06	39	37	35.5
2/06	39	37	35.5
3/06	39	37	35.5
4/06	39	37	35.5
5/06	39	37	35.5
6/06	39	37	35.5
7/06	40.5	37.25	35.5
8/06	40.5	37.25	35.5
9/06	40.5	37.25	35.5
10/06	40.5	37.25	35.5
11/06	40.5	37.25	35.5
12/06	40.5	37.25	35.5

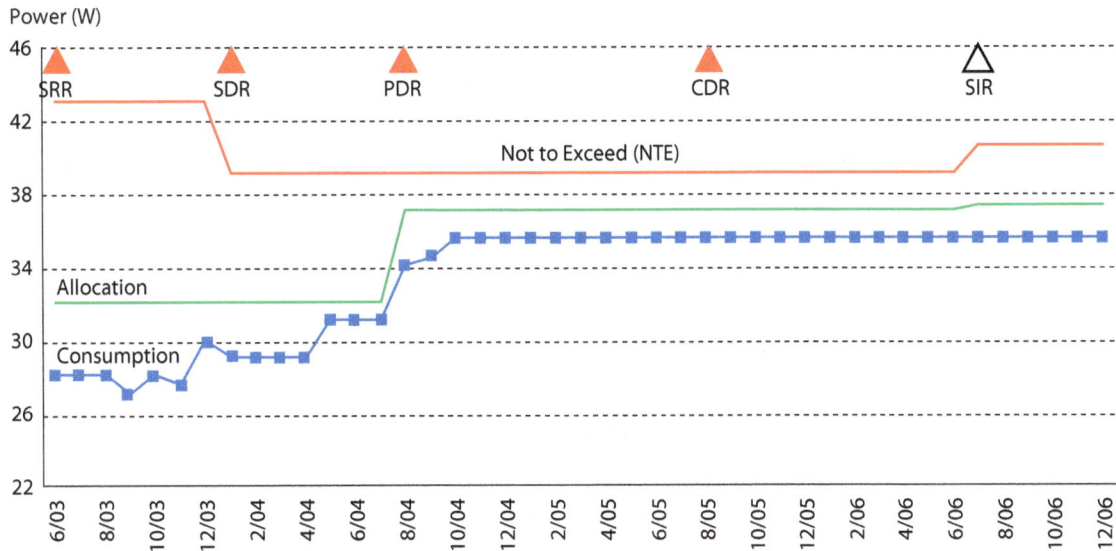

Figure 5-34 **Example Plot for Power Margin Indicator**

Note: The data plotted in this figure are displayed in Table 5-11.

Table 5-11 is an example of a spreadsheet that is tracking the power margin of a project. The margins are estimated prior to the actual manufacturing of the product and measured after it is produced. When reporting at the overall program or project level, the leading indicator may be a combination of estimates for subsystems/components that have not yet been produced and measured power of those that have been produced.

Perhaps a more effective way of displaying the power margin indicator is graphically. Figure 5-34 is an example of displaying the power margins for a particular program or project. Note that symbols are included on the graph to show where the life-cycle reviews were held. This helps put the information into the proper context. Also, enough information needs to be on the graph or in the legend to properly identify what power is being tracked (e.g., total capacity, instrument power, vehicle power).

5.13.4 Other Leading Indicators and Resources

This section provides additional information to aid the program or project team in identifying and tracking leading indicators. An Excel spreadsheet for the three required leading indicators is provided on the NASA Engineering Network (NEN) Systems Engineering community of practice under Tools and Methods at https://nen.nasa.gov/web/se/tools/. The spreadsheet can be used to gather and display these parameters if desired.

Table 5-12 shows examples of other leading indicators that experience has shown to be useful as indicators of whether projects are on track to complete the work within the time and funding that has been allocated. Program- and project-unique indicators are selected based on the type and complexity of the program or project, the key design parameters, and the need for visibility.

Table 5-12 Examples of Other Leading Indicators

Type	Indicator
Requirements Trend	% Requirement Growth
	TBD/TBR Burndown
	Pending Requirement Changes
Interface Trend	% Interface Documentation Approved/Pending
	TBD/TBR Burndown
	Pending Interface Changes
Verification Trend	Verification Closure Burndown
	Number of Deviations/Waivers approved/open
SW Unique Trend*	Number of Requirements verified and validated per Build/Release versus plan
Problem Report/Discrepancy Report Trend	# Open/closed PR/DRs
Manufacturing Trends	# Nonconformance/Corrective Actions
Technical Performance Measures	Trend of Key Design/Performance Parameters
Cost Margin Trend	Expenditure of UFE
	Plan/Actual
	EVM
	NOA
	Burndown/Analysis of closing out threats and liens over time
Schedule Margin Trend	Critical Path Slack/Float
	Critical Milestone Slip
	Schedule Metrics
	EVM
Resource Trends	Number of Key Milestones completed versus planned
	ABC versus Actual Funded Level
	Management Agreement versus Actual Funded Level
Staffing Trend	FTE
	WYE

*Additional non-technical leading indicators for software measurement requirements are also included in *NPR 7150.2, NASA Software Engineering Requirements* (e.g., requirement # SWE-091), which have implementation guidance in NASA-HDBK-2203 (http://swehb.nasa.gov).

5.14 Earned Value Management and Integrated Baseline Reviews

This special topic provides a synthesis of guidance for NASA's Earned Value Management (EVM) requirements for NASA programs, projects, major contracts, and subcontracts. EVM is a disciplined project management process that integrates a project's scope of work with schedule and cost elements. EVM goes beyond simply comparing budgeted costs to actual costs. It is a project management methodology that effectively integrates a project's work scope, schedule, and resources with risk in a single performance measurement baseline plan for optimum planning and control. Progress against the baseline plan can be objectively measured and assessed to determine if the project did what it planned to do for the allocated cost and schedule throughout the duration of the project. This enables management to ask appropriate questions to determine causes and identify corrective actions, along with providing an objective Estimated Actual Cost (EAC) at completion. When properly used, EVM provides an assessment of project progress, early warning of schedule delays and cost growth, and unbiased, objective estimates of anticipated costs at completion.

An Agency-wide EVM Capability Process can be found on the NASA Engineering Network a https://nen.nasa.gov/web/pm/evm. The EVM Capability Process and supporting documentation will be particularly useful in developing the program or project's EVM system.

EVM is required by the Office of Management and Budget (OMB) for compliance with the Federal Acquisition Regulation (FAR) Section 34.2 and guided by industry best practice. OMB Circular A-11 requires EVM for acquisitions with developmental effort and for both in-house government and contractor work using the guidelines in American National Standards Institute (ANSI)/Electronic Industries Alliance (EIA)-748, regarded as the national standard and an industry best practice for EVM systems. There are 32 guidelines that a certified EVM system needs to meet in the areas of organization; planning, scheduling, and budgeting; accounting; analysis and management reports; and revisions and data maintenance. NASA FAR Supplement (NFS), Section 1834.2 requires use of an Earned Value Management System (EVMS) on procurement for development or production work, including flight and ground support systems and components, prototypes, and institutional investments (facilities, IT infrastructure, etc.) when the estimated life-cycle (Phases A–F) costs are $20 million or more. If the program manager applies EVM at the program level, he or she will follow the same process that is used for projects.

EVM requirements apply to NASA projects and contracts with development or production work when the estimated LCC is $20 million or more. Small Category 3/Class D projects with development costs greater than $20 million and a life-cycle cost estimate less than $150 million should reference the EVM guide for applying EVM principles to small projects. The guidance for tailoring 7120.5 requirements for small Cat 3/Class D projects can be found on the OCE tab in NODIS under "Other Policy Documents" at http/nodis3.gsfc.nasa.gov/OCE_docs/OCE_25.pdf. EVM requirements also apply to single-project programs and may, at the discretion of the Mission Directorate Associate Administrator (MDAA), apply to other projects and tightly coupled programs.

For contracts and subcontracts valued from $20 million to $50 million, the EVMS shall be compliant with the guidelines in ANSI/EIA 748, but does not require validation. For contracts and subcontracts valued at $50 million or more, the contractor shall have an EVMS (or a plan to develop a system) that has been formally validated and accepted by the Government. Validation consists of the government testing the contractor's EVMS for compliance through a series of reviews. Compliance means that the contractor must comply with the guidelines listed in ANSI/EIA-748, however, no reviews by the government are required to formally accept the EVMS.

The Performance Measurement Baseline is a time-phased cost plan for accomplishing all authorized work scope in a project's life cycle, which includes both NASA internal costs and supplier costs. The project's performance against the PMB is measured using EVM if EVM is required, or other performance measurement techniques if EVM is not required. The PMB does not include UFE.

EVM planning begins early in project Formulation (Phases A and B) and is applied in project Implementation (Phases C and D). EVM is also applied when modifications, enhancements, or upgrades are made during Phase E when the estimated cost is $20 million or more. *NASA/SP-2012-599, NASA Earned Value Management (EVM) Implementation Handbook* provides detailed guidance on EVM implementation and is maintained electronically at http://evm.nasa.gov/handbooks.html.

During early Formulation, projects need to coordinate with the respective center EVM focal point (http://evm.nasa.gov/fpcpocs.html) to establish the organization and key structures to facilitate effective EVM implementation and usage (e.g., Project Work Breakdown Structure (WBS), Organizational Breakdown Structure, Responsibility Assignment Matrix, control accounts, etc.), and document project-specific tailoring when developing their EVM implementation plans. (See Appendix A for definitions for Organizational Breakdown Structure, Responsibility Assignment Matrix, and control accounts.)

The project's preliminary Performance Measurement Baseline (PMB) is established in Phase B in preparation for Key Decision Point (KDP) C approval and is assessed during the Integrated Baseline Review (IBR) process (see Section 5.14.1). A project-level IBR is completed prior to KDP C, and for this reason it is known as a pre-approval IBR. Project-level EVM reporting begins no later than 60 days after the start of Phase C. However, contract EVM reporting begins no later than 90 days after contract award regardless of the system acquisition phase.

The Program Plan will include the approach for integrating and managing program cost, schedule, and technical performance, including the flow down of EVM requirements to projects. The Project Plan documents the project's approach for meeting the EVM requirements in the Program Plan. Each project flows down EVM requirements to its applicable suppliers (intra-Agency organizations and contractors), ensuring that EVM requirements are included in each Request for Proposal (RFP) and the responses are evaluated for compliance with these requirements. The primary considerations for EVM applicability are the nature of the work and associated risks, and the value of the effort. In the EVM context, there are two basic classifications of the nature of work—discrete and level of effort (LOE). Discrete work is related to the completion of specific end products or services and can be directly planned, scheduled, and measured. LOE is effort of a general or supportive nature that does not produce definite end products. The application of EVM on projects/contracts that are exclusively LOE in nature may be impractical and inefficient and therefore is discouraged. Additionally, EVM is not required or recommended for firm fixed-price contracts. For these contracts, the project manager may implement an alternative method of management control to provide advanced warning of potential performance problems.

5.14.1 Integrated Baseline Review (IBR)

An IBR is required to verify technical content and the realism of related performance budgets, resources, and schedules. It is a risk-based review of a supplier's PMB conducted by the customer (e.g., the Mission Directorate, the program, the project, or even the contractor over its subcontractors). While an IBR has traditionally been conducted on contracts, it can be effective when conducted on in-house work as well. The same principles, objectives, and processes apply for in-house and contract IBRs; however, minor changes may be necessary to the steps in conducting an IBR. See *NASA/SP-2010-3406, Integrated Baseline Review (IBR) Handbook* at http://evm.nasa.gov/handbooks.html for step-by-step instructions on how to conduct an IBR. Center EVM focal points may also be contacted for more information.

The IBR ensures that the PMB is realistic for accomplishing all the authorized work within the authorized schedule and budget, and provides a mutual understanding of the supplier's underlying management control systems.

The IBR is an initialization of the continuous process of analyzing the PMB and will take place periodically any time there are significant changes to the PMB throughout the program or project life cycle.

NASA has many reviews during the program and project life cycles, and some of these reviews share common goals and objectives with the IBR. Therefore, when possible, the IBR can be combined with these other reviews. It is important, however, to ensure that the intent of the IBR is still met and supported by key personnel when reviews are consolidated.

Per NPR 7120.5: "For projects requiring EVM, Mission Directorates shall conduct a pre-approval integrated baseline review as part of their preparations for KDP C to ensure that the project's work is properly linked with its cost, schedule, and risk and that the management processes are in place to conduct project-level EVM." The Mission Directorate and the project will work together to define the process and schedule for the project-level IBR. Subsequent IBRs may be required when there are significant changes to the PMB such as a modification to the project requirements (scope, schedule, or budget) or a project replan.

> An Integrated Baseline Review is not a pass/fail event, an independent review, a time to resolve technical issues, nor a demonstration of EVMS compliance, but a point on the path of a continuous analytical process.

5.14.2 EVM Performance Reporting

The project manager needs to understand and emphasize the importance of the integrated technical, schedule, cost, and risk analyses provided by EVM in conjunction with other project information to formulate an overall project status. NASA projects with EVM requirements will need to integrate and report EVM performance measurement data to various customers. EVM data are obtained from the project team and/or the applicable suppliers by specifying the Contract Performance Report (CPR) as a deliverable and including specific instructions for reporting. CPRs are management reports that provide timely, reliable data that are used to assess the project or supplier's current and projected performance, to quantify and track known or emerging problems, to determine the project or supplier's ability to achieve the PMB, and to assist in decision making. It is important that the CPR is as accurate and timely as possible, so it may be used for its intended purpose, which is to facilitate informed, timely decisions.

EVM reporting requirements normally include explanations of cost, schedule, and (at completion) variances that breach established thresholds. These thresholds can be applied at various levels of the WBS, on a cumulative and/or current basis, and be represented by dollars, percentages, or other customer-specified criteria. For the project, specific reporting requirements and thresholds are defined in a program or project plan or directive. Project EVM reporting begins at Implementation, typically no later than 60 days after KDP C approval.

It is NASA policy that a program or project write a contract requirement for a CPR, Integrated Master Schedule (IMS), and WBS when EVM is required on contracts. Contract reporting requirements are defined in specific Data Requirements Descriptions (DRDs) included in the solicitation and contract. For contracts, the CPR is due no later than 90 days after contract award. When EVM is required on a project but not a contract, selected cost and schedule performance data will be required on those contracts to enable project-level planning, analysis, and EVM reporting. See the *EVM Implementation Handbook, Schedule Management Handbook*, and *WBS Handbook* at http://evm.nasa.gov/handbooks.html for more information on preparing the appropriate DRDs.

EVM data and analysis needs to be included in all management reviews and life-cycle reviews. Project status based on EVM data needs to be reported at the level appropriate for all levels of management and utilized for insight and management actions. Analysis is basically comprised of two major steps, i.e., analyzing past performance and then projecting future performance. NASA's EVM website contains a sample standard analysis package that can be used as a guide http://evm.nasa.gov/reports.html.

While not required, wInsight™[30] meets Agency requirements for analysis and reporting of EVM data, and generally adheres to the American National Standards Institute/Electronic Industries Alliance–748-B EVM standard. wInsight is intended to integrate the scope, schedule, and budget EVM data of NASA's in-house-managed projects as well as contractor data. It graphically displays trends at all levels of the WBS, and produces analyses and reports that can be used to support management reviews. See NASA's EVM website for instructions on how to access wInsight.

5.15 Selecting and Certifying NASA Program and Project Managers

5.15.1 Selecting Program and Project Managers

As part of their many duties, Center Directors are responsible for training, certifying, and providing program and project managers for programs and projects assigned to their Center:

- For Category 3 projects assigned to the Center, the Center Director or designee has the authority to assign a project manager with concurrence from the program manager and MDAA.

- For Category 2 projects assigned to the Center, the Center Director or designee has the authority to assign a project manager with concurrence from the program manager, but the Mission Directorate may also concur in those assignments, particularly for projects with a Life-Cycle Cost (LCC) greater than $250 million.

- For Category 1 projects assigned to the Center, the Center Director (or designee) recommends the project manager candidate to the Mission Directorate Associate Administrator (MDAA) or designee.

- For programs assigned to the Center, the Center Director or designee recommends the program manager candidate to the MDAA or designee.

The MDAA approves the selection of all program managers, all Category 1 project managers, and selected Category 2 project managers.

For very high visibility programs and Category 1 projects, the NASA Administrator and the NASA Associate Administrator may concur in these assignments.

---

[30] wInsight™ is a COTS tool that facilitates effective analysis and reporting of EVM data for management insight and control. wInsight™ is available for use by all NASA programs/projects. See your Center EVMFP for access and training.

5.15.2 Certifying Program and Project Managers

In a letter dated April 25, 2007, the White House Office of Management and Budget (OMB) announced a new set of requirements for program and project management certification that applies to all civilian agencies. OMB's Federal Acquisition Certification for Program and Project Managers (FAC-P/PM) outlines the baseline competencies, training, and experience required for program and project managers in the Federal government. This document may be obtained at http://www.whitehouse.gov/sites/default/files/omb/assets/omb/procurement/workforce/fed_acq_cert_042507.pdf.

To meet these requirements, NASA has established a process to:

- Certify existing experienced program and project managers who manage major acquisitions with LCCs greater than $250 million.
- Ensure certification of future program and project managers assigned to manage major acquisitions with LCCs greater than $250 million.
- Provide an Agency-wide career development framework to support the development of individuals pursuing program or project management career paths.
- Monitor and record the continuous learning achievements of certified program and project managers.
- Manage the process and maintain supporting documentation.

5.15.3 Agency Roles and Responsibilities

- NASA Centers have the responsibility to establish Center review panels that will inventory and validate the capabilities of their program and project managers in accordance with the certification requirements.
- The NASA Office of the Chief Engineer (OCE) has the ultimate responsibility for endorsing certification of NASA employees based on Center-validated career experience and Center reviews and recommendations.
- NASA Mission Directorates maintain an awareness of certified program and project managers within their directorates.
- NASA Academy of Program/Project and Engineering Leadership (APPEL) provides a structured approach to program and project management development through life-long learning at the individual, team, and community level, including on-the-job work experiences, attendance at core and in-depth courses, and participation in knowledge-sharing activities. APPEL also develops the tools and resources for Center implementation.

- The **NASA Acquisition Career Manager** (appointed by the NASA Chief Acquisition Officer) is responsible for oversight of the Agency process for certifying program and project managers.

- **Program and project management practitioners** take the lead in participating in the experiences and training necessary to acquire the competency proficiencies to better perform their job responsibilities and to obtain certification.

5.15.4 Program or Project Manager Certification

The designated point of contact at each Center will establish a Center review panel. This panel will have the responsibility to inventory and validate the capabilities of designated program and project managers (existing or future program and project managers managing major acquisitions) at the FAC-P/PM Senior/Expert certification level. The general roles and responsibilities of the Center review panel are to:

- Validate and approve the satisfaction of certification requirements and attainment of established criteria by the Center program and project management candidates.

- Forward names of recommended candidates to Center Directors for signature and then to the NASA Chief Engineer for final endorsement.

- Ensure candidate records are accumulated and maintained to satisfy OMB tracking requirements.

- Monitor and track workforce members as necessary to ensure training, developmental activities, and experiences are being made available. The System for Administration, Training, and Educational Resources for NASA (SATERN) is used as a resource for tracking development.

- Monitor and track the continuous learning activities of certified program and project managers.

5.15.4.1 Certification Process

OMB requires certification at the senior/expert level for NASA program or project managers who are currently managing major acquisitions with a LCC of more than $250 million. OMB also requires that future program or project managers assigned to projects designated as major acquisitions be certified. Program or project managers assigned to these projects in the future will have one year to become certified if they do not possess the required NASA-awarded FAC-P/PM certification at the time they assume the role.

The Center point of contact verifies the list of existing program and project managers and designates any additional candidates the Center deems eligible for certification at the senior/expert level. Each prospective program or project manager then creates a personal development portfolio (PDP), which documents their experience and development accomplishments. This PDP provides as much information as needed to assess the program or project manager's capabilities relative to OMB's certification requirements. The PDP needs to contain, at a minimum, a current resume, a completed NASA Program/Project Manager (P/PM) Competency Assessment, a supervisory endorsement, a SATERN training record if applicable, and any other supporting documentation the program or project manager deems necessary.

5.15.4.2 The Resume

The resume is a key component that needs to reflect the program or project manager's job history, documenting responsibilities in leading projects and/or programs. To meet OMB requirements for FAC-P/PM Senior/Expert Level certification, the program or project manager needs to complete at least four years of program and project management experience on projects and/or programs. This includes responsibilities such as managing and evaluating Agency acquisition investment performance, developing and managing a program or project budget, building and presenting a successful business case, reporting program or project results, strategic planning, and high-level communication with internal and external stakeholders.

The resume needs to be comprehensive enough that the review panel members and any other reviewers are able to assess the program or project manager's length and types of experiences.

5.15.4.3 The Competency Assessment

The FAC-P/PM requires essential competencies and levels of proficiency for certification. The FAC competency areas are encompassed within the existing NASA P/PM Competency Model which consists of 31 competencies, including 12 common FAC competencies. Table 5-13 shows an example of competencies, which are regularly updated, and their definitions. For each competency, the program or project manager is asked to select the highest degree to which he or she is able to demonstrate a level of proficiency. (More detailed current information on the competencies, including definitions and levels of proficiency, is available on the APPEL website http://appel.nasa.gov/pm-se/project-management-and-systems-engineering-competency-model/.

For senior/expert certification, the program or project manager needs to be able to demonstrate Level 4 proficiency for the 12 common competencies.

He or she needs to be able to demonstrate Level 3 proficiency for at least 80 percent of the remaining 19 NASA competencies.

The program or project manager is also asked to identify how the capability to perform at the specified proficiency level was achieved. Examples include courses, on-the-job training, knowledge-sharing activities, rotational assignments, government or professional organization certification, or other individual assignments. There needs to be some traceability, either on the resume, the training record, or other materials that supports the development experience noted on the competency assessment. For example, if a program or project manager identifies a rotational assignment as a development activity, some information about the rotational assignment (i.e., when, what office, etc.) needs to be referenced on the individual's resume.

Additionally, OMB requires that through acquiring the underlying competencies, senior/expert-level program and project managers possess the capabilities below. Review panels use these as additional guidelines for assessing program and project managers:

- Knowledge and skills to manage and evaluate moderate to high-risk programs or projects that require significant acquisition investment and Agency knowledge and experience.
- Ability to manage and evaluate a program or project and create an environment for program or project success.
- Ability to manage and evaluate the requirements development process, overseeing junior-level team members in creation, development, and implementation.
- Expert ability to use, manage, and evaluate management processes, including performance-based management techniques.
- Expert ability to manage and evaluate the use of Earned Value Management (EVM) as it relates to acquisition investments.

5.15.4.4 The Supervisory Endorsement

The portfolio also needs to include a signed endorsement from the supervisor. This endorsement indicates the supervisor's concurrence that the individual's experience, competency proficiency level, and capabilities meet OMB's requirements for senior/expert-level certification.

For new supervisors who may not be aware of the candidate's capabilities, the Center review panel can use its own discretion in allowing the program or project manager to identify other individuals who can provide validation.

Based on the candidate's PDP, the Center Director recommends the candidate for certification and NASA Chief Engineer provides the final

Table 5-13 NASA Program and Project Management Competencies and Common Competencies

Project Proposal	Conceptualizing, analyzing, and defining program/project plans and requirements and using technical expertise to write, manage, and submit winning proposals. Also involves developing functional, physical, and operational architectures including life-cycle costing.
Requirements Development	Developing project requirements using functional analysis, decomposition, and allocation; finalizing requirements into the baseline; and managing requirements so that changes are minimal. Defining, developing, verifying, reviewing and managing changes to program/project requirements.
Acquisition Management	Developing, implementing, and monitoring acquisition strategies, procurement processes, contract activities, and approval requirements to support flight hardware/software or other project requirements.
Project Planning	Developing effective project management plans and technical integration of project elements for small, moderate, and complex projects including scope definition, schedule and resource estimation and allocation for all project phase activities from concept to launch and tracking.
Cost-Estimating	Developing credible cost estimates to support a variety of systems engineering trade studies, affordability analyses, strategic planning, capital investment decision-making, and budget preparation during project planning. Also, providing information for independent assessments as required.
Risk Management	Risk-Informed Decision Making (RIDM) for selection of program/project alternatives; Continuous Risk Management (CRM) for identifying, analyzing, planning, tracking, controlling, and communicating and documenting individual and aggregate risks for the purpose of meeting program/project objectives within stated risk tolerance levels.
Budget and Full Cost Management	Executing NASA and Center budgeting processes for annual (PPBE) and life-cycle budget projections ensuring consistency between resource availability and project resource needs, including staffing, facilities, equipment, and budget.
Capital Management	Allocating, tracking, and managing funding and other capital resources within a project element, project or program.
Systems Engineering	Integrating technical processes to define, develop, produce, and operate the project's systems in the most technically robust and cost-effective way possible. (See Systems Engineering Competency Model for specific competencies.)
Design and Development	Developing subsystems to meet implementation requirements and producing, integrating, verifying, and testing the subsystem/ system to achieve product quality requirements and optimal technical performance.
Contract Management	Performing acquisition management and monitoring contractor activities to ensure hardware/software components are delivered on time, at projected costs, and meet all performance requirements. Also involves performing variance reporting and change control functions.
Stakeholder Management	Identifying, soliciting, and executing of planning interrelationships with those individuals and organizations that are actively involved in the project, exert influence over the project and its results, or whose interests may be positively or negatively affected as a result of project execution or project completion.
Technology Transfer and Communication	Evaluating the feasibility, development, progression, readiness, cost, risk, and benefits of new technologies so they can be developed and transferred efficiently and effectively to project stakeholders or for possible commercialization.
Tracking/Trending of Project Performance	Monitoring and evaluating performance metrics, project risks, and earned value data to analyze, assess and report program/project status and technical performance.
Project Control	Performing technical and programmatic activities to control cost, schedule, and technical content and configuration to assure the project's performance is within approved baseline and to address performance variances.
Project Review and Evaluation	Planning, conducting and managing internal and external project programmatic and technical reviews that include using metrics to monitor and track the status of the project.

Table 5-13 NASA Program and Project Management Competencies and Common Competencies (continued)

Agency Structure, Mission and Internal Goals	Understanding and successfully adapting work approach and style to NASA's functional, social, cultural, and political structure and interrelationships to achieve Agency, Mission, Directorate, Center, program and project goals. Includes aligning activities with Agency vision, mission, objectives, goals and plans.
NASA PM/SE Procedure & Guidelines	Structuring activities to comply with relevant Agency and Center processes and guidelines, including NPR 7120.5 and NPR 7123.1.
External Relationships	Maintaining cognizance of the policies and procedures of other organizations by participating in professional societies/ organizations, contributing to professional development activities, researching best practices from external sources such as industry standards, procedures, and regulations and Universities, and developing international partnerships and agreements, where applicable, complying with ITAR and as well as international agreements and standards.
Staffing and Performance	All elements of personnel management including, identifying, recruiting, selecting, managing, and evaluating the team members to achieve a coherent, efficient, and effective team. Includes vigorous open communications, decision-making processes, and working relationships.
Team Dynamics and Management	Managing the team aspects of the workforce. This requires: working cooperatively with diverse team members; designing, facilitating, and managing team processes; developing and implementing strategies to promote team morale and productivity; motivating and rewarding team members' performance; managing relationships among team members, customers, stakeholders, and partners; and facilitating brainstorming sessions, conflict resolution, negotiation and problem solving, communication, collaboration, integration and team meetings.
Security	Assuring that all proprietary, classified and privileged information is protected from unauthorized use and dissemination. Also requires identification of information technology (IT) security requirements and developing and implementing an effective IT security plan.
Workplace Safety	Ensuring that workplace safety is an integral part of developing products by applying systems safety analysis techniques throughout the project life cycle and integrating critical hazard elimination/ mitigation measures into risk management and safety plans.
Safety and Mission Assurance	Activities associated with assuring the safety of personnel and property and success of the project. These activities include: Environmental Impact Statements; hazards analyses, elimination, and mitigation; mishap investigations; failure review boards; the flight safety review process; and safety, mission assurance, and risk management plans.
Mentoring and Coaching	Activities designed to help less-experienced members of the team to advance their knowledge and careers by: acting as an advisor, sponsor, or confidant who shares knowledge about NASA's functional, social, cultural, and political aspects or provides counseling to cultivate skills in order to enhance individual, team, and organizational performance and growth.
Communication	Implementing effective strategies for clear and constructive communication both internally within the team and externally to stakeholders, other experts, contractors and others. Also involves communicating decisions in a timely manner.
Leadership	Influencing, inspiring, and motivating individuals and teams to accomplish goals; creating conditions for individuals and teams to be effective; and recognizing and rewarding individual and team achievements.
Ethics	Demonstrating integrity, ethical conduct, and acceptable behavior in all project activities in line with Federal Government principles.
Knowledge Capture and Transfer	Capturing and transferring knowledge in an organized fashion to improve performance and reduce risk associated with future programs and projects.
Knowledge Sharing	Sharing organizational practices and approaches related to generating, capturing, disseminating know-how and other content relevant to NASA's business and processes.

Note: NASA common competencies are in blue.

endorsement. The NASA Chief Engineer signs and sends a letter of endorsement to the acquisition career manager who forwards copies to the program or project manager, a Center review panel representative, and the appropriate Mission Directorate, and ensures update of the program or project manager's SATERN record to reflect certification level, date, etc.

5.15.4.5 Meeting Certification Requirements

In the event a program or project manager does not satisfy the requirements for senior/expert-level certification, the Center review panel, along with the program or project manager and the supervisor, will identify development activities and a timeframe to complete the activities. The program or project manager will complete the identified activities, update his or her portfolio, and resubmit it to the panel for review. This process can be repeated if necessary. The maximum timeframe the program or project manager has to satisfy the requirements is one year.

5.15.4.6 Maintaining Certification

To maintain the certification, certified program and project managers are required to earn 80 continuous learning points (CLPs) of skills currency every two years.

Below are examples of the some of the continuous learning activities for which program and project managers can earn CLPs:

- Serving on NASA boards
- Serving as an instructor or student for APPEL and a Center
- Obtaining other formal education
- Publishing technical papers or other documents
- Rotating jobs
- Attending the PM Challenge, Master's Forum, or Principal Investigator (PI) Forum
- Participating on a Center or Agency team to define policy or improve processes
- Participating in critical NASA board activities (i.e., the Program and Project Management Board (PPMB))
- Participating in critical NASA or other technical Agency reviews
- Serving on a Standing Review Board (SRB), failure review board, or other special-purpose team or committee
- Mentoring or coaching

5.15.4.7 Meeting the Continuous Learning Requirements

If a program or project manager does not meet the continuous learning requirements within the two-year time period, the certification will become conditional. The individual's supervisor and a representative from the Center review panel will meet with the program or project manager and discuss how to satisfy the requirements.

5.15.4.8 Meeting Tracking and Reporting Requirements

Centers are responsible for maintaining all documentation for every reviewed and certified program and project manager. The Center review panel designates a point of contact for records management to maintain copies of PDPs, the recommendation letter, and any documentation or rationale for program and project managers requiring further development.

Appendices

A Definitions

Acquisition. The process for obtaining the systems, research, services, construction, and supplies that NASA needs to fulfill its missions. Acquisition—which may include procurement (contracting for products and services)—begins with an idea or proposal that aligns with the NASA Strategic Plan and fulfills an identified need and ends with the completion of the program or project or the final disposition of the product or service.

Acquisition Plan. The integrated acquisition strategy that enables a program or project to meet its mission objectives and provides the best value to NASA. (See a description in Section 3.4 of the Program Plan and Project Plan templates, Appendices G and H.)

Acquisition Strategy Meeting. A forum where senior Agency management reviews major acquisitions in programs and projects before authorizing significant budget expenditures. The ASM is held at the Mission Directorate/Mission Support Office level, implementing the decisions that flow out of the earlier Agency acquisition strategy planning. The ASM is typically held early in Formulation, but the timing is determined by the Mission Directorate. The ASM focuses on considerations such as impacting the Agency workforce, maintaining core capabilities and make-or-buy planning, and supporting Center assignments and potential partners.

Agency Baseline Commitment. Establishes and documents an integrated set of project requirements, cost, schedule, technical content, and an agreed-to JCL that forms the basis for NASA's commitment to the external entities of OMB and Congress. Only one official baseline exists for a NASA program or project, and it is the Agency Baseline Commitment.

Agency Program Management Council. The senior management group, chaired by the NASA Associate Administrator or designee that is responsible for reviewing Formulation performance, recommending approval, and overseeing implementation of programs and Category 1 projects according to Agency commitments, priorities, and policies.

Agreement. The statement (oral or written) of an exchange of promises. Parties to a binding agreement can be held accountable for its proper execution and a change to the agreement requires a mutual modification or amendment to the agreement or a new agreement.

Allocated Requirements. Requirements that are established by dividing or otherwise allocating a high-level requirement into lower level requirements.

Analysis of Alternatives. A formal analysis method that compares alternative approaches by estimating their ability to satisfy mission requirements through an effectiveness analysis and by estimating their Life-Cycle Costs (LCCs) through cost analysis. The results of these two analyses are used together to produce a cost-effectiveness comparison that allows decision makers to assess the relative value or potential programmatic returns of the alternatives. An analysis of alternatives broadly examines multiple elements of program or project alternatives (including technical performance, risk, LCC, and programmatic aspects).

Announcement of Opportunity. An AO is one form of a NASA Broad Agency Announcement (BAA), which may be used to conduct a type of public/private competition for an R&D effort. NASA solicits, accepts, and evaluates proposers permitted by the terms of the AO. At the discretion of NASA, an AO may permit the following categories to propose: academia, industry, not-for-profits, Government laboratories, Federally Funded Research and Development Centers (FFRDC), NASA Centers, and the Jet Propulsion Laboratory (JPL).

Annual Performance Plan. The Annual Performance Plan (APP) shows the supporting strategic objectives and annual performance goals that are being implemented by one or more program activities for each strategic goal. This plan covers each program activity in the budget and is comprehensive of the strategic objectives. Additionally, the plan addresses the Agency's contributions to Cross-Agency Priority Goals.

Annual Performance Report. NASA's Annual Performance Report (APR) provides the public with key information on whether the performance commitments aligned to the annual budget request were met, and if unmet, plans to address any challenges that were barriers to success. This document also includes progress toward NASA's priority goals.

Approval. Authorization by a required management official to proceed with a proposed course of action. Approvals are documented.

Architectural Control Document. A configuration-controlled document or series of documents that embodies a cross-Agency mission architecture(s), including the structure, relationships, interfaces, principles, assumptions, and results of the analysis of alternatives that govern the design and implementation of the enabling mission systems.

Assure. For the purpose of defining policy in NPD 1000.0B, NPR 7120.5E, and this handbook, "assure" means to promise or say with confidence. It is more about saying than doing.
For example, "I assure you that you'll be warm enough."

Baseline (document context). Implies the expectation of a finished product, though updates may be needed as circumstances warrant. All approvals required by Center policies and procedures have been obtained.

Baseline (general context). An agreed-to set of requirements, cost, schedule, designs, documents, etc., that will have changes controlled through a formal approval and monitoring process.

Baseline Performance Review. A monthly Agency-level independent assessment to inform senior leadership of performance and progress toward the Agency's mission and program or project performance. The monthly meeting encompasses a review of crosscutting mission support issues and all NASA mission areas.

Baseline Science Requirements. The mission performance requirements necessary to achieve the full science objectives of the mission. (Also see Threshold Science Requirements.)

Basis of Estimate. The documentation of the ground rules, assumptions, and drivers used in developing the cost and schedule estimates, including applicable model inputs, rationale or justification for analogies, and details supporting cost and schedule estimates. The BoE is contained in material available to the Standing Review Board (SRB) and management as part of the life-cycle review and Key Decision Point (KDP) process.

Budget. A financial plan that provides a formal estimate of future revenues and obligations for a definite period of time for approved programs, projects, and activities. (See NPR 9420.1 and NPR 9470.1 for other related financial management terms and definitions.)

Center Management Council. The council at a Center that performs oversight of programs and projects by evaluating all program and project work executed at that Center.

Change Request. A change to a prescribed requirement set forth in an Agency or Center document intended for all programs and projects for all time.

Compliance Matrix. The Compliance Matrix (Appendix C in NPR 7120.5) documents whether and how the program or project complies with the requirements of NPR 7120.5. It provides rationale and approvals for waivers from requirements and is part of retrievable program and project records.

Concept Documentation (formerly Mission Concept Report). Documentation that captures and communicates a feasible concept that meets the goals and objectives of the mission, including results of analyses of alternative concepts, the concept of operations, preliminary risks, and potential descopes. It may include images, tabular data, graphs, and other descriptive material. The Concept Documentation is approved at Mission Concept Review (MCR).

Concurrence. A documented agreement by a management official that a proposed course of action is acceptable.

Confidence Level. A probabilistic assessment of the level of confidence of achieving a specific goal.

Configuration Management. A management discipline applied over a product's life cycle to provide visibility into and control changes to performance, functionality, and physical characteristics.

Continuous Risk Management. A systematic and iterative process that efficiently identifies, analyzes, plans, tracks, controls, communicates, and documents risks associated with implementation of designs, plans, and processes.

Contract. A mutually binding legal relationship obligating the seller to furnish the supplies or services (including construction) and the buyer to pay for them. It includes all types of commitments that obligate the Government to an expenditure of appropriated funds and that, except as otherwise authorized, are in writing. In addition to bilateral instruments, contracts include (but are not limited to) awards and notices of awards; job orders or task letters issued under basic ordering agreements; letter contracts; orders, such as purchase orders, under which the contract becomes effective by written acceptance or performance; and bilateral contract modifications. Contracts do not include grants and cooperative agreements.

Control Account. A management control point at which budgets (resource plans) and actual costs can be accumulated and compared to projected costs and resources and the earned value for management control purposes. A control account is a natural measurement point for planning and control since it represents the work assigned to one responsible organizational element (such as an integrated product team) for a single WBS element. Control Accounts are used to set the budget accounts in the finance system (N2).

Control Account Manager. A manager responsible for a Control Account and for the planning, development and execution of the budget content for those accounts.

Convening Authority. The management official(s) responsible for convening a program or project review; establishing the Terms of Reference (ToR), including review objectives and success criteria; appointing the Standing Review Board (SRB) chair; and concurring in SRB membership. These officials receive the documented results of the review.

Cost Analysis Data Requirement. A formal document designed to help managers understand the cost and cost risk of space flight projects. The CADRe consists of a Part A "Narrative" and a Part B "Technical Data" in tabular form, both provided by the program or project or Cost Analysis Division (CAD). Also, the project team produces the project life-cycle cost estimate, (LCCE) schedule, and risk identification, which is appended as Part C.

Decision Authority (program and project context). The individual authorized by the Agency to make important decisions on programs and projects under their authority.

Decision Memorandum. The document that summarizes the decisions made at Key Decision Points (KDPs) or as necessary in between KDPs. The decision memorandum includes the Agency Baseline Commitment (ABC) (if applicable), Management Agreement cost and schedule, Unallocated Future Expenses (UFE), and schedule margin managed above the project (that is, outside of the Management Agreement approved cost), as well as life-cycle cost and schedule estimates, as required.

Decommissioning. The process of ending an operating mission and the attendant project as a result of a planned end of the mission or project termination. Decommissioning includes final delivery of any remaining project deliverables, disposal of the spacecraft and all its various supporting systems, closeout of contracts and financial obligations, and archiving of project/mission operational and scientific data and artifacts. Decommissioning does not mean that scientific data analysis ceases, only that the project will no longer provide the resources for continued research and analysis.

Derived Requirements. Requirements arising from constraints, consideration of issues implied but not explicitly stated in the high-level direction provided by NASA Headquarters and Center institutional requirements, factors introduced by the selected architecture, and the design. These requirements are finalized through requirements analysis as part of the overall systems engineering process and become part of the program or project requirements baseline.

Design Documentation. A document or series of documents that captures and communicates to others the specific technical aspects of a design. It may include images, tabular data, graphs, and other descriptive material. Design documentation is different from the Cost Analysis Data Requirement (CADRe), though parts of the design documentation may be repeated in the latter.

Development Costs. The total of all costs from the period beginning with the approval to proceed to Implementation at the beginning of Phase C through operational readiness at the end of Phase D.

Deviation. A documented authorization releasing a program or project from meeting a requirement before the requirement is put under configuration control at the level the requirement will be implemented.

Directorate Program Management Council. The forum that evaluates all programs and projects executed within that Mission Directorate and provides input to the Mission Directorate Associate Administrator (MDAA). For programs and Category 1 projects, the MDAA carries forward the DPMC findings and recommendations to the Agency Program Management Council (APMC).

Disposal. The process of eliminating a project's assets, including the spacecraft and ground systems. Disposal includes the reorbiting, deorbiting, and/or passivation (i.e., the process of removing stored energy from a space structure at the end of the mission that could result in an explosion or deflagration of the space structure) of a spacecraft.

Dissenting Opinion. A disagreement with a decision or action that is based on a sound rationale (not on unyielding opposition) that an individual judges is of sufficient importance that it warrants a specific review and decision by higher-level management, and the individual specifically requests that the dissent be recorded and resolved by the Dissenting Opinion process.

Earned Value Management. A tool for measuring and assessing project performance through the integration of technical scope with schedule and cost objectives during the execution of the project. EVM provides quantification of technical progress, enabling management to gain insight into project status and project completion costs and schedules. Two essential characteristics of successful EVM are EVM system data integrity and carefully targeted monthly EVM data analyses (e.g., identification of risky WBS elements).

Earned Value Management System. An integrated management system and its related subsystems that allow for planning all work scope to completion; assignment of authority and responsibility at the work performance level; integration of the cost, schedule, and technical aspects of the work into a detailed baseline plan; objective measurement of progress (earned value) at the work performance level; accumulation and assignment of actual costs; analysis of variances from plans; summarization and reporting of performance data to higher levels of management for action; forecast of achievement of milestones and completion of events; forecast of final costs; and disciplined baseline maintenance and incorporation of baseline revisions in a timely manner.

Engineering Requirements. Requirements defined to achieve programmatic requirements and relating to the application of engineering principles, applied science, or industrial techniques.

Ensure. For the purpose of defining policy in NPD 1000.0B, NPR 7120.5E, and this handbook, "ensure" means to do or have what is necessary for success. For example, "These blankets ensure that you'll be warm enough."

Environmental Management. The activity of ensuring that program and project actions and decisions that may potentially affect or damage the environment are assessed during the Formulation Phase and reevaluated throughout Implementation. This activity is performed according to all NASA policy and Federal, State, Tribal Government, and local environmental laws and regulations.

Evaluation. The continual self- and independent assessment of the performance of a program or project and incorporation of the evaluation findings to ensure adequacy of planning and execution according to plans.

Extended Operations. Extended operations are operations conducted after the planned prime mission operations are complete. Extended operations require approval, as determined by the Mission Directorate. Once the extension of operations is approved, program or project documentation must be updated.

Final (document context). Implies the expectation of a finished product. All approvals required by Center policies and procedures have been obtained.

Formulation. The identification of how the program or project supports the Agency's strategic goals; the assessment of feasibility, technology, and concepts; risk assessment, team building, development of operations concepts, and acquisition strategies; establishment of high-level requirements and success criteria; the preparation of plans, budgets, and schedules essential to the success of a program or project; and the establishment of control systems to ensure performance to those plans and alignment with current Agency strategies.

Formulation Agreement. The Formulation Agreement is prepared by the project to establish the technical and acquisition work that needs to be conducted during Formulation and defines the schedule and funding requirements during Phase A and Phase B for that work.

Formulation Authorization Document. This is the document issued by the Mission Directorate Associate Administrator (MDAA) to authorize the formulation of a program whose goals will fulfill part of the Agency's Strategic Plan and Mission Directorate strategies and establish the expectations and constraints for activity in the Formulation Phase. In addition, a FAD or equivalent is used to authorize the formulation of a project. (See Appendix E in NPR 7120.5E.)

Functional Requirements. Requirements that specify what a system needs to do. Requirements that specify a function that a system or component needs to be able to perform.

Funding (budget authority). The authority provided by law to incur financial obligations that will result in expenditures. There are four basic forms of budget authority, but only two are applicable to NASA: appropriations and spending authority from offsetting collections (reimbursables and working capital funds). Budget authority is provided or delegated to programs and projects through the Agency's funds distribution process.

Health and Medical Requirements. Requirements defined by the Office of the Chief Health and Medical Officer (OCHMO).

Implementation. The execution of approved plans for the development and operation of a program or project, and the use of control systems to ensure performance to approved plans and continued alignment with the Agency's goals.

Independent Assessment(s) (includes reviews, evaluations, audits, analysis oversight, investigations). Assessments are independent to the extent the involved personnel apply their expertise impartially and without any conflict of interest or inappropriate interference or influence, particularly from the organization(s) being assessed.

Independent Funding (context of Technical Authority). The funding of Technical Authorities (TAs) is considered independent if funding originating from the Mission Directorate or other Programmatic Authorities is provided to the Center in a manner that cannot be used to influence the technical independence or security of TAs.

Industrial Base. The capabilities residing in either the commercial or government sector required to design, develop, manufacture, launch, and service the program or project. This encompasses related manufacturing facilities, supply chain operations and management, a skilled workforce, launch infrastructure, research and development, and support services.

Information Technology. Any equipment or interconnected system(s) or subsystem(s) of equipment that is used in the automatic acquisition, storage, analysis, evaluation, manipulation, management, movement, control, display, switching, interchange, transmission, or reception of data or information by the Agency.

Institutional Authority. Institutional Authority encompasses all those organizations and authorities not in the Programmatic Authority. This includes Engineering, Safety and Mission Assurance, and Health and Medical organizations; Mission Support organizations; and Center Directors.

Institutional Requirements. Requirements that focus on how NASA does business that are independent of the particular program or project. There are five types: engineering, program or project management, safety and mission assurance, health and medical, and Mission Support Office (MSO) functional requirements.

Integrated Baseline Review. A risk-based review conducted by program or project management to ensure a mutual understanding between the customer and supplier of the risks inherent in the supplier's Performance Measurement Baseline (PMB) and to ensure that the PMB is realistic for accomplishing all of the authorized work within the authorized schedule and budget.

Integrated Center Management Council. The forum used by projects and programs that are being implemented by more than one Center and includes representatives from all participating Centers. The ICMC will be chaired by the director of the Center (or representative) responsible for program or project management.

Integrated Logistics Support. The management, engineering activities, analysis, and information management associated with design requirements definition, material procurement and distribution, maintenance, supply replacement, transportation, and disposal that are identified by space flight and ground systems supportability objectives.

Integrated Master Schedule. A logic network-based schedule that reflects the total project scope of work, traceable to the Work Breakdown Structure (WBS), as discrete and measurable tasks/milestones and supporting elements that are time phased through the use of valid durations based on available or projected resources and well-defined interdependencies.

Integration Plan. The integration and verification strategies for a project interface with the system design and decomposition into the lower-level elements. The integration plan is structured to bring the elements together to assemble each subsystem and to bring all of the subsystems together to assemble the system/product. The primary purposes of the integration plan are: (1) to describe this coordinated integration effort that supports the implementation strategy, (2) to describe for the participants what needs to be done in each integration step, and (3) to identify the required resources and when and where they will be needed.

Interface Control Document. An agreement between two or more parties on how interrelated systems will interface with each other. It documents interfaces between things like electrical connectors (what type, how many pins, what signals will be on each of the pins, etc.); fluid connectors (type of connector or of fluid being passed, flow rates of the fluid, etc.); mechanical (types of fasteners, bolt patterns, etc.); and any other interfaces that might be involved.

Joint Cost and Schedule Confidence Level. (1) The probability that cost will be equal to or less than the targeted cost and schedule will be equal to or less than the targeted schedule date. (2) A process and product that helps inform management of the likelihood of a project's programmatic success. (3) A process that combines a project's cost, schedule, and risk into a complete picture. JCL is not a specific methodology (e.g., resource-loaded schedule) or a product from a specific tool. The JCL calculation includes consideration of the risk associated with all elements, regardless of whether or not they are funded from appropriations or managed outside of the project. JCL calculations include the period from Key Decision Point (KDP) C through the hand over to operations, i.e., end of the on-orbit checkout.

Key Decision Point. The event at which the Decision Authority determines the readiness of a program or project to progress to the next phase of the life cycle (or to the next KDP).

Knowledge Management. A collection of policies, processes, and practices relating to the use of intellectual- and knowledge-based assets in an organization.

Leading Indicator. A measure for evaluating the effectiveness of how a specific activity is applied on a program in a manner that provides information about impacts likely to affect the system performance objectives. A leading indicator may be an individual measure, or collection of measures, predictive of future system and project performance before the performance is realized. The goal of the indicators is to provide insight into potential future states to allow management to take action before problems are realized.

Lessons Learned. Captured knowledge or understanding gained through experience which, if shared, would benefit the work of others. Unlike a best practice, lessons learned describes a specific event that occurred and provides recommendations for obtaining a repeat of success or for avoiding reoccurrence of an adverse work practice or experience.

Life-Cycle Cost. The total of the direct, indirect, recurring, nonrecurring, and other related expenses both incurred and estimated to be incurred in the design, development, verification, production, deployment, prime mission operation, maintenance, support, and disposal of a project, including closeout, but not extended operations. The LCC of a project or system can also be defined as the total cost of ownership over the project or system's planned life cycle from Formulation (excluding Pre–Phase A) through Implementation (excluding extended operations). The LCC includes the cost of the launch vehicle.

Life-Cycle Review. A review of a program or project designed to provide a periodic assessment of the technical and programmatic status and health of a program or project at a key point in the life cycle, e.g., Preliminary Design Review (PDR) or Critical Design Review (CDR). Certain life-cycle reviews provide the basis for the Decision Authority to approve or disapprove the transition of a program or project at a Key Decision Point (KDP) to the next life-cycle phase.

Loosely Coupled Programs. These programs address specific objectives through multiple space flight projects of varied scope. While each individual project has an assigned set of mission objectives, architectural and technological synergies and strategies that benefit the program as a whole are explored during the Formulation process. For instance, Mars orbiters designed for more than one Mars year in orbit are required to carry a communication system to support present and future landers.

Management Agreement. Within the Decision Memorandum, the parameters and authorities over which the program or project manager has management control constitute the program or project Management Agreement. A program or project manager has the authority to manage within the Management Agreement and is accountable for compliance with the terms of the agreement.

Margin. The allowances carried in budget, projected schedules, and technical performance parameters (e.g., weight, power, or memory) to account for uncertainties and risks. Margins are allocated in the formulation process, based on assessments of risks, and are typically consumed as the program or project proceeds through the life cycle.

Metric. A measurement taken over a period of time that communicates vital information about the status or performance of a system, process, or activity.

Mission. A major activity required to accomplish an Agency goal or to effectively pursue a scientific, technological, or engineering opportunity directly related to an Agency goal. Mission needs are independent of any particular system or technological solution.

Non-Applicable Requirement. Any requirement not relevant; not capable of being applied.

Operations Concept (formerly Mission Operations Concept). A description of how the flight system and the ground system are used together to ensure that the mission operations can be accomplished reasonably. This might include how mission data of interest, such as engineering or scientific data, are captured, returned to Earth, processed, made available to users, and archived for future reference. The Operations Concept needs to describe how the flight system and ground system work together across mission phases for launch, cruise, critical activities, science observations, and the end of the mission to achieve the mission. The Operations Concept is baselined at the Preliminary Design Review (PDR) with the initial preliminary operations concept required at the Mission Concept Review (MCR) by the product tables in NPR 7120.5E.

Orbital Debris. Any object placed in space by humans that remains in orbit and no longer serves any useful function. Objects range from spacecraft to spent launch vehicle stages to components and also include materials, trash, refuse, fragments, and other objects that are overtly or inadvertently cast off or generated.

Organizational Breakdown Structure. The project hierarchy of line and functional organizations as applied to the specific project. The OBS describes the organizations responsible for performing the authorized work.

Passback. In the spring of each year, the U.S. Office of Management and Budget (OMB) issues planning guidance to executive agencies for the budget beginning October 1 of the following year. In September, Agencies submit their initial budget requests to OMB. During October and November, OMB staff review the agency budget requests against the President's priorities, program performance, and budget constraints. In November and December, the President makes decisions on agency requests based on recommendations from the OMB director. OMB informs agencies of the President's decisions in what is commonly referred to as the OMB "passback." Agencies may appeal these decisions to the OMB director and in some cases directly to the President, but the timeframe for appeals is small.

Performance Measurement Baseline. The time-phased cost plan for accomplishing all authorized work scope in a project's life cycle, which includes both NASA internal costs and supplier costs. The project's performance against the PMB is measured using Earned Value Management (EVM), if required, or other performance measurement techniques if EVM is not required. The PMB does not include Unallocated Future Expenses (UFE).

Performance Requirement. A performance requirement describes in measurable terms how well a function is to be executed or accomplished. A performance requirement is generally couched in terms of degree, rate, quantity, quality, timeliness, coverage, timeliness or readiness and so on. A performance requirement can also describe the conditions under which the function is to be performed.

Preliminary (document context). Implies that the product has received initial review in accordance with Center best practices. The content is considered correct, though some TBDs may remain. All approvals required by Center policies and procedures have been obtained. Major changes are expected.

Prescribed Requirement. A requirement levied on a lower organizational level by a higher organizational level.

Principal Investigator. A person who conceives an investigation and is responsible for carrying it out and reporting its results. In some cases, Principal Investigators (PIs) from industry and academia act as project managers for smaller development efforts with NASA personnel providing oversight.

Procurement Strategy Meeting. A forum where management reviews and approves the approach for the Agency's major and other selected procurements. Chaired by the Assistant Administrator for Procurement (or designee), the PSM addresses and documents information, activities, and decisions required by the Acquisition Regulation (FAR) and the NASA FAR Supplement (NFS) and incorporates NASA strategic guidance and decisions from the Acquisition Strategy Meeting (ASM) to ensure the alignment of the individual procurement action with NASA's portfolio and mission.

Program. A strategic investment by a Mission Directorate or Mission Support Office (MSO) that has a defined architecture and/or technical approach, requirements, funding level, and management structure that initiates and directs one or more projects. A program implements a strategic direction that the Agency has identified as needed to accomplish Agency goals and objectives. (See Section 2.4.)

Program Commitment Agreement. The contract between the NASA Associate Administrator and the responsible Mission Directorate Associate Administrator (MDAA) that authorizes transition from Formulation to Implementation of a program. (See Appendix D in NPR 7120.5E.)

Program Plan. The document that establishes the program's baseline for Implementation, signed by the Mission Directorate Associate Administrator (MDAA), Center Director(s), and program manager.

Program (Project) Team. All participants in program (project) Formulation and Implementation. This includes all direct reports and others that support meeting program (project) responsibilities.

Programmatic Authority. Programmatic Authority includes the Mission Directorates and their respective programs and projects. Individuals in these organizations are the official voices for their respective areas. Programmatic Authority sets, oversees, and ensures conformance to applicable programmatic requirements.

Programmatic Requirements. Requirements set by the Mission Directorate, program, project, and Principal Investigator (PI), if applicable. These include strategic scientific and exploration requirements, system performance requirements, safety requirements, and schedule, cost, and similar nontechnical constraints.

Project. A space flight project is a specific investment identified in a Program Plan having defined requirements, a life-cycle cost, a beginning, and an end. A project also has a management structure and may have interfaces to other projects, agencies, and international partners. A project yields new or revised products that directly address NASA's strategic goals.

Project Plan. The document that establishes the project's baseline for Implementation, signed by the responsible program manager, Center Director, project manager, and the Mission Directorate Associate Administrator (MDAA), if required. (See Appendix H in NPR 7120.5E.)

Rebaselining. The process that results in a change to a project's Agency Baseline Commitment (ABC).

Reimbursable Project. A project (including work, commodities, or services) for customers other than NASA for which reimbursable agreements have been signed by both the customer and NASA. The customer provides funding for the work performed on its behalf.

Replanning. The process by which a program or project updates or modifies its plans.

Request for Action/Review Item Discrepancy. The most common names for the comment forms that reviewers submit during life-cycle reviews that capture their comments, concerns, and/or issues about the product or documentation. Often, RIDs are used in a more formal way, requiring boards to disposition them and having to get agreements with the submitter, project, and board members for their disposition and closeout. RFAs are often treated more informally, almost as suggestions that may or may not be reacted to.

Reserves. Obsolete term. See Unallocated Future Expenses (UFEs).

Residual Risk. The remaining risk that exists after all mitigation actions have been implemented or exhausted in accordance with the risk management process. (See *NPD 8700.1, NASA Policy for Safety and Mission Success.*)

Resource Management Officer. The person responsible for integrating project inputs for budget planning and execution across many projects or control accounts.

Responsibility Assignment Matrix. A chart showing the relationship between the WBS elements and the organizations assigned responsibility for ensuring their accomplishment. The RAM normally depicts the assignment of each control account to a single manager, along with the assigned budget. The RAM is the result of cross-referencing the OBS with the WBS. Cross-referencing the WBS and OBS creates control accounts that facilitate schedule and cost performance measurement. The control account is the primary point for work authorization, work performance management, and work performance measurement (i.e., where the planned value is established, earned value is assessed, and actual costs are collected).

Risk. In the context of mission execution, risk is the potential for performance shortfalls, which may be realized in the future, with respect to achieving explicitly established and stated performance requirements. The performance shortfalls may be related to any one or more of the following mission execution domains: (1) safety, (2) technical, (3) cost, and (4) schedule. (See *NPR 8000.4, Agency Risk Management Procedural Requirements.*)

Risk Assessment. An evaluation of a risk item that determines: (1) what can go wrong, (2) how likely is it to occur, (3) what the consequences are, (4) what the uncertainties are that are associated with the likelihood and consequences, and (5) what the mitigation plans are.

Risk Management. Risk management includes Risk-Informed Decision Making (RIDM) and Continuous Risk Management (CRM) in an integrated framework. RIDM informs systems engineering decisions through better use of risk and uncertainty information in selecting alternatives and establishing baseline requirements. CRM manages risks over the course of the development and the Implementation Phase of the life cycle to ensure that safety, technical, cost, and schedule requirements are met. This is done to foster proactive risk management, to better inform decision making through better use of risk information, and then to more effectively manage Implementation risks by focusing the CRM process on the baseline performance requirements emerging from the RIDM process. (See *NPR 8000.4, Agency Risk Management Procedural Requirements*.) These processes are applied at a level of rigor commensurate with the complexity, cost, and criticality of the program.

Risk-Informed Decision Making. A risk-informed decision-making process that uses a diverse set of performance measures (some of which are model-based risk metrics) along with other considerations within a deliberative process to inform decision making.

Safety. Freedom from those conditions that can cause death, injury, occupational illness, damage to or loss of equipment or property, or damage to the environment.

Security. Protection of people, property, and information assets owned by NASA that covers physical assets, personnel, Information Technology (IT), communications, and operations.

Signature. A distinctive mark, characteristic, or thing that indicates identity; one's name as written by oneself.

Single-Project Programs. These programs tend to have long development and/or operational lifetimes, represent a large investment of Agency resources, and have contributions from multiple organizations/agencies. These programs frequently combine program and project management approaches, which they document through tailoring.

Space Act Agreements. Agreements NASA can enter into based on authorization from the Space Act. The National Aeronautics and Space Act of 1958, as amended (51 U.S.C. 20113(e)), authorizes NASA "to enter into and perform such … other transactions as may be necessary in the conduct of its work and on such terms as it may deem appropriate…" This authority enables NASA to enter into "Space Act Agreements (SAAs)" with organizations in the public and private sector. SAA partners can be a U.S. or foreign person or entity, an academic institution, a Federal, state, or local governmental unit, a foreign government, or an international organization, for profit, or not for profit.

Stakeholder. An individual or organizational customer having an interest (or stake) in the outcome or deliverable of a program or project.

Standards. Formal documents that establish a norm, requirement, or basis for comparison, a reference point to measure or evaluate against. A technical standard, for example, establishes uniform engineering or technical criteria, methods, processes, and practices. (Refer to *NPR 7120.10, Technical Standards for NASA Programs and Projects.*)

Standing Review Board. The board responsible for conducting independent reviews (life cycle and special) of a program or project and providing objective, expert judgments to the convening authorities. The reviews are conducted in accordance with approved Terms of Reference (ToR) and life-cycle requirements per NPR 7120.5 and NPR 7123.1.

Success Criteria. That portion of the top-level requirements that defines what is to be achieved to successfully satisfy NASA Strategic Plan objectives addressed by the program or project.

Suppliers. Each project office is a customer having a unique, multitiered hierarchy of suppliers to provide it products and services. A supplier may be a contractor, grantee, another NASA Center, university, international partner, or other government agency. Each project supplier is also a customer if it has authorized work to a supplier lower in the hierarchy.

Supply Chain. The specific group of suppliers and the interrelationships among the group members that is necessary to design, develop, manufacture, launch, and service the program or project. This encompasses all levels within a space system, including providers of raw materials, components, subsystems, systems, systems integrators, and services.

System. The combination of elements that function together to produce the capability required to meet a need. The elements include all hardware, software, equipment, facilities, personnel, processes, and procedures needed for this purpose.

Systems Engineering. A disciplined approach for the definition, implementation, integration, and operation of a system (product or service). The emphasis is on achieving stakeholder functional, physical, and operational performance requirements in the intended use environments over planned life within cost and schedule constraints. Systems engineering includes the engineering processes and technical management processes that consider the interface relationships across all elements of the system, other systems, or as a part of a larger system.

Tailoring. The process used to adjust or seek relief from a prescribed requirement to accommodate the needs of a specific task or activity (e.g., program or project). The tailoring process results in the generation of deviations and waivers depending on the timing of the request.

Technical Authority. Part of NASA's system of checks and balances that provides independent oversight of programs and projects in support of safety and mission success through the selection of individuals at delegated levels of authority. These individuals are the TAs. TA delegations are formal and traceable to the NASA Administrator. Individuals with TA are funded independently of a program or project.

Technical Authority Requirements. Requirements invoked by Office of the Chief Engineer (OCE), Office of Safety and Mission Assurance (OSMA), and Office of the Chief Health and Medical Officer (OCHMO) documents (e.g., NPRs or technical standards cited as program or project requirements) or contained in Center institutional documents. These requirements are the responsibility of the office or organization that established the requirement unless delegated elsewhere.

Technical Performance Measure. The set of critical or key performance parameters that are monitored by comparing the current actual achievement of the parameters with that anticipated at the current time and on future dates.

Technical Standard. Common and repeated use of rules, conditions, guidelines, or characteristics for products or related processes and production methods and related management systems practices; the definition of terms, classification of components; delineation of procedures; specification of dimensions, materials, performance, designs, or operations; measurement of quality and quantity in describing materials, processes, products, systems, services, or practices; test methods and sampling procedures; or descriptions of fit and measurements of size or strength. (Source: OMB Circular No. A-119, Federal Participation in the Development and Use of Voluntary Consensus Standards and in Conformity Assessment Activities.) (See *NPR 7120.10, Technical Standards for NASA Programs and Projects.*)

Technology Readiness Level. Provides a scale against which to measure the maturity of a technology. TRLs range from 1, Basic Technology Research, to 9, Systems Test, Launch, and Operations. Typically, a TRL of 6 (i.e., technology demonstrated in a relevant environment) is required for a technology to be integrated into a flight system. (See *Systems Engineering Handbook, NASA/SP-2007-6105 Rev 1*, p. 296 for more information on TRL levels and technology assessment.)

Termination Review. A review initiated by the Decision Authority for the purpose of securing a recommendation as to whether to continue or terminate a program or project. Failing to stay within the parameters or levels specified in controlling documents will result in consideration of a termination review.

Terms of Reference. A document specifying the nature, scope, schedule, and ground rules for an independent review or independent assessment.

Threshold Science Requirements. The mission performance requirements necessary to achieve the minimum mission content acceptable to invest in the mission. In some Announcements of Opportunity (AOs) used for competed missions, threshold science requirements may be called the "science floor" for the mission. (Also see Baseline Science Requirements.)

Tightly Coupled Programs. Programs with multiple projects that execute portions of a mission(s). No single project is capable of implementing a complete mission. Typically, multiple NASA Centers contribute to the program. Individual projects may be managed at different Centers. The program may also include other agency or international partner contributions.

Unallocated Future Expenses. The portion of estimated cost required to meet specified confidence level that cannot yet be allocated to the specific project Work Breakdown Structure (WBS) sub-elements because the estimate includes probabilistic risks and specific needs that are not known until these risks are realized.

Uncoupled Programs. Programs implemented under a broad theme and/or a common program implementation concept, such as providing frequent flight opportunities for cost-capped projects selected through Announcements of Opportunity (AO) or NASA Research Announcements. Each such project is independent of the other projects within the program.

Validation. The process of showing proof that the product accomplishes the intended purpose based on stakeholder expectations. May be determined by a combination of test, analysis, demonstration, and inspection. (Answers the question, "Am I building the right product?")

Verification. Proof of compliance with requirements. Verification may be determined by a combination of test, analysis, demonstration, and inspection. (Answers the question, "Did I build the product right?")

Waiver. A documented authorization releasing a program or project from meeting a requirement after the requirement is put under configuration control at the level the requirement will be implemented.

Work Breakdown Structure. A product-oriented hierarchical division of the hardware, software, services, and data required to produce the program's or project's end product(s), structured according to the way the work will be performed and reflecting the way in which program or project costs and schedule, technical, and risk data are to be accumulated, summarized, and reported.

Work Package. The unit of work established by the control account manager that is required to complete a specific job such as a test, a report, a design, a set of drawings, fabrication of a piece of hardware, or a service.

B Acronyms

ABC	Agency Baseline Commitment
ACD	Architectural Control Document
AI&T	Assembly, Integration, and Test
ANSI/EIA	American National Standards Institute/Electronic Industries Alliance
AO	Announcement of Opportunity
APMC	Agency Program Management Council
APPEL	Academy of Program/Project and Engineering Leadership
ARRA	American Recovery and Reinvestment Act
ASM	Acquisition Strategy Meeting
BAA	Broad Agency Announcement
BoE	Basis of Estimate
BPI	Budget and Performance Integration (scorecard)
BPR	Baseline Performance Review
BY	Budget Year
C3PO	Commercial Crew and Cargo Program
CAD	Cost Analysis Division
CADRe	Cost Analysis Data Requirement
CAIB	Columbia Accident Investigation Board
CAM	Control Account Manager
CDR	Critical Design Review
CERR	Critical Events Readiness Review
CHMO	Chief Health and Medical Officer
CI	Counterintelligence
CLP	Continuous Learning Point
CMC	Center Management Council
CoF	Construction of Facilities
COFR	Certification of Flight Readiness
ConOps	Concept of Operations
COOP	Continuity of Operations
COPV	Composite Overwrapped Pressure Vessel

COSTAR	Corrective Optics Space Telescope Axial Replacement
COTS	Commercial Orbital Transportation Service
CPD	Center Policy Directive
CPR	Center Procedural Requirements
CPR	Contract Performance Report
CR	Continuing Resolution
CRCS-1	Central Resources Control System
CRM	Continuous Risk Management
CRR	Center Readiness Review
CSO	Chief Safety and Mission Assurance Officer
CT	Counterterrorism
CWI	Center Work Instructions
DA	Decision Authority
DCI	Data Collection Instrument
DM	Decision Memorandum
DoD	U.S. Department of Defense
DPMC	Directorate PMC
DR	Decommissioning Review
DRD	Data Requirements Description
DRM	Design Reference Mission
DRR	Disposal Readiness Review
EA	Environmental Assessment
EAC	Estimate at Completion
ECC	Education Coordinating Council
EIS	Environmental Impact Statement
ELV	Expendable Launch Vehicle
EMS	Environmental Management System
EOMP	End of Mission Plan
EPPMD	Engineering Program and Project Management Division
EPO	Education and Public Outreach
EPR	Engineering Peer Review
ESMD/CxP	Exploration Systems Mission Directorate/Constellation Program
ETA	Engineering Technical Authority
EVM	Earned Value Management
EVMS	Earned Value Management System
FAC-P/PM	Federal Acquisition Certification for Program/Project Managers
FAD	Formulation Authorization Document
FAR	Federal Acquisition Regulation
FFRDC	Federally Funded Research and Development Center
FONSI	Finding of No Significant Impact
FRR	Flight Readiness Review

FRR (LV)	Flight Readiness Review (Launch Vehicle)
FTE	Full-Time Equivalent; Full-Time Employee
FY	Fiscal Year
GAO	U.S. Government Accountability Office
GDS	Ground Data System
GFY	Government Fiscal Year
GOES-R	Geostationary Operational Environmental Satellite–R Series
GOS	Ground Operations System
GSE	Government Standard Equipment
HATS	Headquarters Action Tracking System
HEOMD	Human Exploration and Operations Mission Directorate
HMTA	Health and Medical Technical Authority
HQ	Headquarters
HSF	Human Space Flight
HST	Hubble Space Telescope
IAEA	International Atomic Energy Agency
IBPD	Integrated Budget Performance Document
IBR	Integrated Baseline Review
ICA	Independent Cost Assessment
ICB	Intelligence Community Brief
ICD	Interface Control Document
ICE	Independent Cost Estimate
ICMC	Integrated Center Management Council
IG	NASA Inspector General
IIA	Institutional Infrastructure Analysis
ILS	Integrated Logistics Support
IMS	Integrated Master Schedule
INCOSE	International Council on System Engineering
INSRP	Interagency Nuclear Safety Review Panel
IPAO	Independent Program Assessment Office
ISS	International Space Station
IT	Information Technology
IV&V	Independent Verification and Validation
JCL	Joint (Cost and Schedule) Confidence Level
JPL	Jet Propulsion Laboratory
JSC	NASA Johnson Space Center
KDP	Key Decision Point
LaRC	NASA Langley Research Center
LAS	Launch Abort System
LCC	Life-Cycle Cost
LCCE	Life-Cycle Cost Estimate
LDE	Lead Discipline Engineer
LoE	Level of Effort (also LOE)

LRR	Launch Readiness Review
LV	Launch Vehicle
MCP	Mishap Contingency Plan
MCR	Mission Concept Review
MD	Mission Directorate
MDAA	Mission Directorate Associate Administrator
MDCE	Mission Directorate Chief Engineer
MdM	Metadata Manager (database)
MDR	Mission Definition Review
MEL	Master Equipment List
MMT	Mission Management Team
MOA	Memorandum of Agreement
MOS	Mission Operations System
MOU	Memorandum of Understanding
MPAR	Major Program Annual Report
MRB	Mission Readiness Briefing
MRR	Mission Readiness Review
MSFC	NASA Marshall Space Flight Center
MSD	Mission Support Directorate
MSO	Mission Support Office
MSR	Monthly Status Report
N2	(budget formulation system)
NASA	National Aeronautics and Space Administration
NEN	NASA Engineering Network
NEPA	National Environmental Policy Act
NESC	NASA Engineering and Safety Center
NFS	NASA FAR Supplement
NFSAM	NASA Nuclear Flight Safety Assurance Manager
NID	NASA Interim Directive
NIE	National Intelligence Estimate
NOA	New Obligation Authority
NOAA	National Oceanographic and Atmospheric Administration
NODIS	NASA Online Directives Information System
NPD	NASA Policy Directive
NPR	NASA Procedural Requirements
NRA	NASA Research Announcement
NSC	NASA Safety Center
NSM	NASA Structure Management
OBS	Organizational Breakdown Structure
OCE	Office of the Chief Engineer
OCFO	Office of the Chief Financial Officer
OCHMO	Office of the Chief Health and Medical Officer
ODAR	Orbital Debris Assessment Report

OIG	Office of the Inspector General
OIIR	Office of International and Interagency Relations
OMB	U.S. Office of Management and Budget
OoE	Office of Evaluation
OpsCon	Operations Concept
ORR	Operational Readiness Review
OSMA	Office of Safety and Mission Assurance
OSTP	Office of Science and Technology Policy
PAA	Program Analysis and Alignment
PA&E	Program Analysis and Evaluation
PAIG	Programmatic and Institutional Guidance
PBR	President's Budget Request
PCA	Program Commitment Agreement
PCE	Project Chief Engineer
PCLS	Probabilistic Cost-Loaded Schedule
PDLM	Product Data and Life-Cycle Management
PDP	Personal Development Portfolio
PDR	Preliminary Design Review
PE	Program or Project Executive
PFAR	Post-Flight Assessment Review
PI	Principal Investigator
PIR	Program Implementation Review
PLAR	Post-Launch Assessment Review
PM	Program or Project Management
PMB	Performance Measurement Baseline
PMC	Program Management Council
PPBE	Planning, Programming, Budgeting, and Execution
P/PM	Program or project Manager
PPMB	Program and Project Management Board
PPP	Project Protection Plan
PRG	Programming and Resource Guidance
PRR	Production Readiness Review
PSM	Procurement Strategy Meeting
RFA	Request For Action
RFP	Request For Proposal
RID	Review Item Discrepancy
RIDM	Risk-Informed Decision Making
R&M	Reliability and Maintainability
RMO	Resource Management Officer
RMOR	Radioactive Materials On-Board Report
ROD	Record of Decisions
SAA	Space Act Agreement
SAP	Systems, Applications, and Products (in Data Processing)

SAR	System Acceptance Review or Safety Analysis Report
SAS	Safety Analysis Summary
SATERN	System for Administration, Training, and Educational Resources for NASA
SDR	System Definition Review
SEMP	Systems Engineering Management Plan
SER	Safety Evaluation Report
SI	Système Internationale (International System of Units)
SID	NASA OCFO Strategic Investments Division
SIR	System Integration Review
S(&)MA	Safety (and) Mission Assurance
SMA TA	Safety and Mission Assurance Technical Authority
SMC	Strategic Management Council
SMD	Science Mission Directorate
SMSR	Safety and Mission Success Review
SPG	Strategic Planning Guide
SPP	Single-Project Program
SPWG	NASA Space Protection Working Group
SRB	Standing Review Board
SRM	Solid Rocket Motor
SRR	Systems Requirement Review
SSA	Space Situational Awareness
STEM	Science, Technology, Engineering, and Math
TA	Technical Authority
TD	Time Dependent (costs)
TDM	Technology Demonstration Mission
TI	Time Independent (costs)
TLI	Technical Leading Indicator
ToR	Terms of Reference
TPM	Technical Performance Measure
TRL	Technology Readiness Level
TRR	Test Readiness Review
UFE	Unallocated Future Expenses
V&V	Verification and Validation
WBS	Work Breakdown Structure

NPR 7120.5E Requirements Rationale

Para #	Requirement Statement	Rationale for Requirement
1.1.2	NASA Centers, Mission Directorates, and other organizations that have programs or projects shall develop appropriate documentation to implement the requirements of this NPR.	NPRs typically cannot provide adequate Center or Mission Directorate policy, procedural requirements, or instructions, especially when a situation is unique to a particular Center or MD. So, Centers and MDs are required to develop flow-down requirements from this NPR.
1.1.3	The Mission Directorate shall submit their plan for phased tailoring of the requirements of this NPR within 60 days of the effective date of this NPR.	This NPR's requirements apply to existing program's and project's current and future phases as determined by the responsible Mission Directorate, approved by the NASA Chief Engineer (or as delegated), and concurred with by the Decision Authority. The Mission Directorate's plan for phased tailoring of the NPR's requirements is due within 60 days of the effective date of this NPR.
2.1.1	Regardless of the structure of a program or project meeting the criteria of Section P.2, this NPR shall apply to the full scope of the program or project and all the activities under it.	Large projects tend to divide their work into smaller "activities," elements, etc. and these must be managed according to NPR 7120.5 even though they are not listed in a Program or Project Plan.
2.1.4.1	Projects are Category 1, 2, or 3 and shall be assigned to a category based initially on: (1) the project life-cycle cost (LCC) estimate, the inclusion of significant radioactive material, and whether or not the system being developed is for human space flight; and (2) the priority level, which is related to the importance of the activity to NASA, the extent of international participation (or joint effort with other government agencies), the degree of uncertainty surrounding the application of new or untested technologies, and spacecraft/payload development risk classification.	Projects vary in scope and complexity and thus require varying levels of management requirements and Agency attention and oversight. Project categorization defines Agency expectations of project managers by determining both the oversight council and the specific approval requirements. Guidelines for determining project categorization are shown in Table 2-1 of NPR 7120.5, but categorization may be changed based on recommendations by the Mission Directorate Associate Administrator (MDAA) that consider additional risk factors facing the project. The NASA Associate Administrator (AA) approves the final project categorization.
2.1.4.2	When projects are initiated, they are assigned to a NASA Center or implementing organization by the MDAA consistent with direction and guidance from the strategic programming process. They are either assigned directly to a Center by the Mission Directorate or are selected through a competitive process such as an Announcement of Opportunity (AO). For Category 1 projects, the assignment shall be with the concurrence of the NASA AA.	Due to the external visibility and dollar amount of Category 1 projects, it is important that the NASA AA concur that the assignment of the project by the Mission Directorate is consistent with the direction and guidance from the strategic acquisition planning process.

Para #	Requirement Statement	Rationale for Requirement
2.2.1	Programs and projects shall follow their appropriate life cycle, which includes life-cycle phases; life-cycle gates and major events, including KDPs; major life-cycle reviews (LCRs); principal documents that govern the conduct of each phase; and the process of recycling through Formulation when program changes warrant such action.	NASA programs and projects are managed to life cycles, the division of the program's and project's activities over the full lifetime of the program or project, based on the expected maturity of program and project information and products as they move through defined phases in the life cycle. At the top level, this work is divided into two phases, Formulation and Implementation, each of which is divided into subphases. As part of checks and balances, programs and projects must be given formal approval at specific points to progress through their life cycle. This approval is based on periodic evaluation.
2.2.2	Each program and project performs the work required for each phase, which is described in the NASA Space Flight Program and Project Management Handbook and NPR 7123.1. This work shall be organized by a product-based WBS developed in accordance with the Program and Project Plan templates (Appendices G and H).	NASA requires the use of a standard WBS and Dictionary template to ensure that space flight projects define work to be performed and accumulate corresponding costs in a standard manner. This provides uniformity across projects and allows for the accumulation of historical cost data for analysis and comparison.
2.2.3	The documents shown on the life-cycle figures and described below shall be prepared in accordance with the templates in Appendices D, E, F, G, and H.	The purpose of program formulation activities is to establish a cost-effective program that is demonstrably capable of meeting Agency and Mission Directorate goals and objectives. The program Formulation Authorization Document (FAD) authorizes a Program Manager to initiate the planning of a new program and to perform the analyses required to formulate a sound Program Plan. The Program Plan establishes the program's baseline for Implementation, signed by the MDAA, Center Director(s), and program manager. The Program Commitment Agreement (PCA) is the contract between the Associate Administrator and the responsible MDAA that authorizes transition from Formulation to Implementation of a program. The project FAD authorizes a Project Manager to initiate the planning of a new project and to perform the analyses required to formulate a sound Project Plan. (The Formulation Agreement represents the project's response to the FAD). The Project Plan establishes the project's baseline for Implementation, signed by the responsible program manager, Center Director, project manager, and the MDAA, if required. The templates are designed to ensure all content necessary is addressed.
2.2.4	Each program and project shall perform the LCRs identified in its respective figure in accordance with NPR 7123.1, applicable Center practices, and the requirements of this handbook.	LCRs provide a periodic assessment of the program's or project's technical and programmatic status and health at key points in the life cycle. An LCR is complete when the governing Program Management Council (PMC) and Decision Authority complete their assessment and sign the Decision Memorandum. The maturity tables identify the expected program/project maturity state for each major review specified by the following six assessment criteria: Agency Strategic Goals and Outcomes, Management Approach, Technical Approach, Budget Schedule, Resources other than Budget, and Risk Management.

Para #	Requirement Statement	Rationale for Requirement

2.2.5	The program or project and an independent Standing Review Board (SRB) shall conduct the SRR, SDR/MDR, PDR, CDR, SIR, ORR, and PIR LCRs in Figures 2-2, 2-3, 2-4, and 2-5.	The Governance model provides an organizational structure that emphasizes mission success by taking advantage of different perspectives that different organizational elements bring to issues. The organizational separation of the Mission Directorates and their respective programs and projects (Programmatic Authorities) and the Headquarters Mission Support Offices, the Center organizations that are aligned with these offices, and the Center Directors (Institutional Authorities) is the cornerstone of this organizational structure and NASA's system of checks and balances. Independent assessments provide:
		
		 1. The program/project with a credible, objective assessment of how they are doing
		
		 2. NASA senior management with an understanding of whether
		
		    a. The program/project is on the right track,
		
		    b. Is performing according to plan, and
		
		    c. Externally imposed impediments to the program's/ projects' success are being removed
		
		 3. A credible basis for a decision to proceed into the next phase
		
		 4. The independent review also provides additional assurance to external stakeholders that NASA's basis for proceeding is sound.

2.2.5.1	The conflict of interest procedures detailed in the *NASA Standing Review Board Handbook* shall be strictly adhered to.	NASA accords special importance to the policies and procedures established to ensure the integrity of the SRB's independent review process and to comply with Federal law.
2.2.5.2	The portion of the LCR conducted by the SRB shall be convened by the convening authorities in accordance with Table 2-2.	The convening authorities are the heads of the organizations principally responsible for authorizing, overseeing, supporting and evaluating the programs and projects. The SRB is convened by these individuals, as part of the Agency's checks and balances, to provide an independent assessment that addresses each organization's perspective. This approach minimizes the burden on programs and projects by using only one review team to meet the needs of multiple organizations.
2.2.5.3	The program or project manager, the SRB chair, and the Center Director (or designated Engineering Technical Authority representative) shall mutually assess the program's or project's expected readiness for the LCR and report any disagreements to the Decision Authority for final decision.	Life-cycle reviews are important in determining program or project readiness to proceed to the next phase. Conducting a life-cycle review before a program or project is ready would waste the time of both the program/project and the SRB.
2.2.6	In preparation for these LCRs, the program or project shall generate the appropriate documentation per the Appendix I tables of this document, NPR 7123.1, and Center practices as necessary to demonstrate that the program's or project's definition and associated plans are sufficiently mature to execute the follow-on phase(s) with acceptable technical, safety, and programmatic risk.	The documents are the tangible evidence of the work performed by the program/project during the current life-cycle phase and a concrete way to demonstrate readiness to proceed to the next phase.

Para #	Requirement Statement	Rationale for Requirement
2.2.8	Projects in Phases C and D (and programs at the discretion of the MDAA) with a life-cycle cost estimated to be greater than $20 million and Phase E project modifications, enhancements, or upgrades with an estimated development cost greater than $20 million shall perform earned value management (EVM) with an EVM system that complies with the guidelines in ANSI/EIA-748, Standard for Earned Value Management Systems.	The Office of Management and Budget (OMB) Circular A-11 (Part 7 Planning, Budgeting, Acquisition, and Management of Capital Assets and the Capital Programming Guide) sets forth the policy, budget justification, and reporting requirements that apply to all agencies of the Executive Branch of the government that are subject to Executive Branch review for major capital acquisitions. It requires that Earned Value Management (EVM) be consistent with the guidelines in the American National Standards Institute/Electronic Industries Alliance 748 (ANSI/EIA-748), Earned Value Management Systems, for developmental efforts for both government and contractor work, and that in-house work be managed with the same rigor as contract work. While a Project Plan or Intra-Agency Work Agreement replaces the contract for NASA in-house work, the other requirements for good project management, including the use of EVM in accordance with the ANSI/EIA-748 standard, are applicable for developmental and production efforts.
2.2.8.1	EVM system requirements shall be applied to appropriate suppliers, in accordance with the NASA Federal Acquisition Regulation (FAR) Supplement, and to in-house work elements.	To comply with NASA FAR Supplement 1834.201, and thus, OMB requirements in Circular A-11 as described above.
2.2.8.2	For projects requiring EVM, Mission Directorates shall conduct a preapproval integrated baseline review as part of their preparations for KDP C to ensure that the project's work is properly linked with its cost, schedule, and risk and that the management processes are in place to conduct project-level EVM.	Cost control is essential to meeting the commitments in the Project Plan. Implementation of project EVM is needed to measure and assess project performance in Phase C/D. Performance is measured and assessed through the integration of technical scope with schedule and cost objectives during the execution of the project. A project-level IBR is completed prior to KDP C, and for this reason it is known as a preapproval IBR.
2.2.10	Each program and project shall complete and maintain a Compliance Matrix (see Appendix C) for NPR 7120.5 and attach it to the Formulation Agreement for projects in Formulation and/or the Program or Project Plan. The program or project will use the Compliance Matrix to demonstrate how it is complying with the requirements of this document and verify the compliance of other responsible parties.	The program or project uses the Compliance Matrix to demonstrate how it is complying with the requirements of NPR 7120.5 and verify the compliance of other responsible parties.
2.3.1	Each program and project shall have a Decision Authority who is the Agency's responsible individual who determines whether and how the program or project proceeds through the life cycle and the key program or project cost, schedule, and content parameters that govern the remaining life-cycle activities.	The Agency's Governance model requires that there be a single approving authority for all program/project phase transitions.
2.3.1.1	The NASA AA shall approve all Agency Baseline Commitments (ABCs) for programs requiring an ABC and projects with a life-cycle cost greater than $250 million.	The ABC for these programs and projects is required to be externally reported to OMB and the Congress. Thus they are inherently highly visible to our stakeholders. The NASA Associate Administrator is responsible for the technical and programmatic integration of these programs and projects at the Agency level and serves as the Decision Authority for them. The NASA Associate Administrator, as chair of the Agency PMC, ensures that projects are subjected to an appropriate level of Agency oversight.

Para #	Requirement Statement	Rationale for Requirement
2.3.2	Each program and project shall have a governing PMC.	A KDP is an event where the Decision Authority determines the readiness of a program/project to progress to the next phase of the life cycle. As such, KDPs serve as gates through which programs and projects must pass. Within each phase, the KDP is preceded by one or more reviews, including the governing PMC review. These reviews enable a disciplined approach to assessing programs and projects. Per NPD 1000.3 charter, the Agency Program Management Council (APMC) serves as the Agency's senior decision-making body to baseline and assess program/project performance and ensure successful achievement of NASA strategic goals. This role is delegated to the DPMC for projects as specified in NPR 7120.5E or as delegated by the NASA AA.
2.3.3	The Center Director (or designee) shall oversee programs and projects usually through the Center Management Council (CMC), which monitors and evaluates all program and project work (regardless of category) executed at that Center.	The Center Director has a unique role as the only person who can ensure proper planning and execution of activities requiring constructive integration across Programmatic, Technical, and Institutional Authorities. The Center Director is therefore responsible and accountable to the Administrator for the safe, effective, and efficient execution of all activities at his/her Center. As part of the Institutional Authority, Center Directors are responsible for establishing, developing, and maintaining the Center's institutional capabilities (such as processes, competency development and leadership, human capital, facilities, and independent review) required for the execution of programs, projects, and missions assigned to the Center. The Center Directors work closely with the AA for Mission Support in this role. Center Directors have specifically delegated Technical Authority (TA) responsibilities for work performed at the Center and are responsible for establishing and maintaining Center technical authority policies and practices, consistent with Agency policies and standards. The Center Directors work closely with the Chief Engineer, Chief Safety and Mission Assurance Officer, and Chief Medical Officer in this role. While the Center Directors do not exercise Programmatic Authority over programs and projects (i.e., do not make programmatic cost and schedule decisions), they work closely with the Mission Directorate (MD) AAs to balance the specific needs of individual programs and projects alongside thoughtful compliance with applicable priorities, policies, procedures, and practices. The summation of the "balanced" agreements between the Program/Project Manager and Center Directors of participating NASA Centers are documented in the Program/Project Plan, consistent with the Mission Directorate's requirements and with Agency policy and the Center's best practices and institutional policies. The Center Director is a convening authority for SRBs and uses their assessment along with lower level review teams and his/her Center leadership team such as the CMC or ICMC (Integrated Center Management Council) in forming his/her assessment and affecting the plans as necessary. When the Center Director sees an issue which, in his/her judgment, may require programmatic direction, he/she engages the MDs or Program Office as needed to cooperatively identify solutions, including cases where resolution of a Technical Authority issue might impact top level programmatic requirements.

Para #	Requirement Statement	Rationale for Requirement
2.3.4	Following each LCR, the independent SRB and the program or project shall brief the applicable management councils on the results of the LCR to support the councils' assessments.	It is important for the Governing PMC and the Decision Authority to hear the assessments and feedback from the both the SRB and the program or project in order for them to make the best informed decision possible. In this way they will hear issues and plans to address where the program/project agrees or they will disagreements and rationales for those disagreements so that decisions can be made.
2.4.1	After reviewing the supporting material and completing discussions with concerned parties, the Decision Authority determines whether and how the program or project proceeds into the next phase and approves any additional actions. These decisions shall be summarized and recorded in the Decision Memorandum signed at the conclusion of the governing PMC by all parties with supporting responsibilities, accepting their respective roles.	As stated in the sentence preceding the requirement, after reviewing the supporting material and completing discussions with concerned parties, the Decision Authority determines whether and how the program or project proceeds into the next phase and approves any additional actions, which are documented in the Decision Memorandum.
2.4.1.1	The Decision Memorandum shall describe the constraints and parameters within which the Agency, the program manager, and the project manager will operate; the extent to which changes in plans may be made without additional approval; any additional actions that came out of the KDP; and the supporting data (i.e., the cost and schedule datasheet) that provide further details.	The Decision Memorandum describes the Decision Authority's decisions. Within the Decision Memorandum, the parameters and authorities over which the program or project manager has management control constitute the program or project Management Agreement. A program or project manager has the authority to manage within their Management Agreement and is accountable for compliance with the terms of the agreement. The Management Agreement is established at every KDP but may be changed between KDPs as the program or project matures and in response to internal and external events. The Program Plan or Project Plan is updated and approved during the life cycle, if warranted, by changes in the stated Management Agreement commitments.
2.4.1.2	A divergence from the Management Agreement that any party identifies as significant shall be accompanied by an amendment to the Decision Memorandum.	The purpose is to document rationale for changes as part of the permanent record and that all signatories in the original Decision Memorandum have an opportunity to review and agree or disagree.
2.4.1.3	During Formulation, the Decision Memorandum shall establish a target life-cycle cost range (and schedule range, if applicable) as well as the Management Agreement addressing the schedule and resources required to complete Formulation.	The cost/schedule range during Formulation (end of Phase A) is required by Congress. It is important for the project to assess and request the resources it needs during Formulation so that expectations by the project and resources provided by the Mission Directorate are aligned and agreed to in the Management Agreement.
2.4.1.5	All projects and single-project programs shall document the Agency's life-cycle cost estimate and other parameters in the Decision Memorandum for Implementation (KDP C), and this becomes the Agency Baseline Commitment (ABC).	The ABC is the baseline against which the Agency's performance is measured during the Implementation Phase. The ABC for projects with a life-cycle cost of $250 million or more forms the basis for the Agency's external commitment to the Office of Management and Budget (OMB) and Congress.
2.4.1.6	Tightly coupled programs shall document their life-cycle cost estimate, in accordance with the life-cycle scope defined in the FAD or PCA, and other parameters in their Decision Memorandum and ABC at KDP I.	Tightly coupled programs can be viewed as very large projects and KDP I is where the program ends formulation and begins implementation (KDP C for projects). Like projects, a tightly coupled program needs to provide their life-cycle cost estimate at KDP I for external commitment purposes. Since tightly coupled programs generally have very long life cycles that exceed normal planning horizons, the life cycle to be used is documented in the FAD or PCA.

Para #	Requirement Statement	Rationale for Requirement
2.4.1.7	Programs or projects shall be rebaselined when: (1) the estimated development cost exceeds the ABC development cost by 30 percent or more (for projects over $250 million, also that Congress has reauthorized the project); (2) the NASA AA judges that events external to the Agency make a rebaseline appropriate; or (3) the NASA AA judges that the program or project scope defined in the ABC has been changed or the tightly coupled program or project has been interrupted.	For (1), per the NASA Appropriation Act and the 2005 NASA Authorization Act, NASA is required to notify Congress of significant cost growth. For (2) and (3), performance is not to be assessed against the original baseline when significant events outside the program/project control occur. Therefore a new baseline is generated for the program/project to perform against.
2.4.2	All programs and projects develop cost estimates and planned schedules for the work to be performed in the current and following life-cycle phases (see Appendix I tables in NPR 7120.5E). As part of developing these estimates, the program or project shall document the Basis of Estimate (BoE) in retrievable program or project records.	The BoE is documentation of the ground rules, assumptions, and drivers used in developing the cost or schedule estimates including applicable model inputs, rationale or justification for analogies, and details supporting cost and schedule estimates. The basis of estimate is contained in material available to the SRB and management as part of the LCR and KDP process.
2.4.3	Tightly coupled and single-project programs (regardless of life-cycle cost) and projects (with an estimated life-cycle cost greater than $250 million) shall develop probabilistic analyses of cost and/or schedule estimates to obtain a quantitative measure of the likelihood that the estimate will be met in accordance with the following requirements.	See 2.4.3.1 and 2.4.3.2
2.4.3.1	Tightly coupled and single-project programs (regardless of life-cycle cost) and projects (with an estimated life-cycle cost greater than $250 million) shall provide a range of cost and a range for schedule at KDP 0/KDP B, each range (with confidence levels identified for the low and high values of the range) established by a probabilistic analysis and based on identified resources and associated uncertainties by fiscal year.	A cost/schedule range is required by Congress by KDP 0/KDP B.
2.4.3.2	At KDP I/KDP C, tightly coupled and single-project programs (regardless of life-cycle cost) and projects (with an estimated life-cycle cost greater than $250 million) shall develop a resource-loaded schedule and perform a risk-informed probabilistic analysis that produces a JCL.	A cost/schedule estimate is required by Congress by KDP I/KDP C. The JCL is required to enable the Agency to assert that the programs/projects have executable plans. The Agency has moved to this because external stakeholders have demanded better cost/schedule performance. This is required for all tightly coupled programs and single-project programs regardless of LCC, and projects with an LCCE greater than $250 million.
2.4.4	Mission Directorates shall plan and budget tightly coupled and single-project programs (regardless of life-cycle cost) and projects (with an estimated life-cycle cost greater than $250 million) based on a 70 percent joint cost and schedule confidence level or as approved by the Decision Authority.	In response to GAO cost growth concerns, NASA surveyed the best practices in the industry and determined that all projects need to be estimated using probabilistic cost estimating methods and budgets need to reflect a 70 percent probability the project could be completed for that budget request or lower.
2.4.4.1	Any JCL approved by the Decision Authority at less than 70 percent shall be justified and documented.	This is required to ensure the agency has a record of and rationale for deviating from our 70 percent policy.
2.4.4.2	Mission Directorates shall ensure funding for these projects is consistent with the Management Agreement and in no case less than the equivalent of a 50 percent JCL.	Any JCL under 70 percent is considered a management challenge and it is understood that any JCL less than 50 percent significantly diminishes a project's chances of being successful in meeting cost and schedule commitments.

Para #	Requirement Statement	Rationale for Requirement
2.4.5	Loosely coupled and uncoupled programs are not required to develop program cost and schedule confidence levels. These programs shall provide analysis that provides a status of the program's risk posture that is presented to the governing PMC as each new project reaches KDP B and C or when a project's ABC is rebaselined.	Because these programs are principally made up of unrelated or loosely related projects, confidence levels at the program level are not particularly meaningful. However, it is important the program provide an analysis of its risk posture to management so that management can assess for itself the likelihood the program can meet its commitments and to understand if any of the risks might impact other programs within the Mission Directorate or the Agency.
3.3.1	Programs and projects shall follow the Technical Authority process established in Section 3.3 of this NPR.	NASA established the Technical Authority process as part of its system of checks and balances to provide independent oversight of programs and projects in support of safety and mission success through the selection of specific individuals with delegated levels of authority. These individuals are the Technical Authorities. The responsibilities of a program or project manager are not diminished by the implementation of Technical Authority. The program or project manager is ultimately responsible for the safe conduct and successful outcome of the program or project. This includes meeting programmatic, institutional, technical, safety, cost, and schedule commitments.
3.4.1	Programs and projects shall follow the Dissenting Opinion process in this Section 3.4.	NASA teams need to have full and open discussions, with all facts made available, to understand and assess issues. Diverse views are to be fostered and respected in an environment of integrity and trust with no suppression or retribution. In the team environment in which NASA operates, team members often have to determine where they stand on a decision. In assessing a decision or action, a member has three choices: agree, disagree but be willing to fully support the decision, or disagree and raise a Dissenting Opinion. Unresolved issues of any nature (e.g., programmatic, safety, engineering, health and medical, acquisition, accounting) within a team need to be quickly elevated to achieve resolution at the appropriate level.
3.5.1	Programs and projects shall follow the tailoring process in this Section 3.5.	It is NASA policy to comply with all prescribed directives, requirements, procedures, and processes unless relief is formally granted. Tailoring is the process used to adjust or seek relief from a prescribed requirement to accommodate the needs of a specific task or activity (e.g., program or project). Additional details regarding the tailoring process are in the NASA Space Flight Program and Project Management Handbook.
3.5.3	A request for a permanent change to a prescribed requirement in an Agency or Center document that is applicable to all programs and projects shall be submitted as a "change request" to the office responsible for the requirements policy document unless formally delegated elsewhere.	This requirement alleviates the need for programs and projects at a Center to continually request deviations or waivers to requirements that no longer apply to the programs and projects at that Center.
3.6.1	A Center negotiating reimbursable space flight work with another agency shall propose NPR 7120.5 as the basis by which it will perform the space flight work.	It is understood that outside agencies come to NASA for its expertise and approach to program/project management. Therefore it is only natural that NASA propose the management policy that it uses itself.
3.7.1	Each program and project shall perform and document an assessment to determine an approach that maximizes the use of SI.	Federal policy requires agencies to use the International System of Units (SI) to the extent possible without incurring a substantial increase in cost or unacceptable delays in schedule. Documentation of the project's assessment and approach to its system of measurement demonstrates why the program or project is or is not fully using SI and is also needed to ensure that all members of the project team understand and use consistent units of measure.

D Roles and Responsibilities Relationships Matrix

Table D-1 provides a summary of the roles and responsibilities for key program and project management officials. The table is informational only. Implementation of specific roles and responsibilities is determined on a case-by-case basis and is documented in the Program or Project Plan.

Table D-1 Roles and Responsibilities Relationships Matrix

	Office of the Administrator	Administrator Staff and Mission Support Offices	Mission Directorate Associate Administrator	Center Director — Institutional	Center Director — Technical Authority	Program Manager	Project Manager
Strategic Planning	Establish Agency strategic priorities and direction Approve Agency Strategic Plan and programmatic architecture, and top-level guidance Approve implementation plans developed by Mission Directorates	Lead development of Agency Strategic Plan Lead development of Annual Performance Plan	Support Agency strategic planning Develop directorate implementation plans and cross-directorate architecture plans consistent with Agency strategic plans, architecture, and top-level guidance	Support Agency and Mission Directorate strategic planning and supporting studies		Support Mission Directorate strategic implementation plan	
Program Initiation (FAD/ASM)	Approves programs and assigns them to MDs and Centers and validates partnerships Approve Acquisition Strategy	Approve Program Chief Engineers' (Technical Authority) (OCE) When applicable, approve program's approach to Health and Medical Technical Authority based on Center's HMTA infrastructure (OCHMO) Approves procurement strategy (AA for Procurement)	Implement new programs via FAD Recommend assignment of programs to Centers Approve appointment of Program Managers Approve Acquisition Strategy	Provide human and other resources to execute FAD Recommend Program Managers to MDAA Ensure Acquisition Strategy and the program's plans are executable	Appoint Program Chief Engineers* and SMA TA (Technical Authority) in consultation with and after approval by OCE and OSMA, respectively Appoint Center Lead Discipline Engineers (LDEs)	Establish the program office and structure to direct/monitor projects within program Develop Acquisition Strategy Develop Formulation Agreement (only combined SPP)	

Table D-1 Roles and Responsibilities Relationships Matrix

	Office of the Administrator	Administrator Staff and Mission Support Offices	Mission Directorate Associate Administrator	Center Director		Program Manager	Project Manager
				Institutional	Technical Authority		
Project Initiation (FAD/ASM)	Approves Category 1 projects and assigns them to MDs and Centers and validates partnerships Approve Acquisition Strategy	Approve Project Chief Engineers[1] (Technical Authority) appointment to Category 1 projects (OCE) Is notified of Project Chief Engineers[1] (Technical Authority) assigned to Category 2 and 3 projects (OCE) When applicable, approve project's approach to Health and Medical Technical Authority based on Center's HMTA infrastructure (OCHMO) Approves procurement strategy (AA for Procurement)	Implement new projects via FAD Recommend assignment of Category 1 projects to Centers Assign Category 2 and 3 projects to Centers. Approve appointment of Category 1 and selected Category 2 Project Managers Approve Acquisition Strategy Approve Formulation Agreement	Provide human and other resources to execute FAD Recommend Category 1 Project Managers to MDAA Appoint Category 2 and 3 Project Managers Ensure Acquisition Strategy and the project's plans are executable Approve Formulation Agreement	Appoint Project Chief Engineers[1] (Technical Authority) on Category 1 projects in consultation with and after approval by OCE Appoint Project Chief Engineers[1] (Technical Authority) on Category 2 and 3 projects with OCE concurrence Appoint project CSO with OSMA concurrence	Concur with appointment of Project Managers Define content with support of Project Approve Formulation Agreement	Establish the project office and structure to direct and monitor tasks/activities within project Develop Formulation Agreement Develop acquisition strategy.

Table D-1 Roles and Responsibilities Relationships Matrix

	Office of the Administrator	Administrator Staff and Mission Support Offices	Mission Directorate Associate Administrator	Center Director — Institutional	Center Director — Technical Authority	Program Manager	Project Manager
Policy Development	Approve policies	Establish Agency Institutional policies and ensure support infrastructure is in place: this includes: Technical Authority (OCE), SMA functions (OSMA), Health and Medical functions (OCHMO) Develop and maintain Agency-wide engineering (OCE), health and medical (OCHMO), and safety and mission assurance (OSMA) standards applicable to programs and projects	Establish Directorate policies (e.g., guidance, risk posture, and priorities for acquisition) applicable to program, projects, and supporting elements	Ensure Center policies are consistent with Agency and Mission Directorate policies Establish policies and procedures to ensure program and projects are implemented consistent with sound technical and management practices	Establish institutional engineering design and verification/validation best practices for products and services provided by the Center Develop implementation plan for technical authority at the Center		

	Office of the Administrator	Administrator Staff and Mission Support Offices	Mission Directorate Associate Administrator	Center Director		Program Manager	Project Manager
				Institutional	Technical Authority		
Program/Project Concept Studies		Provide technical expertise for advanced concept studies, as required (OCE/NESC)	Develop direction and guidance specific to concept studies for formulation of programs and non-competed projects	Develop direction and guidance specific to concept studies for Formulation		Initiate, support, and conduct program-level concept studies consistent with direction and guidance from MDAA	Initiate, support, and conduct project-level concept studies consistent with direction and guidance from program (or Center for competed projects)
Development of Programmatic Requirements			Establish, coordinate, and approve high-level program requirements Establish, coordinate, and approve high-level project requirements, including success criteria	Provide support to program and project requirements development Provide assessments of resources with regard to facilities	Approves changes to and deviations and waivers from those requirements that are the responsibility of the TA and have been delegated to the CD for such action	Originates requirements for the program consistent with the PCA Approve program requirements levied on the project	Originates project requirements consistent with the Program Plan

Table D-1 Roles and Responsibilities Relationships Matrix

	Office of the Administrator	Administrator Staff and Mission Support Offices	Mission Directorate Associate Administrator	Center Director		Program Manager	Project Manager
				Institutional	Technical Authority		
Development of Institutional Requirements	Approve Agency-level policies and requirements for programs and projects	Develop policies and procedural requirements for programs and projects and ensure adequate implementation (OCE, OCHMO, OSMA, MSOs) Approve/disapprove waivers and deviations to requirements under their authority	Develop crosscutting Mission Directorate policies and requirements for programs and projects and ensure adequate implementation Approve/disapprove waivers and deviations to requirements under their authority	Develop Center policies and requirements for programs and projects and ensure adequate implementation Approve/disapprove waivers and deviations to requirements under their authority	Develop TA policies and requirements for programs and projects and ensure adequate implementation Approve/disapprove waivers and deviations to requirements under their authority		
Budget and Resource Management	Determine relative priorities for use of Agency resources (e.g., facilities) Establish budget planning controls for Mission Directorates and Mission Support Offices Approve Agency Budgets	Manage and coordinate Agency annual budget guidance, development, and submission (OCFO) Analyze Mission Directorate submissions for consistency with program and project plans and performance (OCFO) Develop Agency operating plans and execute Agency budget (OCFO).	Develop workforce and facilities plans with implementing Centers Provide guidelines for program and project budget submissions consistent with approved plans Allocate budget resources to Centers for assigned projects Conduct annual program and project budget submission reviews	Confirm program and project workforce requirements Provide the personnel, facilities, resources, and training necessary for implementing assigned programs and projects Support annual program and project budget submissions, and validate Center inputs	Provide resources for review, assessment, development, and maintenance of the core competencies required to ensure technical and program/project management excellence Ensure independence of resources to support the implementation of technical authority	Implement program consistent with budget Develop cost estimates for program components Develop workforce and facilities plans Manage program resources Provide annual program budget submission input	Develop mission options, conduct trades, and develop cost estimates Maintain up-to-date estimated costs at completion Develop workforce and facilities plans Implement project budget Manage project resources Provide annual project budget submission input

Table D-1 Roles and Responsibilities Relationships Matrix

	Office of the Administrator	Administrator Staff and Mission Support Offices	Mission Directorate Associate Administrator	Center Director — Institutional	Center Director — Technical Authority	Program Manager	Project Manager
PCA	Approve Program Commitment Agreement (NASA AA)	Concur with PCA (OCE)	Develop and approve Program Commitment Agreement			Support development of the Program Commitment Agreement	
Agency Baseline Commitment	Approve Category 1 and 2 projects greater than $250 million	Concur with ABC (OCFO)	Develop and approve ABC			Support development of the ABC	Support development of the ABC
Program Plans			Approve Program Plans	Approve Program Plans	Approve the implementation of Technical Authority	Develop and approve Program Plan. Execute Program Plan	
Project Plans			Approve Project Plans	Approve Project Plans	Approve the implementation of Technical Authority	Approve Project Plans	Develop and approve Project Plan. Execute Project Plan
Program/Project Performance Assessment	Assess program and Category 1 project technical, schedule, and cost performance through Status Reviews. Conduct Agency PMC (NASA AA). Chair monthly BPR (NASA AA)	Conduct special studies for the Administrator (IPAO). Provide independent performance assessments (OCE, OCFO, OSMA). Administer the BPR process(OCE, Office of Agency Council Staff)	Assess program technical, schedule, and cost performance and take action, as appropriate, to mitigate risks. Conduct Mission Directorate PMC. Support the Agency BPR process.	Assess program and project technical, schedule, and cost performance against approved plans as part of ongoing processes and forums and the Center Management Council. Provide summary status to support the Agency BPR process and other suitable forums		Assess program and project technical, schedule, and cost performance and take action, as appropriate, to mitigate risks. Provide data to support the monthly BPR process and report on performance.	Assess project technical, schedule, and cost performance and take action, as appropriate, to mitigate risks. Support monthly BPR reporting.

Table D-1 Roles and Responsibilities Relationships Matrix

	Office of the Administrator	Administrator Staff and Mission Support Offices	Mission Directorate Associate Administrator	Center Director		Program Manager	Project Manager
				Institutional	Technical Authority		
Program/ Performance Issues	Assess project programmatic, technical, schedule and cost through Agency PMC and BPR	Maintain issues and risk performance information (OSMA) Track project cost and schedule performance (OCFO) Manage project performance reporting to external stakeholders (OCFO)	Communicate program and project performance issues and risks to Agency management and present plan for mitigation or recovery	Monitor the technical and programmatic progress of programs and projects to help identify issues as they emerge Provide support and guidance to programs and projects in resolving technical and programmatic issues and risks Proactively work with the Mission Directorates, programs, projects, and other Institutional Authorities to find constructive solutions to problems Direct corrective actions to resolve performance issues, if within approved plans Communicate program and project technical performance and risks to Mission Directorate and Agency management and provide recommendations for recovery		Communicate program and project performance issues and risks to Center and Mission Directorate management and present recovery plans	Communicate project performance, issues and risks to program, Center, and Mission Directorate management and present recovery plans

Table D-1 Roles and Responsibilities Relationships Matrix

	Office of the Administrator	Administrator Staff and Mission Support Offices	Mission Directorate Associate Administrator	Center Director		Program Manager	Project Manager
				Institutional	Technical Authority		
Termination Reviews	Determine and authorize termination of programs and Category 1 projects through Agency PMC	Provide status of program or project performance, including budgetary implications of termination or continuation (OCFO) Support Termination Reviews as requested (IPAO)	Determine and authorize termination of programs and Category 2 and Category 3 projects through DPMC and coordinate final decision with Administrator	Support Termination Reviews Perform supporting analysis to confirm termination, if required		Conduct program and project analyses to support Termination Reviews	Support Termination Reviews
Life-Cycle Reviews	Authorize implementation of programs and Category 1 projects through PMC	Convene and support life-cycle reviews for programs and Category 1 and 2 projects (IPAO) Support life-cycle reviews or technical assessments, as required (OCE/NESC, OSMA, and OCHMO) Provide project budget and performance status to SRB (OCFO)	Convene and support life-cycle reviews	Ensure adequate checks and balances (e.g., technical authority) are in place Convene and support life-cycle reviews requiring an independent Center SRB Conduct independent life-cycle reviews that do not require an SRB	Convene and support life-cycle reviews	Prepare for and conduct/support LCRs Provide assessment of program and project readiness to enter next phase	Prepare for and conduct/support LCRs Provide assessment of project readiness to enter next phase

	Office of the Administrator	Administrator Staff and Mission Support Offices	Mission Directorate Associate Administrator	Center Director		Program Manager	Project Manager
				Institutional	Technical Authority		
KDPs (all)	Authorize program and Category 1 and Category 2 above $250 million projects to proceed past KDPs (NASA AA)	Provide Executive Secretariat function for APMC KDPs, including preparation of final decision memorandum	Authorize program and Category 2 and 3 projects to proceed past KDPs (MDAA may delegate Category 3 project KDPs as documented in the Program Plan) Provide recommendation to NASA AA for programs and Category 1 projects at KDPs, including proposing cost and schedule commitments	Perform supporting analysis to confirm readiness leading to KDPs for programs and Category 1, 2, and 3 projects Conduct readiness reviews leading to KDPs for Category 1, 2, and selected Category 3 projects Present Center's assessment of readiness to proceed past KDPs, adequacy of planned resources, and ability of Center to meet commitments Engage in major replanning or rebaselining activities and processes, ensuring constructive communication and progress between the time it becomes clear that a replan is necessary and the time it is formally put into place	Present TA assessment of readiness to proceed past KDPs	Conduct readiness reviews leading to KDPs for program Conduct readiness reviews leading to KDPs for Category 1, 2, and 3 projects Present program and project readiness to proceed past KDPs Provide proposed program management agreement, cost and schedule estimates for KDPs	Conduct readiness reviews leading to KDPs for projects Present readiness to proceed past KDPs Provide proposed Management Agreement, cost and schedule commitments

Table D-1 Roles and Responsibilities Relationships Matrix

	Office of the Administrator	Administrator Staff and Mission Support Offices	Mission Directorate Associate Administrator	Center Director		Program Manager	Project Manager
				Institutional	Technical Authority		
Decision Memorandum (DM)	As DA, approve program and Cat 1 project DM, approving P/p plans and accepting technical and programmatic risks for the Agency (AA)	Approve DM (OCE, OSMA, OCHMO), certifying policies and standards have been followed and risks are deemed to be acceptable Approve DM (OoE), certifying policies and standards have been followed and independent analysis was used to inform Agency decision processes Approve DM (OCFO), certifying policies and standards have been followed and funding; obligations; and commitments on budget, schedules, LCC, and JCL estimates are accurate and consistent with previous commitments	Approve DM (either as DA for Cat 2 & 3 projects or as MDAA), certifying P/p can execute mission within resources and committing funding for mission at proposed levels, approving P/p plans and accepting technical and programmatic risks for the MD	Approve DM, concurring with P/p Plans as approved by GPMC and committing to oversee P/p; providing necessary institutional staffing and resources to make P/p successful; making sure that policies, requirements, procedures, and technical standards are in place and are being properly implemented, and accepting technical and programmatic risks for the Center		Approve DM to commit to execute P/p plan approved at GPMC	Approve DM to commit to execute project plan approved at GPMC
International and Intergovernmental Agreements	Concur on agreements	Support the development and negotiate international and intergovernmental agreements (OIIR)	Negotiate content of agreements with international and other external organizations			Support development of content of agreements with international and other Government Agencies	Support development of content of agreements with international and other Government Agencies

Table D-1 Roles and Responsibilities Relationships Matrix

	Office of the Administrator	Administrator Staff and Mission Support Offices	Mission Directorate Associate Administrator	Center Director		Program Manager	Project Manager
				Institutional	Technical Authority		
Launch Readiness	Approve launch request Forward request for nuclear launch approval to OSTP as required	Validate, certify, and approve human rating and launch readiness to Administrator (OCE, OSMA, and OCHMO)	Validate, certify, and approve human rating	Certify that programs and/or projects assigned to the Center have been accomplished properly as part of the launch approval process		Develop program launch readiness criteria Sign the COFR	Develop program/project launch readiness criteria Sign the COFR
Program Operations	Approve continued program implementation, when desired Ensure program performance through periodic reviews	Support Program Implementation Reviews, when requested (IPAO) Provide assessments of program performance at periodic reviews (OCE, OSMA, and OCHMO)	Recommend continued implementation when requested Ensure program performance through periodic reviews	Recommend continued implementation when requested Ensure program performance against approved plans through periodic reviews	Maintain continuous insight into program performance	Support Program Implementation Reviews, when requested Execute the Mission Operations Plan	
Project Operations	Ensure project performance through periodic reviews	Provide assessments of project performance at periodic reviews (OCE, OSMA, and OCHMO)	Ensure project performance through periodic reviews	Ensure project performance against approved plans through periodic reviews	Maintain continuous insight into project performance	Ensure project performance through periodic reviews	Execute the Mission Operations Plan
Decommissioning	Approve program and project Decommissioning Plans	Support development and assessment of program and project Decommissioning Plans	Approve program and project Decommissioning Plans	Approve program and project Decommissioning Plans		Approve project Decommissioning Plan Develop program Decommissioning Plan Execute approved Decommissioning Plan	Develop project Decommissioning Plan Execute approved Decommissioning Plan

¹ Centers may use an equivalent term for these positions, such as Program/Project systems engineer.

E

Addressing the Six Life-Cycle Review Criteria

For the following tables, note that the life-cycle review entrance and success criteria in Appendix G of NPR 7123.1 and the life-cycle phase and KDP requirements in NPR 7120.5 provide specifics for addressing the six criteria required to demonstrate that the program or project has met the expected maturity state.

Table E-1 Expected Maturity State Through the Life Cycle of Uncoupled and Loosely Coupled Programs

KDP Review	Associated LCR and LCR Objectives	Expected Maturity State by Assessment Criteria						Overall Expected Maturity State at KDP
		Agency Strategic Goals	Management Approach	Technical Approach	Budget and Schedule	Resources Other Than Budget	Risk Management	
KDP 0'	SRR—To evaluate whether the program functional and performance requirements are properly formulated and correlated with the Agency and Mission Directorate strategic objectives; to assess the credibility of the program's estimated budget and schedule.	The program has merit and is within the Agency scope; program requirements reflect Mission Directorate requirements and constraints, and are approved.	Program Formulation Authorization Document (FAD) has been approved and a preliminary Program Plan is appropriately mature; the management framework is in place with key interfaces and partnerships identified; and preliminary acquisition strategy is defined.	Functional and performance requirements have been defined, and the requirements satisfy the Mission Directorate needs; a feasible set of program implementation options has been identified that broadly addresses the functional and performance requirements.	Credible risk-informed program implementation options exist that fit within desired schedule and available funding profile.	Preliminary staffing and essential infrastructure requirements have been identified and documented; preliminary sources have been identified.	The driving risks associated with each identified program implementation option have been identified; approaches for managing these risks have been proposed and are adequate.	Overall KDP 0: Program addresses critical NASA needs and can likely be achieved as conceived.
KDP I	SDR—To evaluate the proposed program requirements/ architecture and allocation of requirements to initial projects, to assess the adequacy of project pre-Formulation efforts, and determine whether the maturity of the program's definition and associated plans are sufficient to begin implementation.	Program requirements, program approaches, and initial projects reflect Mission Directorate requirements and constraints, and fulfill the program needs and success criteria.	Program Plan and Program Commitment Agreement (PCA) are complete and management infrastructure, including interfaces and partnerships, are in place; initial project(s) have been identified and project pre-Formulation is ready to be (or already) started; technology development plans are adequate, and acquisition strategy is approved.	Driving program and project requirements have been defined, and program architectures, technology developments and operating concepts respond to them; initial project pre-Formulation responds to program needs and appears feasible.	Credible cost/ schedule estimates are supported by a documented basis of estimate (BoE) and are consistent with driving assumptions, risks, system requirements, conceptual designs, and available funding and schedule profile.	Availability, competency and stability of staffing, essential infrastructure and additional resources other than budget are adequate for remaining life-cycle phases.	Significant program and project safety, development, cost, schedule, and safety risks are identified and assessed; mitigation plans have been defined; a process and resources exist to effectively manage or mitigate them.	Program is in place and stable, addresses critical NASA needs, has adequately completed Formulation activities, has an acceptable plan for Implementation that leads to mission success, has proposed projects that are feasible within available resources, and has risks that are commensurate with the Agency's expectations.

Table E-1 Expected Maturity State Through the Life Cycle of Uncoupled and Loosely Coupled Programs

KDP Review	Associated LCR and LCR Objectives	Expected Maturity State by Assessment Criteria						Overall Expected Maturity State at KDP
		Agency Strategic Goals	Management Approach	Technical Approach	Budget and Schedule	Resources Other Than Budget	Risk Management	
KDP II to KDP n	PIR—To evaluate the program's continuing relevance to the Agency's Strategic Plan, assess performance with respect to expectations, and determine the program's ability to execute the implementation plan with acceptable risk within cost and schedule constraints.	Program's goals, objectives, and requirements remain consistent with the Agency strategic goals; requirements are complete and properly flowed down to projects.	Program Plan and PCA are up-to-date and management infrastructure, including interfaces and partnerships, are working efficiently; program/project relationships are good; technology development plans remain adequate; and acquisition strategy is working properly.	Program's technical approach and processes are enabling project mission success; and technology development activities (if any) are enabling improved future mission performance; projects are proceeding as planned.	Credible cost/schedule estimates are supported by a documented BoE and are consistent with driving assumptions, risks, project implementation, and available funding and schedule profile.	Availability, competency and stability of staffing, essential infrastructure and additional resources other than budget are adequate for continuing program acquisitions and operations.	Significant program and project safety, development, cost, schedule and safety risks are identified and assessed; mitigation plans have been defined; a process and resources exist to effectively manage or mitigate them.	Program still meets Agency needs and is continuing to meet Agency commitments as planned.

[1] KDP 0 may be required by the Decision Authority to ensure major issues are understood and resolved prior to formal program approval at KDP I.

NOTE: LCR entrance and success criteria in Appendix G of NPR 7123.1 and the life-cycle phase and KDP requirement in NPR 7120.5 provide specifics for addressing the six criteria required to demonstrate the program or project has met expected maturity state.

Table E-2 Expected Maturity State Through the Life Cycle of Tightly Coupled Programs

KDP Review	Associated LCR and LCR Objectives	Expected Maturity State by Assessment Criteria						Overall Expected Maturity State at KDP
		Agency Strategic Goals	Management Approach	Technical Approach	Budget and Schedule	Resources Other Than Budget	Risk Management	
	SRR—To evaluate whether the functional and performance requirements defined for the system are responsive to the Mission Directorate requirements on the program and its projects and represent achievable capabilities.	KDP 0 may be required by the Decision Authority to ensure major issues are understood and resolved prior to formal program approval at KDP I.						
KDP 0¹	SDR—To evaluate the credibility and responsiveness of the proposed program requirements/ architecture to the Mission Directorate requirements and constraints, including available resources, and allocation of requirements to projects. To determine whether the maturity of the program's mission/system definition and associated plans are sufficient to begin preliminary design.	Program requirements, program approaches, and initial projects incorporate Mission Directorate requirements and constraints, and fulfill the program needs and success criteria; and allocation of program's requirements to projects is complete.	Draft Program Plan and PCA are appropriately mature and management infrastructure, including interfaces and partnerships, are in place; project Formulation is underway: technology development plans are adequate, and acquisition strategy is approved and initiated.	Driving program and project requirements have been defined, and program architectures, technology developments and operating concepts respond to them; initial project Formulation responds to program needs and appears feasible.	Credible cost and schedule range estimates and associated confidence levels are supported by a documented BoE and are consistent with driving assumptions, risks, system requirements, conceptual design, and available funding.	Availability, competency and stability of staffing, essential infrastructure and additional resources other than budget are adequate for remaining life-cycle phases.	Significant mission, development, cost, schedule, and safety risks are identified and assessed; mitigation plans have been defined; a process and resources exist to effectively manage or mitigate them.	Program addresses critical NASA needs, and projects are feasible within available resources.

Table E-2 Expected Maturity State Through the Life Cycle of Tightly Coupled Programs

KDP Review	Associated LCR and LCR Objectives	Expected Maturity State by Assessment Criteria						Overall Expected Maturity State at KDP
		Agency Strategic Goals	Management Approach	Technical Approach	Budget and Schedule	Resources Other Than Budget	Risk Management	
KDP I	PDR—To evaluate the completeness/ consistency of the program's preliminary design, including its projects, in meeting all requirements with appropriate margins, acceptable risk, and within cost and schedule constraints and to determine the program's readiness to proceed with the detailed design phase of the program.	Program requirements and program/ project preliminary designs satisfy Mission Directorate requirements and constraints, mission needs and success criteria.	Program Plan and PCA are complete; external agreements and infrastructure business case are in place; contractual instruments are in place; and execution plans for the remaining phases are appropriate; projects have successfully completed their PDRs per the Program Plan.	Program and project preliminary designs satisfactorily meet requirements and constraints with acceptable risk; projects are properly integrated into the larger system.	The integrated cost/schedule baseline has a sound basis and is consistent with driving assumptions; reflects risks; is fully supported by a documented BoE; fits within the available funding and schedule profile; and cost/schedule management tools/ processes are in place.	Adequate agreements exist for staffing, essential infrastructure and additional resources, as appropriate, for remaining life-cycle phases.	Mission, development and safety risks are addressed in designs and operating concepts; a process and resources exist to effectively manage or mitigate them.	Program is in place and stable, addresses critical NASA needs, has adequately completed Formulation activities, and has an acceptable plan for Implementation that leads to mission success. Proposed projects are feasible within available resources, and the program's risks are commensurate with the Agency's tolerances.

Table E-2 Expected Maturity State Through the Life Cycle of Tightly Coupled Programs

KDP Review	Associated LCR and LCR Objectives	Expected Maturity State by Assessment Criteria						Overall Expected Maturity State at KDP
		Agency Strategic Goals	Management Approach	Technical Approach	Budget and Schedule	Resources Other Than Budget	Risk Management	
KDP II	CDR—To evaluate the integrity of the program integrated design, including its projects and ground systems. To meet mission requirements with appropriate margins and acceptable risk, within cost and schedule constraints. To determine if the integrated design is appropriately mature to continue with the final design and fabrication phase.	Changes in program scope affecting Mission Directorate requirements and constraints have been approved and documented and have been or will be implemented.	Acquisitions, partnerships, agreements and plans are in place to complete the remaining life-cycle phases; projects have successfully completed their CDRs per the Program Plan.	Detailed program and project design satisfactorily meets requirements and constraints with acceptable risk.	Driving ground rules and assumptions are realized; adequate technical and programmatic margins and resources exist to complete the remaining life-cycle phases of the program within budget, schedule, and risk constraints.	Infrastructure and staffing for final design and fabrication are available/ready; adequate agreements exist for remaining life-cycle phases.	Accepted risks are documented and credibly assessed; a process and resources exist to effectively manage or mitigate remaining open risks.	Program is still on plan. The risk is commensurate with the projects' payload classifications. The program is ready for AI&T with acceptable risk within its ABC.
	SIR—To evaluate the readiness of the program, including its projects and supporting infrastructure, to begin system Assembly, Integration, and Test (AI&T) with acceptable risk and within cost and schedule constraints.	Changes in program scope affecting Mission Directorate requirements and constraints have been approved and documented and implemented.	Acquisitions, partnerships, agreements, and plans are in place to complete the remaining phases; projects have successfully completed their SIRs per the Program Plan.	The hardware/software systems, processes, and procedures needed to begin system AI&T are available.	AI&T and remaining life-cycle phases can be completed within budget, schedule, and risk constraints.	Infrastructure and staffing for start of system AI&T are available and ready; adequate agreements exist for remaining life-cycle phases.	Accepted risks are documented and credibly assessed; a process and resources exist to effectively manage or mitigate remaining open risks.	

Table E-2 Expected Maturity State Through the Life Cycle of Tightly Coupled Programs

KDP Review	Associated LCR and LCR Objectives	Expected Maturity State by Assessment Criteria						Overall Expected Maturity State at KDP
		Agency Strategic Goals	Management Approach	Technical Approach	Budget and Schedule	Resources Other Than Budget	Risk Management	
KDP III[2]	ORR—To evaluate the readiness of the program, including its projects, ground systems, personnel, procedures and user documentation, to operate the flight system and associated ground systems in compliance with program requirements and constraints during the operations phase.	Any residual shortfalls relative to the Mission Directorate requirements have been identified to the Mission Directorate and documented and plans are in place to resolve the matter.	Acquisitions, partnerships, agreements, and plans are in place to complete the remaining phases; projects have successfully completed their ORRs per the Program Plan.	Certification for mission operations is complete, and all systems are operationally ready.	Mission operations and sustainment can be conducted within budget, schedule, and risk constraints.	Infrastructure support and certified staff on which the mission relies for nominal and contingency operations are in an operationally ready condition.	Accepted risks are documented and credibly assessed; a process and resources exist to effectively manage or mitigate remaining open risks.	Program is ready for launch and early operations with acceptable risk within Agency commitments.
	FRR—To evaluate the readiness of the program and its projects, ground systems, personnel, and procedures, for a safe and successful launch and flight/mission.	Any residual shortfall relative to the Mission Directorate requirements has been resolved with the Mission Directorate and documented.	Acquisitions, partnerships, agreements, and plans are in place to complete the remaining phases; projects have successfully completed their FRRs per the Program Plan.	Certification for flight is complete, and all systems are operationally ready.	Launch and subsequent operations can be conducted within budget, schedule, and risk constraints.	Infrastructure support and certified staff on which the launch and the mission rely are in an operationally ready condition.	Accepted risks are documented, credibly assessed and communicated; acceptable closure plans, including needed resources, exist for any remaining open risks.	

Table E-2 Expected Maturity State Through the Life Cycle of Tightly Coupled Programs

KDP Review	Associated LCR and LCR Objectives	Expected Maturity State by Assessment Criteria						Overall Expected Maturity State at KDP
		Agency Strategic Goals	Management Approach	Technical Approach	Budget and Schedule	Resources Other Than Budget	Risk Management	
	PLAR—To evaluate the in-flight performance of the program and its projects. To determine the program's readiness to begin the operations phase of the life cycle and transfer responsibility to the operations organization.	Any newly discovered shortfalls relative to the Mission Directorate requirements have been identified to the Mission Directorate and documented; plans to resolve such shortfalls are in place.	Acquisitions, partnerships, agreements, and plans are in place to complete the remaining phases; projects have successfully completed their PLARs per the Program Plan.	All systems are operationally ready and accommodate actual flight performance; anomalies have been documented, assessed and rectified or plans to resolve them are in place.	Full routine operations and sustainment, including accommodation of actual flight performance, can be conducted within budget, schedule, and risk constraints.	Infrastructure support and certified staff on which the mission relies, including accommodation of actual flight performance, are in an operationally ready condition.	Accepted risks are documented, credibly assessed and communicated; acceptable closure plans, including needed resources, exist for any remaining open risks.	PLAR Expected State: Project is ready to conduct mission operations with acceptable risk within Agency commitments.
Non-KDP Mission Operations Reviews	CERR—To evaluate the readiness of the program and its projects to execute a critical event during the flight operations phase of the life cycle.	Critical event requirements are complete, understandable and have been flowed down to appropriate levels for implementation.	Program and project agreements needed to support the Critical Event are in place; projects have successfully completed their CERRs per the Program Plan.	Critical event design complies with requirements and preparations are complete, including V&V.	Planned Critical Event can be conducted within budget, schedule, and risk constraints.	Infrastructure support and certified staff on which the Critical Event relies, including accommodation of actual flight performance, are in an operationally ready condition.	Accepted risks are documented, credibly assessed and communicated; acceptable closure plans, including needed resources, exist for any remaining open risks applicable to the Critical Event.	Mission CERR Expected State: Project is ready to conduct critical mission activity with acceptable risk.
	PFAR—To evaluate how well mission objectives were met during a human space flight mission. To evaluate the status of the flight and ground systems, including the identification of any anomalies and their resolution.	Any newly discovered shortfalls relative to the Mission Directorate requirements have been identified to the Mission Directorate and documented; plans to resolve such shortfalls are in place.	Acquisitions, partnerships, agreements, and plans are in place to support future flights; projects have successfully completed their PFARs per the Program Plan.	All anomalies that occurred in flight are identified; actions necessary to mitigate or resolve these anomalies are in place for future flights.	Future flights and missions operations can be conducted within budget, schedule, and risk constraints.	Infrastructure support and certified staff on which future flights and missions rely, including accommodation of actual flight performance, are in an operationally ready condition.	Risks to future flights and missions, identified as a result of actual flight performance, are documented, credibly assessed, and closed or acceptable closure plans, including needed resources, are in place.	PFAR Expected State: All anomalies that occurred in flight are identified, and actions necessary to mitigate or resolve these anomalies are in place.

Table E-2 Expected Maturity State Through the Life Cycle of Tightly Coupled Programs

KDP Review	Associated LCR and LCR Objectives	Expected Maturity State by Assessment Criteria						Overall Expected Maturity State at KDP
		Agency Strategic Goals	Management Approach	Technical Approach	Budget and Schedule	Resources Other Than Budget	Risk Management	
KDP IV to KDP n-1	PIR—To evaluate the program's continuing relevance to the Agency's Strategic Plan, assess performance with respect to expectations, and determine the program's ability to execute the implementation plan with acceptable risk within cost and schedule constraints.	Program's goals, objectives and requirements remain consistent with the Agency's strategic goals; requirements are complete and properly flowed down to projects.	Program Plan and PCA are up-to-date and management infrastructure, including interfaces and partnerships, are working efficiently; program/project relationships are good; technology development plans remain adequate; and acquisition strategy is working properly.	Program's technical approach and processes are enabling project mission success; and technology development activities (if any) are enabling improved future mission performance; projects are proceeding as planned.	Credible cost/schedule estimates are supported by a documented BoE and are consistent with driving assumptions, risks, project implementation, and available funding and schedule profile.	Availability, competency and stability of staffing, essential infrastructure and additional resources other than budget are adequate for continuing program acquisitions and operations.	Significant program and project safety, cost, schedule, and safety risks are identified and assessed; mitigation plans have been defined; a process and resources exist to effectively manage or mitigate them.	Program still meets Agency needs and is continuing to meet Agency commitments as planned.
KDP n	DR—To evaluate the readiness of the program and its projects to conduct closeout activities, including final delivery of all remaining program/project deliverables and safe decommissioning/disposal of space flight systems and other program/project assets.	Decommissioning is consistent with Agency and Mission Directorate objectives and requirements; decommissioning requirements are complete, understandable and have been flowed down to appropriate levels for implementation.	Acquisitions, partnerships, agreements, and plans are in place to support decommissioning, disposal, data analysis and archiving and contract closeout; projects have successfully completed their DRs per the Program Plan.	The flight hardware, and software and all associated ground systems are ready for decommissioning, including deorbit (if appropriate), and disposal.	Planned decommissioning and disposal operations can be completed within budget, schedule, and risk constraints.	Infrastructure support and certified staff on which decommissioning, deorbit and disposal rely are in an operationally ready condition.	Risks associated with decommissioning, deorbit or disposal are documented, credibly assessed and closed, or acceptable closure plans, including needed resources, are in place.	Program decommissioning is consistent with program objectives, and program is ready for final analysis and archival of mission and science data and safe disposal of its assets.

[1] KDP 0 may be required by the Decision Authority to ensure major issues are understood and resolved prior to formal program approval at KDP I.

[2] See Section 4.4.4 for a detailed description of the reviews associated with KDP III, the launch approval process, and the transition to operations for human and robotic space flight programs and projects.

Table E-3 Comprehensive Expected Maturity State Through the Life Cycle of Projects and Single-Project Programs

KDP Review	Associated LCR and LCR Objectives	Expected Maturity State by Assessment Criteria							Overall Expected Maturity State at KDP
		Agency Strategic Goals	Management Approach	Technical Approach	Budget and Schedule	Resources Other Than Budget	Risk Management		
KDP A	MCR—To evaluate the feasibility of the proposed mission concept(s) and its fulfillment of the program's needs and objectives. To determine whether the maturity of the concept and associated planning are sufficient to begin Phase A.	The proposed project has merit, is within the Agency/Program scope, and initial objectives and requirements are appropriate.	The Project FAD and Formulation Agreement are ready for approval and the management framework is in place; key interfaces and partnerships have been identified; and appropriate plans for Phase A are in place.	One or more technical concepts and attendant architectures that respond to mission needs are identified and appear feasible. Driving technologies, engineering development, payload, heritage hardware and software needs and risks have been identified.	Credible risk-informed options exist that fit within desired schedule and available funding profile.	Infrastructure and unique resource needs, such as special skills or rare materials, have been identified and are likely available.	The driving risks associated with each identified technical concept have been identified; approaches for managing these risks have been proposed and are adequate.		Project addresses critical NASA need. Proposed mission concept(s) is feasible. Associated planning is sufficiently mature to begin Phase A, and the mission can likely be achieved as conceived.
KDP B	SRR—To evaluate whether the functional and performance requirements defined for the system are responsive to the program's requirements on the project and represent achievable capabilities.	Project requirements reflect program requirements and constraints, and are responsive to mission needs.	Project documentation is appropriately mature to support conceptual design phase and preliminary acquisition strategy is defined.	Conceptual design documented; spacecraft architecture baselined; functional and performance requirements have been defined, and the requirements satisfy the mission.	Credible preliminary cost and schedule range estimates and associated confidence levels are supported by a documented BoE and are consistent with driving assumptions, risks, system requirements, design options, and available funding.	Preliminary staffing and essential infrastructure requirements have been identified and documented; preliminary sources have been identified.	Significant mission safety, technical, cost, and schedule risks have been identified; viable mitigation strategies have been defined; a preliminary process and resources exist to effectively manage or mitigate them.		Proposed mission/system architecture is credible and responsive to program requirements and associated constraints including resources. The maturity of the project's mission/system definition and associated plans is sufficient to begin Phase B, and the mission can likely be achieved within available resources with acceptable risk.

Table E-3 Comprehensive Expected Maturity State Through the Life Cycle of Projects and Single-Project Programs

| KDP Review | Associated LCR and LCR Objectives | Expected Maturity State by Assessment Criteria ||||| Overall Expected Maturity State at KDP |
		Agency Strategic Goals	Management Approach	Technical Approach	Budget and Schedule	Resources Other Than Budget	Risk Management	
KDP B	SDR/MDR— To evaluate the credibility and responsiveness of the proposed mission/system architecture to the program requirements and constraints, including available resources. To determine whether the maturity of the project's mission/ system definition and associated plans are sufficient to begin Phase B.	Mission/System requirements, design approaches, and conceptual design incorporate program requirements and constraints, and fulfill the mission needs and mission success criteria.	Preliminary Project Plan is appropriately mature to support preliminary design phase, technology development plans are adequate, acquisition strategy is approved, and U.S. partnerships are baselined. Formulation Agreement for Phase B is ready for approval.	Driving requirements have been defined, and system architectures and operating concepts respond to them. Inheritance assumptions identified, verified, and assessed for risk; components and subassemblies with significant engineering development prototyped.	Credible cost/schedule estimates are supported by a documented BoE and are consistent with driving assumptions, risks, system requirements, conceptual design, and available funding and schedule profile.	Availability, competency and stability of staffing, essential infrastructure, and additional resources other than budget are adequate for remaining life-cycle phases.	Significant mission, development, cost, schedule and safety risks are identified and assessed; mitigation plans have been defined; a process and resources exist to effectively manage or mitigate them.	Proposed mission/ system architecture is credible and responsive to pro-gram requirements and constraints including resources. The maturity of the project's mission/ system definition and associated plans is sufficient to begin Phase B, and the mission can likely be achieved within available resources with acceptable risk.
KDP C	PDR— To evaluate the completeness/ consistency of the planning, tech-nical, cost, and schedule baselines developed during Formulation. To assess compliance of the preliminary design with applicable requirements and to determine if the project is sufficiently mature to begin Phase C.	Project require-ments and pre-liminary designs satisfy program requirements and constraints, mission needs and mission success criteria.	Project Plan is complete; external agree-ments and infrastructure business case are in place; contractual instruments are in place; and execution plans for the remaining phases are appropriate.	Performance, cost, and risk trades completed; preliminary design satis-factorily meets requirements and constraints with acceptable risk; subsystem interfaces defined and evaluated for complexity and risk; assemblies with moderate to significant engineering development prototyped.	The integrated cost/schedule baseline has a sound basis and is consistent with driving assumptions; reflects risks; is fully supported by a documented BoE; fits within the available funding and schedule profile; and cost/schedule management tools/processes are in place.	Adequate agree-ments exist for staffing, essential infrastructure and additional resources, as appropriate, for remaining life-cycle phases.	Mission, development, and safety risks are addressed in designs and operating concepts; a process and resources exist to effectively manage or mitigate them.	Project's planning, technical, cost and schedule baselines developed during Formulation are complete and consistent. The preliminary design complies with its requirements. The project is sufficiently mature to begin Phase C, and the cost and schedule are adequate to enable mission success with acceptable risk.

Table E-3 Comprehensive Expected Maturity State Through the Life Cycle of Projects and Single-Project Programs

KDP Review	Associated LCR and LCR Objectives	Expected Maturity State by Assessment Criteria						Overall Expected Maturity State at KDP
		Agency Strategic Goals	Management Approach	Technical Approach	Budget and Schedule	Resources Other Than Budget	Risk Management	
KDP D	CDR—To evaluate the integrity of the project design and its ability to meet mission requirements with appropriate margins and acceptable risk within defined project constraints, including available resources. To determine if the design is appropriately mature to continue with the final design and fabrication phase.	Changes in project scope affecting program requirements and constraints have been approved and documented and have been or will be implemented.	Acquisitions, partnerships, agreements, and plans are in place to complete the remaining life-cycle phases.	Detailed project design satisfactorily meets requirements and constraints with acceptable risk.	Driving ground rules and assumptions are realized; adequate technical and programmatic margins and resources exist to complete the remaining life-cycle phases of the project within budget, schedule, and risk constraints.	Infrastructure and staffing for final design and fabrication are available/ ready; adequate agreements exist for remaining life-cycle phases.	Accepted risks are documented and credibly assessed; a process and resources exist to effectively manage or mitigate remaining open risks.	Project is still on plan. The risk is commensurate with the project's payload classification, and the project is ready for AI&T with acceptable risk within its ABC.
KDP D	PRR— To evaluate the readiness of system developer(s) to produce the required number of systems within defined project constraints, for projects developing multiple similar flight or ground support systems. To evaluate the degree to which the production plans meet the system's operational support requirements.	Changes in project scope affecting program requirements and constraints have been approved documented and have been implemented in the design.	Acquisitions, partnerships, agreements, and plans are in place to complete the remaining phases.	Project design is sufficiently mature to proceed with full-scale production and is consistent with requirements and constraints.	Production and remaining life-cycle phases can be completed within budget, schedule, and risk constraints.	Infrastructure and staffing for conducting production are available and ready; adequate agreements exist for remaining life-cycle phases.	Accepted risks are documented and credibly assessed; a process and resources exist to effectively manage or mitigate remaining open risks.	Project is still on plan. The risk is commensurate with the project's payload classification, and the project is ready for AI&T with acceptable risk within its ABC.

Table E-3 Comprehensive Expected Maturity State Through the Life Cycle of Projects and Single-Project Programs

KDP Review	Associated LCR and LCR Objectives	Expected Maturity State by Assessment Criteria						Overall Expected Maturity State at KDP
		Agency Strategic Goals	Management Approach	Technical Approach	Budget and Schedule	Resources Other Than Budget	Risk Management	
KDP D	SIR— To evaluate the readiness of the project and associated supporting infrastructure to begin system AI&T, evaluate whether the remaining project development can be completed within available resources, and determine if the project is sufficiently mature to begin Phase D.	Changes in project scope affecting program requirements and constraints have been approved, documented and implemented.	Acquisitions, partnerships, agreements, and plans are in place to complete the remaining phases.	The hardware/software systems, processes and procedures needed to begin system AI&T are available.	AI&T and remaining life-cycle phases can be completed within budget, schedule, and risk constraints.	Infrastructure and staffing for start of system AI&T are available and ready; adequate agreements exist for remaining life-cycle phases.	Accepted risks are documented and credibly assessed; a process and resources exist to effectively manage or mitigate remaining open risks.	Project is still on plan. The risk is commensurate with the project's payload classification, and the project is ready for AI&T with acceptable risk within its ABC.
KDP E[1]	ORR—To evaluate the readiness of the project to operate the flight system and associated ground system(s) in compliance with defined project requirements and constraints during the operations/sustainment phase of the project life cycle.	Any residual shortfalls relative to the program requirements have been identified to the program and documented and plans are in place to resolve the matter.	Acquisitions, partnerships, agreements, and plans are in place to complete the remaining phases.	Certification for mission operations is complete, and all systems are operationally ready.	Mission operations and sustainment can be conducted within budget, schedule, and risk constraints.	Infrastructure support and certified staff on which the mission relies, for nominal and contingency operations, are in an operationally ready condition.	Accepted risks are documented and credibly assessed; a process and resources exist to effectively manage or mitigate remaining open risks.	Project and all supporting systems are ready for safe, successful launch and early operations with acceptable risk within ABC.

Table E-3 Comprehensive Expected Maturity State Through the Life Cycle of Projects and Single-Project Programs

KDP Review	Associated LCR and LCR Objectives	Expected Maturity State by Assessment Criteria						Overall Expected Maturity State at KDP
		Agency Strategic Goals	Management Approach	Technical Approach	Budget and Schedule	Resources Other Than Budget	Risk Management	
KDP E[1]	MRR/FRR— To evaluate the readiness of the project and all project and supporting systems for a safe and successful launch and flight/mission.	Any residual shortfall relative to the program requirements has been resolved with the program and documented.	Acquisitions, partnerships, agreements, and plans are in place to complete the remaining phases.	Certification for flight is complete, and all systems are operationally ready.	Launch & subsequent operations can be conducted within budget, schedule, and risk constraints.	Infrastructure support and certified staff on which the launch and the mission rely are in an operationally ready condition.	Accepted risks are documented, credibly assessed and communicated; acceptable closure plans, including needed resources, exist for any remaining open risks.	Project and all supporting systems are ready for safe, successful launch and early operations with acceptable risk within ABC.
KDP En[2]	PIR — To evaluate the program's continuing relevance to the Agency's Strategic Plan, assess performance with respect to expectations, and determine the program's ability to execute the implementation plan with acceptable risk within cost and schedule constraints.	Program's goals, objectives and requirements remain consistent with the Agency's strategic goals; requirements are complete and properly flowed down to the project if there is one.	Program Plan and PCA are up-to-date and management infrastructure, including interfaces and partnerships, are working efficiently; program/project relationships are good; technology development plans remain adequate; and acquisition strategy is working properly.	Program's technical approach and processes are enabling program/project mission success; and technology development activities (if any) are enabling improved future mission performance; program/projects are proceeding as planned.	Credible cost/schedule estimates are supported by a documented BoE and are consistent with driving assumptions, risks, program/project implementation, and available funding and schedule profile.	Availability, competency and stability of staffing, essential infrastructure and additional resources other than budget are adequate for continuing program acquisitions and operations.	Significant program and project technical, cost, schedule, and safety risks are identified and assessed; mitigation plans have been defined; a process and resources exist to effectively manage or mitigate them.	Program still meets Agency needs and is continuing to meet Agency commitments as planned.

Table E-3 Comprehensive Expected Maturity State Through the Life Cycle of Projects and Single-Project Programs

KDP Review	Associated LCR and LCR Objectives	Expected Maturity State by Assessment Criteria						Overall Expected Maturity State at KDP
		Agency Strategic Goals	Management Approach	Technical Approach	Budget and Schedule	Resources Other Than Budget	Risk Management	
Non-KDP Reviews	PLAR—To evaluate in-flight performance of the flight system early in the mission and determine whether the project is sufficiently prepared to begin Phase E.	Any newly discovered shortfalls relative to the program requirements have been identified to the program and documented; plans to resolve such shortfalls are in place.	Acquisitions, partnerships, agreements, and plans are in place to complete the remaining phases.	All systems are operationally ready and accommodate actual flight performance; anomalies have been documented, assessed and rectified or plans to resolve them are in place.	Full routine operations and sustainment, including accommodation of actual flight performance, can be conducted within budget, schedule, and risk constraints.	Infrastructure support and certified staff on which the mission relies, including accommodation of actual flight performance, are in an operationally ready condition.	Accepted risks are documented, credibly assessed and communicated; acceptable closure plans, including needed resources, exist for any remaining open risks.	Project still meets Agency needs and is continuing to meet Agency commitments as planned.
Non-KDP Reviews	CERR—To evaluate the readiness of the project and the flight system for execution of a critical event during the flight operations phase of the life cycle.	Critical event requirements are complete, understandable and have been flowed down to appropriate levels for Implementation.	Project agreements needed to support the Critical Event are in place.	Critical event design complies with requirements and preparations are complete, including Verification and Validation (V&V).	Planned Critical Event can be conducted within budget, schedule, and risk constraints.	Infrastructure support and certified staff on which the Critical Event relies, including accommodation of actual flight performance, are in an operationally ready condition.	Accepted risks are documented, credibly assessed and communicated; acceptable closure plans, including needed resources, exist for any remaining open risks applicable to the Critical Event.	PLAR Expected State: Project is ready to conduct mission operations with acceptable risk within ABC.

Table E-3 Comprehensive Expected Maturity State Through the Life Cycle of Projects and Single-Project Programs

KDP Review	Associated LCR and LCR Objectives	Expected Maturity State by Assessment Criteria						Overall Expected Maturity State at KDP
		Agency Strategic Goals	Management Approach	Technical Approach	Budget and Schedule	Resources Other Than Budget	Risk Management	
Non-KDP Reviews	PFAR—To evaluate how well mission objectives were met during a human space flight mission and to evaluate the status of the returned vehicle.	Any newly discovered shortfalls relative to the program requirements have been identified to the program and documented; plans to resolve such shortfalls are in place.	Acquisitions, partnerships, agreements, and plans are in place to support remaining flights.	All anomalies that occurred in flight are identified; actions necessary to mitigate or resolve these anomalies are in place.	Continuing flights and missions operations can be conducted within budget, schedule, and risk constraints.	Infrastructure support and certified staff on which continuing flights and missions rely, including accommodation of actual flight performance, are in an operationally ready condition.	Risks to future flights and missions, identified as a result of actual flight performance, are documented, credibly assessed, and closed or acceptable closure plans, including needed resources, are in place.	Mission CERR Expected State: Project is ready to conduct critical mission activity with acceptable risk.
KDP F	DR—To evaluate the readiness of the project to conduct closeout activities, including final delivery of all remaining project deliverables and safe decommissioning of space flight systems and other project assets. To determine if the project is appropriately prepared to begin Phase F.	Decommissioning is consistent with Agency and program objectives and requirements; decommissioning requirements are complete, understandable and have been flowed down to appropriate levels for implementation.	Acquisitions, partnerships, agreements, and plans are in place to support decommissioning.	The flight hardware, software, and all associated ground systems are ready for decommissioning.	Planned decommissioning can be completed within budget, schedule, and risk constraints.	Infrastructure support and certified staff on which decommissioning rely are in an operationally ready condition.	Risks associated with decommissioning are documented, credibly assessed and closed, or acceptable closure plans, including needed resources, are in place.	PFAR Expected State: All anomalies that occurred in flight are identified. Actions necessary to mitigate or resolve these anomalies are in place.

Table E-3 Comprehensive Expected Maturity State Through the Life Cycle of Projects and Single-Project Programs

KDP Review	Associated LCR and LCR Objectives	Expected Maturity State by Assessment Criteria						Overall Expected Maturity State at KDP
		Agency Strategic Goals	Management Approach	Technical Approach	Budget and Schedule	Resources Other Than Budget	Risk Management	
Non-KDP Reviews	DRR—To evaluate the readiness of the project and the flight system for execution of the spacecraft Disposal Event.	Disposal event requirements are complete, understandable and have been flowed down to appropriate levels for implementation.	Project agreements needed to support the Disposal Event are in place.	Disposal event design complies with requirements and preparations are complete, including V&V.	Planned Disposal Event can be conducted within budget, schedule, and risk constraints.	Infrastructure support and certified staff on which the Disposal Event relies, including accommodation of actual flight performance, are in an operationally ready condition.	Accepted risks are documented, credibly assessed and communicated; acceptable closure plans, including needed resources, exist for any remaining open risks applicable to the Disposal Event.	Project decommissioning is consistent with program objectives and project is ready for safe decommissioning of its assets and closeout of activities, including final delivery of all remaining project deliverables and disposal of its assets.

[1] See Section 4.4.4 for a detailed description of the reviews associated with KDP E, the launch approval process, and the transition to operations for human and robotic space flight programs and projects.

[2] Applies only to single-project programs.

Table E-4 Objectives for Other Reviews

Review Name	Review Objective
System Acceptance Review (SAR)	To evaluate whether a specific end item is sufficiently mature to be shipped from the supplier to its designated operational facility or launch site.
Safety and Mission Success Review (SMSR)	To prepare Agency safety and engineering management to participate in program final readiness reviews preceding flights or launches, including experimental/test launch vehicles or other reviews as determined by the Chief, Safety and Mission Assurance. The SMSR provides the knowledge, visibility, and understanding necessary for senior safety and engineering management to either concur or nonconcur in program decisions to proceed with a launch or significant flight activity.
Launch Readiness Review (LRR)	To evaluate a program or project and its ground, hardware, and software systems for readiness for launch.

F
Control Plan Description and Information Sources

Control Plan	Description	For Additional Information
Technical, Schedule, and Cost Control Plan	Describes how the program or project plans to control program or project requirements, technical design, schedule, and cost to achieve its high-level requirements.	• Section 3.3.3.5, 4.3.4.2.2, 4.3.6.2.2 and 4.5.3.2.2 of this handbook
Safety and Mission Assurance Plan	The SMA Plan addresses life-cycle SMA functions and activities, including SMA roles, responsibilities, and relationships.	• NPD 8730.5 and 8720.1 • NPR 8715.3, 8705.2, 8705.6 and 8735.2 • NASA Standards 8719.13 and 8739.8 • Section 3.3.4, 4.3.4.3 and 4.4.3.3 of this handbook
Risk Management Plan	Summarizes how the program or project will implement the NASA risk management process.	• Section 3.3.3.5 and 4.3.4.2.2 of this handbook • NPR 8000.4
Acquisition Plan	This plan documents an integrated acquisition strategy that enables the program or project to meet its mission objectives.	• Section 3.3.3.5, 4.3.4.2.2 and 4.3.6.2.2 of this handbook
Technology Development Plan	Describes the technology assessment, development, management, and acquisition strategies needed to achieve the program or project's mission objectives. Also describes opportunities for leveraging ongoing technology efforts, transitioning technologies and commercialization plans.	• NPD 7500.2 • NPR 7500.1 • Section 3.3.4 and 4.3.4.3 of this handbook
Systems Engineering Management Plan	Describes the overall approach for systems engineering including the system design and product realization processes as well as the technical management processes.	• NPR 7123.1 • SP 6105 (Systems Engineering Handbook) • Section 3.3.4 of this handbook
Product Data and Life-cycle Management Plan	Identifies how the product data and life-cycle management capabilities will be provided and how authoritative data will be managed effectively.	• NPR 7120.9 • Section 3.3.3.5 of this handbook
Verification and Validation Plan	Summarizes the approach for performing verification and validation of the program or project products including the methodology to be used.	• NPR 7123.1 • SP 6105 (Systems Engineering Handbook) • Section 3.3.4, 4.3.4.3 and 4.3.6.3 of this handbook
Information Technology Plan	Describes how the program or project will acquire and use information technology including IT security requirements.	• NPR 2830.1 • Section 3.3.4 and 4.3.4.3 of this handbook

Control Plan	Description	For Additional Information
Review Plan	Summarizes the program or project's approach for conducting a series of reviews including internal reviews and program life-cycle reviews.	• Section 3.3.4 and 4.3.4.3 of this handbook
Mission Operations Plan	Describes the activities required to perform the mission. Discusses how the program or project will implement the associated facilities, hardware, software, and procedures required to complete the mission.	• Section 3.3.4 and 4.4.1.3 of this handbook
Environmental Management Plan	Describes the program's NEPA strategy at all affected Centers, including decisions regarding programmatic NEPA documents.	• NPR 8580.1 • Section 3.3.4 and 4.3.4.3 of this handbook
Integrated Logistics Support Plan	Describes how the program or project will implement a maintenance and support concept, enhancing supportability, supply support, maintenance planning, packaging, handling and transportation, training, manpower, required facilities, and logistics information systems for the life of the program or project.	• NPD 7500.1B • Section 3.3.4, 4.3.4.3 and 4.3.6.3 of this handbook
Science Data Management Plan	Describes how the program will manage the scientific data generated and captured by the operational mission and any samples collected and returned for analysis. Describes how the data will be generated, processes, distributed, analyzed and archived.	• Section 3.3.4, 4.3.6.3 and 4.4.3.3 of this handbook
Configuration Management Plan	Describes the approach that the program or project team will implement for configuration management. Describes the organization, tools, methods, and procedures for configuration identification, configuration control, traceability, and accounting/auditing.	• SP 6105 (*Systems Engineering Handbook*) • Section 3.3.4 and 4.3.4.3 of this handbook
Security Plan	Describes the program or project plans for ensuring security and technology protection. Describes the approach for planning and implementing requirements for information, physical, personnel, industrial, and counterintelligence/counter terrorism security.	• NPD 1600.2 • NPR 1600.1, 1040.1 and 2810.1 • Section 3.3.4, 4.3.4.3 and 4.4.6.3 of this handbook
Threat Summary	Documents the threat environment the system is most likely to encounter as it reaches operational capability. May contain Top Secret/Sensitive Compartmented Information. For more information on Threat Summary and Project Protection Plan specifics, go to the Community of Practice for Space Asset Protection at https://nen.nasa.gov/web/sap.	• Section 3.3.4 of this handbook
Technology Transfer Control Plan	Describes how the program or project will implement the export control requirements.	• NPR 2190.1 • Section 3.3.4 and 4.3.4.3 of this handbook
Education Plan	Planned activities to enhance science, technology, engineering or math education using the program or project science and technical content.	• Section 3.3.3.5, 4.3.4.2.2 and 4.4.3.3 of this handbook
Communications Plan	Describe plans to implement a diverse, broad, and integrated set of efforts and activities to communicate with and engage target audiences, the public, and other stakeholders in understanding the program or project, its objectives, elements, and benefits.	• Section 3.3.3.5, 4.3.4.2.2, 4.3.6.2.2 and 4.4.3.3 of this handbook
Knowledge Management Plan	Describes the program's approach to creating the knowledge management strategy and processes, examining the lessons learned database for relevant lessons, and creating the plan for how the program continuously captures and documents lessons learned throughout the program life cycle.	• NPD 7120.4 • NPD 7120.6 • Section 3.3.4 and 4.4.4.3 of this handbook

Control Plan	Description	For Additional Information
Human Rating Certification Package	Focuses on the integration of the human into the system, preventing catastrophic events during the mission, and protecting the health and safety of humans involved in or exposed to space activities, specifically the public, crew, passengers, and ground personnel.	• NPR 8705.2 • SP 6105 (*Systems Engineering Handbook*) • Section 3.3.3.5, 4.3.4.3, 4.3.6.3, 4.4.1.3 and 4.4.3.3 of this handbook
Software Management Plan	Summarizes how the project will develop and/or manage the acquisition of software required to achieve project and mission objectives. Plan should be coordinated with the SEMP.	• NPR 7150.2 • NASA Standard 8739.8 • Section 4.4.4.3 and 4.3.6.3 of this handbook
Integration Plan	This plan defines the integration and verification strategies. It is structured to show how components come together in the assembly of each subsystem and how the subsystems are assembled into the system/product. Also describes the participants and required resources and when they will be needed.	• SP 6105 (*Systems Engineering Handbook*) • Section 3.4.1.1, 4.3.4.3, 4.3.6.3 and 4.4.1.3 of this handbook
Project Protection Plan	This plan is based on the program Threat Summary, which documents the threat environment the project is most likely to encounter as it reaches operational capability and recommends potential countermeasures. For more information on Threat Summary and Project Protection Plan specifics, go to the Community of Practice for Space Asset Protection at https://nen.nasa.gov/web/sap.	• Section 3.3.4, 4.3.4.3 and 4.3.6.3 of this handbook
Planetary Protection Plan	Planetary protection encompasses: (1) the control of terrestrial microbial contamination associated with space vehicles intended to land, orbit, fly by, or otherwise encounter extraterrestrial solar system bodies and (2) the control of contamination of the Earth by extraterrestrial material collected and returned by missions. The scope of the plan contents and level of detail will vary with each project based upon the requirements in NASA policies.	• NPD 8020.7 • NPR 8020.12 • Section 4.3.4.3 and 4.3.6.3 of this handbook
Nuclear Safety Launch Approval Plan	Needed for any U.S. space mission involving the use of radioactive materials. Describes potential risks associated with a planned launch of radioactive materials into space, on launch vehicles and spacecraft, and during flight.	• NPR 8715.3 • Section 4.3.4.3 of this handbook

G References

External References

Air Force Space Command Manual 91-710, *Range Safety User Requirements Manual Volume 3 - Launch Vehicles, Payloads, and Ground Support Systems Requirements (for ELVs).*

American National Standards Institute (ANSI) Electronic Industries Alliance/(EIA)-748-C, *Earned Value Management Systems,* 2013. This document is regarded as the national standard and an industry best practice for EVM systems. OMB Circular A-11 requires EVM for acquisitions with developmental effort and for both in-house government and also contractor work using these guidelines.

BAE Presentation "Technical Performance Measure" Jim Oakes, Rick Botta and Terry Bahill, 2005.

Consolidated and Further Continuing Appropriations Act, 2013 (P.L. 113-6), Section 522. Congress requires NASA to report on projects greater than $75 million that encounter a 10 percent LCC growth.

Commerce, Justice, Science, and Related Agencies' appropriations acts (annual), General Provisions, Section 505. Lays out reprogramming requirements and requires NASA to notify the House and Senate Committees on Appropriations of a decision to terminate a program or project 15 days in advance of the termination of a program or project. The NASA Office of Legislative and Intergovernmental Affairs (OLIA) is responsible for notifying the Committees on Appropriations pursuant to this reprogramming requirement.

Executive Order 12114, Environmental Effects Abroad Of Major Federal Actions, Jan. 4, 1979.

National Security Presidential Directive (NSPD) 49, National Space Policy of the United States of America, June 28, 2 010.

Federal Acquisition Regulation (FAR), 48 C.F.R., Chapter 1.

FY 2008 House Appropriations Report H.R.2764 (P.L.110-161), "Audit of NASA large-scale programs and projects," Refer to the House report for the details. All appropriations since FY 2008 have included direction for the Government Accountability Office (GAO) to "identify and gauge the progress and potential risks associated with selected NASA acquisitions. This has resulted in GAO's annual "Assessment of Large-Scale NASA Programs and Projects," the audit known internally as the Quick Look Book.

Garvey, P.R., *Probability Methods for Cost Uncertainty Analysis: A Systems Engineering Perspective*, New York, Marcel Dekker, 2000.

Government Performance Reporting and Accountability (GPRA) Modernization Act of 2010, Public Law 111–352—Jan. 4, 2011, 124 STAT. 3866. Additional information on the GPRA Modernization Act can be found at http://www.gpo.gov/fdsys/pkg/PLAW-111publ352/pdf/PLAW-111publ352.pdf.

"Metric Guidebook for Integrated Systems and Product Development," INCOSE, 1995.

NASA Authorization Act of 2005 (P.L. 109-155), Section 103. Congress created an external reporting requirement in this Act, i.e., the Major Program Annual Report (MPAR) for projects in development (whether or not they are space flight projects) with an estimated life-cycle cost exceeding $250 million.

National Environmental Policy Act (NEPA), 42 U.S.C. § 4321 et seq.

NASA FAR Supplement (NFS), 48 C.F.R., Chapter 18.

- The contents of written acquisition plans and PSMs are delineated in the FAR in Subpart 7.1—Acquisition Plans, the NFS in Subpart 1807.1—Acquisition Plans.
- EVM is required by OMB for compliance with the FAR (section 34.2) and guided by industry best practice. NASA FAR Supplement, section 1834.2 requires use of an Earned Value Management System (EVMS) on procurement for development or production work, including flight and ground support systems and components, prototypes, and institutional investments (facilities, IT infrastructure, etc.) when their estimated life-cycle (Phases A–F) costs $20 million or more.

"Systems Engineering Leading Indicators for Assessing Program and Technical Effectiveness," Donna H. Rhodes, Ricardo Valerdi, Garry J. Roedler. Massachusetts Institute of Technology and Lockheed Martin Corporation. Reprint of an article accepted for publication in Systems Engineering, 2008.

"Systems Engineering Leading Indicators Guide," Version 2.0, MIT, INCOSE and PSM joint paper, 2010.

"Technical Measurement," a Collaborative Project of PSM, INCOSE and Industry, 2005.

White House, Office of Management and Budget (OMB) Circular A-11, *Preparation, Submission and Execution of the Budget*, Section 6. (7/26/2013). See http://www.whitehouse.gov/sites/default/files/omb/assets/a11_current_year/s200.pdf. Some reporting requirements, such as the Annual Performance Plan (APP), are Government-wide to meet guidance in OMB Circular A-11.

White House, OMB Circular A-119, Federal Participation in the Development and Use of Voluntary Consensus Standards and in Conformity Assessment Activities. (02/10/1998). For technical standards, see http://www.white-house.gov/omb/circulars_a119/.

White House, OMB, Office of Federal Procurement Policy (OFPP), *The Federal Acquisition Certification for Program and Project Managers (FAC-P/PM)*, April 25, 2007, sets requirements for project management certification that apply to all civilian agencies and outlines the baseline competencies, training, and experience required for program and project managers in the Federal government. See http://www.whitehouse.gov/sites/default/files/omb/procurement/workforce/fed_acq_cert_042507.pdf

U.S. Department of Commerce, National Institute of Standards and Technology (NIST), Special Publication 330, 2008 Edition, *International System of Units* (United States version of the English text of the eighth edition (2006) of the International Bureau of Weights and Measures publication *Le Système Internationale d' Unités* (SI), commonly known as the metric system of measurement).

U.S. Government Accountability Office (GAO), GAO-04-642, *NASA: Lack of Disciplined Cost-Estimating Processes Hinders Effective Program Management*, 2004.

GAO, *"Assessment of Large-Scale NASA Programs and Projects,"* annual audit known internally as the Quick Look Book. GAO has generally chosen to review projects already required to file MPAR or NSPD-49 reports and publishes its results annually at http://gao.gov/search?search_type=Solr&o=0&facets=&q=NASA+Assessments+of+Selected+large-scale+projects&adv=0

NPDs Referenced

The latest versions of these policy documents can be found in the NASA Online Directives Information System library at http://nodis3.gsfc.nasa.gov.

- *NPD 1000.0, NASA Governance and Strategic Management Handbook*
- *NPD 1000.5, Policy for NASA Acquisition*, also referenced as *NPD 1000.5A, Policy for NASA Acquisition*
- *NPD 1001.0, NASA Strategic Plan*
- *NPD 1440.6, NASA Records Management*
- *NPD 1600.2, NASA Security Policy*
- *NPD 2200.1, Management of NASA Scientific and Technical Information*
- *NPD 7120.4, NASA Engineering and Program/Project Management Policy*
- *NPD 7120.6, Knowledge Policy on Programs and Projects*
- *NPD 7500.1, Program and Project Life Cycle Logistics Support Policy*
- *NPD 7500.2, NASA Innovative Partnerships Program*
- *NPD 8010.3, Notification of Intent to Decommission or Terminate Operating Space Missions and Terminate Missions*
- *NPD 8020.7, Biological Contamination Control for Outbound and Inbound Planetary Spacecraft*

- *NPD 8610.7, Launch Services Risk Mitigation Policy for NASA-Owned and/or NASA-Sponsored Payloads/Missions*
- *NPD 8610.12, Human Exploration and Operation Mission Directorate (HEOMD) Space Transportation Services for NASA and NASA-Sponsored Payloads*
- *NPD 8610.23, Launch Vehicle Technical Oversight Policy*
- *NPD 8700.1, NASA Policy for Safety and Mission Success*
- *NPD 8720.1, NASA Reliability and Maintainability (R&M) Program Policy*
- *NPD 8730.2, NASA Parts Policy*
- *NPD 8730.5, NASA Quality Assurance Program Policy*
- *NPD 8820.2, Design and Construction of Facilities*

NPRs Referenced

The latest versions of these policy documents can be found in the NASA Online Directives Information System library at http://nodis3.gsfc.nasa.gov.

- *NPR 1040.1, NASA Continuity of Operations (COOP) Planning Procedural Requirements*
- *NPR 1441.1, NASA Records Retention Schedules*
- *NPR 1600.1, NASA Security Program Procedural Requirements*
- *NPR 2190.1, NASA Export Control Program*
- *NPR 2200.2, Requirements for Documentation, Approval, and Dissemination of NASA Scientific and Technical Information*
- *NPR 2810.1, Security of Information Technology*
- *NPR 2830.1, NASA Enterprise Architecture Procedures*
- *NPR 7123.1, NASA Systems Engineering Processes and Requirements*
- *NPR 7120.5E, NASA Space Flight Program and Project Management Requirement*
- *NPR 7120.9, NASA Product Data and Life-Cycle Management (PDLM) for Flight Programs and Projects*
- *NPR 7120.10, Technical Standards for NASA Programs and Projects*
- *NPR 7120.11, NASA Health and Medical Technical Authority (HMTA) Implementation*
- *NPR 7150.2, NASA Software Engineering Requirements.*
- *NPR 7500.1, NASA Technology Commercialization Process*
- *NPR 7900.3, Aircraft Operations Management Manual*
- *NPR 8000.4, Agency Risk Management Procedural Requirements*
- *NPR 8020.12, Planetary Protection Provisions for Robotic Extraterrestrial Missions*
- *NPR 8580.1, NASA National Environmental Policy Act Management Requirements*
- *NPR 8621.1, NASA Procedural Requirements for Mishap and Close Call Reporting, Investigating, and Recordkeeping*
- *NPR 8705.2, Human-Rating Requirements for Space Systems*

- *NPR 8705.4, Risk Classification for NASA Payloads*
- *NPR 8705.6, Safety and Mission Assurance (SMA) Audits, Reviews, and Assessments*
- *NPR 8715.3, NASA General Safety Program Requirements*
- *NPR 8715.5, Range Flight Safety Program*
- *NPR 8715.6, NASA Procedural Requirements for Limiting Orbital Debris*
- *NPR 8715.7, Expendable Launch Vehicle Payload Safety Program, and local range requirements*
- *NPR 8735.1, Procedures for Exchanging Parts, Materials, Software, and Safety Problem Data Utilizing the Government-Industry Data Exchange Program (GIDEP) and NASA Advisories*
- *NPR 8735.2, Management of Government Quality Assurance Functions for NASA Contracts*
- *NPR 8820.2, Facility Project Requirements*
- *NPR 9250.1, Property, Plant, and Equipment and Operating Materials and Supplies*
- *NPR 9420.1, Budget Formulation*
- *NPR 9470.1, Budget Execution*

NASA Standards Referenced

NASA Technical Standards can be found under the "Other Policy Documents" menu in the NASA Online Directives Information System library at http://nodis3.gsfc.nasa.gov.

- *NASA-STD-0005, NASA Configuration Management (CM) Standard*
- *NASA-STD-8709.20 – Process for Limiting Orbital Debris*
- *NASA-STD-8719.13, NASA Software Safety Standard*
- *NASA-STD 8719.14, Process for Limiting Orbital Debris*
- *NASA-STD-8739.8, Software Assurance Standard*

Handbooks Referenced

NASA Special Publications (SP) handbooks can be found in the Scientific and Technical Information (STI) library at http://www.sti.nasa.gov. All the other handbooks, except the Science Mission Directorate handbook can be found on the Office of the Chief Engineer tab under the "Other Policy Documents" menu in the NASA Online Directives Information System library at http://nodis3.gsfc.nasa.gov. Contact the Science Mission Directorate for a copy of their handbook.

- *NASA Cost Estimating Handbook*
- *NASA/SP-2012-599, NASA Earned Value Management (EVM) Implementation Handbook*
- *NASA/SP-2010-3406, Integrated Baseline Review (IBR) Handbook*

- *NASA/SP-2011-3422, NASA Risk Management Handbook*
- *Schedule Management Handbook*
- *SSSE MH2002, The Science Mission Directorate Enterprise Management Handbook.*
- *NASA-HDBK-2203 NASA Software Engineering Handbook*, https://swehb. nasa.gov/display/7150/Book+A.+Introduction
- *NASA Standing Review Board Handbook*
- NASA/SP-2007-6105, *Systems Engineering Handbook,* Rev 1, 2007.
- *NASA Work Breakdown Structure (WBS) Handbook*

H Index

Italicized page numbers indicate material in boxes, figures, or tables.

System Assembly, Integration and Test, Launch and Check-out, 91–93
technical activities and products of, 68–79
program and project managers
certification of, 372–377
competencies of, 376–377
competency assessment of, 374–375
maintaining certification, 378–379
matrix of roles and responsibilities, *418–428*
selection of, 371–372
Program Commitment Agreement (PCA), 10, 46, 80, 396
Program Implementation Review (PIR), 94, 340, 344
Programmatic Authority, 11, *12*, 13, 230, 232–379, 239
See also dissenting opinion resolution process.
and derived requirements, 268–269
definition of, 396
organizational hierarchy of, *5*
Program Plan, 61, 274, 276, 368, 396
Program/Project Chief Engineer (PCE), 245
project, 15–17, 109–228
and Program Management Councils (PMCs), 128
categorization of, 109–110, *110*
definition of, 6, 109, 396
evolution and recycling of, 122
Formulation activities of, 114–115, 137–189, 274–275
Implementation activities of, 114
initiation of, 110–111
internal reviews of, 117
life cycle of, 112–115
life-cycle reviews of, 115–120, 127
milestone product maturity matrix for, 223–226
other review of, 120–122
oversight and approval for, 126–136
Phase A activities of, *156*, 153–176
Phase B activities of, *180*, 178–189
Phase C activities of, *191*, 189–198
Phase D activities of, *199*, 198–211
Phase E activities of, *213*, 211–215
Phase F activities of, *220*, 219–221

Pre–Phase A activities of, *140*, 137–148
products by phase, 221–228
resources, 121–122
tailoring of, 122–126
Project Plan, 396
Project Protection Plan (PPP), 78, 173, *449*

Q

Quality Audit, Assessment, and Review, 343
quarterly data call
See external reporting.
Quick Look Book
See external reporting; Government Accountability Office (GAO).

R

radioactive material, 15, 109, 110, 174–176, 176, 247
See also nuclear safety launch approval.
Radioactive Materials On-Board Report (RMOR), 175
Rebaseline Review, 31, 120, 282
replanning and rebaselining, 281, 278–282, 396, 397
Requests For Action (RFAs)
See leading indicators.
Requirement Flow Down and SMA Engineering Design Audits and Assessments, 343
requirements
of NPR 7120.5E and rationale for, *409–416*
program level, 58–59
requirements, types of, 14, 265
allocated, 14, 265, 384
baseline science, 58, 146, 385
derived, 14, 265, 268–269, 269, 387
engineering, 389
infrastructure, 64
institutional, 13, 14, 230, 233, 265, 391, *422*
programmatic, 13, 14, 230, 231, 265, 396, *421*
Technical Authority, 14, 265, 269–270, 400
threshold science, 401
Review Plan, 75–76, 171, *448*

reviews
See internal reviews, special reviews; specific reviews by name.
risk, 297, 397
Risk Management Plan, 66–67, 143, 163–164, *447*
robotic space flight, 92, 93, 112, 113, 115, 163, 183, 198, 203, 204, 251
flow chart for, 208
launch approval process for, 207–210

S

Safety and Mission Assurance (SMA) compliance/verification review, 340, 343
Safety and Mission Assurance (SMA) Plan, 73–74, 169–171, *447*
Safety and Mission Success Review (SMSR), 93, 343, *446*
safety data package
requirements for, 71
Science Data Management Plan, 76, 186–187, *448*
science mission, 58, 134, 146
Science, Technology, Engineering, and Math (STEM) education, 67, 164
Security Plan, 77, 173, *448*
Shuttle Program, xiii, 41, 204
See also Challenger, Columbia, Discovery.
single-project program
control plan maturity matrix for, 106–107
definition of, 21, 398
evolution and recycling of, 33
Formulation activities of, 51, 54–56, 274–275
Implementation activities of, 87, 88, 276–277
internal reviews of, 81–83
life cycle of, 24–25, 26
milestone product maturity matrix for, 101–105
Software Management Plan, *449*
Space Act Agreement (SAA), 111, 398–399

special reviews, 30–31, 120, 340
  See also Rebaseline Review,
  Termination Review.
Standing Review Board (SRB), 22,
  330–339, 399
  See also life-cycle review.
  conflict of interest procedures
    for, 333–334
  forming, 333
  life-cycle reviews that require,
    25, 26, 113, 117
  participation in various reviews
    by, 82, 282, 340, 341, 344
  role and responsibilities, 331–333
  terms of reference for, 334
strategic acquisition process, 13,
  137
Strategic Implementation Planning
  (SIP), 46, 154, 313
Strategic Programming Guidance
  (SPG), 231, 313, 314, 318
Sustainment and Sustainment
  Engineering, 94, 212
System Acceptance Review (SAR),
  446
System Assembly, Integration and
  Test, Launch and Checkout,
  91–93
System Definition Review/Mission
  Definition Review (SDR/
  MDR), 47–48
Système Internationale (SI), 71–72,
  167
System for Administration,
  Training, and Educational
  Resources for NASA
  (SATERN), 373, 374, 378
System Requirements Review
  (SRR), 47, 58
Systems Engineering Management
  Plan (SEMP), 74, 148, *447*

T

tabletop reviews
  See internal reviews.
tailoring, 263–270
  See also Compliance Matrix.
  and delegation of approval au-
    thority, 263–264
  and request for a permanent
    change, 270

considering other stakeholders
  in, 266
documentation of, 267–268
of a derived requirement,
  268–269, 269
of a non-applicable prescribed
  requirement, 270
of a Technical Authority requirement, 269–270
of NPR 7120.5 requirements,
  264–266, 267
of projects, 124–125, 122–126
tracking data for, 267
Technical Authority, 12, 400
  and Columbia Accident Investi-
    gation Board (CAIB), 238
  and dissent, 242
    See also dissenting opinion
    resolution process.
  and Governance, 239
  Engineering Technical Authority
    (ETA), 241, 244, 243–246
  Health and Medical Technical
    Authority (HMTA), 241,
    248–252
  illustration of flow, 250, 251
  origin of process, 238–239
  overview of, 237–238
  roles and responsibilities of,
    240–242, 243–252
  Safety and Mission Assurance
    Technical Authority (SMA
    TA), 241, 247–248
Technical, Schedule, and Cost
  Control Plan, 61–63, 143,
  157–160, *447*
Technology Development Plan,
  71–72, 147, *447*
Technology Transfer Control Plan,
  78–79, 173, *448*
termination decision
  See termination review.
Termination Review, 31, 121–122,
  340–343
Terms of Reference (ToR), 334, 401
Threat Summary, 78, *448*
Threshold Report
  See external reporting.
tightly coupled program
  control plan maturity matrix for,
    100
  definition of, 21–22, 401

evolution and recycling of, 33
Formulation activities of, 50,
  52–54, 275–276
Implementation activities of, 86,
  88, 276–277
internal reviews of, 81–83
life cycle of, 24, 25
milestone product maturity
  matrix for, 99

U

Unallocated Future Expenses
  (UFE), 41, 134, 271, 281,
  277–282, 284, 308
uncoupled program
  definition of, 21, 401
  Formulation and Implementa-
    tion activities of, 48, 49, 85,
    277
  life cycle of, 23
  milestone product and control
    plan maturity matrix for, 98

V

Verification and Validation (V&V)
  Plan, 74, 171, *447*

W

waiver, 263, 402
Work Breakdown Structure (WBS),
  319–330, 402
  and Metadata Manager System
    (MdM), 328–329
  dictionary, 321, 322, 330
  example of, 323
  for programs, 63, 321
  for projects, 160, 321–322
  generic structure of, 320
  Level 3 and subsequent elements
    of, 326–327
  program work elements of, 330
  standard elements of, 322–326
  standard template for space
    flight projects, 160, 319
  translating work breakdown into
    funds, 327
Work Instruction Management
  System
  funds distribution systems, 328
work package, 160, 402

Mars Rover

To the right is an artist's concept of NASA's Mars 2020 rover. This new rover capitalizes on the design and engineering done for Curiosity, but with new science instruments for the 2020 mission. It will continue the search for signs of life on Mars.

NASA's Mars rover Curiosity landed on Mars in August 2012. A self-portrait panorama (below) combines dozens of exposures taken by the rover's Mars Hand Lens Imager (MAHLI) in February 2013 during the 177th Martian day, or sol, of Curiosity's work on Mars. In the picture, the rover rests in the

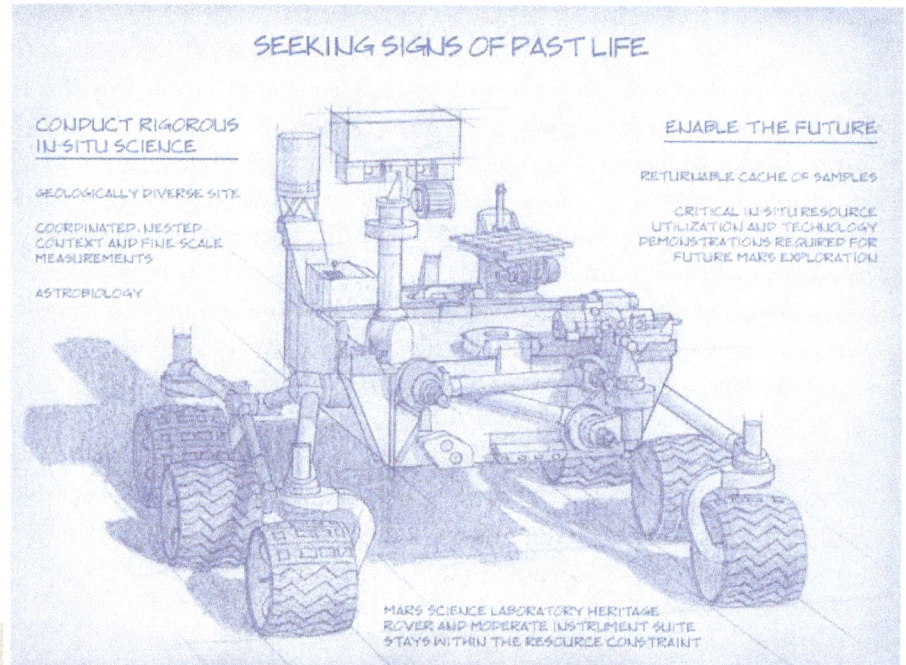

SEEKING SIGNS OF PAST LIFE

CONDUCT RIGOROUS IN-SITU SCIENCE

GEOLOGICALLY DIVERSE SITE

COORDINATED, NESTED CONTEXT AND FINE-SCALE MEASUREMENTS

ASTROBIOLOGY

ENABLE THE FUTURE

RETURNABLE CACHE OF SAMPLES

CRITICAL IN-SITU RESOURCE UTILIZATION AND TECHNOLOGY DEMONSTRATIONS REQUIRED FOR FUTURE MARS EXPLORATION

MARS SCIENCE LABORATORY HERITAGE ROVER AND MODERATE INSTRUMENT SUITE STAYS WITHIN THE RESOURCE CONSTRAINT

Yellowknife Bay region of the red planet's Gale Crater at a patch of flat outcrop called "John Klein."

NASA's Jet Propulsion Laboratory (JPL), a division of the California Institute of Technology (Caltech) in Pasadena, California, designed and assembled the rover and manages the Mars Science Laboratory Project for NASA's Science Mission Directorate in Washington, D.C.

Kepler Space Telescope

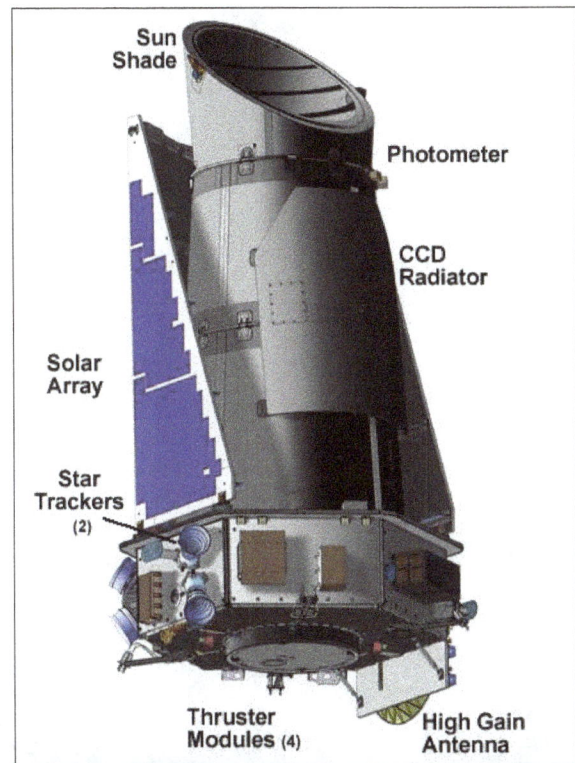

Planet Kepler-186f is the first validated Earth-size planet in the habitable zone (a distance from a star where liquid water might pool on the planet's surface). Kepler-186f and four inner planets line up in orbit around a host star that is half the size and mass of our sun in the solar system Kepler-186, about 500 light-years from Earth in the constellation Cygnus. Kepler-186f orbits its star once every 130 days. Less than 10 percent larger than Earth, Kepler-186f receives one-third as much energy from its sun as Earth, putting it at the edge of the habitable zone. This artist's conception (below) depicts a life-friendly version of the planet, but the composition, density, and actual conditions on the planet are not known. The Kepler Space Telescope, launched in March 2009 to search for Earth-like planets, infers the existence of a planet by the amount of starlight blocked when it passes in front of its star. The discovery of Kepler-186f confirmed that Earth-size planets exist in the habitable zones of other stars, one significant step closer to finding habitable worlds.

Kepler Space Telescope

Credit: NASA

Sun Shade

Photometer

CCD Radiator

Solar Array

Star Trackers (2)

Thruster Modules (4)

High Gain Antenna

EARTH

KEPLER-186f

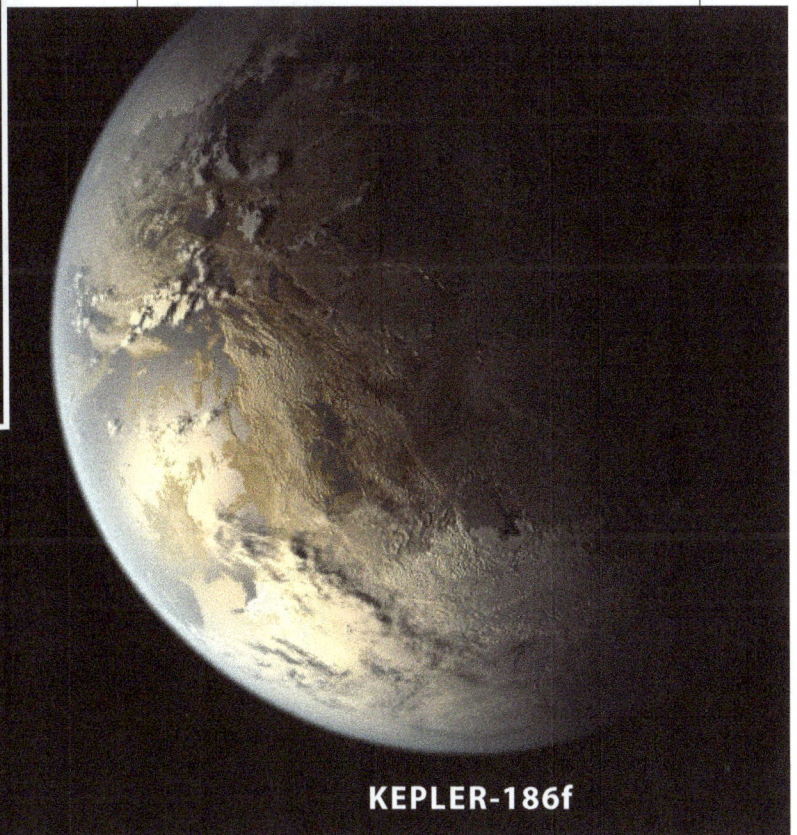

Credit: NASA Ames/SETI Institute/JPL-Caltech

Orion Capsule and SLS Rocket

NASA's Space Launch System (SLS) will be the largest launch vehicle ever built. It is designed for missions beyond Earth's orbit deep into space. It is envisioned that the Orion, SLS, and a modernized Kennedy spaceport will support missions to multiple deep space destinations extending human existence beyond our Moon, to Mars, and across our solar system.

NASA's SLS heavy-lift booster is shown in a computer-aided design image of SLS and Orion being stacked inside the Vehicle Assembly Building high bay at the Kennedy Space Center and in the line-art diagram (left).

The photo (above) shows NASA's Orion crew module ready to be mated to the largest heat shield ever built, which will protect the crew module from the extreme 4000-degree heat of reentry.

Image courtesy Lockheed Martin

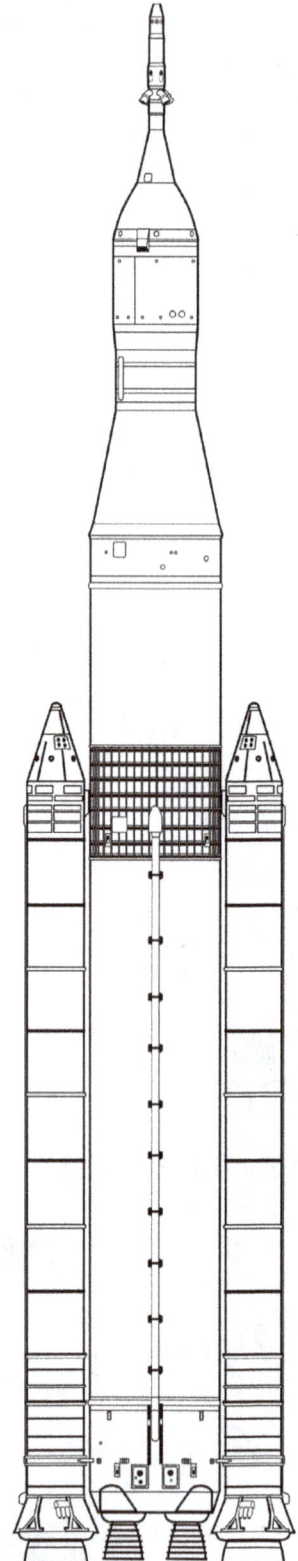

Hubble Space Telescope

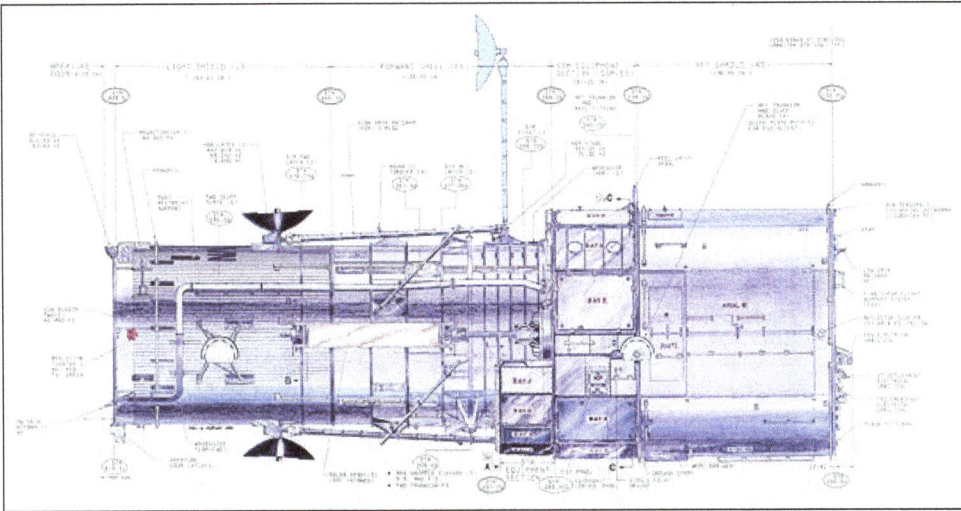

Diagram of Hubble from 1981
Credit: Lockheed Martin.
Image provided by ESA/Hubble.

The photo below shows Hubble floating free in orbit against the background of Earth after a week of repair and upgrade by Space Shuttle Columbia astronauts in March 2002.

Credit: NASA/ESA

Back Cover:
One of the first photos released of the Hubble Space Telescope launched in April 1990, the back photo shows Hubble being grappled by the robotic arm of the Space Shuttle. Of NASA's four Great Observatories—powerful, space-based telescopes designed to examine different regions of space at different wavelengths—Hubble was the first to reach orbit. In its decades of operation, Hubble has revolutionized knowledge of the cosmos.
Credit: NASA.